SOUTHERN AFRICAN PREHISTORY AND PALEOENVIRONMENTS

SOUTHERN AFRICAN PREHISTORY AND PALEOENVIRONMENTS

Edited by
RICHARD G.KLEIN
Department of Anthropology, University of Chicago

A.A.BALKEMA / ROTTERDAM / BOSTON / 1984

Dedicated to the memory of our friend and colleague:
Franz Richard (Frank) Schweitzer (1925-1979)

ISBN 90 6191 097 8

Distributed in USA & Canada by: A.A.Balkema Publishers, P.O.Box 230, Accord, MA 02018

Printed in the Netherlands

Table of contents

Preface

Scientific research on the rich archeological and associated fossil record of southern Africa began in the latter part of the last century. It has continued vigorously ever since, stimulated especially by the discovery of the australopithecines in the 1920's and 1930's and by the fundamental culture-stratigraphic synthesis that A.J.H. Goodwin formulated at about the same time.

We now know that human beings, broadly defined, have inhabited southern Africa for more than two million years. Only in East Africa is the prehistoric record ostensibly longer. Since 1959, the repeated and often spectacular discoveries of australopithecines, of other early hominids, and of very early stone artifacts in East Africa have drawn much specialist and lay attention away from southern Africa. Yet, the southern African australopithecine sample remains larger than the East African one, and it continues to play a crucial role in our understanding of australopithecine biology and ecology. Equally important, southern Africa has provided unique information on human biological and cultural evolution after the australopithecines. Indeed, with regard to the late Quaternary, from perhaps 200,000 years ago to historic times, the southern African record is as interesting and informative as any in the world. In addition, paleoenvironmental studies have shown that southern Africa contains data that are vital to a fuller understanding of late Cenozoic climatic/environmental change and of past man-environment relationships.

It is the purpose of this book to provide an up-to-date and authoritative summary of southern African prehistory and related paleoenvironmental studies. In the opening chapter, K.W. Butzer addresses Quaternary environmental reconstruction and change from the perspective of a geo-archeologist with extensive experience in southern Africa. His field studies have included the Plio/Pleistocene australopithecine caves, as well as a remarkable variety of Middle and Late Quaternary sites in alluvial, lacustrine, spring, and cave contexts. Many of the data and conclusions he presents about these sites are published here for the first time. His tables provide especially succinct summaries of the Quaternary climatic/environmental events he has reconstructed from observations at several key southern African localities.

In Chapter 2, L. Scott summarizes the information that fossil pollens have provided on Quaternary environments in southern Africa. Much of this information was collected by E.M. van Zinderen Bakker and J.A. Coetzee, Scott's colleagues at the University of the Orange Free State, and more recently by Scott himself. Scott's emphasis on modern surface sampling points up the difficulty of interpreting the pollen record, let alone of reconstructing vegetation for a particular period. The still fragmentary data base presently suggests a complex sequence of climatic change that varied importantly from region to region within southern Africa and that in no way resembles the once-popular notion of 'pluvials' and 'interpluvials' corresponding neatly to 'glacials' and 'interglacials' in middle and upper latitudes.

In Chapter 3, Q.B. Hendey summarizes present knowledge of Miocene and early Pliocene vertebrates, especially mammals, from southern Africa. It is Hendey's own research on fossils from the mid-Miocene site of Arrisdrift in Namibia and the early Pliocene site of Langebaanweg in the southwestern Cape Province of South Africa that has provided most of the available information. His work at Langebaanweg should be of especial interest to paleoanthropologists and paleoecologists. The volume of the Langebaanweg faunal sample probably exceeds that of the combined total from East African Pliocene localities, yet it totally lacks hominids. It seems probable that they had simply not penetrated this far south in the early Pliocene. Because the Langebaanweg sample is so large and so well studied, it is uniquely suitable for elucidating the composition of an African animal community only shortly before the appearance of hominids.

In Chapter 4, R.G. Klein surveys the fossil record of southern African mammals from late Pliocene to historic times, emphasizing the paleoenvironmental and paleoecological implications of faunal change through time. This emphasis reflects Klein's particular research interests, as illustrated in his published analyses of various southern African Quaternary faunas. Hominids were present in southern Africa throughout the period with which Klein is concerned, and it is possible that they were in part responsible for some of the faunal turnover he discusses.

In Chapter 5, G.P. Rightmire summarizes the contributions of southern Africa to our knowledge of hominid evolution. As a specialist reader would expect, Rightmire considers the southern African australopithecines at some length, but he also devotes considerable space to later hominid fossils, virtually all of which he has seen and studied in detail. Among the various problems he addresses, one of particular interest is the possibility that Homo sapiens sapiens originated in southern Africa.

In Chapter 6, T.P. Volman summarizes the archeology of the Early and Middle Stone Age peoples who lived in southern Africa from perhaps 2,000,000 to 40,000-30,000 years ago. He focuses on the nature and meaning of chronological and regional variation in the stone artifacts they left behind. This is a subject he is particularly well qualified to address as a result of his detailed studies of artifact assemblages from several pertinent southern African sites, including key Middle Stone Age sites in the southern Cape Province of South Africa. Besides artifacts, he also considers the evidence for Early and Middle Stone Age man-environment relationships. Some of this evidence is so far without parallel anywhere else in the world.

In Chapter 7, J. Deacon reviews the prehistory of the Later Stone Age people who appeared in southern Africa 40-30,000 years ago and who were still present in some parts of the subcontinent in historic times. Deacon draws on her broad experience in Later Stone Age studies, gained from directing or participating in the excavation of several vital sites and from analyzing the artifact assemblages afterwards. She makes extensive use of ethnohistoric and ethnoarcheological observations both to interpret the archeological record and to fill in the inevitable blanks.

Finally, in Chapter 8, T. Maggs deals with the last episode in southern African prehistory, the appearance of Iron Age mixed agriculturalists, beginning at or shortly after the time of Christ. Maggs' own research in the Orange Free State and Natal has not only helped to define the chronology of the Iron Age, but also the manner in which Iron Age people adapted to southern African environments. Maggs' chapter shows how Iron Age studies comprise both prehistory, as it is understood by archeologists, and history, as it might be understood by many of the modern inhabitants of southern Africa.

The contributors know that many readers will want to pursue particular topics more deeply. Consequently, each contributor has cited numerous pertinent primary sources. The result is a unique, composite bibliography of approximately 800 items on southern African prehistory and paleoenvironments.

Unlike many multi-authored books, the present one was not born of a conference. Instead, it owes its origin to a suggestion by Mr. A.T. Balkema and its existence to the contributors' shared belief that an authoritative, comprehensive, and up-to-date survey of southern African prehistory and paleoenvironments would be welcomed by professional archeologists, their colleagues in allied Quaternary sciences, and their students.

KARL W. BUTZER
Departments of Geography and Anthropology, University of Chicago, Ill., USA

1

Archeogeology and Quaternary environment in the interior of southern Africa

INTRODUCTION

In attempting to elucidate past environments, several approaches are possible. The traditional, geomorphologic method was to identify erosional surfaces and river terrace levels, to postulate specific climatic trends explaining their development, and then to collate the results into a relative chronology of environmental change. With the advent of absolute dating and specialized analytical techniques, the emphasis switched to detailed evaluation of long sediment sections and to interpret these according to sedimentary, geo-chemical, or paleobiological criteria. Whether the example be a deep-sea core or a pollen profile, the apparent implications of a single time-line were extrapolated to entire regions, perhaps an ocean or a continent.

Neither approach is satisfactory to those of us who believe that past environments were spatially complex, and that temporal trends were always accompanied by equally complex, spatial interactions. Paleoclimatic reconstruction, both temporal and spatial, must be primarily inductive, if we are to avoid fitting the data to a hypothesis, true or false. Deductive research phases have their place, in formulating research programs, and in the final phases of synthesis and interpretation. The primary work must however be inductive, at the level of data-gathering and analysis.

In the case of southern Africa, the 1970s have been a decade of data-gathering and analysis, despite repeated attempts at premature, deductive synthesis. In accepting the task of writing a chapter on Quaternary environments in southern Africa, I felt that I had only two options, firstly, a gloss of the key published sequences with some very general conclusions or, secondly, an interim assessment of the mass of data that I had gathered during the course of 13 months of fieldwork on 8 occasions between 1969 and 1982. My work had focused on some 3 dozen archeological or faunal sites, and almost 4000 sediment samples were tediously analyzed between periods in the field. I chose the time-consuming option, because I feel that too many meaningless generalizations have already been offered.

Again I had two choices, I could provide summaries of the conclusions derived from each local or regional study, but these would not be self-explanatory, and to specialists they would appear unconvincing, even speculative, because the bulk of the background data remains unpublished. I could also present several sets of information in some detail, outlining the criteria for low-level interpretation, and suggesting the first stages of synthesis. I chose the latter alternative, focusing on four categories of data that each represent a coherent set of interrelated archeological and paleoenvironmental issues; (a) the australopithecine breccias of the Transvaal and northern Cape; (b) the valleys of the lower Vaal River and its tributaries; (c) the lakes, pans, and springs of the semiarid to arid interior basins; and (d) a selection of long cave sequences that illustrate the contrast between the arid and the humid extremes of southern Africa. Not all of the information so generated is readily reconciled within any one hypothesis, and I feel that a broader synthesis would be premature at this point in time. At any rate, I am presently unwilling to offer one, and the reader would be better served by being able to personally confront the persistent problems as they emerge in the course of this paper.

Ultimately, I am convinced that logical spatial and temporal configurations will emerge, patterns suitable for paleoclimatic interpretation. In the meantime I will forego the reconstructions of rainfall belts and pressure fields that have become a popular pastime for some scientists, who should know their own qualifications as well as the limitations of their data better.

Before embarking on this explicitly analytical study, a few words are appropriate in regard to the strategy of archeo-sedimentological research that happens to be my own preferred technique in establishing a particular data base for the study of prehistoric human ecology.

One of the most immediate records available for productive study of fossil or archeological strata is the sediment matrix that embeds the site components. This matrix commonly consists of multiple, event-specific or process-specific lenses that document depositional micro-environments. In their turn, such micro-environments reflect the natural moisture regime, the balance of soil formation and erosion, as well as the vegetation cover of the external, topographic matrix. At an even larger scale, this meso-environment forms part of a physical and biotic landscape, the regional matrix (Butzer 1982).

The interpretation of field and laboratory analyses devoted to sequences of site deposits consequently involves a scale-translation between sedimentary, topographic, and regional contexts that is or should be explicit. Clues are provided by (1) the proportions of clay, silt and sand, which reflect sandy or clayey soils in the meso-environment, as well as transport energy in the micro-environment; (2) the sorting of size-grades into distinct or blurred, and single or multiple peaks, reflecting the efficiency of one or more transport modes, so helping in their identification; (3) the arrangement of sediment into beds or laminae, subhorizontal or inclined, reflecting minute episodes of deposition and/or erosion, frequently characteristic of one particular transport mode; (4) presence or absence of cementing material, such as calcium bicarbonate, reflecting surface or subsoil water movement during or after sedimentation, in relation to specific evaporation properties within the micro-or meso-environment; and (5) erosional evidence, within or between horizons, reflecting mechanical abrasion or chemical corrosion as a result of processes such as running water, wind, or human disturbance.

THE AUSTRALOPITHECINE BRECCIAS

Introduction

The late Pliocene and early Pleistocene in southern Africa are probably best documented in the richly fossiliferous deposits of the Transvaal and northern Cape that yielded australopithecine remains since the 1920s. These 'bone breccias' are exposed as complex, dolomite cavern sequences at: (1) Swartkrans, Sterkfontein, and Kromdraai, in the Sterkfontein or Bloubankspruit valley 9 km north of Krugersdrop; (2) Makapansgat, in the upper Dorpsrivier, 18 km east of Potgietersrus; and (3) the Norlim quarries, near Taung, 64 km north of Warrenton. Collectively these sedimentary suites record the details of Plio-Pleistocene landscape evolution, while at the same time providing a context for some of Africa's most important hominid fossils.

Today the Sterkfontein Valley averages 715 mm annual precipitation, 85% during the summer half-year, with a dry-subhumid climate (South African Weather Bureau 1965); it is a degraded grassland, with deciduous trees limited to valley bottoms, spring emergences, and sinkholes. The Makapansgat area has 625 mm of mainly summer rainfall (88%), a dry-subhumid climate, and a grassy thornbush vegetation. Taung, on the other hand, has 450 mm of predominantly summer rainfall (85%), a semiarid climate, with thorn-tree savanna.

In the Sterkfontein and Makapansgat areas the fossiliferous caverns were originally dissolved below karstic landscapes, subsequently incised by 30-40m. The caves were enlarged through solution and roof collapse, and repeatedly filled in with flowstones, collapse rubbles, and external soil; at intervals these fills were partly dissolved and eroded, before fresh sediment formed in place or through external derivation. The interactive geomorphic processes responsible were controlled by local groundwater tables, by the distance and diameter of vents to the surface outside, by the local topography, and by the regional hydrological and geomorphic balance as reflecting climatic regime, vegetative cover, soil type and sediment supply. The available data speak for alternating periods of active and passive slope balance, the former with soil attrition and valley or cavern filling, the latter with re-establishment of sub-continuous soil mantles. As a result of such periodic landscape evolution, episodes of accentuated erosion favoured accelerated landform sculpture and net, long-term soil attrition, exemplified by active erosion, transport and deposition. Such morphodynamic episodes alternated with intervals of comparative stability and slow geomorphologic change, during which the rate of weathering equals or exceeds the rate of erosion (Butzer 1976).

Swartkrans

The deposits of Swartkrans are preserved as a truncated cave 'root' on the side of a hill, and now lie 22-30m above the valley floor. An evolutionary model of the hillside is given by Butzer (1976), of the cavern complex by Brain (1976, 1981). Several distinct sedimentary units can be recognized that change in facies from below the former external vent (Outer Cave) to what was an interior cavern (Inner Cave). They are as follows:

a) Resting on over 2m of clean travertine (Unit IA), arranged in stalagmitic bosses and flowstone cascades below the former vent, is a 1m mass of crude dolomite rubble, with a matrix of reddish brown (5 YR 5/3), clayey sand-silt, very poorly sorted, with 74% calcium carbonate equivalent (C.C.E.) (Unit IB1).

b) Some 1.5-2.0m of thin-bedded, reddish brown (5 YR/5/4) stoney sandy silt or clayey sand-silt, poorly to very poorly sorted (67% C.C.E.), with concentrations of medium-grade rubble, and interbedded flowstones in lower half, dipping 5-20° to interior. Terminated by flowstone within Inner Cave (Unit IB2).

c) About 1m of light reddish to reddish brown (5 YR 5.5/4), clayey sand-silt, poorly to very poorly sorted (64% C.C.E.), with local concentrations of medium-grade rubble in Outer Cave; thin-bedded, dipping 5-15° to interior in Outer Cave, 20-40° in Inner Cave. Terminated by thick flowstone in Inner Cave (Unit IC).

d) The Outer Cave has over 6m of reddish brown (5 YR 4.5/3), clayey sand-silt, very poorly sorted (61% C.C.E.), arranged in gravelly lenticles (8-15° dips); thin-bedded (4-8° dips), sheets of finer material; or rubbles with basal, laminated flowstones (8-27° dips). The Inner Cave equivalent measures 10.5m thick and consists of reddish brown (5 YR 4.5/3), sandy silt or clayey sand-silt, moderately to poorly sorted (57% C.C.E.); laminated and thin-bedded (3-10° dips), with rare, shallow cross laminae; some fine gravel zones and traces of a terminal flowstone interdigitation (Unit II).

e) Several generations of younger deposits, with differing degrees of cementation and sorting, in part rubbles, in part soilwash, mainly clayey sand-silts, are found in solution conduits or larger fissures opened in units I or II by subsequent, large-scale corrosion. Two sediment suits were studied from the Lower Cave (under the Outer Cave). The earlier of these is rich in microbone, and consists of alternating flowstone-rich lenses, with a residual of very poorly sorted, clayey sand-silt, and compact sandy silt, poorly sorted. The later suite has less flowstone development and is dominated by gritty to clayey sand-silts with better and more consistent sorting. These suites are here informally labelled units IIIa and IIIb, but stratigraphic distinctions will be difficult to apply to the maze of younger solution pipe fillings.

The 16m of sediment in Units I and II were defined as the Swartkrans Formation, with two members, by Butzer (1976). A more recent paper by Brain (1982) introduces two new stratigraphic elements that require clarification here. Firstly, Member IB was originally defined with respect to the 'Hanging Remnant' of Brain (1982 and pers. comm.); his new 'Lower Bank' represents a steep debris cone, eroded and then indurated by stalagmite. This 'Lower Bank' may be older than unit IB and requires reevaluation of the position of the 'basal' travertine (IA). Secondly, Brain (1982) writes of an older and younger component to Member II, with reference to the fauna rather than to a specific cave profile. For the Inner Cave, Member II was originally defined by Butzer (1976) on the basis of a fully conformable suite of deposits exposed along most of the north walls; for the Outer Cave this member was defined by a 6m profile on the west wall that contains several depositional breaks but no erosional or corrosional unconformity. It is probable that Brain's 'younger component,' rich in *Antidorcas bondi* but lacking hominid remains, refers to intrusive fillings younger than Member II as defined in Butzer (1976); they should presumably be relegated to the informal unit III complex.

The robust australopithecine fossils came primarily from unit IB, but also from IC, as well as (the base of?) II, where they may be in a reworked context (Brain 1982). Three individuals of *Homo* cf. *habilis* (SK 847 composite, SK 27, SK 2635) come from IB, while the *Homo erectus* (SK 15) and several related teeth appear to come from II (Brain 1982). At least several of the Oldowan artifacts (Leakey 1970) came from unit IB; none of the hand axes I have seen in place came from unit II, but were in later, intrusive deposits, so that the Acheulian assignment to that member must be considered dubious; I have not seen Middle Stone Age artifacts from Swartkrans, although such may be present in younger fissure fills (see Vrba 1976). Finally, the older Swartkrans fauna (SKa) comes from Member I, while the 'older component' of the younger fauna (SKb), as originally used by Vrba (1976, 1982), is representative of Member II.

Attempts to isolate remnant magnetism for the different stratigraphic units at Swartkrans were not particularly successful: 8 of 14 samples were unstable or inconsistent, 4 were intermediate, and 2 normal–both of these from Unit II (outer) (Brock *et al.* 1977).

The basal flowstone (IA) was deposited over an essentially clean cavern floor, but contains some external debris in its final lenses. Unit IBI is a minimally-bedded mass of blocks and residual soil, with reduction of size and improvement of stratification away from the former vent. Rapid, if not penecontemporaneous accumulation is suggested by the lack of flowstone interlaminations. This collapse breccia includes soil, bone, and rare artifacts probably derived from an older, higher-lying surface depression or external cave. The matrix includes minimal proportions of sand but considerable clay. This implies derivation from a deep, clayey surface soil, unlike modern stoney slope soils. The closest external analogs are provided by the relict B-horizon preserved on the upland flat north of Swartkrans Hill or the dissected, red colluvium on the lower slopes below Sterkfontein.

Unit IB2 suggests a much slower, incremental build-up of sediments, interrupted by local travertine episodes, and including fewer and fewer crude scree components. Stratigraphic differentiation is impeded by the rapid vertical and lateral changes within the successive sweeps of sediment constituting this complex of minor beds. Eventually the topmost strata develop greater horizontal uniformity, with a composition remarkably similar to gritty, modern slope soils. Although some soil and organic debris will have washed into the doline overhead after each protracted rainstorm, the presence of thin-bedded, multiple sediment increments argues for deposition in the wake of unusually strong surface runoff, following major 1-in-10 or 1-in-100 year downpours. In between such sporadic surges of water, loose sediment, bone and artifacts probably accumulated slowly on the sides of the sinkhole above.

Ultimately the vent area of the Outer Cave was largely filled with sediment, and accretion was restricted to thin flowstones. At about or after this time, the huge dolomite block on the cave floor appears to have settled, to open a new vent of at least 60cm height between the Outer and Inner caverns, allowing a new generation of deposits to accumulate in both cave sectors.

Unit IC is relatively shallow but extensive and homogeneous, except where crude detritus is interbedded in former vent proximity. Erosion from clayey soils by surging, concentrated runoff waters and soil creep is suggested, with some rubble picked up within the overhead sinkhole. Except for their high grit level, modern slope soils are textually similar; this implies the former presence of B-horizons within the catchment of the doline, i.e., a deeper external soil mantle than at the time unit IB2 was accumulating. After unit IC had filled most of the Outer Cave, thick flowstones were deposited in the Inner Cave.

Member II in the Outer Cave suggests a cavern with direct access to a doline in process of active enlargement. Periodic rock collapse complicates a record of alternating deposition and erosion. The lower third of this sequence includes distinct erosional disconformities, and one large block of already-cemented sediment appears to have been dislodged (after substantial corrosion) and rotated backwards by at least 20°. The topmost 2m indicate mass movements, possibly slow earthflows. The bulk of the sediment implies gradual accretion in response to periodic, surging runoff combined with soil creep, but the frequent flowstone interbeds argue for repeated episodes with very limited sediment transfer. The closest modern analog for this overall suite is provided by the B-horizon on the upland flat north of Swartkrans Hill, so that a deep soil mantle must be posited beyond the margins of the stoney doline.

The Inner Cave facies of Member II represents a 'filtered,' i.e., sorted variant of the Outer Cave, lacking both the coarsest fraction (trapped in the Outer Cave) and much of the clay (flushed out via the Lower Cave). There are no external analogs that are texturally similar, but comparable sediments have been laid down in small depressions of the cave during the last few years: these are laminated silts, swept together from the miners' back-dirt into small muddy pools during downpours. The 10m column of Member II in the Inner Cave indicates a very protracted interval of periodic, high but uniform-velocity influx from one or more openings, with silt and sand accumulation in the ponded waters of the Inner Cave.

The unit III sediments, found in the Lower Cave, have also been subject to a filter effect, and tend to be intermediate in character between those of unit II in the Outer and Inner Caves. They lack evidence of mass movements but generally suggest similar conditions of accumulation; they span a somewhat shorter period.

Sterkfontein

Like at Swartkrans, the australopithecine deposits of Sterkfontein form a truncated cave 'root,' at 28-41m above the modern valley. The total sediment accumulation of the Sterkfontein cavern complex has been assessed at over 30m by Partridge (1978), the lower half of which is limited to the (now) subterraneuan Silberberg ('Day light') Grotto. Wilkinson (1983) has shown that the lower fillings of the cavern complex cannot be directly linked to the higher ones, because much of the initial fill was dissolved and removed after a 15m fall of the water table, also allowing younger sediment to penetrate deep into the various caverns. Consequently, Member 1 of the Sterkfontein Formation of Partridge (1978) only starts halfway up the stratigraphic succession, while Member 3 is split by a 30m gap, posing serious problems for resolution of the early history of the cavern complex. Since the fossils of these units have not been systematically investigated, I prefer not to discuss these deposits, focusing instead on those in the surface exposures.

Partridge (1978) presents summary sedimentological data for the key exposed units (his 4A-4D, 5, 6A and 6B) which unfortunately does not match that of my own analyses of 4 and 5. His acid-soluble percentages are substantially lower, as are his silt and clay fractions, indicating procedural differences, possibly related to incomplete decalcification of his samples. I also obtained substantial contrasts of sorting, between 4A on the one hand and 4B to 5 on the other, that are not apparent in his descriptions. My detailed analyses are summarized here and will be published elesewhere.

a) The base of the exposed sequence at the Sterkfontein Type Site is formed by over 4m of dolomite slabs and blocks, with an indurated matrix of light reddish brown (5 YR 6/4), clayey sand-silt, averaging 75% C.C.E., and very poorly sorted. This lower block breccia, Partridge's unit 4A, is essentially sterile and probably reflects one or more rapid and massive rock collapses around the sinkhole; the matrix is almost certainly younger than the rubble, washing in slowly from above, to fill in the interstitial voids. Several laminated calcite horizons within the breccia probably reflect hydrological changes during the infilling process.

b) The gracile australopithecine fossils and terminal Pliocene fauna (STS, Vrba 1976) essentially all come from a second, finer accumulation between this block breccia and a travertine bank prominently visible under a remnant of dolomite roof. This unit has been almost completely removed in the Type Site, but is preserved in the 'Mrs. Ples' pedestal and can be traced to the west and south walls, where it assumes the configuration of a cone coming in from the south. The sediment is a relatively well-sorted sandy silt, of variable color (averaging light reddish brown to reddish yellow, (5 YR 5.8/5.1) and C.C.E. (averaging 66%), and a thickness of up to 3m. It is equivalent to Partridge's (1978) 4B and 4C units, as separated by the impact of corrosion and minor erosion. This is a soil wash, carried into the cave periodically by surface runoff, and variably intermixed with some roof rubble or external ferricrete or cherty sand-sized debris. Relatively rapid water movements, the conspicuous bedding (7-15%) with minor reversed lenses, the limited clay, and presence of some roof rubble all suggest that the exterior surface preserved a measure of doline topography at this time, channelling water in. The terminal travertine of 20-70cm suggests temporary closing of the cavern, after several large roof blocks had collapsed upon unit 4B/4C.

c) After a long period of corrosion, a new suite of sediments washed into the cave (Partridge's unit 5). These contain some basal subfacies that reflect slow, interior accumulations, such as pockets of 'rodent breccia' derived from cave owl pellets or rubble-rich zones. This subfacies is a reddish brown (5 YR 4.5-6/3-4), poorly to very poorly sorted, stoney clayey sand-silt, whereas the dominant facies is a reddish brown (5 YR 5-6/3-4), better-sorted sandy silt with conspicuous, low-angle bedding (10-15°) in the Extension Site, steeper (15-28°) in the Type Site; C.C.E. averages 62%. Within the Extension Site, unit 5 is a classic colluvium, deposited immediately inside the cave entrance; only further inside is there evidence of minor collapse rubbles and older soil derivatives. A minor intervening flowstone indicates two phases of deposition. This is the unit containing the artifacts described by Leakey (1970) as Oldowan, together with very fragmentary fossil material suggesting a gracile australopithecine or *Homo* aff. *habilis* (Hugher & Tobias 1977). The fauna is believed to be late early Pleistocene (SE, Vrba 1982).

Sediments exposed within the base and on the walls of the 'Lower Cave' are reddish to light reddish brown (5-7.5 YR 4-6/2-4), poorly to very poorly sorted, sandy clay-silts with an average 79% C.C.E. To me they appear to be intrusive, and bedded against and under the block breccia. But they correspond closely to the description given by Partridge (1978) for his member 3.

I did not find a satisfactory exposure of Partridge's (1978) youngest unit, member 6, and cannot provide independent confirmation for the derivation of the younger late Pleistocene fauna (D16) posited by Vrba (1976). The modern surface soil I sampled on top the Sterkfontein hillock is dolomite and chert 'gravel' with a matrix of poorly sorted, sandy silt (7% C.C.E.); it resembles nothing in the cave suite.

There are no good modern valley or slope analogs for any of the Sterkfontein units. None of the Sterkfontein units match any of those at Swartkrans exactly, but Sterkfontein 4A has similarities with Swarktkrans IBI, while Sterkfontein 4B-4C has some limited similarities with Swartkrans IB2. In any event, the temporal sequence of sedimentary events was decidedly different in both caves.

Kromdraai

At Kromdaai there are three excavated trenches (see Brain 1981: 250), located on a bedrock platform at 40-48m above the valley floor.

The key hominid site (KB) represents a well-stratified fissure infilling generally inclined at 14-24°. Several units can be recognized, representing discrete sediments without major erosional breaks or changes of dip.

a) Heavily oxidized, yellow (10 YR 8/5), partly decalcified (12% C.C.E), sandy silt with dispersed rubble; thickness unknown. Large voids, commonly refilled with much younger, decalcified breccia (g).

b) Some 1.5m of light reddish brown (5 YR 6/5), clayey silt-sand, with dispersed gravel and several levels of crude rubble, suggesting repeated but minor vent-wall collapses. Bedding (16-23°) improves from discontinuous, wavy, parallel to continuous, parallel and even in the upper third, and ultimately to laminated units; locally, near a former sluice, the final beds change abruptly in dip and direction, and there are knobs of white stalagmite. Partial decalcification of the upper third was followed by recementation, with abundant manganese and iron (47% C.C.E.). The upper contact marks a depositional break, not a true erosional unconformity (contra Partridge 1983).

c) A stoney marker horizon, with 35cm of reddish yellow (5 YR 6/6), clayey sand-silt (54% C.C.E.), was preceded and terminated by laminated calcretes. This unit is further distinguished by its feathered bedding (16-27°) and its irregular, dark gray (5 YR) discoloration due to heavy concentrations of manganese dioxide (pyrolusite), with some iron compounds and phosphates. Faunal concentration, with a robust australopithecine and abundant cercopithecoids similar in age to Swartkrans I (Vrba 1981, 1982).

d) Almost 1m of pink (7.5 YR 7/4), clayey sand-silt with dispersed rubble (60% C.C.E.); thin-bedded (parallel, wavy, 16°) or laminated, with flowstone lenses or pockets.

e) Some 1.5m of reddish brown to yellow (5 YR 5/4, 7/5), clayey sand-silt (56-63% C.C.E.) with dispersed rubble and even, non-parallel bedding (16-30°).

f) Almost 4m of reddish brown (5 YR 5-6/3-5), clayey sand-silt or sandy silt (48-73% C.C.E.) as matrix to medium to boulder-grade, angular to subangular rubble; thin-bedded throughout, typically non-parallel and 13-14°, but with finer terminal beds parallel-bedded, with 2° reversed interbeds, as the fissure was infilled and sedimentation came to a halt.

g) Inhomogeneous, contiguous mass of decalcified breccia (2% C.C.E.), with a matrix of dark brown (7.5 YR 4/4), silty clay-sand, with much manganese discoloration. The chaotic bedding and the abundance of boulder-grade gravel distinguish this as a new deposit, filling a large solution or erosional vent near the eastern end of KB, intruding unit (a), and in part now encountered under the older beds. The microbone points to an interior facies.

The KB fissure continues as a small, former underground passage (KB-West) with laminated, reddish brown (5 YR 5/3), gritty, sandy silt (55% C.C.E.), terminating in a laminated pool fill some 60cm in diameter, with a residual of sandy silt (80% C.C.E.); dips are 5-15° and manganese stains are present. These minor beds may represent a lateral extension of unit KB (f).*

The faunal site (KA), with an early Pleistocene fauna (Brain 1981; Vrba 1982), is not directly connected with the 15m-distant hominid site but once formed part of the same underground fissure complex. Both sites parallel the strike of the local dolomite. At KA there now are vestiges of at least 1.5m of deposit resting on basal travertines. First is a 15cm lense of crude rubble with a matrix of reddish brown (5 YR 4/3), sandy silt (58% C.C.E.), with abundant bone. This is followed by laminated sandy silts (57% C.C.E.), with microbone. Finally there are alternating thin beds of sandy silt and gritty lenticles with microbone, terminating in a laminated deposit (56% C.C.E.).

The third trench (KC) is 5m east of KB. It exposes 75cm of reddish brown (5 YR 4/2.5), sandy silt (65% C.C.E.), with gritty basal interbeds, over horizontal travertine and under a thin cap of the same. Vrba (1981) suggests correlations with KB units (c) and (d).

The modern footslope deposit below Kromdraai is a dark brown (7.5 YR 4/4), very stoney, organic sandy silt (1.5% C.C.E.).

Overall, the KB site and the basal rubble of KA indicate free access to external sediment, and these are the beds that contain macro-fauna. Conspicuous thin-bedding and laminations are notable for most of the surviving outcrops. Laminated, subhorizontal, deep passage beds are reasonable, but the abundance of clay and moderate carbonate levels in the systematically inclined beds of KB are more difficult to explain. However, in detail, the KB bedding, whether even or wavy, and parallel or non-parallel, coincides with minute carbonate enrichments. This indicates repeated, cyclical deposition, with each graded set beginning with coarser pebbly silts, terminating with lime muds. Episodes of vigorous flushing are also recorded by the fine, prograding pool deposits in KB-West. The closest parallels for the KB sediments are provided by units IC and II in the Outer Cave of Swartkrans. Good modern analogs for the KB fill are provided by the stoney surface soils above Swartkrans (for KB rubbly units) and by the relict B-horizon on the 15m platform south of Swartkrans (for KB fine fill). Moderately clayey soils were evidently washing into the KB fissure, together with bone fragments or concentrations, in response to discrete runoff events outside. Contrary to the claim of Partridge (1982), the KB facies does not record a

*The proposed status of the KB-West deposits as a separate formation, with 3 members, as contrasted with a KB-East formation, with 5 members (Vrba 1981; Partridge 1983), is quite unreasonable. In fact, any formal stratigraphic definition of the relatively simple fissure fills at KB serves no practical purpose.

unique geomorphologic environment, and it indicates a relatively moist, not aberrantly wet setting. Partridge's (1982: 637) rationale for linking the KB unit with 'the Great Patagonian Glaciation' of 1.0-1.2 million years ago eludes me.

The KA facies are not dissimilar to the modern surface sediment at Kromdraai, and indicate somewhat coarser surface sediment washing into some type of karstic fissure, the original morphology of which is now largely destroyed. In combination with the subsequent laminated facies, the KA sediment package has fundamental similarties with units 4C and 5 at Sterkfontein.

Interpretation of the Sterkfontein Valley sequence

The sedimentary units from Swartkrans, Sterkfontein, and Kromdraai that include hominids and/ or fauna can now be summarized, in the sequence suggested internally, as well as by the faunal stratigraphy of Vrba (1976, 1982) (Table 1).

Table 1. Sediment interpretation for fossiliferous units in Swartkrans, Sterkfontein and Kromdraai (fauna after Vrba 1976, 1982, and Brain 1981). Note: in all tables, the youngest units or phases are at the top, despite various numbering conventions.

Swartkrans III (complex of several fills) (mid or late Pleistocene fauna [SKb, early and/ or late component]). Moderately long accumulation of clayey surface soils by surging runoff, with interruptions marked by minor flowstones. Implies deep soils and periodic heavy rains, with relatively effective ground vegetation outside. External conditions similar to those of today, possibly a little moister.

Swartkrans II. (Oldowan or Acheulian artifacts and *Homo*; late Lower Pleistocene fauna [SKb, early component], with 71% alcelaphines and antilopines). Protracted accumulation of clayey surface soils by surging runoff or mass movements, with repeated interruptions marked by minor flowstones or corrosion. Implies deep soils and periodic heavy rains of considerable duration, with relatively effective ground cover. Moister than today.

Sterkfontein 5. (Oldowan artifacts and *Homo*; mid-Lower Pleistocene fauna [SE], with 82% alcelaphines and antilopines). Moderately long accumulation of stoney surface soils by surging runoff in two phases, interrupted by minor flowstone. Implies shallow soils and high intensity rains, with incomplete ground cover. Conditions probably similar to today.

Kromdraai KA. (mid-Lower Pleistocene fauna, with 81% alcelaphines and antilopines). Relatively brief accumulation under conditions similar to those applying to Sterkfontein 5; possible partial overlap with Sterkfontein 5.

Swartkrans IC. (robust australopithecine; early Lower Pleistocene fauna [SKa]). Relatively brief accumulation of clayey surface soils by surging runoff and limited mass movements. Implies deep soils and periodic heavy rains of considerable duration, with relatively effective ground cover. Moister than today.

Kromdraai KB. (robust australopithecines, essentially no artifacts; early Lower Pleistocene fauna, with at least 88% alcelaphines and antilopines). Identical sedimentary record to Swartkrans IC, except for absence of mass movements; possible overlap with Swartkrans IC, alternatively with IB1.

Swartkrans IB2. (robust australopithecines and Oldowan artifacts; early Lower Pleistocene fauna [SKa], with 78% alcelaphines and antilopines). Protracted accumulation of sandy surface soils by surging runoff, repeatedly interrupted by breaks with minor flowstone development. Implies thin soils and periodic high-intensity rains, with incomplete ground cover. Conditions similar to today.

Swartkrans IB1 (robust australopithecines, possible *Homo,* and Oldowan artifacts). Rapid accumulation of collapse breccia with clayey matrix soil. Implies deep soils and probably a relatively effective ground cover. Similar to or moister than today.

Sterkfontein 4B-4C (gracile australopithecines; terminal Pliocene fauna [STS], with 5% alcelaphines and antilopines). Moderately long accumulation of stoney surface soils by surging runoff, in two phases, interrupted by some erosion. Implies shallow soils and high intensity rains, with incomplete ground cover. Conditions probably similar to today.

Sterkfontein 4A (some fauna). Relatively rapid collapse of block breccia, followed by longer infilling of voids with clayey soil derivatives, interrupted by minor flowstone developments. Implies deep soils and periodic heavy rains, probably with relatively effective ground cover. Possibly moister than today.

The record of Table 1 is biased in favor of these time spans leading to net cave sedimentation, as a result of active slope balance and soil attrition. Long periods with less intensive rains and a climate slightly drier than today would have favored sedimentary breaks, with little or no soil erosion. On the other hand, protracted intervals of intensive corrosion would require a ground vegetation sufficiently dense to impede tangible soil erosion or, alternatively, would need lower-intensity rains but acidic soil environments and considerable effective moisture. Such corrosion was notable between Sterkfontein 4 and 5, as well as between Swartkrans IC and II. The flowstone record is more ambiguous. Massive travertines, such as Swartkrans IA or Sterkfontein 4D, reflected more or less closed cave microclimates; on the other hand, minor flowstone interbeds within detrital sediment packets indicate periods of more moderate rainfall, but not necessarily drier than today.

Whereas the various detrital sediments indicate climatic ranges from at least as dry as modern Makapansgat (ca. 600mm) to wetter than today (ca. 800mm), the corrosion phases must have been so wet as to preclude supersaturation of groundwaters with calcium bicarbonates (areas with over 900mm rainfall in the eastern Transvaal today). All of the detrital sediments, reflecting as they do high-intensity precipitation, are likely to have been eroded by summer rains - which today increase in intensity from southwest to northeast across the South African interior (South African Weather Bureau 1965: Fig. 155). Cooler climates may have favored less intensive rainfall or denser grass swards or both.

The bovid fauna of the Sterkfontein Valley deposits carries potential implications for grass versus bush or woodland cover (Vrba 1976), with the alcelaphines and antilopines representing specialized grazers, at least during Pleistocene time (Klein, this volume). The correspondence between the alcelaphine/antilopine proportion and the geomorphic evidence in Table 1 is unsatisfactory, presumably because both criteria are indirect reflections of a primary condition. The faunal record is complicated by temporal trends, with a less competitive and less specialized grassland fauna during Pliocene times (Klein, this volume); by the question of which kinds of predators accumulated the bone and from what distance (Brain 1981; Klein, this volume); and by the complex mosaic of grass and woodland to be expected in relation to topographic details and surface hydrology. I feel that the sedimentary record, by explicitly attempting to translate microfacies variability into external geomorphic and hydrological controls, is more likely to be a reliable reflection of the regional landscape. And, detailed faunal statistics aside, a rainfall span of 600-800mm would not change the basic mix of grassland and local thickets-in physiognomic rather than floristic terms-with which all the faunal assemblages are readily reconciled.

Makapansgat Breccias, Central Transvaal

The australopithecine site of Makapansgat is located in the uppermost Dorpsrivier valley, within a mountainous dolomite landscape that goes over into other, quartzitic rocks in the headwaters. Solution processes have roughed up the topography, created jagged, micro-karstic (*lapiès*) slopes, and opened several large, complex dolines (uvalas). The Limeworks Cave is part of one such uvala, the nearby Cave of Hearths represents another large solution cavern, while the configuration of the valley floor is locally controlled by the outlines of another breached uvala. The Limeworks Cave is below a dolomite platform now at 40m above the modern floodplain, with the key deposits at 24-44m above the stream bed.

Partridge (1979) has described a stratigraphic scheme with 5 members to a Makapansgat Formation of up to 50m original thickness*. I reexamined the Limeworks exposures with Partridge in 1974, when we were able to resolve the fundamental stratigraphy to our mutual satisfaction. However, as in the case of the Sterkfontein sediments, there are methodological problems in their analysis, and I fail to follow the vague geomorphologic arguments on which the paleoclimatic conclusions of Partridge (1982) are based. The summary description that follows presents my own observations and data, with complementary empirical information from Partridge (1979) and Brain (1958) for the final two members.

a) Member 1. A continuous sheet of white travertine originally covered the uneven dolomite base of Makapansgat, with typical thickness of 3-10m. This unusually extensive and massive flowstone is dominated by parallel, even bedding and lamination, with a few basal, wavy-bedded cascades (15° dips) and local, lenticular wedges of vertically crystallized material (2-7°). The upper third of the unit is interbedded with 10-30cm lenticles of reddish silts, and eventually goes over conformably into the next unit via wavy, non-parallel beds. Locally there also are roof blocks at the base and which are covered by convex, dripstone bosses with upper contacts of 40-65°, while in another area roof rubble is interbedded with the later flowstone accumulation.

b) Member 2. The northwestern cave sector has 2-6m of well stratified (even to wavy, non-parallel), light reddish brown to red (2.5 YR 6/4-8) and reddish yellow (5 YR 6/8) detrital material with flowstone laminae. The non-calcic residual is a poorly sorted clayey silt (average C.C.E. 73%). There is frequent cross-stratification, including low-angle, simple or planar crossbeds (to 15°), low-angle trough crossbeds (to 10°), and high-angle, planar crosslaminations (to 25°) (terminology after McKee & Weir 1953; Campbell 1967). These document the accretion of small silt lobes, interrupted by occasional scour-and-fill structures, across a surface originally dipping as much as 22° down from a former entrance ("Rodent Corner"), where coarser, sandy facies are represented (Turner 1980). In basal irregularities, Member 2 begins with pockets of dipping, interlaminated, reddish silty calcite and buff flowstones, with bone and calcite grit; and in other areas it grades up conformably into the next unit, despite local erosional contacts.

c) Member 3. The key fossiliferous unit or 'Gray Breccia' was originally best developed in the modern entrance and is now fragmentarily preserved in shallow outcrops just under the dolomite roof, with a modal thickness of 0.5-1.5m. Typical development is in the form of multiple lenses, 1-1.5m long and 5-20cm thick, with curved parallel or wavy non-parallel bed and laminasets, and dips averaging up to 5° but locally attaining 30°. Such lenses tend to consist of reddish yellow or light reddish brown (5 YR 6/4-6), silty calcite (54-98% C.C.E.), with stalactite rubble of sand to gravel grade, and massive bone concentrations. The non-calcic residual tends to be a poorly sorted, sandy clay silt, with abundant opaline silica and traces of shale and

*The advantages of a formal stratigraphic framework at Makapansgat are considerable, given the number of poorly chosen or imprecisely defined lithological designations used by various researchers over the years.

quartzite sand. The lenses are interbedded with white to yellow (10 YR 8/4-6) or pink (5 YR 8/3) flowstones with pyrolusite dendrites, or thin lenticles of reddish (2.5 YR 4/2, 5 YR 6/5) clayey silt of clayey sand silt, with abundant opaline silica and stalactitic grit, and possibly bone. The irregular basal surface on which Member 3 was deposited has been three-dimensionally plotted by Maguire *et al.* (1980). These authors show, by orientation diagrams, how bone originally derived from hyena dens was concentrated in localized sediment traps by current action. In various areas Member 3 was terminated by up to 3m of white, wavy and parallel-bedded flowstone, with laminae rich in pyrolusite. A protracted hiatus followed.

d) Member 4. Deposition was renewed with masses of light reddish brown or brownish red (5 YR 5-6/4, 2.5 YR 6/5), sandy silt or clayey sand-silt (70-94% C.C.E.), very poorly sorted, positively skewed, with chaotic dolomite rubble. Stratification is limited to certain zones with flowstone laminae (3-10° dips, in various directions), and many areas of unlaminated, massive-bedding suggest later infilling of voids in roof or vent rubbles with external colluvium. Sands of shale, quartzite, and metamorphic minerals, as well as quartzite pebbles indicate exotic materials. Typically 2-15m thick, Member 4 is subdivided by Partridge (1979) into Beds A and B, the lower subaquatic, the upper part of a debris cone coming in from near the modern entrance; the distinction is frequently difficult to make. The important fossils, including cercopithecoid and rodent concentrations, preferentially came from the lower facies or the distal parts of the cone (Partridge 1979). A long break separates Member 4 from the next unit, during which the entrance was considerably enlarged, new small openings created, and parts of the cone compacted and sagged.

e) Member 5. Renewed deposition began with 2m of reddish yellow (5 YR 6/5), clayey sand silt, with abundant subangular to subrounded rubble, dipping up to 22° in thin, discontinuous bedsets. This is followed by about 5-8m of reddish yellow (5 YR) 6/6), gravelly silty sand or packed, coarse-grade dolomite and quartzite gravel, in thick bedsets. Finally there are reddish brown, pink, or red (5 YR 5/3-4, 8/4; 2.5 YR 4-5/6), gritty and sandy clay silts with gravel pockets, some 3-9m thick. This suite, divided by Partridge (1979) into Beds A-E, is very poorly sorted, negatively skewed, and C.C.E. values are highly variable (46-87%): originally the deposit was calcified *per descensum* but in more recent times it has been partly decalcified from the top down. The overall accumulation marks a detrital cone derived from the outside, terminating with wavy, non-parallel laminasets (3° dips), that continue as a surface veneer on top of the platform over the Limeworks Cave. There is some fragmentary bone.

Discussion of the Makapansgat sequence

It is readily apparent that Members 1, 2 and 3 are distinct from any facies in the Sterkfontein Valley, whereas 4 and 5 are more comparable. The bedding properties cited are as essential to comprehension as the suite of modern analog samples studied. The latter are substantially better sorted than almost all the cave samples, and generally have a much larger coarse silt to fine sand component-precluding arguments for eolian admixture in the Makapansgat Formation. Further, the strong coarse sand component in the modern analogs is only replicated in members 4B and 5. The relatively low clay levels of the surface analogs-channel, floodplain, footslope, or midslope-are matched solely in 5A-5D, and the high clay and fine silt ratio of the

Table 2. Sediment interpretation for fossiliferous units in Makapansgat.

Makapansgat 5D, 5E. Protracted accumulation of clayey surface soils by surging runoff into open cave, with interruptions marked by flowstone laminae. Implies deep soils and periodic heavy rains, with relatively effective ground cover. Moister than today. Possibly overlaps with Swartkrans IB1.

Makapansgat 5A, 5B, 5C. Protracted accumulation of sandy surface soils by surging runoff into open cave. Rapid decline of clay level indicates destruction of soil mantle, with periodic, high-intensity rains and incomplete ground cover. Conditions apparently similar to today. Probably coeval with Sterkfontein 4B, 4C.

Makapansgat 4 (Gracile australopithecine; terminal Pliocene, cercopithecoid and rodent faunas). Protracted accumulation of clayey surface soils into small cave opening, in combination with internal collapse rubble. Implies deep soil and periodic heavy rains, with relatively effective ground cover. Moister than today, but possibly transitional. Probably overlaps with Sterkfontein 4A.

Makapansgat 3 (Gracile australopithecines; late Pliocene, bovid fauna; paleomagnetic interval ca. 2.9-3.2 million years). Repeated accumulation of fossiliferous marl lenses, interbedded with silts and flowstones, in shallow interlinked underground ponds; eventually cavern was sealed and extensive flowstones developed. Abundant subterranean waters linked to hydrology of a humid karst river. Climate substantially moister than today. May antedate Sterkfontein 4A.

Makapansgat 2 (some fossiliferous units, not differentiated from unit 3; paleomagnetic interval ca. 3.1-3.35 million years). Accumulation of silt lobes in an underground cavern by periodic lateral ingression of river discharge. Preponderance of fine suspended sediment infers hydrology of humid karst landscape, with reduced seasonal contrasts. Climate substantially moister than today.

Makapansgat 1. Long-term accumulation of massive, continuous flowstone under several meters of subterranean water seeping through cavern. During later stages, inruption of suspended sediment via underground fissures. Climate substantially moister than today with reduced seasonality.

fossil deposits is largely responsible for their very poor sorting. One must therefore conclude that a different soil and hydrological environment obtained during most of the time represented by the Makapansgat Formation: slope soils were clayey and river discharge was perennial rather than episodic.

The bedding characteristics of unit 1 are incompatible with an "emergent" cave floor, which would develop shorter bedsets, with prominent wavy stratification, and evidence of cascades. The dominance of continuous, even and parallel bedsets and laminasets through such a massive deposit argues for a fair depth of water with limited fluctuation. Under such circumstances CO_2 can best be lost during movement from water-saturated rock joints and fissures into the ventilated Limeworks cavern; rapid movement is excluded by the absence of planar cross-stratification. This suggests a meter or two of water, slowly seeping through the cavern as part of the standing water table. Since the dolomite is structurally complex, underground water behaviour was probably linked to perennial subsurface flow beneath a river. Although that watercourse need not have been perennial, such quantities and reliability of circulating water are incompatible with an arid or semiarid karst environment.

The initial appearance of external clayey silt within the upper third of unit 1, and the dominance of such sediments in unit 2, indicate enlargement of one or more horizontal fissures to permit detritus from the river channel to enter the underground water system. The complex laminasets indicate periodic ingression of suspended sediment with considerable velocity, presumably during the season of peak discharge. If bedload sediment (see Turner 1980) had been abundant, it would soon have choked off the entrance; this, plus the very limited clay and fine silt fraction of the rather varied spectrum of modern floodplain analogs sampled, argues for sediment transfer in the

Dorpsrivier similar to that in modern humid karst environments: a heavy preponderance of fine suspended load.

Unit 3 does not indicate a change of hydrology or stream sediment caliber so much as a change in water level relative to cavern floor. The Limeworks Cave had been largely but unevenly filled in, so that less water spilled in via externally-open conduits, with a higher proportion of subterranean water, free of detrital sediment. Ultimately the external conduits were sealed off, and cascade flowstones became characteristic of a wet but emergent cavern floor.

Thus, units 1 to 3 record a temporal sequence not obviously linked to climatic shifts. The abundant flow of subterranean waters, the prominence of derivatives from red clay soils, the minimal role of sandy or gravelly sediment fractions, the absence of torrential influx symptomatic of high peak discharge, combined with planar and trough stratification, all point to a substantially wetter and more equitable climate, with a distinct but reduced measure of seasonality, and an effective vegetative mat. This context is relevant to all the gracile australopithecine materials other than specimen MLD 37/38 (see Partridge 1979) and to much of the Makapansgat fauna, specifically the bovid assemblage which appears to have washed or floated into the cavern. Unfortunately this rich fauna has not been systematically analyzed, precluding a paleo-ecological interpretation. It corresponds most closely with but is a little older than the fauna from Sterkfontein 4 (Cooke H.B.S. 1978a; Vrba 1982). Paleomagnetic work indicates polarity switches at the base of units 2 and 4, with most of unit 2 and 4 normal (McFadden *et al.* 1979); this, in combination with the African faunal stratigraphy, indicates a minimum age of 3.3 million years for the base of 2, of greater than 2.4 million for the lower half of 4, and a time span between 2.9 and 3.2 million for the fossiliferous 'Grey Breccia' of unit 3. In other words, units 1 to 3 reflect late Pliocene conditions.

Deposition of Makapansgat 4 followed after an interval, but still within the Sterkfontein 4 faunal span, since the rich cercopithecoid fauna of unit 4 is matched by species in Sterkfontein 4 (see Collings *et al.* 1975; Cooke H.B.S. 1978a). The decrease in clay and increase in coarse sand, coupled with the invasion of a debris cone from above, indicate a fundamental microenvironmental change, but our five sediment samples from unit 4, and the data reported by Partridge (1979), are inadequate to determine whether temporal trends were more important than distance from the entrance sinkhole.

The fact that 4 of 5 samples are strongly positively skewed indicates that the small opening was selecting against external sands. After another depositional break, probably substantial, external detritus progressively filled the cavern, and persistent negative skewing shows that a sandy matrix was characteristic of the dominant transport mode, despite prominent clays in the basal and upper subunits. Our and Partridge's (1979) textural data corroborate that Beds A, D and E have finer, somewhat clayey matrices, B and C coarser, and non-clayey. It is therefore probable that two sedimentation cycles are recorded: the first began with rubble and clayey soil derivatives (Bed A), probably from a B-horizon, and then continued with persistent erosion of incompletely weathered debris from skeletal slope soils (Beds B, C); after a longer interval of soil reconstitution and non-deposition, an initial flush of rubble (Bed D) was followed by protracted and much slower accumulation of clayey soil derivatives, with interlaminated flowstones. Akin to our interpretations in the

Sterkfontein Valley, we would argue for high-intensity rains, slope erosion, and ineffective ground cover during the first of these cycles, and reduced erosion and a good vegetative mat during the second.

The geo-environmental sequence at Makapansgat is summarized in Table 2. The tentative correlations with the Sterkfontein Valley are based on the assumption that environmental trends were similar across the Transvaal, but require confirmation from as yet inadequately published and differentiated faunal assemblages. Translating effective moisture from an inferred geomorphological environment is hazardous, but suggestions can be made by way of illustration. The earlier part of the record suggests a humid climate, perhaps comparable to that of the eastern Transvaal Escarpment today (over 800mm precipitation), while Makapansgat 4 and 5D-5E suggest a rainfall at least as great as that of the modern Sterkfontein Valley (over 750mm); units 5A-5C speak for precipitation values of about 600mm or less. These estimates clearly presume that other variables be held reasonably constant. The point to underscore is that the sedimentary record speaks for substantial climatic variation, but not of a magnitude sufficient to change vegetation physiognomy in a dramatic way.

Taung

The so-called Taung fossil came from the former Norlim quarries along the Gaap Escarpment, near Buxton. This escarpment runs 275 km along the western margin of the Vaal and Hartz valleys, and represents the rim of a vast dolomite plateau extending back towards the Kalahari. During the course of the later Cenozoic, springs that emerged along these 70-120m high cliffs were periodically able to precipitate dissolved bicarbonates to form "secondary limestones," resembling cave travertines, but deposited as surface sheets or waterfalls of calc-tufa. Such tufas form large tongues or plugs of complex origin at discrete points along the escarpment, either below shallow drainage lines coming off the plateau, or where the cliffs intersect larger, subterranean aquifers.

Closer examination shows that the tufas commonly are part of a regular sequence of deposition and erosion, and include several facies. Collectively they record a geomorphic cycle that provides considerable environmental information. Six major cycles can in fact be identified, each recording as many as 6 sequential stages (Butzer, Stuckenrath, *et al.* 1978):

Stage 1. Deposition begins as surface runoff increases to erode soils and surface sands, collecting them into local colluvial sheets or fissure fillings. Climate semiarid, but characterized by a greater incidence of high-intensity, protracted rains.

Stage 2. Cliff faces are subject to intense mechanical weathering, producing footslope rubbles of crude rock and wearing back the scarp face. Increased severity of frost.

Stage 3a. The angular footslope rubbles are partly swept into the beds of intermittent streams, characterized by violent flash floods sufficiently competent to move crude debris and remodel it into water-laid fans extending out from the foot of the cliffs. Repeated, high-intensity rains converge on the rim of the escarpment, to spill down onto the lowlands with unsual energy.

Stage 3b. Stream flow becomes attenuated and seasonal, with reduced flood peaks, as the aquifers begin to carry a higher proportion of the rainfall, that now emerges from great springs at the edge of the escarpment. Less and less debris is transported; instead, calcium carbonate begins to accumulate as sheets along the footslope. Climate increasingly moist.

Stage 4. Stream flow becomes perennial, but subject to seasonal fluctuation, with deposition limited to waterfall tufas ('carapaces') that build out over slightly older sheets ('aprons'). Episodes of maximum discharge lead to reduced bicarbonate saturation of waters, favoring temporary corrosion of existing tufas. A less variable climate, with 150% the modern rainfall average (i.e. 600-650mm), would allow the

Table 3. Depositional units at Buxton-Norlim.

Buxton-10. Several minor tufas in the bottom of the north (main) valley downstream of Buxton-8; as yet not analyzed, these deposits appear similar to ones dated between 3200 and 400 B.P. at other sites of the Gaap Escarpment. Minor increases of spring discharge.

Buxton-9. Channel Alluvium. 5m of calcified silty sand, with three subdivisions, derived from residual soils and reworked eolian sands, resting on manganese-stained deposits downstream of Buxton-6 in the abandoned south valley. Base contemporary with Buxton-8, the last phase possibly coeval with tufas dated 9700-7600 B.P. at other Gaap sites. Accelerated surface runoff, converging in south valley stream.

Buxton-8. Blue Pool Tufa II. 20m tufa apron within north valley; center of lobe has 15,980 B.P. date, top 14,010 B.P. Significant spring discharge, more generally dated >21,000-14,000 B.P. for Gaap Escarpment, where locally interbedded with or followed by frost-weathered breccias. Late Upper Pleistocene.

(Downcutting into Buxton-7 tufa [10m]; development of deep, reddish paleosol as well as widespread, black manganese patina on exposed surfaces of older deposits)

Buxton-7. Blue Pool Tufa I. 9m tufa apron at plateau level and choking top of north valley; ^{14}C minimum date of 32,700 B.P. at -2m. Fossiliferous beds with Middle Stone Age at *Equus* and Witkrans caves predate the black patina; *Equus* was disolved within Buxton 5 and 6, the 2m of sandy silt fill washing in from a karst pipe as a result of accelerated surface runoff; the Witkrans deposits are interbedded with travertine that has a ^{14}C minimum date of 30,700 B.P. Significant spring discharge, preceded by strong runoff (*Equus* Cave). Early Upper Pleistocene.

(Downcutting or reexcavation of 55m-deep north valley)

Buxton-6. Oxland Tufa II. 30m, steeply-dipping carapace tufa with lenses of partly-crossbedded rubble seals off south valley; older than 43,000 B.P. (^{14}C minimum date), with basal contact to Buxton-5 approximately 250,000 B.P. (Th/U, J.C. Vogel, unpulished). Major spring discharge. Late Middle Pleistocene.

(Development of brown paleosol, preserved in fissures below unit 6)

Buxton-5. Oxland Tufa I. 10m tufa carapace and aprons (5b), over 5m of frost-weathered footslope rubble (5a), and partly filling south valley. Initial cold, followed by major spring discharge. Mid-Middle Pleistocene.

(Major bedrock and tufa downcutting; surface corrosion)

Buxton-4. Norlim Tufa. Massive, 60m-thick tufa carapace and apron, blocking south valley. *Australopithecus africanus* fossil comes from interbedded travertine, clay and sand, over 'Baboon sands' within fissure dissolved in Buxton-2, and therefore dates early in accretion of Buxton-4. Very long period of major spring discharge. Mid or late Lower Pleistocene.

Buxton-3. Norlim Conglomerate. Coarse to cobble-grade gravel, 5m thick, filling bottom of south valley as a high-competence alluvium, probably derived from slope rubbles. 'Baboon Sands,' with an early Lower Pleistocene fauna, probably contemporary. Period of very intensive rainfalls and effective runoff, as well as discharge. Early Lower Pleistocene.

(Some downcutting into Buxton-2; fissure corrosion opens up *Australopithecus* cave)

Buxton-2. Thabaseek Tufa. Major, 40m-thick tufa carapace and apron, formed in front of bedrock valley below escarpment edge. Very long period of major spring discharge. Late Pliocene.

Buxton-1. Thabaseek Breccia Complex. Massive, 20m- thick, crude frost-weathered rubble with thick tufa interbeds, inclined along footslopes of escarpment. Early Pliocene?

requisite dolomite solution without excessive dilution of bicarbonates. A good ground-cover is implied by the lack of colluvial debris.

Stage 5. The drainage network disintegrates and tufa accretion becomes limited to a few, local spring seeps as rainfall decreases again. Rainwater corrosion attacks existing tufas and enlarges jointing systems, favoring slow but cumulative surface cavitation, fissure enlargement, and stream downcutting.

Stage 6. Eventually even corrosion becomes rare and sporadic, and a long interval with limited geomorphic change ensues. Climate and ground cover comparable to those today.

For the specific case of the Buxton-Norlim tufa-breccia complex, the geomorphic evolution is outlined in Table 3 after Butzer (1974a), Butzer, Stuckenrath *et al.* (1978), and Butzer (1980a). The Taung australopithecine fits in Buxton-4 of this sequence and is younger than a fauna intermediate in age between Sterkfontein 4 and Swartkrans I, i.e. ca. 2.0 million years (Vrba, 1982). Depending on how much younger, the *Australopithecus africanus* specimen could well be partly coeval with Swartkrans I. Of greater significance than the stratigraphic position is the association of the fossil with a relatively moist climate and a well-vegetated meso-environment, perhaps not significantly different than that of the Makapansgat area during the mid-19th century.

The Buxton-Norlim sequence extends beyond the immediate setting of the Taung australopithecine, to provide a representative environmental history for the entire Gaap Escarpment and one that spans the Pliocene to the Holocene.

ALLUVIAL TERRACES OF THE VAAL ORANGE DRAINAGE

The Older Vaal Gravels and the Rietputs Formation

The australopithecine breccias form a set of discrete sedimentary sequences that collectively record a broad segment of late Pliocene to early Pleistocene time, perhaps from 3.5 to as late as 1.0 million years. Later archeological and paleontological records come primarily from cave infillings, small lake basins, or spring and karst accumulations, few of which are older than 200,000 years. The intervening span of time is poorly recorded throughout southern Africa and, other than the Gaap tufa cycles, the best connecting framework between the Plio-Pleistocene, on the one hand, and the late Pleistocene on the other, is provided by the alluvial gravels of the Vaal River. These deposits have provided important faunas, in part helping to define the mid-Pleistocene Vaal-Cornelia 'faunal span' (see Klein, this volume), as well as collections of Acheulian artifacts that figured significantly in the archeological exploration of southern Africa.

The Vaal alluvial record is best developed between about Warrenton and the Orange River confluence, an area of significant Paleozoic glacial scour and fluvial sculpture (Helgren 1979a, 1979b). Three discontinuous erosional platforms (Nooitgedacht, Holpan, Proksch Koppie, in descending order of relative elevation) at 30-110m above the modern Vaal appear to have been stream cut during Paleozoic times, then partly mantled with Karoo sediments, and later exhumed and remodeled by a Mio-Pliocene river (Helgren 1979a: 108-17, 123). The subsequent alluvial deposits can be outlined as follows:

a) The oldest alluvium consists of patches of calcreted gravel that have been largely destroyed by over a century of mining activities (Helgren 1979a: chap. 4). They are found on the Paleozoic erosional platforms, and appear to have averaged less than 2m in thickness, rarely attaining more than 5m. These deposits (but not the various cover mantles of later detritus, mainly unconsolidated) appear to lack fauna or artifacts. They may or may not represent the remnants of distinct alluvial terraces and certainly can no longer be separated into distinct stages. Generally found downstream of Paleozoic tillites, such conglomerates commonly represent locally reworked, till-derived cobbles, suggesting that finer, sandy deposits may once have been more representative (Helgren 1979a: 167-71).

b) The first morphostratigraphic marker is provided by the Wedburg Terrace, a sediment-capped fluvial platform at 22-24m above the Vaal near Windsorton and that rises to +52m near the Orange confluence (Helgren 1979a: 156-67, 171-82). In detail, the deposits comprise a meter or two of sands and coarse to cobble-grade gravels, variably calcreted, and mark a river stage after the Vaal had cut down into shales below the Proksch Koppie platform. There is reason to believe that the most archaic fossils, always found in abraded and battered condition, were derived from the Wedburg Terrace, even though recovered from younger deposits. Based on correlations with East Africa, these forms *(Notochoerus capensis, Mammuthus subplanifrons)* suggest a time span of 4.2-3.2 million years (Helgren 1977). The Wedburg sediments represent the youngest unit of what have been called the Older Gravels of the Vaal. They suggest that even higher conglomerates may be of Miocene age.

c) Valley incision resumed as the Wedburg terrace was abandoned, the river cutting down 25 or 30m into bedrock, over a protracted interval during which the Wedburg Terrace was extensively undermined by lateral erosion (Helgren 1979a: 219, 234, 245, 257).

d) Aggradation resumed with 10-30m of interfingering lateral and channel bars of a braided, gravelly Vaal. These deposits include tributary gravels from coeval hillslope erosion and, in the Harts River, extensive sands. Along both the Harts and the lower Vaal (Harts to Orange confluence) the gravel and sand facies interdigitate on their western margins with a massive tufa sheet that extends back 10km and more to the Gaap Escarpment. This link is important because it argues that this particular Vaal unit, the Rietputs Formation, Member A (Helgren 1979a: chaps. 5-8), was coeval with the last major Gaap tufa (Buxton-4 of Table 3). Acheulian artifacts are found sporadically within the Rietputs-A, and the older artifactual unit at Canteen Koppie pertains to this alluvial body, implying that Member A can be no older than 1.5 million years (Helgren 1978).

e) The Rietputs-A deposits were partly removed as the rivers cut back down to their previous beds, incising their fills, and, near the Orange confluence, cutting at least another 5m into bedrock. During this general time span the adjacent tufa surfaces were extensively and intensively corroded, a reddish paleosol developed on top of the Rietputs-A alluvium, and eroded segments thereof were locally calcreted (Helgren 1979a). The Acheulian Larsen Site dates from this interval (Helgren 1978).

f) Renewed aggradation, of Rietputs Formation, Member B, comprises 12-18m or more of interbedded gravels and sands in channelways or along the floodplain margins, and includes the products of local slope erosion (Helgren 1979: chaps. 5-8). Calcification accompanied and also followed the terminal phases of accumulation.

Occasional Acheulian artifacts are present within the Rietputs-B, but no major concentrations have been identified.

g) Between Windsorton and the Orange confluence the Rietputs-B was subsequently dissected 12-18m, down to below the modern channelway. But erosion was accompanied or rapidly followed by the long-term reworking and gradual aggradation of 4-6m of coarse to cobble gravels as part of a gleyed channel-lining with local confluence fans (Helgren 1979a: chaps, 5-8). This is the Rietputs Formation, Member C, which is rare in the major tributary valleys, where the calcrete surface of Rietputs-B was heavily corroded at about this time. The Acheulian Power's and Pniel sites pertain to the Rietputs-C (Helgren 1978), as does the great bulk of the 'Younger Gravels' (Rietputs Fm.) faunal material (Helgren, 1977). This fauna is roughly contemporary with that of Olduvai Bed IV, i.e. younger than 0.7 million years.

h) Eventually the Vaal again cut down to its modern channelway, and the surface segments of the Rietputs-C were locally calcreted.

The history of the so-called Older and Younger Vaal gravels, more specifically the Wedburg Terrace and the several members of the Rietputs Formation, is summarized in Table 4. Subsequent alluvial history is recorded in the Riverton Formation, outlined below, prior to discussion of paleoclimatic interpretations.

The Riverton Formation

The five members of the Riverton Formation consist primarily of flood silts and sands that mark an incisive change in the regime of the Vaal River. These sediments may form anywhere between one and four morphostratigraphic terraces, but most frequently are present as two complex bodies, an Older Fill (including members I, II and III) of Pleistocene age and a Younger Fill (members IV and V) of Holocene age. The type site at Riverton-on-Vaal is described in detail and illustrated in longitudinal profile by Butzer, Helgren et al. (1973).

At Riverton, Member I includes at least 6m of massive-bedded, gleyed flood silts, interdigitated with tributary sands or gravels. There are derived Acheulian hand axes at the base. Calcification preceded erosional truncation. Member II is 7m thick and ranges from flood silts to channel sands, capped by a weakly-organic, prismatic-structured (cracking) paleosol. After a period of downcutting, Member III accumulated as 9m of channel and reworked slope deposits, primarily sands and gravels, their top cemented in a siliceous calcrete. Several Middle Stone Age collections have been eroded from this unit (Humphreys 1973), and a reworked concentration of flakes of Acheulian type (made of andesite, instead of the indurated shale typical of the Middle Stone Age) has been recovered from its base. Major cutting (15m) ensued, then deposition of Member IV, with 9m of flood silts and limited tributary sands, with a basal ^{14}C date of 8640 B.P. on ostrich eggshell. Finally, Member V follows another 11m of fill cutting, and consists of flood silts interdigitating with up to 6m of tributary sands; the capping, dark organic soil has a cracking structure, with carbonate enrichment at 1-1.5m below the surface. In the lower Harts Valley this unit contained a hippo skull dated 1960 B.P. by ^{14}C. Younger events include 6m downcutting by the Vaal, a striking, brown sandy colluvium on the interfluves, and local wash in many channels.

The contacts of the Riverton and Rietputs formations are better documented at Windsorton (e.g. profiles I and II of Helgren 1979a: 208-16), where the younger sequence rests directly on 4m or more of Rietputs-C (with a surface calcrete), or upon 1-1.5m of unconsolidated gravel reworked from Rietputs-C, with relatively fresh, late Acheulian bifaces. Member I here consists of 1.5-7m flood silts and sands, or channel sands and gravels with local crossbeds. Member II is mainly recorded by lenses of sandy silts, with some pebble lines. The three key, terminal Acheulian (Fauresmith) sites of Muirton, Riverview VI, and Newman's Pont either fall between Members I and II or were coeval with Member II (Helgren 1978). Member III comprises 1-3m of partly calcified silt and sand, interbedded with channel gravels, which, together with a capping sheet of yellowish red (5 YR) sand, form a prominent 8-13m terrace body ('Older Fill') that locally abuts Rietputs-B alluvium and rests on Rietputs-C. Member IV forms an 8m terrace, beginning with reddish sands and gravelly slopewash lenses, followed by brown sandy-silts.

Further precisions on the Riverton Formation are provided in the Riet River (Butzer, Fock *et al.* 1979; Butzer, unpublished). At the base of this tributary sequence, 2.5m of calcrete and gleyed conglomerate with worn Acheulian bifaces rest on bedrock, probably representing Rietputs-C. Member I begins with 7m of partly calcified and well stratified silty sand, followed by 3m of clayey sand-silt with gravel lenses near the valley shoulder and including fresh Acheulian bifaces or flakes. This +7m terrace body is capped by Member II, 5m of calcrete sand, gleyed at the base, with dispersed gravel, fresh flakes, and worn bifaces, both typical and terminal (Fauresmith-type) Acheulian. After 7m of downcutting, Member III accumulated as a +13-15m terrace, comprising 3-4m of silty flood sands with fresh Middle Stone Age blades and points; after formation of a weak soil, a channel was cut through the top of this deposit, leaving behind 1.5m of fresh silty sand with a shell horizon (*Corbicula,* Unionidae) that is dated greater than 37,400 B.P. by ^{14}C, and followed by another paleosol. Soil carbonates of this latter paleosol have local ^{14}C dates of 8855, 15,980 and 18,480 B.P., which compare with dates of 17,240 and 17,100 B.P. on land snails (*Achatina*) instrusive in the calcrete atop Member III at the Riverton type site (Butzer, Helgren *et al.* 1973). This terminal Pleistocene paleosol represents an important stratigraphic marker in the area and records a relatively moist but geomorphically stable environment.

Member IV of the Riet River is represented by another, stratified channel silt-sand, with a carbonate date of 13,740 B.P., forming a +10m terrace. After a little cutting, some 5m of sandy deposits accumulated in the valley bottom, grading into dark, organic silty fills in minor tributaries, with a date of 2665 B.P. on soil carbonates; local concentrations of mint Later Stone Age (Smithfield) implements are found within this sediment. This first phase of Member V was followed by 13m of downcutting, then by another +5m, minor fill of similar material. Both subunits are veneered by brown sands, observed at Riverton as well as in several lake basins.

The units and basic features of the Riverton Formation are included in Table 4.

Environmental interpretation of the fluvial record

The fact that the Lower Vaal sequence is repeated, with minor variations, in the Riet Valley, matches the direct evidence that floodplain and slope facies interfingered

Table 4. Alluvial history of the Lower Vaal River. (Environmental conditions refer to direct sediment evidence; for geomorphic interpretation see text).

Riverton 5. Channel fill of 5-6m silt or sand, commonly altered by a dark floodplain soil, forming before 2665 until after 1960 B.P. (Late Holocene). This general aggradation is recorded even in minor tributaries. Locally a second phase of infilling is indicated (Riet River), with intervening downcutting (13m). Low energy, moist floodplain environment. Smithfield occupation.

(River downcutting in fill [2-11m])

Riverton 4. Channel fill of 8-10m silt or sand, with slope interbedded wash. locally beginning by 13,740 B.P. Mainly Early Holocene. Low energy, flood deposits of trunk streams.

(River downcutting in fill [8-15]; soil formation and local calcretion from before 18,500 to after 13,750 B.P.)

Riverton 3. Channel and tributary fill of 1-9m sand, with interbedded slope wash, forming floodplains 5 to 10 times as wide as those of Riverton 4 or 5. Terminal channel older than 37,400 B.P.; probably early Upper Pleistocene. Low-energy environment, with superabundant sand supply. Middle Stone Age occupation.

(River downcutting in fill [1-7m]; local calcretion)

Riverton 2. Channel and floodplain accumulation of 3-7m sand and silt, sometimes altered by a weak floodplain soil, and interfingered with slope wash. Probably late Middle Pleistocene. Low energy, relatively moist floodplain environment. Terminal Acheulian (Fauresmith) occupation possible.

(Limited erosion; Fauresmith occupation)

Riverton 1. Channel and floodplain sands and silts, the latter waterlogged; typically 6-7m thick, and locally resting on reworked gravel bars. Late Middle Pleistocene. Relatively moist floodplain environment, generally low energy. Late Acheulian or Fauresmith occupation probable.

(River downcutting in fill [10-18m]; local calcretion)

Rietputs C. Channel reworking of coarse/cobble gravel, with net aggradation of 4-6m in trunk streams. Abundant fauna, coeval with Olduvai Bed IV, i.e. Middle Pleistocene. High energy, probably braided and torrential Vaal floodplain, with a high water table, accompanied by calcrete corrosion on uplands. Major Acheulian occupation.

(River downcutting in fill [12-18m] in part coeval with Rietputs C; local calcification)

Rietputs B. Channel and floodplain aggradation of 12-18m gravels and sands, with interfingering slope wash and tributary deposits, eventuallyaccompanied by contemporaneous calcretion. Probably early Middle Pleistocene. Moderate to high energy floodplain. Some Acheulian occupation.

(Complex and extended interval with 10-30m downcutting in fill, and locally 5m or more into bedrock; intensive corrosion of upland calcretes, formation of reddish paleosol, and local calcretion, exact sequence unknown)

Rietputs A. Channel aggradation of 10-30m coarse/cobble gravels and sandy interbeds, with tributary rubbles, but sandy Harts River fill; interdigitate with massive tufa sheet from Gaap Escarpment. Probably late Lower Pleistocene. High energy, braided floodplain with extensive, seasonal sheetfloods on interfluves. Acheulian occupation.

(Protracted interval of 25-30m bedrock incision of Vaal, spanning late Pliocene and early Lower Pleistocene; formation of reddish paleosols as well as calcretion, temporal sequence unknown)

Wedburg Terrace. Stream planation of platform capped by 1-2m of sand and coarse/cobble gravel. Probable faunal associations suggest 4.2-3.2 million year age, i.e. mid-Pliocene. Moderate energy floodplain (?).

Intermittent river planation, with capping, high-level conglomerates, alternating with cumulative bedrock incision of 5-60m along different river stretches; one or more phases of soil formation or calcretion.

during most, but not all aggradation episodes. This argues that such records commonly reflect envrionmental changes within the Lower Vaal Basin, rather than impulses derived from the more distant catchment in the southeastern Transvaal and eastern Free State.

The climate of the lower basin is semiarid, with 320-480mm precipitation, increasing from southwest to northeast (South African Weather Bureau 1965). Seasonality of rainfall increases in the same direction, shifting from 79 to 83% concentration during the summer half year. Strong nightly frosts are not uncommon, and 4-5 months of the year have average monthly minimum temperatures below the freezing point, while the average daily minimum of the coldest month varies from –1.8 to +2.4° C. Vegetation ranges from Karoo semidesert grass and shrub at the Orange confluence, to thorn-tree savanna (on well-drained ground) or low bush (on poorly-drained surfaces) in the north. The lower basin consequently represents an environmental transition sensitive to changes of climate. In particular, colder winters should rapidly increase the effectiveness of frost weathering, while wetter summers and greater rainfall intensities should increase periodic stream competence.

Helgren (1979a: 309-26) has provided an interpretative model for the Vaal record in which the major variables are frequency and magnitude of fluvial events, sediment calibre and supply, and slope-soil-vegetation stability. By Helgren's criteria, (a) the Rietputs and older gravel terraces would be explained by the 'arid aggradation' principle, i.e. infrequent, high-magnitude events that supplied abundant slope debris to braided rivers prone to sporadic but competent flood peaks; (b) the phases of bedrock downcutting between the Miocene and mid-Pleistocene would be examples of 'humid incision,' i.e. frequent, moderate-magnitude events linked with low-sinuousity rivers and stable, soil-covered slopes; (c) the accumulation of increasingly fine, sandy and silty alluvia since the mid-Pleistocene would represent a trend toward 'humid aggradation,' i.e frequent, moderate-magnitude events that entrained only limited amounts of generally fine sediment from relatively stable slopes, to be deposited on floodplains and in channels of sinuous rivers; and (d) the repeated dissection of older alluvium since the mid-Pleistocene would represent 'arid incision,' as a consequence of infrequent, high-magnitude events at times of minimal sediment supply.

Such model variants are best applied in the context of all available information. In the case of Rietputs-A aggradation (Helgren 1979a: 326-34), it is known that:

a) Thick colluvial rubbles interfinger with alluvium along the valley margins, indicating substantial sediment supply as well as unusual competence in the mobilization of surficial debris.

b) Tufa accretion across the entire piedmont of the Gaap Escarpment requires extraordinary quantities of sustained runoff, while frost-rubble breccias interbedded with the lobes and carapaces of Buxton-4 point to accelerated frost-weathering and, possibly, frost-assisted mass movements (Butzer, Stuckenrath, *et at.* 1978).

c) Reconstruction of energy gradient, particle size, and channel depth, roughness and width, by means of empirical data and equations generally established for modern rivers, allows an estimate of the average annual discharge of the Vaal above the Vaal-Harts confluence. It is 1.42×10^{10} m^3/year, compared with 3.095×10^9 m^3/year for the modern Vaal at Riverton (1909-48), i.e. 4 to 5 times greater.

d) The almost exclusive dominance of coarse to cobble, even boulder-grade gravel within the Rietputs-A, as well as the characteristic bar and lag structures argue for a braided channel, activated by intermittent episodes of phenomenally high, peak discharge. This in turn suggests that annual discharge estimates would be misleading, because discharge was not only highly seasonal in character but also prone to discrete, "catastrophic" events. Such regimes are now best represented in seasonally-dry subtropical environments with very high variability, such as Texas, although no adequate, modern geomorphic analogs can be cited there.

In total, the Rietputs-A aggradation marks a phenomenal era in South African paleoclimatic history, the intensity of which was unparalleled before or after. A cold, dry winter, combined with periodic rainstorms of exceptional intensity and duration during the summer or transitional seasons, is suggested. The Mio-Pliocene Older Gravels, seen in conjunction with the massive tufas and interbedded breccias and conglomerates of the Gaap (Buxton-1 Table 3) on the one hand, and the diamondiferous gravel outcrops along the southwestern perimeter of the Transvaal (Cooke H.B.S. 1947) on the other, may indicate a similar pattern, but on a more subdued scale. The Rietputs B and C aggradations lack both substantial valley-margin rubbles and coeval Gaap breccias, while tufa accretion no longer extended across the Gaap piedmont; this speaks for a reduction of coarse sediment supply and for far less spectacular rainstorms in the lower basin, implying that frost-weathering was ineffective and that peak discharges were lower and derived to a greater extent from the upper basin, Yet substantial, if sporadic peak discharges were necessary, for example, to entrain the cobble bars of the Rietputs-C, spread over a channel width 10 times that of the modern river at the Vaal-Harts confluence (see Helgren 1979a 227, 241, 243). A strongly seasonal climate is further suggested for Rietputs-B by multiple, calcareous horizons, whereas the heavy oxidation and reduction of the Rietputs C beds suggests a more persistently high watertable.

Interpretation of the Riverton Formation can only be attempted in terms of the new sediment supply, of abundant quartz sand, first verified by important, red, sandy interbeds within the Rietputs-B. The origin of these sands, either from the Kalahari Basin via the Harts drainage, or from the middle and upper Vaal, through weathering of sandy Karoo lithologies, is uncertain; both sources may be relevant (Helgren 1979a: 335-37). More important is that such sands, now displaying relatively deep red soils, are widely distributed across the interfluves of the northern Cape, the south-western Transvaal, and the western half of the Free State. Their dunal topography south of the middle Vaal, and their linear disposition over tens of kilometers along the vector of effective prevailing winds (NNW-SSE), demonstrates eolian distribution, probably after deflation from the Vaal floodplain. Such sands dominate all units of the Riverton Formation along the Riet, as well as Member III at Windsorton, Riverton, and the Vaal-Harts confluence. In addition, the minor tributaries interdigitated with the Riverton Formation at the type site also deposited similar sands during all aggradation phases.

The Riverton units basically match the model of 'humid aggradation' in terms of floodplain processes, namely flood silt or sand accretion beyond a defined channel, and local but repetitive slackwater environments. But the persistent supply of sand from the interfluves is more compatible with the 'arid aggradation' model. Member V, with its wet floodplain mesoenvironment, and Member IV, interbedded with a local (not escarpment) tufa (Helgren 1979a: 245), qualify for the 'humid' model, but Member III represents a giant, valley sand-plug and is better referred to the 'arid' model; Members I and II were intermediate in character. The first three units of the Riverton Formation each define floodplains an order of magnitude larger than those of members IV and V; substantial differences in time depth and sediment supply are implied. Regional interpretation of Members III-V must however be done in conjunction with the broadly coeval Buxton stages 7, 9 and 10 (Table 3).

Bedrock entrenchment operates best in the channels of warm, perennial streams that allow hydrolytic weathering of subjacent rock, in combination with (a) turbulent discharge that favors plucking, saltation and potholing, and (b) a relatively low sediment supply, that ensures persistent contact between water and bedrock along the channel. Such conditions presuppose at least moderate energy gradients and relatively straight channels that commonly, but not necessarily, relate to morphotectonic environments. 'Humid incision,' so defined, would be most effective in a perennial river prone to strong seasonal floods, but receiving little or no tributary sediment or colluvial influx, either because of stable, well-vegetated slopes or insignificant mechanical weathering on the interfluves. Applied to the Vaal, the long intervals of bedrock incision between the later Miocene and the mid-Pleistocene could either reflect discharge almost exclusively from the upper basin, with negligible surface runoff in the lower basin, or a moist, if distinctly seasonal environment throughout the basin, with effective groundcover stabilizing slope soils as far as the Orange confluence. Paleosols are too poorly preserved to provide positive support for the second alternative.

Alluvial entrenchement subsequent to the Rietputs-A also requires strong, periodic floods but limited coarse sediment, if cobble and boulder gravels are to be 'removed'. This point is supported by coeval soil formation and bedrock cutting. Similar conditions applied, at least at times, while the Rietputs-B bars were being eroded and the Rietputs-C cobbles emplaced.

On the other hand, downcutting into the various Riverton fills involved the sporadic but cumulative effects of flash floods operating in the manner of gully (donga) erosion, in environments with stable interfluves or limited surface runoff. This pattern has characterized the Vaal Basin during the last century or so, and conforms well to the 'arid incision model.

By way of overview, the later Pleistocene to Holocene alluvial record differs substantially from that of the late Tertiary to mid-Pleistocene counterpart. Frost-weathered rubbles provided the major sediment supply prior to the Rietputs-B, but eolian or eolian-reworked sands were the key sediment source thereafter. Phenomenal flood discharges culminated with the Rietputs-A, and tapered off with later alluvial units. Bedrock incision during the earlier period probably indicates rather moist conditions. The Vaal floodplain was broadly similar in behaviour to that about 1850 during the various Riverton aggradations, some of which however imply wetter local conditions. The Rietputs and older floodplains were substantially different; a somewhat comparable mobilization of gravel shoals in contemporary southern Africa can only be observed along parts of the lower Fish and Cunene rivers.

Other alluvial sequences

The alluvial terraces of the Lower Vaal Basin outlined here provide relatively detailed information on a specific environment. Without implying simplistic correlations, it can be shown briefly that similar alluvial trends were characteristic of a much wider area in the interior of southern Africa.

Along the Orange River, between the Vaal and Caledon confluences, Riverton Formation equivalents are repeatedly recorded by a prominent terrace of Member III-like deposits found at +11-13m in more open valley sectors, as much as +30m in constricted areas. In former channel proximity there is a sandy facies, with a deep reddish soil, while away from the channel, silts or clays are characteristic, with gray cracking floodplain soils. The Middle Stone Age sites of Elandskloof 13, Orangia I, and Zeekoegat 27 and 27a (Sampson 1968a) all come from this unit:

a) Elandskloof 13 represents a thin channel bar on a bedrock bench, and has the characteristic reddish soil;

b) Orangia I comes from the top of a 3m tributary terrace (Bergspruit), with a similar reddish soil; and

c) The Zeekoegat sites both come from the clay-silt facies, calcified and under a strong floodplain soil.

To the extent that the archeological materials are primary, these sites are of similar age.

The younger Riverton equivalents are less conspicuous, due in part to erosion, in part to burial under more recent stream and eolian sands. The Later Stone Age site of Zeekoegat 13 (Sampson 1967b) comes from a calcareous sandy silt, locally grading into tributary rubble and representing Member IV. Older gravels outcrop at +22m near Bethulie, Aliwal North, and near the Kraai confluence; they consist of several meters of quartzite, basalt, and dolerite gravel, somewhat smaller in size than those of the Vaal, but also diamondiferous. Traces of higher-lying gravels at +35, 45-50, 65 and 90m are additionally present (Macfarlane 1945).

Riverton III as well as fragmentary records of the younger units are invariably sandy along the Caledon River, with Member III typically found at +13-16m near Wepener and Maseru and characterized by the reddish soil variant.

Archeological links are provided in the headwaters, along the Little Caledon River, where Visser and Van Riet Lowe (1956) developed an alluvial sequence reminiscent of that of the Vaal. The problems are also similar, in that many of the stages are recorded by isolated, gravel-veneered, bedrock spurs that may or may not represent broad fluvial surfaces of the Little Caledon. Focusing on the less problematic units, the basic stratigraphic elements can be summarized as follows:

a) +56-59m gravel-veneered platform, with possible pebble tools, younger than isolated high terrace spurs at +70 and +112m;

b) +36-40m gravel-covered platform with Acheulian, rolled as well as fresh;

c) +29-32m gravel terrace with rolled Acheulian; older than three terrace spurs, with thin gravels and fresh 'Fauresmith' or rolled Acheulian artifacts, at +10-24m;

d) Valley floor filling of coarse to cobble-grade gravel and cross-bedded grit, with Middle Stone Age;

e) +15-17m silt terrace, resting on valley floor gravel, and interstratified with gravel lines; Later Stone Age artifacts in topmost sediments;

f) Erosional terraces at +10-11m, +7.5m, +4.5m and +1.8m cut into the silt terrace.

Field reexamination indicates that stages (d), (e), and (f) represent the Riverton Fm., the main silty alluvium being Member III. The complex sediments, soils, and pollen (L. Scott) of two younger sedimentary units are currently being studied.

Downvalley of the Vaal confluence, the Orange also provides fragmentary evidence of the Riverton Fm. and, in some reaches, of higher gravel terraces. A good example of the former is provided near Vioolsdrif, where Member III forms a 10m terrace of sandy silt, resting on basal gravel and covered by further channel fills, with an intervening calcareous paleosol, much as along the Riet River, 1100 km upstream (Butzer, unpublished). But the sands are mineralogically of more local origin. A +6-7m flood silt resembles Member IV or V. The multiple gravel terraces at +20-50m are coarse to cobble-grade, with sandy interbeds, and considerable evidence of multiple weathering events, including red soils and calcretes. Even further downvalley, at Arrisdrift, one such terrace body, at +35m, includes a mid-Miocene vertebrate fauna (Corvinus & Hendey 1978).

Turning to the Middle Vaal Basin, Riverton III is conspicuously developed at +9-10m between Christiana and Bloemhof, with corresponding +12-14m alluvial bodies in the adjacent Free State tributaries, viz. the Vet and Sand rivers. The as yet unpublished Munro Site – Acheulian, with abundant fauna (Mason 1969b), comes from the base of the Member III Vaal terrace, possibly from an earlier unit of the Riverton Fm.

At Virginia, in the upper Sand drainage, the +10m Member III terrace exposes complex clay and silt beds with an intermediate paleosol. At +34m there are remnants of an ancient river, represented by 10m of crossbedded sands and gravels, deposited on the former bed of a large stream, initially with deep and perennial flow (Butzer 1973c). These Virginia beds included remains of *Mammuthus subplanifrons*, indicating a late Pliocene age for this period of substantially more vigorous hydrology.

The Upper Vaal record is best represented in tributary valleys near Cornelia, in the northeastern Free State. The nearby Uitzoek locality has provided a key fauna of the Vaal-Cornelia faunal span, as well as an Acheulian industry (of 'Developed Oldowan' facies) (Butzer, Clark & Cooke 1974). For this reason the sequence is described in Table 5 as comprising two members (and 5 beds) of the Cornelia Formation. Correlations with the Riverton Formation cannot be offered, because shale bedrocks, the absence of sand veneers, and a moister climate (690mm, 82% summer concentrations) have shifted both the sediment facies and geomorphic thresholds. Faunistically, unit 1b is more likely to be equivalent to Rietputs B or C than

Table 5. Alluvial history of the Cornelia Area, Upper Vaal drainage.

Aggradation of 1-6m of alluvium and colluvium, now cut by deep dongas. Late Holocene. Later Stone Age.
(Downcutting in fill [5-25m], accompanied by sheet erosion on slopes)
Development of Cornelia Soil on a seasonally very wet, poorly-drained floodplain. Late Pleistocene or early Holocene.
Aggradation of Cornelia 2b, some 4-8m of gleyed valley clays, interfingering with hillslope loams. A moist environment with good ground cover but intensive rainfall. Late Pleistocene.
Aggradation of Cornelia 2a, up to 6m of colluvium with local valley-bottom clays; enrichment of soil carbonates. Relatively dry with seasonal rainfall and incomplete soil cover.
(Downcutting in fill [over 8m])
Aggradation of Cornelia 1c, some 1-6m of gleyed clayey to silty valley-floor deposits, including channel sand or gravel, and some slope rubble. Relatively moist with marshy floodplain, but downcutting initiated near Vaal.
Aggradation of Cornelia 1b, some 1-8m gleyed valley clay. Moist, with poorly-drained bottomlands and well-vegetated slopes. Major faunal horizon, equivalent to Olduvai Beds III/IV*. Middle Pleistocene. Acheulian.
Aggradation of Cornelia 1a, a meter of gleyed, valley-floor rubble. Increasingly moist, with high-intensity runoff and incomplete ground cover. Acheulian.
Intermittent river planation, alternating with cumulative bedrock incision of 40-60m.

*Bone from this level has a Th/U date of 290,000 (+100,000, -40,000) years, to be regarded as a minimum age (Szabo 1979).

Riverton 1. More significant than stratigraphic comparison is the repetitive cycle of initial, arid aggradation, followed by a long interval of humid aggradation, terminated by arid incision. Humid incision has been insignificant since mid-Pleistocene time. Similar aggradation of valley-floor clays with minimal channel development can be observed today in Zambian environments with more than 900mm rainfall.

A final point of reference for evaluation of the Lower Vaal sequence is given by the Zambezi River, near Victoria Falls (Clark 1950), some 1100 km north of the Vaal. The point to be made here is that there are high-level 'plateau gravels' without artifacts, 'older' and 'younger' gravels with Acheulian, reworked eolian sands before, after, and between the 'younger gravel' units, and a complex, late Pleistocene sand-clay fill with several facies. Without implying correlation, it is noteworthy that gross geomorphic evolution had broad similarities in these two distant basins. Further research in the Upper Zambezi drainage should help define the nature of convergences and divergences in the alluvial histories represented within the interior of southern Africa.

PAN, LAKE, AND SPRING RECORDS

Introduction

In contrast to alluvial histories, which involve many variables operating over large basins, the sedimentary sequences of pans or lakes, as well as spring sites, tend to be less difficult to interpret. For one, the record is commonly localized in a smaller area or even within a single stratigraphic column, facilitating lateral correlations of facies variants. For another, such records tend to reflect a local hydrological regime, with the changes in the volume of stored surface water in that lake basin providing the empirical data necessary to reconstruct a hydrological budget. Even where such budgets have not yet been attempted, changes in lake volume can be overwhelmingly attributed to climatic variables, such as precipitation and evaporation, as a function of temperature, radiation and windiness – without recourse to geomorphic and biological variables such as ground cover, sediment supply, and concentration of surface runoff, or tectonic inpulses.

In outlining and reviewing the pan or lake records it soon becomes apparent that such research in southern Africa is still in the data-gathering phase. But the longer term prospects for more sophisticated evaluation are good.

Springs pose a special problem. Whether we are dealing with karstic springs on the Gaap Escarpment, or deep-seated, often thermal springs issuing from complex intrusive structures, the additional variables render hydrological interpretations difficult, if not impossible at this time. Nonetheless springs tend to accumulate long records at or near a single point, and commonly allow organic preservation that permits palynological and other biological research, as well as ^{14}C dating. For these reasons spring deposits provide a critical component in the Pleistocene data base.

Research at lake and spring sites increased exponentially during the 1970s and now spans most of the semiarid and arid environments of southern Africa.

Florisbad and Vlakkraal

One of the oldest study areas is provided by the Florisbad springs (28° 46′S, 26° 04′Ė), near the Modder River, 46 km north of Bloemfontein. The water emerges at a constant temperature of 29.6° C, some 13° warmer than the mean surface air temperature, so that it could derive from Precambrian Basement rocks. However, the sodium chloride content (1gm/liter) and the steady release of gas, that is 71.5% methane, instead indicate intermediate primary derivation from Karoo marine rocks and decomposing coal seams, in contact with the local dolerite intrusion (Fourie 1970). Most probably, therefore, the water emerges under pressure from a moderately deep aquifer in the carbonaceous Ecca shales, which are only a few hundred meters thick in the area. This is supported by the fact that after excavation at the turn of the century had removed most of the overburden, a very minor earthquake in 1912 was able to increase the water flow as much as 8-fold, after an initial, violent burst.

The extensive mound at Florisbad has a relief of 7m and consists of quartz sands, with interbedded organic horizons and local spring-eye intrusions, that reflect violent spring eruptions. The heavy mineral spectrum, dominated by ilmenite, with subsidiary epidote, pyroxene and zircon (Fourie 1970), derives from the breakdown of dolerite along the pipeways of the rising water. The Florisbad sequence consequently records the behaviour of a regional aquifer, rather than a typical thermal spring of volcanic type.

The sedimentary units are outlined in Table 6, as exposed in R.J. Clarke's new excavation in 1982 (Butzer, in preparation), but using Fourie's (1970) descriptions for the units now below watertable. The position of the fauna and artifacts was carefully compiled by Fourie (1970) from information provided by the National Museum staff after the 1953 excavations (see also Meiring 1956; Sampson 1972: 101-103). Clarke recovered unretouched Middle Stone Age material from FL-7A, contrasting with the retouched Middle Stone Age assemblage obtained from FL-9 and FL-10 during earlier excavations. The artifacts from FL-11 and FL-12 lack Middle Stone Age diagnostic types (Meiring 1956; Sampson 1972) and may represent a 'Developed Oldowan' type of Acheulian (see Butzer, Clarke & Cooke 1974). The archaic *Homo sapiens* cranium comes from a spring eye deposit ('eastern eye'), intrusive into units FL-12 through FL-9 or 10, and must therefore have been churned up from FL-12 or FL-11; the related vent sands included both Middle Stone Age artifacts and doleritic spheroids, the latter from FL-12 or FL-11 (Meiring 1956). A later vent ('western eye') intruded FL-10 through FL-6, the related sands blending with FL-5B (Flourie 1970: Fig. 5).

The great bulk of the Florisbad fauna, including 6 extinct species or genera, comes from FL-11 and 12 and belongs to the early Florisian (Cooke H.B.S. 1963; Klein 1980, and this volume); it includes such aquatic forms as lechwe, water mongoose, and clawless otter, suggesting extensive waters as well as dense vegetation. The description of Peat IA suggests a semiaquatic accumulation similar to Peat II, and formerly extensive marsh terrain is documented by 7m of clayey sediments along and under the stream now draining from the Florisbad spring, via a shallow donga, towards the Hagenstadt salt pan (Table 7, units FD-5 and 6, recorded 1970 and analyzed by Butzer). Two exposed calcretes, that preserve abundant aquatic snails (Fourie 1970), indicate periods of non-deposition probably coeval with major soil horizons in the

Table 6. Sediment stratigraphy of Florisbad Spring Eye.

FL-1A. Light gray (10 YR) silty sand (25-35cm), partly laminated. Under disturbed or removed surface soil (FL-0).

FL-1B. Light gray (10 YR) silty sand (5-10cm), partly laminated.

FL-2. Gray (10 YR), organic silty sand (15-25cm), partly laminated, with light grayish brown (10 YR) filling of 3-4mm root voids in lower half. Buried A-horizon. Locally converges with FL-4.

FL-3. White (2.5 Y) silty sand (10-20cm), partly laminated, within root zone of FL-2.

FL-4A. Gray (10 YR), organic, silty sand (5-8cm), with moderate soil structure and light grayish brown (10 YR) sandy filling of 5mm root voids. Buried A-horizon. Locally separated from FL-4B by white silty sand (5-8cm)

FL-4B. (PEAT IV). Dark gray (10 YR), organic clayey silt sand (10-25cm) with strong, cracking soil structure and abundant organic voids. Buried A1-horizon of major paleosol. ^{14}C 5530 B.P. (Pta-1128).

FL 4C. Mixed gray (10 YR) and light gray (10 YR/2.5Y), organic silty sand (7-15cm), with organic intrusions penetrating from 4B along decomposing vegetation structures. A3-horizon of overlying paleosol.

FL-5A. Light gray (2.5Y) silty sand (25-30cm), with abundant, light brownish gray (10 YR), organic mottles related to root zone from 4B.

FL-5B. Light gray (10 YR) silty sand (20cm), with vertical structure and weak organic mottling; basal 5cm laminated sand.

FL-6A. Light brownish gray (10 YR) silty sand (0-8cm), with vertical structure and abundant carbonized root fragments (1-4mm) in lower part. Weak A-horizon.

FL-6B. Light gray (2.5Y) laminated sand (7-25cm), with abundant root fragments from above, and non-calcareous concretions.

FL-6C. (PEAT III). Dark gray (10 YR), organic silty sand (5-12cm), laminated, as matrix for web of carbonized vegetation mat and 5mm roots, extending 25-60cm down into lower units. Major A1-horizon of paleosol, forming undulating surface. Locally converges with FL-4 and attains 60cm in thickness. The ^{14}C date of 19,350 B.P. (L-271D) may well be on intrusive, younger organics and was determined by the solid carbon techniques in 1954; a minimum age only.

FL-7A. Loose light gray to very pale brown (10 YR), sand (110-115cm), with multiple organic laminae in thin-bedded (5-20cm) sets of mainly wavy, parallel bedding, dipping 4° toward modern donga. Top third of deposit is mainly a medium-grade sand, lower two-thirds are coarse sands. Interrupted midway by lense (0-10cm) of grayish brown (10 YR) organic sand. Major episode of spring flush flow. Middle Stone age horizons up to 20cm below top; rare fossil bone.

FL-7B. Light gray (10 YR), organic silty sand (0-15cm) partly laminated, with small-scale, organic 'flame' structures above and below.

FL-7C. Light gray (10 YR), laminated coarse sand (7-20cm) with 'flame' structures above and below.

FL-8. (PEAT II). Very dark gray (10 YR), highly organic silty sand (20-25cm), interbedded with dark gray (10 YR), organic clayey silt sand and pockets of light gray (2.5 Y) sandy silt. Laminated but disposed in convoluted 'sweeps', with vegetation structures, root impressions, and both upper and lower 'flame' structures. A-horizon of semiaquatic soil. Older than 42,500 B.P. (Pta-1108). Some fossil bone.

FL-9A. Light gray (10 YR), laminated sand (15-20cm), with fine (1mm) carbonized roots. Middle Stone Age occupation.

FL-9B. Light brownish gray (2.5 Y), laminated sandy silt (10-15cm), with vegetation structures and fine vertical rooting. Weak A-horizon.

FL-9C. Light gray (2.5 Y), laminated and reduced silty sand (>25cm), with prominent vertical root impressions (1-4mm).

(Watertable February 1982. Units below after Fourie [1970])

FL-10. Silty sand (60cm). Middle Stone Age artifacts.

FL-11A (PEAT IA) (38cm), a black 'waxy clay', and FL-11B (PEAT IB) (43cm), a gray and white organic sand. Older than 48,900 B.P. (GrN-4208). Some fossil bone and abundant artifacts.

FL-12. Brown to yellowish, mottled silty sand with basal gravel (70cm). Major faunal horizon; some artifacts near base.

FL-13 (BASAL PEAT) (2cm) over dolerite bedrock.

Table 7. Sediment stratigraphy of the Florisbad Donga.

1.5m Low Terrace

FD-1A. Dark gray (5Y) sandy clay silt (30cm), poorly sorted, with cracking soil structure. Represents active floodplain, prior to mound excavation to emplace the bath houses.

FD-1B. Grayish brown (10 YR) clayey silt sand (over 90cm), moderately well sorted.

4m High Terrace

FD-2. Light yellowish brown (10 YR), thin bedded and laminated sand (60-70cm), moderately sorted. Surface colluvium, derived in part from erosion of eolian sands.

FD-3A. Brown (10 YR) silty sand (60cm), moderately sorted, with cracking soil structure; land snails. Buried A-horizon, developed in eolian derivatives.

FD-3B. Yellowish red (5-7.5 YR) sand (80cm), moderately sorted. Buried B-horizon.

FD-4. Light yellowish brown (10 YR), silty clay sand (30cm), thin-bedded, and moderately sorted, with reworked shell and Middle Stone Age artifacts. Colluvium derived in part from older, fine alluvium and calcrete.

(Erosional break)

FD-5A. Light gray to white (2.5Y), platey to wavy-bedded calcrete (20cm) with residue of moderately sorted, silty sand clay; with aquatic snails (*Succinea*, *Lymnaea*, *Bulinus*) and *Helix*.

FD-5B. Gray (5Y), clayey silt sand (100cm), poorly sorted, with strong cracking soil structure, concretions, and yellow (2.5Y) mottling.

(Erosional break)

FD-6A. Light gray (10 YR), platey and salty calcrete (15cm), with residue of poorly sorted, sandy clay silt; with *Succinea*, *Lymnaea*, and *Pupilla*.

FD-6B. Pale yellow (5Y), reduced and calcified clayey silt sand (40cm exposed), moderately sorted, with cracking soil structure and rich in opaline silica. Boreholes penetrated another 4m of similar, fine, reduced sediment before reaching Ecca shale (R.J. Clarke, pers. comm.).

Florisbad mound. The derived Middle Stone Age horizon in FD-4B, reworked from FD-5, suggests that FD-5A marks the Peat III land surface, FD-6A the Peat II level, and the FD-3 soil peat IV.

This direct relationship of the Florsibad sequence to local alluvial histories is further demonstrated along the nearby Modder River, and with reference to the spring site of Vlakkraal (9km due south, also known as the Prinsloo site), seasonally submerged in the Abrahamskraal Dam since 1971. The important fauna of Vlakkraal has the same six extinct forms as Florisbad (Cooke H.B.S. 1963), and is therefore of similar age. The associated, large artifact assemblage, recovered when the spring eye was deepened, remains unpublished in the National Museum, but has a low incidence of retouch, as does the Middle Stone Age of FL-7A. The Vlakkraal sequence is outlined in Table 8, as recorded in 1970 (Butzer, unpublished). The Rietputs section described is exposed in a Modder north-bank quarry, where a complex calcrete is developed in flood silts and sands; fragmentary exposures on the south bank include channel facies with medium-grade pebbles, mainly exotic quartzite, chert/jasper, and chalcedony/agate derived from now-eroded Stormberg strata of the uppermost Modder drainage.

A good case can be made that Peat III corresponds to FD-5A, FV-3A, and the top of Riverton III. It probably is beyond the effective range of ^{14}C dating. Peat II and FD-6A may correlate with FV-5, but this case is tenuous. The Florisian fauna of Peat I leaves no doubt that it is younger than the Cornelian fauna of Rietputs-C, so that the classic faunal horizon of Florisbad and its archaic industry must be placed near the base of the

Table 8. Sediment Stratigraphy of the Modder River Terraces at Vlakkraal.

6m Low Terrace–(Riverton 4)

FV-1A. Brown (7.5 YR), silty coarse sand (2.5-4.0m), well-stratified and poorly sorted, with lenses of fine to coarse dolerite gravel; vertical soil structure in upper half. Mantled by colluvium at terrace shoulder.

FV-1B. Brown (7.5 YR) coarse sand (0.5m exposed), thin-bedded and laminated, moderately well sorted.

12m Middle Terrace–(Riverton 1 or 2 and 3, under Riverton 5 colluvium)

FV-2. Pale brown (10 YR) silty coarse sand (1.5m), thin-bedded with soil structure and moderately well sorted. Capping colluvium.

FV-3A. White (10 YR), concretionary calcrete (0-10cm) with residue of clayey silt and opaline silica, poorly sorted, intermixed with dolerite rubble and fossils (*Megalotragus*,? *Damaliscus*, H.B.S. Cooke, pers. comm.).

FV-3B. Grayish brown (2.5 Y), sandy clayey silt (20-100cm), with calcareous concretions, cracking structure, and poor sorting.

FV-4A. Pale brown (10 YR), clayey silt sand (5-7m), thinbedded and moderately well sorted, and soil structure in upper 1.5m which are more reddish.

FV-4B. Brown (10 YR) to light gray (2.5 Y), partly organic silty sand, moderately sorted, with yellow (10 YR) mottling, forming spring eye of Vlakkraal, intrusive within top of unit 4A. Abundant Middle Stone Age and a Florisian fauna.

FV-5. Light gray (5 Y), reduced silty clay sand (2.5-3m exposed), with cracking structure, poor sorting, and strong brown (7.5 YR) mottling.

20-22m High Terrace–(Rietputs Formation)

FV-6. Brown (10 YR) silty sand (0.5m), with cracking structure, moderate sorting and reworked calcrete rubble. Colluvial soil, capping terrace.

FV-7A. White (10 YR, 5 Y) platey calcrete (30-40cm), with residue of poorly sorted silty sand, with extensive reddish yellow (7.5 YR) staining. Truncated BCa-horizon of ancient soil.

FV-7B. Very pale brown (10 YR), silty sand (25-30cm), with vertical structure and poor sorting.

FV-7C. As 7A (20cm).

FV-7D. As 7B (20cm).

FV-7E. As 7A (60cm).

FV-7F. Pale yellow (5Y) clayey silt sand (30-40cm), with cracking structure and poorly sorted.

FV-7G. White (10 YR), platey and chalky calcrete (60cm exposed), with residue of moderately sorted, silty sand clay.

Riverton Formation, in the later Middle Pleistocene. The affinity of the skull to that Elandsfontein (Rightmire 1978a, and this volume), rather than to the early Upper Pleistocene hominids from Border Cave, supports such a conclusion.

Interpreted in the context of the local alluvial sediments, the Florisbad sequence evidently reflects regional environmental changes. The composite history of these major sites is detailed in Table 9. Interesting is the fact that the Holocene aggradations were 'drier' than their Pleistocene counterparts, reversing the pattern noted previously for the Vaal Basin. This may reflect the fact that Florisbad lies near the semiarid-subhumid boundary, with a rapid transition from 480mm precipitation at the spring (1272m elevation) to 558mm at Bloemfontein (1387m), only 46km away. The older pollen results from Florisbad (Van Zinderen Bakker 1956) suggest a treeless environment at all times, and imply that Peats I and II should have reflected drier macro- and microenvironments than Peat III, contradicting the geomorphological and faunal framework developed here. The latent problems of the palynological

Table 9. Environmental interpretation for the Florisbad-Vlakkraal sequence.

Florisbad-5B. Marshy clay floodplain (FD-1A) and surface soil of mound (FL-0) mark geomorphic stability during low-discharge phase prior to excavation of 1912.

Florisbad-5A. Moderately strong spring discharge from eye at top of mound, with accumulation of FL-1 and FD-1B (silty medium-grade sand). Riverton 5B.

(Incision of donga [over 4m])

Florisbad-4D. Geomorphic stability with organic soil development on mound (FL-2) and on terrace colluvia (soil structure in FD-2 and FV-2).

Florisbad-4C. Moderately strong spring discharge (FL-3, 2: silty medium sand) and mobilization of sandy interfluve materials by repeated surges of runoff, to accumulate liminated colluvia (FD-2, FV-2). Intensive rains and arid aggradation. Riverton 5A.

Florisbad-4B. Geomorphic stability with major organic soil development on mound, interrupted by intervals of spring activation (Peat IV and FL-4A); deep but modest soil alteration of terrace surfaces (FD-3, FV-1). Relatively moist. 5530 B.P.

Florisbad-4A. Moderately strong spring discharge (FL-5, 4C: silty medium sand) preceded by minor flush episode (FL-5B base), initiated by eruption of Western Eye; aggradation of sandy alluvium (FD-3) preceded by colluvium (FD-4) in donga; aggradation of gravelly fill (FV-1A) over initial laminated sands (FV-1B) in Modder valley. Periodic strong rains and arid aggradation. Anomalously young ^{14}C dates, of 4370 B.P. (Pta-1127) on FL-6A and 10,000 B.P. (Pta-1125) on FL-6C, suggest bracketing dates of 10,000-4300 B.P. for Florisbad 4 and the subsequent period of stability and soil development. Riverton 4.

(Incision of Modder River [12m] and erosion of donga)

Florisbad-3B. Geomorphic stability with major organic soil development on mound, interrupted by intervals of spring activation (Peat III and FL-6B, 6A), with formation of calcretes under stable floodplain surfaces (FD-5A, FV-3A). Terminal organic accumulation no younger than 19,350 B.P., with an anomalously young date of 28,450 B.P. (L-271C) on Peat II possibly indicating some 10 millennia of soil development for the Peat III interval.

Florisbad-3A. Very strong Florisbad spring discharge (FL-7C, 7A; coarse sands), with some periods of reduced flow (FL-7B; silty medium sand), and periodic vegetative recolonization (organic laminae), and increasing moderation towards close of FL-7A; later Middle Stone Age occupations. Eruption of Vlakkraal spring (FV-4B) (mid-way FL-7A?), preceded by Middle Stone Age occupation around quiescent seep (FV-4A), and followed by reduced spring activity impinging a clayey and marshy floodplain (FV-3B). Clayey marsh with aquatic molluscan fauna in Florisbad donga (FD-5B). Humid aggradation with lechwe present. Riverton 3.

(Erosion of unknown dimensions)

Florisbad-2B. Geomorphic stability with semiaquatic soil development around extensive Eastern Eye zone (Part II), and calcrete formation on donga floodplain (FD-6A). Certainly older than 42,500 B.P., probably beyond 50,000 years.

Florisbad-2A. Moderately strong spring discharge (FL-10, 9C: silty medium sand), then weakening (FL-9B: sandy silt), later accelerating abruptly (FL-9A: medium-grade sand); eruption of Eastern Eye either at beginning of FL-10 or of FL-9A. Earlier Middle Stone Age occupation. Slow accretion of clayey flood silts on Modder and Florisbad floodplains (FV-5, FD-6B). Humid aggradation. Riverton 2.

Florisbad-1B Geomorphic stability with semiaquatic soil development at Florisbad (Peat I).

Florisbad-1A. Moderately strong spring discharge (FL-12: silty medium sand), becoming increasingly quiescent (FL-11B organic zone). Archaic *Homo sapiens*, Acheulian(?) occupation. Wet meso-environment, with lechwe, water mongoose, and clawless otter in early Florisian fauna. Perhaps coeval with massive throughflow calcretes on interfluve. Probably Riverton I.

(Downcutting in Modder valley [22m])

Florisbad-0. Aggradation of channel beds (gravelly sands) and flood silt/sand by Modder River, flowing strong from Lesotho border to Vaal. Arid aggradation. Rietputs B or C.

interpretation are hinted at by Scott (this volume). The modern analytical and interpretative framework recently applied by Scott (1982a) to the Wonderkraterspring site, central Transvaal, suggests that the renewed palynological effort underway at Florisbad will yield substantially more informative results.

Alexandersfontein, Voightspos, and Deelpan

The western Orange Free State is characterized by innumerable closed depressions, generally floored by sodium salts, calcium carbonate, or gypsum. Called pans, they are invariably developed in Karoo shales, situated between more resistant dolerite watersheds: local converging drainage favors salt accumulation, which in turn accelerates hydration weathering of underying shales, the debris of which is removed by wind or water (DeBruiyn 1972). Unlike the pans of the Kalahari, those of the Free State are primarily bedrock-controlled and show no orientation related to effective wind vectors.

Three such salt pan basins have been studied:

1) Alexandersfontein Pan (floor elevation 1134m, average rainfall 400mm), 28° 50′S, 24° 48′E, with three key sites on the farms Benfontein 442, Uitzigt 124, and Mauritsfontein 26.

2) Liebenbergspan (1224m, 400mm), 29° 08′S, 25° 20′E, on farm Voightspos 537.

3) Deelpan (1279m, 500mm), 29° 10′S, 25° 47′E, on the farms Panfontein 185 and Heuningkoppie 354.

These sites provide 5 informative sedimentary columns that, together, delineate a good environmental record for the late Pleistocene and Holocene, for comparison with the Florisbad-Vlakkraal sequence.

The Liebenbergspan sediments are the most shallow, and consequently provide direct shoreline interdigitations. The earlier stratigraphic study of Horowitz *et al.* (1978) was found to be inaccurate, and a new composite profile was drawn up in 1977 by Butzer and L. Scott (unpublished), with corresponding sediment analyses (Table 10). Dates were provided by J.C. Vogel. A relatively complete Holocene record is indicated, with two lake phases dated roughly 6900-6300 B.P. and 3500-2000 B.P. The second lake was coeval with more luxuriant vegetation in the area, judging by the prominent organic soil. After the pan floor was deflated, another but modest soil began to form, prior to about 1300 B.P., when eolian processes became dominant in the region.

Deelpan sediments span the Holocene and later Pleistocene, and are exposed in semicontinuous sections of much greater thickness. A composite profile is given in Table 11, after Butzer and L. Scott. Although ^{14}C dates are still pending, units 2-5 closely parallel the Liebenbergspan sequence, but the basal calcrete complex at Deelpan is substantially older than the units of Table 10. There is no trace of sediments between DP-1E and DP-2, probably because of repeated wind erosion. The material deflated from the pan was deposited within a kilometer of the southeastern shoreline, where it weathered to a reddish yellow soil, before being swept together as the colluvial facies of DP-2. The lowest exposures of DP-1, and projections of the bedrock contacts, indicate that a total of 1.5-2m of shale were removed from the pan floor since initial sedimentation of DP-1A.

Table 10. Sediment stratigraphy on Voightspos, Liebenbergspan, near Petrusberg.

VP-4A. Very pale brown to light brownish gray (10 YR), medium-grade sand (55cm), moderately sorted and well stratified, with some crossbedding. Eolian sand. ^{14}C date on ostrich eggshell: 1220 B.P. (Pta-1483) with Later Stone Age material (Smithfield) (Horowitz *et al.* 1978). Subsequent development of marshy soils between dunes (VP-4B).

VP-3A. Brown (10 YR), silty medium sand (35cm), moderately sorted, in part laminated. Grades laterally in swales to light gray (10 YR), clayey silty sand (20cm), poorly stratified, with strong soil structure and partly calcified, representing the buried A-horizon (VP-3B) of a wet zone near the lakeshore. Eolian deposition, terminated as lake was recreated and rose to about +lm.

VP-2B. Dark brownish gray (10 YR), silty medium sand (20-30cm), poorly sorted with cracking structure. Grades laterally into white (10 YR) calcified silty sand with sodium bicarbonate, and eventually into white (10 YR) laminated calcrete (15cm), with residue of clayey silt sand and sodium chloride. Represents a strong, buried A-horizon which, in lowerlying areas, developed on lake marls that record at +2.2m water level. ^{14}C date on carbonate: 2020 B.P. (Pta-2428), with earlier part of phase probably dated 3500 B.P. (carbonates in upper part of VP-2A). Traces of artifacts.

VP-2A. Light gray or grayish brown (10 YR), or mottled, very pale to light yellowish brown (10 YR), silty sand, clayey silt sand, or clayey sand silt (50cm), poorly sorted and partly calcified, with sodium salts. Represents transition from subaerial sands, in part eolian, to lake sediment, in part reduced, in part oxidized, with water level fluctuating up to +2.0m. ^{14}C date on eggshell: 6350 B.P. (Pta-1520) with Later Stone Age material (Albany) (Horowitz *et al.* 1978); traces of bone.

VP-1B. White (10 YR/ 2.5 Y) laminated calcrete (30cm), with residue of silty medium to coarse sand, moderately sorted and positively skewed. Carbonates derived from subsequent (VP-2A) lake phase in view of ^{14}C date of 6830 B.P. (Pta-2461). Shore-near colluvium, with strong eolian component.

VP-1A. Pale brown (10 YR), silty medium to coarse sand (75cm), poorly sorted and positively skewed, with weak pink mottling (7.5-10 YR). Colluvium, with strong eolian component, resting on shale and dolerite bedrock with a palimpsest of worn Middle Stone Age. Lake initially dry, then rising a little.

Table 11. Sediment stratigraphy at Deelpan, near Brug.

DP-5. Pale to very pale brown (10 YR), sand (60-180cm), in part laminated and crossbedded.

DP-4. Very pale brown (10 YR) sand (25-140cm), laminated, with calcareous root casts; locally interbedded with partly calcified lenses of light gray (2.5 Y-10 YR), silty sand with *Succinea*, and truncated by wave action at +3m.

DP-3. Light brown, brown, or grayish brown (10 YR), silty sand (60-75 cm), with moderate cracking structure (upper half) and partial calcification (lower half); local proliferations of *Succinea* or traces of bone. At northeastern end of pan, the sedimentary equivalent of this buried Al-ACa soil thickens to 170cm of sand, with little structure or humification, and interbedded with dispersed grit or lines of fine gravel, as a result of persistent colluvial transport. Lake level mainly below +2.5m during formation. ^{14}C date, on soil organics: 450 B.P. (Pta-3354). Several Later Stone Age (Smithfield) concentrations.

DP-2. Very pale brown, pink, or reddish yellow (5-10 YR), calcified and heavily mottled (oxidized), silty sand (40-380cm), with cracking structure, root casts, ostrich eggshell and bone; inshore, changes rapidly from such marly lake sands to non-calcareous silty sand with interbedded gravel lenses, a distinct colluvial facies. At northeastern end of lake, unit begins with gritty or gravelly colluvium, that grades up into lacustrine facies. Lake level fluctuating both above and below +3.5m. ^{14}C date on ostrich eggshell: 7650 B.P. (Pta-3345).

(Protracted erosion, with at least 50cm deepening of pan in bedrock).

DP-1E. White (2.5Y) cracking calcrete (40-90cm), platey towards top, where it incorporates recemented calcrete rubble concentrations of Middle Stone Age artifacts in topmost 15cm (DP-1F); residues vary from silty sand, clayey sand, clayey sand silt, to clayey sand silt. This formerly marly sand or sandy marl has a gently sloping surface at 3.5-4.2m above the pan and records a lake fluctuating near +4m. The artifacts are in secondary position, part of a colluvium transported from backshore at a later date.

DP-1D. Pale brown (10 YR), partly decalcified, silty sand (0-10cm). representing a weak, reworked soil; the underlying 20-25cm of intact calcrete is partly decalcified.

DP-1C. White (5Y), massive to platey calcrete (110cm), increasingly gritty towards base, with root casts.

DP-1B. Pink (5-7.5 YR), partly calcified, silty sand (45cm), as matrix to medium or coarse-grade shale rubble.

DP-1A. White (10 YR) massive, gritty calcrete (35cm), resting on shale.

Examination of the floor showed up to 15cm of saline and oxidized, laminated, sandy silt or clayey sand silt, over a white (5 Y), vertically structured, silty residue (25-30cm) derived from shale; the shale itself is partly decomposed to at least 1.5m depth, forming a honeycomb of pale yellow (5 Y), platey, sandy silt and gray to olive gray (5 Y) intact rock. The pan is normally dry but may have up to 50cm of water (the approximate relief of the floor) during an average rainy season. In March, 1974, 300mm of rain fell over the basin during 3 weeks, leading to a sudden flood surge from the southern tributaries that rapidly filled the pan to a maximum depth of 4.2m (Mrs. G.R. De Villiers, pers. comm.): this water body only dried out in 1980, and had been reduced to +2.5m in 1976 and +1.2m in 1978, by which time the calcium carbonate concentration was sufficiently high to create a lime-coated shore zone still visible today. The receding lake also left shallow mud ridges (relief up to 15cm) below+1.5m, as well as a bay bar with 50cm relief at the southern end of the pan; there is no evidence of any direct or indirect water action above +2.8m.

Located 1.3km north of the Alexandersfontein Pan, the Benfontein sequence is exposed as a series of tufas, clays, and soils resting against and over an old wave-cut nip at +19m. It was therefore mistakenly identified as a set of low-energy shore deposits interbedded with colluvium, by Butzer, Fock *et al.* (1973); in fact, the deposits record a mix of spring and colluvial deposits, as detailed in Table 12 (Butzer, unpublished). Units BT-1 to 9 are present in a 100m section trenched in the course of well-digging, while BT-10 and 11 are located just west of the nearby stream channel, in the same topographic situation. In addition to the Later Stone Age (Albany) artifacts, lightly patinated and edge-fresh, in BT-1 to 4, there also are two generations of derived Middle Stone Age artifacts, one edge-fresh with medium patina, the other edge-worn with medium to heavy patina.

Noteworthy in the Benfontein sequence is (a) the accelerated spring discharge, undersaturated in calcium bicarbonate prior to 13,600 B.P., culminating in organic soils; (b) the subsequent dehydration of the deposits, with penetration of reworked, eölian sands; (c) the renewal of spring discharge, now saturated in calcium, from somewhat before 11,500 B.P. to a little after 7000 B.P.; and (d) a final episode of spring activity, undersaturated in calcium, beginning a little after 4000 B.P. Colluvial material was interbedded at the very beginning, then about 14,500, after 13,600, and again before 11,500 B.P. The distinction between saturated and undersaturated discharge is significant.

The Uitzigt sequence is exposed in a valley head, 2.6km beyond the eastern end of the pan. Two calcrete terraces and one unconsolidated fill all converge on the edge of the shale slope, at which point there has long been a persistent seep or active spring. Deposits younger than all three terraces are embanked against the spring, or veneer the surface above it, in part calcreted by persistent throughflow waters. The composite section of Table 13 is based on study of the local exposures and borings at the spring, a 1:1000 mapping of the spring area, as well as detailed, continuous profiles of the 3km valley between the spring and the pan (Butzer, unpublished, in conjunction with A.J. Bruzewicz and L. Scott). Because of the 3-dimensional stratigraphy, the same deposits may be found in two or even three parts of Table 13.

The pollen from US-4B and 4C (Scott 1976) indicates two environmental patterns, one with grasses, compositae and sedges, but essentially no trees, the other with chenopodiaceae, sedges, and trees. The former pattern comes exclusively from one pit,

Table 12. Sediment stratigraphy of the Benfontein Well Trench, Alexandersfontein Basin. (CCE: calcium carbonate equivalent; OC: organic carbon).

BT-11B. Black (5 YR), crumb-structured silty sand clay (5% CCE, 7.8% OC) (12cm), with dispersed dolerite rubble; poorly sorted. Soil horizon of organic spring mud.

BT-11A. Very dark gray (7.5 YR), sandy clay silt (2.5% CCE, 8.2% OC) (40cm), with low-density organic packing, and poorly sorted. Organic spring deposit. High concentration of Later Stone Age (Smithfield) artifacts in top of unit or along 11A/11B contact.

BT-10. Very dark gray (7.5 YR), sandy clay silt (7% CCE, 3.2% OC), (over 40cm), with vertical structure, poor sorting, and sodium bicarbonate efflorescence. ^{14}C date 3450 B.P. on humus (SI-2049) near top. Organic spring muck.

(Erosion and long depositional break)

BT-9. White (10 YR), banded 'calcrete' (20-70cm) (76%, CCE), with residue of salty clay sand, moderately sorted. ^{14}C dates of 7035 and 7905 B.P. on carbonate (SI-3569, 3050). A recrystallized spring tufa on convex slope.

BT-8. Light grayish brown (10 YR-2.5 Y), vertically crystallized tufa (10-20cm) (98% CCE, 0.7% OC), with residue of silty clay sand, and abundant opaline silica. Snail proliferations: *Pupilla fontana, Pupilla tetrodus, Vertigo antivertigo*, and *Succinea striata* (Butzer, Fock, *et al.* 1973). ^{14}C dates of 11,025 and 11,230 B.P. on carbonate (SI-1116, 3048). Spring-seep tufa on concave slope.

BT-7. Light gray (10 YR-2.5 Y), partially calcified (38% CCE, 0.3% OC) clayey silt sand (30 cm), poorly sorted. Colluvium.

BT-6. Very pale brown (10 YR), calcified (27% CCE, 0.3% OC) and laminated, clayey silt sand (0-40 cm), poorly sorted. A marly spring deposit on concave slope.

BT-5. Pink (7.5-10 YR) sand (8% CCE, 0.2% OC), moderately sorted, filling fissures and pockets of underlying sediments to 30cm maximum thickness. Eolian-derived colluvium.

BT-4. Grayish brown (10 YR), partly calcified (24% CCE, 0.7% OC), clayey silt sand (10-30cm) with vertical structure, poorly sorted. ^{14}C dates: 13,800 and 13,950 on humus (SI-2583B, 2583). Organic soil developed in colluvium around spring seep. Fresh Later Stone Age (Albany) artifacts dispersed through this and all lower units.

BT-3. Light gray (10 YR), partly calcified and mottled (yellow, 10 YR) (19% CCE, 0.3% OC), sand (0-15cm), moderately sorted, Colluvium.

BT-2B. Dark and light gray (10 YR), irregularly calcified (13% CCE, 0.8% OC), silty sand clay (0-60cm), poorly sorted. ^{14}C date of 14,900 on humus (SI-2584). Organic soil developed in spring clays.

BT-2A. Dark gray (10 YR), cracking, silty clay sand (10% CCE, 1.0% OC) (20-70cm), moderately sorted. Organic soil developed in spring muds.

BT-1C. White (2.5Y), silty clay sand (CCE 6%, 0.2% OC) (100-200cm), poorly sorted. Spring muds, interbedded with colluvium at base (1A, 1B), mantling footslope.

BT-1B. Pale yellow (2.5Y), calcified lense (CCE 33%, 0.1% OC) (0-30cm) of silty clay sand, poorly sorted. Marly clay, interbedded in lower spring muds.

BT-1A. Pale yellow (5Y), partly calcified (CCE 14%, 0.1% OC) and reduced, silty sand clay (50-180cm), with gypsum microconcretions; dispersed dolerite cobbles in upper part, lines of shale rubble in lower part; poorly sorted. Intermixed spring muds and colluvium, resting on shale and dolerite.

contradicting the radiocarbon dates and sediment properties used to distinguish US-4B from 4C. However, the dates are fallible due to millennia of tree-rooting, while the sediment criteria are weak; since no sediment boundaries can be identified, the distinction made in Table 13 may require revision. The pollen from US-3 (Scott 1976) indicates a much drier micro-environment (almost exclusively chenopodiaceae) than the sediments and soils would infer (grassy swales with sedge stands), possibly due to younger pollen subsequently penetrating dehydration cracks. Unpublished pollen from US-2 belong almost exclusively to Compositae, which is compatible with the semieolian character of the calcified sands.

Unit US-3, with its capping 14,700-year soil, provides a direct counterpart to BT-2B,

Table 13. Spring sediments and alluvial terraces at Uitzigt, Alexanderfontein Basin.

Below spring face

US-4C. Gray (10 YR-2.5Y), calcified (35-41% CCE), clayey silt sand (60cm), poorly sorted and variably organic (1.7-3.5% OC). Forms narrow tongue of sediment grading down into modern, artificially deepened seep. ^{14}C dates of 1555, 1880 and 2220 B.P. on humus, as well as one inverted date of 1070 on charcoal (carbonized root?) from older organic sediment (SI-2233, 2228, 2231, 3564).

US-4B. Gray (10 YR), calcified (19-26% CCE), clayey silt sand (175 cm), poorly sorted and less organic (0.5-1.3% OC). Forms spurs of sediment around seep, embanked against various older beds. ^{14}C date: 4075 B.P. on humus (SI-2232).

US-4A. White (10 YR-2.5 Y), chalk (12-32% CCE), with residue of silica-rich, silty clay sand (5-15cm), poorly sorted. Forms thin sheet under US-4B.

Spring surface and face

US-5A. Dark grayish brown (7.5-10 YR), clayey silt sand (10cm) with reworked calcrete grit, poorly sorted. Modern A-horizon.

US-4B, C. Light gray (10 YR), laminated calcrete (56-62% CCE), with residue of silica-rich, silty medium sand (5-15cm), moderately sorted. Embeds rich Later Stone Age (Smithfield) industry. ^{14}C dates of 1395 and 3290 B.P. on ostrich eggshell, and 2520 B.P. on inorganic carbonate (SI-2581, 2580, 2582). Forms thin surface sheet, cemented by throughflow waters.

US-2D. Pink (7.5 YR), calcified (47% CCE), silica-rich silty medium sand (25cm), moderately sorted, with soil structure. Weak B-horizon of buried paleosol, secondarily enriched with calcium from above.

US-2C. Light brown (7.5-10 YR), partly calcified (30% CCE), silica-rich silty medium sand (25cm), locally with *Succinea*. B-horizon of paleosol.

US-2B. Very pale brown (10 YR), chalky (32% CCE), medium-to-coarse sand (20cm). CCa-horizon of paleosol. ^{14}C minimum date of 12,450 B.P. on carbonates (SI-3565).

US-2A. Very pale brown (10 YR), partly calcified (24% CCE), medium-to-coarse sand (over 200cm), moderately sorted. Unit US-2 was largely derived from eolian transport, with sand fraction better sorted (Folk σ0.68-0.74) than modern eolian sand sheet upslope (σ1.14), and containing no local grit. Sand probably trapped in wet valley bottoms fed by bicarbonate-saturated spring waters.

Alluvial Terraces

US-5B. (+0.9m, Low Terrace). Pale brown (10 YR), silty clay sand (50cm), well stratified, locally current-bedded, with lenses of medium to coarse-grade calcrete gravel, poorly sorted. Developed recently as narrow bench within gullied Middle Terrace, from which it is derived.

US-4C. (Wash on top of Middle Terrace). Grayish brown (2.5Y), silty clay sand (35cm), weakly organic (0.4% OC), often gritty, and poorly sorted. Once a continuous valley floor sediment, becoming a more organic (3.4% OC), clayey sand silt a little below the spring.

US-3B. (+1-1.7m, Middle Terrace Soil). Gray (5Y), calcified (38% CCE), clayey silt (15cm), moderately organic (0.7% OC), moderately sorted, with cracking structure and occasional proliferations of *Succinea*. A-horizon of paleosol. ^{14}C date 14,670 B.P. on humus (SI-3563).

US-3A. (Body of Middle Terrace). Light olive gray (5Y), clayey sand silt (over 1.5m), poorly sorted, with derived calcrete sand and grit, weak cracking structure, and some organic matter (0.5% OC). C-horizon of paleosol in fine alluvium, with surface gradient of 1.38%, merging at 1.1km from pan with classic, cracking lake beds of yellowish to grayish brown (2.5Y), oxidized and calcified (15-42% CCE) or silicified (3-31%), clayey sand silt, poorly sorted. This facies is identical to that of the extensive lake plain adjacent to the modern pan, and is preserved to +12.5m.

(Fill and bedrock incision [2-2.5m])

US-2A. (+2-2.5m, Lower Calcrete Terrace). White (10 YR-2.5Y), chalky calcrete (42-50% CCE), with silica-rich residue of silty clay sand (1.5m), moderately sorted. Sands identical in mean size and sorting to those of US-2 at the spring. Along valley margins, the upper part, possibly recemented, contains some major concentrations of Middle Stone Age (in lydianite), transported only short distances from slope; occasional edge-worn Acheulian bifaces and cleavers (in andesite) occur within body of terrace. Aberrant ^{14}C date of 4395 B.P. on inorganic carbonates (SI-2027) indicates crystallization during mid or

(Contd.)

Table 13 *(Contd.)*

late Holocene spring phase, when a veneer of clayey silt sand was swept over the surface from higher slopes. Forms terrace once connecting directly with US-2 at spring face, inclined at 1.8%, dipping below Middle Terrace level 1.6km from pan, suggesting a closer shore position but at a broadly similar level.

(Bedrock incision [4-5m])

US-1. (+4-6m Upper Calcrete Terrace). White (10 YR), laminated or hardpan calcrete (over 80% CCE), with residue of silica-rich sandy clay silt, moderately sorted. Poorly exposed mantle on bedrock-lined spurs, continuous with projection of slopes above spring, with gradient of 1.65% and projected to lake floor at +19m; later truncated by erosion at 1.2km from pan.

---

while US-4 parallels BT-10 and 11. The Uitzigt data serves to tie BT-1 to 4 with a high lake in the basin. Similar lake clays are found west of the pan to +12m and southwest of the pan at +15.5m. In the western sector these lake beds predate 12,755 B.P. (SI-2028), on calcium carbonate that intruded the fissures of a calcified, sandy clay silt (under a considerable overburden); a spring chalk (sandy to clayey silt residue), laterally equivalent to the lake clay, has a date of 16,010 B.P. (SI-1115) (Butzer, Fock *et al.* 1973). In the southwestern sector, gleyed, sandy clay silts are interbedded with laminated calcrete or chalk (silty sand residue) upslope, and then prograded by a spring tufa (sandy silt clay residue) dated 15,400 B.P. (SI-2029). Relatively fresh, wave-cut platforms and nips in bedrock are common in the basin at 11-13m and 15-16.5, and they argue that a 12-16m lake persisted from before 16,000 to about 13,500 B.P., after which the pan rapidly dried out.

The US-1 terrace projects to a lake at +18-19m. Low beach cliffs or platforms are repeatedly found at 18-21m above the pan. Although distinct, these are eroded and very much older. Calcrete mantles, but no unambiguous, associated sediments are preserved.

Mauritsfontein 26 (28° 52′ 21″ S, 24° 47′ 58″ E) represents a locus of rich archeological occurrences, 3.2km southeast of the modern Alexanderfontein Pan. A 4.3 hectare surface was mapped at 1:1000, with sediments and soil types, then archeologically sampled and tested by controlled surface collection or excavation in 1974-75. 110 sediment samples were analyzed to control for facies variability. Relationships to the pan and the basin margins were established by one direct, 4km transect as well as 10 further transects along other basin quadrants. The depositional sequence is outlined in Table 14 (Butzer, unpublished). The ranges of sand, calcium carbonate equivalent (CCE), and opaline silica (as percentage of non-carbonate residue) represent the 1σ variation for each unit and help define distinctive lithologies. Mean size, sorting, and skewness (see indices of Folk [1966]) are as follows for the fraction coarser than 0.37mm (with standard deviations in parentheses):

Unit	Mean (in φ [log] units)		Sorting (σ)	Skewness
MS-5A	2.85	(0.10)	0.602 (0.047)	+0.15 (0.11)
MS-4	2.85	(0.145)	0.656 (0.038)	0.12 (0.13)
MS-3C	3.06	(0.150)	0.694 (0.042)	0.07 (0.10)
MS-3B	3.12	(0.114)	0.708 (0.054)	0.075(0.10)
MS-3A	3.03	(0.125)	0.735 (0.062)	0.10 (0.12)
MS-2 (fine)	3.06	(0.155)	0.730 (0.043)	0.06 (0.10)
MS-2 (coarse)	2.97	(0.141)	0.700 (0.041)	0.08 (0.11)

Table 14. Shore deposits and eolian sands at Mauritsfontein 26, Alexandersfontein Basin.

MS-5A. Pale brown (10 YR), superficially humified sand (90-92% sand fraction), partly stratified and laminated, with reworked calcrete grains (5-10% CCE). Eolian sand, thinning from 60 to 10cm, west to east. ^{14}C date: 805 B.P. on wildebeest horncore at base (SI-2024), and 310 B.P. on humus of marshy fill (MS-5B) in depression subsequently cut into MS-3C to 5 (SI-3566). Some Later Stone Age occupation.

(Erosion of 20m wide, 50cm deep gullies, exposing Middle Stone Age artifacts then used as quarry by Later Stone Age (Smithfield) occupants; ^{14}C date: 3040 B.P. on associated ostrich eggshell [SI-2587])

MS-4. Pale brown (10 YR), weakly humified sand (78-88% sand), partly calcified (8-20% CCE). Eolian sand, forming uniform 10 to 20cm cover over site. ^{14}C date: 4475 B.P. on ostrich egg resting on basal contact (SI-2022); 3370 B.P. on quagga and warthog teeth found on or in top of sediment (SI-2025); 3805 and 2225 B.P. on soil carbonates near base suggesting span of soil formation (SI-2586, 3568).

(Local soil erosion [$>$30cm], after end of soil development. Surface occupation, without diagnostic artifacts, made from rejuvenated Middle Stone Age pieces)

MS-3C. Light to pale brown (7.5-10YR) sand, silty sand, or clayey silt sand (67-80% sand), partly calcified (15-27% CCE), with weak soil structure. B-horizon (25-30cm) of buried paleosol developed in shore or shore-near silty sand and sand. Lake fluctuating at and below +13m. Dispersed, edge-worn, Middle Stone Age artifacts.

MS-3B. Very pale brown (10 YR), silty sand or clayey silt sand (60-70% sand), calcified (24-35% CCE), with opaline silica enrichment (2.5-25%), and moderate soil structure. Rare *Succinea* shells. BCa-horizon (10-20cm) of paleosol. ^{14}C dates: 9310, 13,680, 13,895 and 15,175 B.P. on calcareous concretions of soil (SI-3567, 2021, 2018, 2020) date second half of high lake phase and subsequent leaching. Dispersed, edge-worn, Middle Stone Age artifacts.

MS-3A. White (10 YR), silty sand (55-73% sand), heavily calcified (34-38% CCE), with opaline silica enrichment (6-33%). Ca-horizon of paleosol. ^{14}C date: 10,875 B.P. on calcareous concretion (SI-2017). Dispersed but locally abundant, edge-worn, Middle Stone Age artifacts.

(Wave cutting of smooth platform across calcrete at +11.5-13.5m, responsible for reworking of Middle Stone Age artifacts originally embedded in MS-2)

MS-2. White (10 YR) massive calcrete with two facies: (a) moderately calcareous (30-65% CCE) and highly siliceous (19-60% opaline silica), with a residue of clayey silt sand (37-63% sand); and (b) highly calcareous (50-81% CCE) and moderately siliceous (11-47% opaline silica), with a residue of sand or silty sand (63-82% sand). Locally contains one or two horizons of concentrated, mainly edge-worn, Middle Stone Age artifacts. Mantles undulating shale surface with thickness of 20-100cm, representing highly altered shore and shore-near deposits related to lake at or a little below +12.5m. ^{14}C date: 11,285 B.P. on calcrete (SI-2585) represents recrystallization of much older lake and soil carbonates, during terminal Pleistocene submergence and subsequent soil formation.

MS-1. White (10 YR), wavy-bedded to laminated calcrete (77-84% CCE, 51-66% opaline silica); the lower 80-120cm have a residue of clayey sand (74% sand) and are truncated by an irregular, erosional surface; the upper 100-50cm have a residue of silty sand (83% sand) and incorporate fine, reworked calcrete pebbles. Forms a conspicuous bar culminating at +12.5m, resting on a broad bedrock-cut surface at +10-10.5m, situated 200m in front of the Mauritsfontein site. Either contemporary with or older than MS-2. ^{14}C date: 25,250 B.P. on the lower calcrete (SI-2019) represents a minimum age only. No artifacts.

The size and sorting statistics demonstrate two kinds of background sands, one coarser and better sorted – the eolian sands of MS-4 and 5A, the other finer and less well sorted – the mixed beach sands of MS-2 and 3. The former represent extensive eolian remodelling of the basin at a time of minimal lake expanse, with equivalents of MS-5A continuing 2km down towards the pan, to just above a +2m shoreline. The latter facies complex is represented within and below younger unconsolidated beach ridges at +7m that preserve the full transition of lake plain to shoreline. The heavy calcification of MS-2 and 3 in part reflects primary, marly sands, in part calcareous throughflow waters from slope seepage, activated after longer periods of heavy rains, e.g. as observed in 1977 after 4 summers with well above-normal rainfall that began in January 1974.

The MS-3 deposits represent the +12-16.5m, terminal Pleistocene lake already discussed. The subsequent soil formation and leaching, before 9000 B.P., relate to accelerated spring discharge at Benfontein from before 11,500 to after 7000 B.P., and probably date the extensive sediments of a +5.5-7.5m lake that only contains reworked, older artifacts in its sandy shorelines, and nothing datable in its cracking clay flats. During wetter parts of the late Holocene, the lake level fluctuated near or below +2m; again, there are no direct means to date this youngest lake stage. The MS-2 beach at +12.5m, broadly coeval with Middle Stone Age activities in the area, finds confirmation in the Uitzigt and Deelpan sequences.

It is now possible to synthesize the information from the pan basins of the western Free State, as in Table 15. A realistic picture of interblending, minor environmental changes emerges for the last 4500 years or so, with an overall tendency to a little more water and vegetation until about 1300 B.P. The details, particularly of the fluctuations and their dating, still requires greater precision. The period spanned by Alex 6 and 7 is represented by the alluvial fills included in Riverton 5, and at Florisbad includes phases 4C to 5B, while Peat IV appears to be a little earlier.

The period from 6000-4500 B.P. represents an apparent gap at each site, probably as a result of limited geomorphic activity during a time of stable climate, with few high-energy climatic events. This break also marks a lacuna in regional occupation, after the Albany and prior to the Smithfield-Wilton technocomplexes of the Later Stone Age.

The lake episode 11,500-6000 B.P. (Alex 5) is widely documented, including the Klipfontein basin (6920 B.P., Butzer, Fock *et al.* 1979), and corroborated by tufas of the Benfontein and Gaap Escarpment springs (Buxton 9). Water was generally abundant, and saturated with bicarbonates, but there is no evidence for organic marsh accumulations. This may explain the enigmatic pollen spectra from Voightspos (Horowitz *et al.* 1978), Klipfontein (Butzer, Fock *et al.* 1979), and the Riverton 4 deposits of the Vaal River (L. Scott, unpublished), which suggest that grass cover was comparatively poor but bush more expansive. Florisbad 4A and Riverton 4 evidently document this period in other settings and areas.

The end of the Pleistocene, between 13,600 and 11,500 B.P., was characterized by several rapid and incisive climatic shifts, including one or more episodes of total pan desiccation and spring quiescence, at a time when reddish Kalahari sands were widely distributed, including emplacement on the slopes east of Florisbad. This same variability is documented in the pollen profile of the spring at Aliwal North between about 13,200 and 9600 B.P. (Coetzee 1967; Scott, this volume). Again, the dating requires substantial refinement.

The Alexanderfontein lake phase of 16,100 to 13,600 B.P. (Alex 2), complemented by good spring records, is spectacular. To maintain lakes at +12-16.5m, even with a 6°C temperature reduction, would require about 670 to 860mm rainfall over the basin (compared with 400mm today) (Butzer, Fock *et al.* 1973). This is a compatible model for the presence of Cape elements (*Myrica*, Ericaceae) among the sparse pollen preserved in BT-2 (L. Scott, unpublished). The moist environment was reflected in the prominent weathering of Riverton 3 surfaces in the nearby Riet valley about 18,600 to 13,600 B.P.(root casts and Ca-horizon nodules with dates of 18,480, 15,980 and 13,740 B.P. [SI-3385, 3384, 3384], the last in a lateral channel cutting through Riverton 3). The absence of lake beds at Voightspos and Deelpan can best be explained by expanded lacustrine plains, beyond the modern exposed littoral perimeter. The

Table 15. Environmental interpretation for the Alexandersfontein, Deelpan, and Voightspos basin records.

Alex 7B. Minor filling of small gullies with clayey alluvium (US-5B) or marshy sediment (MS-5B, 310 B.P.; VP-4B) marks return to geomorphic stability prior to recent disturbance. About 350-75 B.P.

(Local donga incision [US and MS, 1.5m])

Alex 7A. Eolian activity in basins, with deflation of pans and with low dunes formed downwind (DP-5; VP-4A, 1220 B.P.; MS-5A, 805 B.P.). Drier than today. Local Later Stone Age occupation. About 1300-500 B.P.

Alex 6B. Some eolian activity in basins, reduced as low lakes temporarily rise (DP-4,+3m lake maximum; VP-3,+1m lake); accelerated spring discharge (late US-4C, 1555-1395 B.P.; BT-11B). Slightly moister than today, but also fluctuating to the drier side. Local Later Stone Age occupation. About 1800-1300 B.P.

Alex 6A. Some eolian activity in basins, with fluctuating lake levels (DP-3, lake mainly under +2.5m; VP-2B,+2.2.m lake, 3500-2200 B.P.; MS-4, sand deposition 4475-3370 B.P., with soil decalcification 3805-2225 B.P., interrupted by shallow gully cutting before 3040 B.P. Alexandersfontein lake probably up to +2m maximum); accelerated spring discharge initially undersaturated in bicarbonates (US-4B, early 4C, 4075-1880 B.P., soil carbonates in US-2A, 4395 B.P.; BT-10, 3450 B.P., BT-11A). Moister than today, but fluctuating. Considerable Later Stone Age (Smithfield) occupation after 3500 B.P. Began about 4500 B.P., with major dry spells ca. 3100 and 1800 B.P., and maximum moisture about 3500 B.P.

(Depositional break, probably relatively dry, about 6000-4500 B.P.)

Alex 5. Accelerated surface runoff and colluvial sedimentation at time of high lake levels (DP-2, lake +3.5m; VP-2A, lake +2.0m, 6830-6350 B.P.; Alexandersfontein probably +7m); accelerated spring discharge (US-4A; BT-8, 9, 11,230-7035 B.P.) saturated with bicarbonates; decalcification of surface soils (development of MS-3 profile until after 9310 B.P.). Mainly moister than today. Local Later Stone Age (Albany) occupation. About 11,500-6000 B.P.

Alex 4B. Local erosion and colluvial activity (VP-1, BT-7?) Drier.

Alex 4A. Some accelerated spring activity (BT-6, and 12,755 B.P. calc-filling of dehydration cracks west of Alexandersfontein; mobilization of carbonates in US-2B, 12,450 B.P.). Slightly moister, perhaps 12,900-12,300 B.P.

Alex 3. Erosion and redistribution of reddish eolian sands (BT-5). Drier, from a little after 13,600 to about 12,900.

Alex 2. Moist valley aggradation (US-3, 14,670 B.P.); organic soils developing around strong springs, in part with water undersaturated in bicarbonates (BT-1, 2, 4, 14,900, 13,950, 13,800 B.P., separated by drier interval with spring quiescence and colluvial activity, BT-3; other, carbonate-rich Alexandersfontein spring activity dated 16,010 B.P., at time of maximum lake extent, and 15,400 B.P., after the lake level had dropped from +15.5m to under +13.5m); high lake in Alexandersfontein Basin, fluctuating between +16 and +12m (MS-3; oldest concretions of MS-3B dated 15,175 B.P.). Quite moist environment. Local Later Stone Age (Albany) occupations. From before 16,100 to 13,600 B.P. or a little later, with a sharp dry spell about 14,500 B.P.

(Long depositional break and some episodes of erosion followed by recementation of calcrete caps; environment mainly as dry or drier than today)

Alex 1. High lake levels (MS-2,+12.5m lake; DP-1, +4m lake); arid valley aggradation, with accelerated eolian activity (MS-2); accompanied or followed by calcrete formation due to water seepage. Abundance of water, yet effective wind action and semiarid vegetation (US-2 Compositae). Middle Stone Age occupation. Early Upper Pleistocene.

marshy Uitzigt terrace (US-3) and a lateral Riet channel fill of the same age both suggest that aggradation of Riverton 4 may have begun at about this time, at least in some valleys. At Florisbad, the Peat III complex appears to continue into this period, with the tufas of Buxton 8 at least parly coeval.

A major lacuna, of many tens of thousands of years, preceded Alex 2. This also represents a remarkable archeological gap in the region, between the Middle Stone

Age and Albany, that is only bridged by certain cave sequences of the eastern montane zone and in the Kuruman Hills.

The early Upper Pleistocene lake phase (Alex 1), linked to Middle Stone Age settlement, appears to be coeval with Riverton 3, Florisbad 3A and Buxton 7. Around Florisbad this aggradation was decidedly 'humid,' but at Alexandersfontein and in the Vaal valley quintessentially 'arid,' despite the abundance of water. The differences may reflect regional contrasts in sediment supply, complex vegetation associations as yet not understood, and unusual combinations of climatic variables.

Of comparative interest is that the wettest intervals recorded near Florisbad, in the western Free State basins, and on the Gaap Escarpment all date to the later Pleistocene, rather than to the Holocene. This contrasts with the East African Rift, where the Holocene lakes were highest (Butzer 1978a, 1980b); so for example, Lake Chilwa in southern Malawi (15° 15′S, 35° 40′E) reached peak levels ca. 9140 and 5660 B.P. (Lancaster 1981a).

Doornlaagte and Rooidam

The Alexandersfontein record is carried backward in time by two former pans, west of Kimberley, on the farms Doornlaagte and Rooidam, known on account of their archeological excavation. Although not adjacent, Doornlaagte (28° 47′S, 24° 21′E; 1115m elevation) is older than Rooidam (28° 47′S 24° 31′E; 1165m elevation) in terms of upstream position in a series of stepped depressions, potentially draining towards the Riet River. In effect there is a record of four cycles of pan excavation and sedimentation in the Kimberley region: Alexandersfontein belongs to the youngest of these, Rooidam to the two next older, and Doornlaagte to the oldest.

The Doornlaagte sequence, with almost 6m of deposit exposed, has been published and is presented as a summary interpretation in Table 16 (Butzer 1974b). Similarly, the Rooidam sequence, comprising 9.5m of deposit over dolerite, is given in Table 17 (Butzer 1974b), incorporating minor revisions of 1977 (Szabo & Butzer 1979) and the

Table 16. Sedimentary and environmental sequence at Doornlaagte, Kimberley District.

DO-7. Corrosion of calcrete surface and leaching, with subsequent penetration of reddish soil wash, with strong eolian components, into soil pipes. Initially wetter, then semiarid.

DO-6B. Formation of calcrete hardpan on the old lake deposits. Seasonally dry, but with abundant throughflow water.

DO-6A. Seasonal pan, with strong eolian components (Unit Eiii: 50cm). Seasonally moist, but becoming drier.

DO-5. Fluctuating but shallow lake, recorded by freshwater limestones, interbedded with sandy lenses representing fine wash of conspicuous eolian derivation (Units Ei, Eii: 140cm). Environment more humid than today, but subject to long dry intervals with desiccation.

DO-4. Seasonal pan, with colluvial and eolian sedimentation (Unit D; 30cm). Seasonally moist, with stronger surface runoff and long dry spells.

DO-3. Fluctuating but deep lake with freshwater limestone and some aquatic snails (Unit C: 50cm). Subhumid climate, interrupted by dry periods. Sporadic Acheulian presence.

DO-2. Seasonal lake, with freshwater limestone (including proliferations of aquatic snails) interbedded with coarse colluvium (Unit B: 60cm). Seasonally moist, with strong runoff. Major Acheulian occupation.

DO-1. Deep, possible permanent lake with freshwater, limestone (Unit A: over 240cm). Long period of subhumid climate.

Table 17. Sedimentary and environmental sequence at Rooidam, Kimberley District.

RO-8. Corrosion of calcrete surface and leaching, with subsequent penetration of reddish soil wash, with strong eolian components and Middle Stone Age artifacts, into fissures and bedding planes. Initially wetter, then semiarid.

RO-7B. Formation of calcrete hardpan on the old lake deposits. Cements Middle Stone Age artifacts. Seasonally dry but with abundant throughflow waters.

RO-7A. Fluctuating but deep lake, recorded by freshwater limestone with some eolian components (Unit I: 130cm). Subhumid climate, interrupted by dry spells.

RO-6. Seasonal lake, with eolian and colluvial sedimentation; Acheulian artifacts (Unit H: 30-60cm). Seasonally moist, with surface runoff and long dry intervals.

RO-5. Fluctuating but shallow lake, with freshwater limestone containing eolian components (Unit G: 75-140cm). Mainly wetter climate, but interrupted by long dry periods. Th/U date: 174,000 B.P.

RO-4. Fluctuating, shallow lake or seasonal pan, with laminated marl and cracking clays; some colluvial components (Unit F: 40-60cm). Acheulian artifacts. Seasonally moist but subject to long dry spells with desiccation.

(Palesosol: corrosion and carbonate mobilization, with root cast formation and solution fissures. Watertable low, but seasonally wet)

RO-3. Seasonal pan, with laminated marl, cracking clays, and coarse colluvial lenses (Units C, D, E: 40-130cm). Concentrations of Acheulian artifacts. Seasonally moist, with protracted dry periods and episodes of strong surface runoff.

RO-2. Seasonal pan, with aquatic snails; mainly recorded by eolian shore ridge that grades into mixed eolian-colluvial beds with shallow channel structures at Rooidam 2 (unit B: 50-150cm). Major Acheulian occupations. Seasonally moist, with strong runoff and long dry spells.

RO-1. Mainly colluvial environment, with reworked eolian sands and interrupted by shallow channels. (Units O, A: 450cm). Long period of semiarid climate, with strong but intermittent runoff.

1982 recording of new quarry exposures ('Rooidam 2'). It is apparent that the two sites do not overlap in time: the Doornlaagte sequence begins with a deep lake and ends with a shallow, fluctuating one. The Rooidam sequence begins with a dry pan floor that becomes a fluctuating lake; midway, this lake dries out and soil development ensues, before a new lake cycle begins, terminating with a deep lake, prior to rapid desiccation. An even younger cycle is incompletely exposed in a lower basin downvalley of Rooidam, at Biesiesput (28° 48′s, 24 30′E; 1136m elevation): almost 2m of sandy-chalky calcrete record a seasonal pan with strong eolian components (Butzer, unpublished). Wheras DO-2 and 3 have Upper Acheulian sites or occurrences, RO-2 to 6 have terminal Acheulian (Fauresmith), and RO-7B and Biesiesput have Middle Stone Age artifacts. RO-5 has an approximate Th/U age of 174,000 B.P. on aragonite (Szabo & Butzer 1979), while RO-3 has an aberrant date of 108,000 B.P. on recrystallized calcite. The last phase of higher watertable, that reset the radiometric 'clock' in RO-3 about 104,000 years ago (concordant thorium and protactinium ages), in all probability provides a generalized date for the Biesiesput sequence.

Seen in conjunction, there are 4 lake cycles recorded here: one at Doornlaagte; two at Rooidam (separated by a soil), of which the second dates roughly 174,000; and one at Biesieput, probably about 104,000. The last cycle presumably correlates with Riverton 3 and Alex 1, which both have similar sandy facies and an identical Middle Stone Age variant. The penultimate cycle, recording the Fauresmith to Middle Stone Age transition or break, after 174,000 B.P., may well parallel the sedimentation of Riverton 2 (Table 4). The antepenultimate lake cycle, with the key Fauresmith site at Rooidam, provides a reasonable parallel to Riverton 1. Finally, the first, Doornlaagte cycle could be correlated with the Rietputs C. In effect these records appear to provide a lake history spanning the mid-Middle to early Upper Pleistocene, complementing that of the western Free State pans.

Kathu Vlei and the Southern Kalahari

Perhaps the best paleoenvironmental sequence from the Kalahari Basin comes from Kathu Vlei (or Pan), a broad surface of organic marshland next to the new city of Kathu (27 40′S, 23 01′E), 15km north of Sishen. Here a smooth surface between the Kuruman Hills and Koranaberg dips toward the Molopo drainage; this surface is developed on some 50m of Tertiary calcrete and up to 25m of gravel, sand or shale that mantle an erosional surface cut across Precambrian rocks. The alternating sandy, porous, and indurated facies of the gently-inclined calcrete complex (ISCOR borings 1971) allow waters from the 25-50km distant mountains to rise under pressure during wet periods, so that the vlei is primarily maintained by artesian seepage rather than surface waters. Recent groundwater pumping has revealed a covered karst in the calcrete substrate of Kathu Vlei. Although calcrete is commonly found 2-3m below the surface, an 8m drop of the watertable has led to compaction of numerous doline fills, with collapse and partial exposure of the sedimentary sequence.

One such collapse was excavated by P.B. Beaumont and showed the doline base to be at −11.7m. The sedimentary column is described in detail in Table 18 (Butzer, unpublished), employing the units and thicknesses of Beaumont; it can be interpreted from top to bottom as follows:

a) 160cm of organic, clayey silt sand (Unit 1) deposited during the last 7500 B.P. or so (Pretoria ^{14}C date), under marshy conditions similar to those persisting into the 1970s. Two beds (at 15-56cm and 80-160cm) record very wet conditions, interrupted by a spell of accelerated surface runoff (coarse sand at 95-110cm). Assuming that the base of the unit dates to 7500 B.P. and that sedimentation rates were constant, maximum moisture is indicated from 7500-3750 B.P. and 2700-2400 B.P., with accelerated surface runoff 5000-4500 B.P.; conditions were drier than today 3750-2700 and 2400-600 B.P., since then they have been intermediate. This chronology has a probable error of 200 to 500 years, either way, and will be refined by further ^{14}C dating as well as by pollen data (samples collected by E.M. Van Zinderen Bakker).

b) Some 180cm of calcareous silty sands (Unit 2) accumulated during the terminal Pleistocene, and these are increasingly clayey and calcified towards the top, suggesting a decrease in sedimentation rates as a pan environment was established. The well sorted, fine to medium-grade sands speak for a strong, primary or derivative eolian component, underscoring the case for relatively dry conditions.

c) The 80cm of Middle Stone Age rubble (Unit 3A) represents a horizontal, poorly sorted detritus, swept across the Kathu area at the level of the cavernous calcrete surface by intermittent flash floods. At this time the wet season watertable was at least 4.5m lower than it was during the past few centuries (annual variability ca. 1971 was 2m); rainfall, which today averages 340-440 in the basin, must have been substantially less. The high concentration of artifacts and debitage, which are distinctly edge-worn, within a gravelly sand, speaks clearly for a derivative lag. The actual occupation may have been much older. A mid-Upper Pleistocene age seems probable.

d) The 360cm of Upper Acheulian deposit (Unit 3B) represents two temporal phases, possibly with a partial overlap. Initially, medium grade sands or silty sands

Table 18. Sedimentary sequence at Kathu Vlei, near Sishen (with depths below surface; OC: organic carbon; CCE: calcium carbonate equivalent).

Unit KV-1A (Iron Age or Pottery, Later Stone Age)

0-15cm. Grayish brown (10 YR), silty sand (OC 10.4%, CCE 32%).

KV-1B (Low-density late, Later Stone Age?)

15-56cm. Gray to light gray (2.5 Y), calcified clayey silt sand, with concretions towards top (OC 2. 8-8. 3%, CCE 48-85%),

56-62cm. Dark gray (5 Y), peaty, clayey silt sand with concretions (OC 14.0%, CCE 32%).

KV-1C (Essentially sterile)

62-80cm. Light gray (10 YR), calcified, silty clay sand, with concretions (OC 8.5%, CCE 58%),

80-95cm. Dark gray (5 Y), peaty, clayey silt sand, with microconcretions (OC 23.3%, CCE 33%),

95-110cm. Gray (5 Y), peaty, silty coarse sand, with concretions (OC 19.8% CCE 36%),

110-160cm. Dark gray (5 Y), clayey silt and, with detrital peat and concretions (OC 17.7% CCE 35%). ^{14}C on basal peat 7350 B.P. (Pta-3073).

KV-2 (Essentially sterile)

160-180cm. Light gray (10 YR), calcified clayey silt sand, with concretions (OC 0.3%, CCE 67%),

180-340cm. Light grayish brown or brown (10 YR), calcified silty fine sand (OC 0.5-2.3%, CCE 18-31%).

KV-3A (Relatively fresh Middle Stone Age, worn Acheulian)

340-370cm. Light brownish gray (10 YR), gravelly, silty coarse sand, with microconcretions and abundant artifacts (OC 0%, CCE 2.4%),

370-420cm. Light gray (10 YR), gravelly sand with some concretions; abundant artifacts (OC 0-0.2%, CCE 0-3.1%).

KV-3B (Locally, Middle Stone Age at top, otherwise Acheulian with Middle Pleistocene fauna)

420-430cm. Pinkish gray (7.5 YR) laminated sand (OC and CCE 0),

430-540cm. Inhomogeneous complex of pink to very pale brown (7.5-10 YR), silty sand, sand, coarse sand, and gritty or gravelly sand, with extensive reddish yellow (7.5 YR) oxidation mottling, and artifact concentrations (OC 0, CCE 0-2.5%). Arranged as background sands, multiply intruded by gritty to gravelly spring vents from below,

540-670cm. Pink to light brown (7.5 YR), moderately oxidized, stratified silty sand and coarse sand, with ash laminae and concentrations, multiply intruded by gritty to gravelly sands with artifacts (OC 0-0.1%, CCE 0-1. 7%),

670-780cm. Pink (5-7.5 YR), moderately oxidized, gravelly coarse sand, and sand with ash; local artifacts (OC 0, CCE 0.2-5.1%).

KV-4 (Sterile)

780-980cm. Reddish yellow (7.5 YR), heavily oxidized, stratified sand, silty sand, and clayey silt sand, with ash concentrations and laminae (OC 0, CCE 1.3-2.1%),

980-1140cm. Reddish yellow (5-7.5 YR), heavily oxidized, silty clay sand, rich in ash (OC 0, CCE 0.8-1.1%),

1140-1170cm. Reddish yellow 5-7.5 YR), heavily oxidized, silty clay sand, with ash and organic laminae, in part with carbonized fibers (OC 4.6% CCE 1.0%). Rests at base of doline dissolved in hard calcrete.

accumulated together with some concentrations of fibrous organic matter (subsequently reduced to ash) and, probably, carbonates (now leached); eolian components are present, but fewer than in Unit 2. Subsequently the aquifer became prone to strong, sudden recharges, leading to repeated "spring-eye" eruptions that squeezed coarse sands and gravelly sands up through the existing sediments. The Acheulian artifacts are linked to such vent facies but, since I was not able to observe their excavation, no explanation is offered for their derivation or concentration. The faunal remains, mainly durable enamel parts of teeth, are also concentrated in such vents, and are not only differentially preserved but also derived. Many of the artifacts and some of the tooth fragments are heavily coated by a glossy, light brown and transparent silicate patina that requires special study. The Acheulian sands and gravels suggest conditions initially drier than those of the Holocene, but followed by a notable increase of groundwater on a highly periodic basis, a geomorphic trend similar to that which could have emplaced the Rietputs-C gravels of the Vaal. The Kathu fauna is of comparable age (Olduvai Bed IV-type, Klein, this volume and pers. comm.). At some point, pan conditions ceased (terminal, laminated bed), the water table dropped to at least 9m below that of the last few centuries (and 4m below that during the time of original deposition), to allow the organic matter of the desiccated beds to burn out as a result of protracted fires.

e) The lowest 4m or so of Unit 4 originally comprised peaty, silty clay sands, judging by 4.5% of organic carbon preserved in the unvolatized basal bed, and by the abundance of stratified ash throughout the deposit. Conditions resembled those of 8000-4000 B.P. These beds could have been volatized by peat fires after sedimentation of Unit 3B had been completed, but it is more likely that this happened shortly after Unit 4 had been deposited and the watertable dropped about 3.5m: the deposits at -10.3m include some concentrations of ash that still preserve traces of carbon.

Pending further precisions on Unit 1, the Kathu climatic record is summarized in Table 19. The significance of this sequence is that it primarily reflects long-term groundwater trends, rather than short-term geomorphic processes. It therefore gives a filtered climatic record, and one that provides unique evidence for protracted intervals during the Pleistocene that were substantially drier than any part of Holocene time. Broad

Table 19. The environmental record of Kathu Vlei and its aquifer.

Kathu 5E. Relatively moist with vlei conditions.
Kathu 5D. Drier, with pan conditions and seasonal water.
Kathu 5C. Wet, with peat marsh and perennial water. (3100-2800 B.P.?).
Kathu 5B. Drier, with pan conditions and seasonal water.
Kathu 5A. Wet, with peat marsh, but some surface runoff (8000-4000 B.P. ?).
Kathu 4. Xeric deposition of eolian sands, culminating with pan conditions and intermittent water (KV-2). (Terminal Pleistocene).
(Depositional break)
Kathu 3. Xeric deposition of colluvial rubble (KV-3A). Low watertable. (Mid-Upper Pleistocene).
(Depositional break; Middle Stone Age occupation in area).
(Watertable drops at least 4m [9m below recent level] and peat fires destroy organic matter of KV-3B).
Kathu 2. At first, pan with seasonal water; later, strong periodic rises of watertable with spring eruptions (KV-3A). Acheulian; Cornelian fauna (Mid-Middle Pleistocene).
(Watertable drops 3.5m [11m below recent level], with protracted peat fires)
Kathu 1. Wet, with peat marsh and perennial water; slow accumulation of fresh sediment at base of doline (KV-4). (Late Lower or early Middle Pleistocene).
(Long period of subsurface, karstic solution creating Kathu doline [9-10m below calcrete surface, with maximum area of 6 by 10m], followed by draining out of residual sediment through underground sluices at a time of very low watertable)

similarities with the Gaap and Vaal records also seem indicated, suggesting that each provides complementary evidence that must be evaluated in a holistic context.

In comparison with Kathu Vlei, other depressions of the southern Kalahari Basin provide only a superficial record of late Pleistocene remodelling. In particular, the pattern is one of deflation of dunes, and eolian deepening of existing pans, prior to a lake phase and a subsequent interval of eolian intervention (Lancaster 1979a, 1981b). There are few opportunities to date the lake phase but, at Urwi Pan (23° 29′S, 20° 30′E), the end of one lacustrine interval is dated 17,000-15,500 B.P. by algal stromatolites that preceded saline clays and renewed deflation (Lancaster 1979b). Another fossil lake near the Orange River (but not further identified) has a date of 24,890 B.P. on lacustrine chalk (Heine 1979), possibly representing an early part of this same phase.

The information from the Molopo drainage, presented by Heine (1981), is very cursory. The lower terrace (TI) has dates of 13,000-12,200 (3 ^{14}C determinations) and 16,800-15,200 B.P. (3 further determinations) on ponded silts with *Corbicula*, Unionidae, *Bulinus*, *Planorbis* and *Xerocerastus*. The top of the next older terrace (T2) of the lower Molopo has an assay of greater than 18,100 B.P. on a silt-sand fill with mollusca. In the tributary Nossob drainage, Heine confused two different alluvial terraces, the 17-20m high terrace (HT) and the 5-6m early middle terrace (MTI, with Middle Stone Age) of Leser (1971: 174-79), providing minimum dates of 25,000 and 17,500 B.P. that do little to resolve later Pleistocene geochronology.

It appears that the Kalahari lakes were more responsive to climatic changes than Kathu Vlei. The absence of Holocene lacustrine beds suggests that reduced late Pleistocene evaporation was at least as significant as increased precipitation in sustaining lakes (Urwi) and surface runoff (Molopo, Buxton), without discernible impact on deeper aquifers (Kathu). Similar problems are apparent in the northern Kalahari basin.

Gi Pan and the Northern Kalahari Basin

The only longer sedimentary sequence from the northern Kalahari comes from Gi Pan (19° 37′S, 23° 01′E), in the Dobe Valley of northwesternmost Botswana. Dates, fauna, and prehistoric occupations add significance to this succession, outlined in Table 20, after Helgren and Brooks (1983).

The pan complex of the Makgadikgadi depression (21° S, 26° E) remains surprisingly poorly studied, despite evidence for multiple shorelines (Grey & Cooke 1977) and a suite of ^{14}C dates (Heine 1978a, 1978b). The major basin-wide shorelines are at 920-925m and 940-950m, the higher level commonly obscured by eolian features or long colluvial slopes with Acheulian artifacts (D.M. Helgren, pers. comm.). Heine's dates were apparently not collected from specific shorelines but from vaguely identified points on the extensive 910-920m lake plains (e.g., north of the Sua Pan, from the lower Nata river); they include calcretes, lake chalks or marls, and stream silts with mollusca. The dates basically serve to verify extended lakes or perennial stream flow about 30,750-18,500 B.P. (7 dates) and ca. 12,000 B.P. (1 date), with less reliable assays on pedogenic calcretes suggesting soil water mobilization ca. 31,750, ca. 13,150-14,700 (3 dates) and ca. 8650-9450 B.P. (2 dates). Possibly these meagre facts can be interpreted to date the last regional ('mega') lake with 920m shorelines, from >31,000-18,500 B.P. After this lake dried out (a formidable event in its own right), subsequent

Table 20. Pan, lake and stream beds at Gi Pan, Dobe Valley, northwestern Botswana.

GI-1A. Black, clayey silt sand (18-35cm), with dates of 110, 495, and 810 B.P., representing organic, colluvial pan infilling. Semiarid local environment. Later Stone Age occupation.

GI-1B. Light gray, sandy silt clay (7-10cm); colluvial pan fill. Arid.

GI-2A. Light gray to white, sandy clay slit (0-40cm); pan sediment. Arid. Later Stone Age occupation.

GI-2B. Light gray to white lake limestone (5-40cm), also local spring and channel tufas, with dates of 22,250 and 23,980 B.P. Subhumid.

GI-2C. White silty sand and stratified, light gray, clayey silt sand (0-40cm), representing stream and colluvial beds. Arid. Early Later Stone Age occupation.

(Depositional hiatus and limestone corrosion)

GI-3. Light brownish gray, lake limestone (20-60cm), also local spring and channel tufas; minimum date of 20,450 B.P. Subhumid.

GI-4. Interbedded, light gray calcrete conglomerates, gray silty sands, and calcified, white clayey silt sands, forming a 50-75cm package of alluvial and colluvial deposits. Older than 43,000 B.P. Semiarid. Middle Stone Age occupation with abundant fauna, with 3 extinct species (Early Florisian, i.e. early Upper Pleistocene).

GI-5. Crossbedded white and gray stream sands with dispersed gravel (75cm). Semiarid. Traces of artifacts, possibly Middle Stone Age.

GI-6. Indurated conglomerates (base not seen), including Acheulian in other exposures. Probably mid-Pleistocene.

lacustrine phases were more local in occurrence and possibly also in their impetus (e.g., different drainage sectors, river progradation); more water is again suggested very roughly 14,500-8500 B.P. Deflation later carved out large scallops of the 910-920m lake plains, and evaporite veneers then accumulated on the floors of these new pans (Helgren, unpublished); a Later Stone Age site with a modern fauna was occupied on the 906-908m shores of a prodeltaic lake of the Botletle River before 2960 B.P.

The available Makgadikgadi information seems compatible with the Gi sequence, with intermediate or even drier conditions suggested since almost 20,000 B.P. Phase Gi-2B recalls the 920m Makgadikgadi stage of >31,000-18,500 B.P., while Gi-3 may or may not parallel the 940-950m stage.

Comparison of the cave and stream deposits of the Kwihabe Hills, only 65km southeast of Gi, appears more difficult. Here, H.J. Cooke (1975, Grey & Cooke 1977; Cooke, H.J. & Verhagen 1977) has a large suite of dates on cave travertines (flowstone and dripstone) or 'sinter.' The composite scheme of geomorphic evolution for Kwihabe is presented in Table 21. Modifications include separation of the intervals of major cave enlargement from those of travertine formation, since the latter demand a higher level of bicarbonate saturation and marked moisture periodicity, i.e., a less humid environment. With some logic, the episodes of maximum (ground) watertable should be contemporary, namely Kwihabe-1 and 7 with Gi-3 and 2B. But the residual problems of correlation are too difficult and suggest that, within this single region, geomorphic thresholds and climatic variables were too complex to presently allow a synthesis between Gi, the Kwihabe Hills, and the Makgadikgadi.

Further regional data come from the Etosha Pan (18 40° S, 10° 30° E) of northern Namibia. Here Heine (1979) cites three dates of 13,800-10,200 B.P. for extensive lacustrine chalks and for mollusca related to an unspecified shoreline, without further context.

Table 21. Valley calcretes and cave travertines, Kwihabe Hills, northwestern Botswana.

Kwihabe-11. Intermittent stream flow, with accumulation of calcified, gravelly sand (Calcrete IV); travertine growth in Drotsky's Cave (Sinter IV, ca. 2500-500 B.P.), but periodic ingression of eolian sands. Climate intermittently wetter than today. Late Holocene.

Kwihabe-10. Repeated flooding of some dune areas by strong stream flow, with partial submergence and corrosion of cave interior; beginning of new travertine phase (Sinter IV, earliest date 4350 B.P.). Climate wetter. Early to mid-Holocene.

Kwihabe-9. Intermittent stream flow, with calcification of light gray, gravelly sand (Calcrete III), dated 11,000-9,800 B.P. (4 dates); downfaulting within cave, and no corrosion or travertines. Climate intermittently dry. Terminal Upper Pleistocene.

Kwihabe-8. Major travertine development within cave as watertable falls (Sinter III), dated 17,300-13,000 B.P. (16 dates). Climate relatively wet. Later Upper Pleistocene.

Kwihabe-7. Cave flooded, with extensive solution and corrosion; valley incision in dolomite. Climate wet. Mid-Upper Pleistocene.

Kwihabe-6. Dune formation, eolian remodelling, and sand ingression in cave. Faulting. Climate dry.

Kwihabe-5. Intermittent stream flow, with accumulation of calcified, light buff yellow, gravelly sand (Calcrete II, minimum dates of 30,000-45,000 B.P.). Climate intermittently dry.

Kwihabe-4. Surface corrosion of Calcrete I; travertine development possibly resumed inside cave. Climate wetter.

Kwihabe-3. Intermittent stream flow, with accumulation of calcified, light buff yellow, gravelly sand (Calcrete I); eolian sand penetrates cave. Climate intermittently dry.

Kwihabe-2. Travertine forms in cave (Sinter II), as watertable falls. Climate relatively wet.

Kwihabe-1. Cave flooded, with extensive solution and cavern enlargement; valley incision in dolomite. Climate wet.

The Namib and Namaqualand

Evidence for climatic change in Namibia is provided by several rivers draining the high western escarpment, only to disappear amid the dune fields of the piedmont area.

In Sossus Vlei (24° 45′S), the Tsauchab River deposited fossil silts from waters once extensively ponded in swales between dunes. Two such depositional zones indicative of greater runoff have dates of 24,800 and of 9600 and 9460 B.P. (Vogel & Visser 1981).

In the Tsondab valley (23° 53′S), similar extensive fossil silts under later dunes have dates of 14,300, 13,300 and 8640 B.P., while 110km downstream from the present terminus, fossil root casts at Narabeb are dated 28,500 B.P. (Seeley & Sandelowsky, 1974; Vogel & Visser 1981). An even older fossil lake at Narabeb has Th/U dates of 210,000 and 260,000 B.P. (Selby *et al.* 1979). The calcrete surfaces underlying the dunes between the Tsondab and Kuiseb rivers have 4 dates between 32,700 and 28,100 B.P., while calcareous root casts and termite nests in the dune swales along the Kuiseb have 9 dates between 31,900 and 20,900 B.P. (Vogel & Visser 1981).

In the Kuiseb River gorge (23° 38′S), there are 27m of well-stratified silty 'ponded' deposits that now form a 40m terrace; the 6 dates range from 25,000 to 18,100 B.P.; at another point, a +40m fluvial conglomerate, capping silts and sands, has 4 dates of 35,600-28,900 B.P., while silts possibly embanked against this terrace have dates of 12,700 and 6830 B.P.; finally, some 3m of silts on a 12m terrace have a 10,600 B.P. date (Vogel & Visser 1981; Vogel 1983). These dates from a now high-energy sector of the Kuiseb help pinpoint the 'Homeb silts,' more characteristic of humid aggradation.

The Namib dates indicate protracted river discharge, with extensive ponding as well as soil carbonate mobilization by throughflow water, between 33,000 and 18,000 B.P. (with a possible interruption about 26,000 B.P.). Less significant evidence for greater floods is available ca. 14,400-12,600 and 10,700-8600 B.P. Particularly interesting is verification of a late Middle Pleistocene river-end lake. However, all the Namibian evidence requires more detailed analysis and publication.

A particularly promising situation is available in the Koa valley of northernmost Namaqualand, where this former south-bank tributary of the Orange was drowned by eolian sands during the earlier Pleistocene. Trenches by the Anglo-American Corp. near Henkries reveal extensive lenses of diatomaceous earth, interbedded with organic sediments and cracking clays, as well as reworked eolian sands, reduced under anaerobic conditions. These lacustrine units occupy interdunal swales or deflation scars. Unpublished Th/U dates (R.S. Harmon, pers. comm.) from the major lacustrine level are patterned so as to suggest early Upper Pleistocene formation, with renewed hydrological activity about 40,000 B.P. and again 8000 B.P. Eriksson (1978) has mapped the subsurface geology from 38 boreholes and identified 4 lacustrine beds, subject to large lateral and vertical changes of shape and position. The oldest of these lake phases may date to the Middle Pleistocene. Pollen work appears promising for the organic facies.

In sum, the Kalahari-Namib evidence is both patterned and inchoate. No distinctive interregional contrasts emerge, and different categories of data are often difficult to reconcile within one area. Yet there also are repeated commonalities across wide areas, or with squences further east – along the Gaap, in the Vaal valley and Kimberley area, or in the western Free State. The problem in part is one of incompletely preserved records, in part of differential thresholds for response and, finally, of field and laboratory work insufficiently detailed to discern the true complexity of local records.

ARCHEOLOGICAL CAVE FILLS

The major advantage of cave infillings is that they potentially provide long, perhaps unbroken records at a single point, simplifying the perennial problem of collating lateral stratigraphies. But as was evident in our discussion of the australopithecine breccias, cave deposits are inherently difficult to interpret, because of their blend of cultural, cave-internal, and cave-external sedimentary components (Butzer 1982: 79ff.). But these problems also bring advantages: facies mixtures and interdigitations can provide broader insights as to the environmental variables influencing geomorphic processes, while the interplay of occupation residues with ongoing sedimentation gives a more realistic impression of occupation spans and their intensity.

Four major cave sites, together providing three long sequences, are presented here: Wonderwerk, on the margin of the Kalahari; Rose Cottage, on the southeastern montane fringe of the Free State High Veld; and Bushman Rock and Heuningneskrans, just above the escarpment to the Transvaal Low Veld. All are as yet unpublished. They are complemented for coastal South Africa by the published cave stratigraphies from Border Cave, above the Swazi-Kwazulu Low Veld (Butzer, Beaumont & Vogel 1978), Klasies River Mouth and Nelson Bay, on the Cape south coast (Butzer 1973a; Butzer & Helgren 1972; Helgren & Butzer 1977; Butzer 1978b), Die Kelders, in the southwestern Cape (Tankard & Schweitzer 1976; Butzer 1979b), and Elandsbaai, on the western Cape coast (Butzer 1979a).

Wonderwerk cave

The great solution cavity at Wonderwerk (27° 50′ 45″S, 23° 33′ 19″E) is located near the base of the Kuruman Hills (peak elevation 1815m, approximate rainfall 450mm, 80% during the summer half year) near 1680m. The cavern is 140m long, 10-20 wide, and typically 6-8m from the irregular bedrock floor to the domed roof. It follows bedding planes of Precambrian stromatolitic dolomite that dip about 1° into the mountainside. Two composite sequences are outlined here: one is from the front of the cave, based on two sample suites, collected by F.E. Peabody in 1949 and by Beaumont (1979b) from his 1978 excavations (Table 22); the other is from the deeper interior, based on existing

Table 22. Sediment stratigraphy of Wonderwerk Cave, Front Sections (Unit WB-1 to 4A at 18-24m from entrance, units WB-8 to 16 at 28-32m, units WB-4B to 7 from both sections) (P: phosphorus in parts per million, K: potassium, p.p.m.; OC: organic carbon).

Modern

WB-1. 0-5cm. Dung (150-350 P, 3700-4220 K, 7.7-9.5% OC)

WB-2A. 5-10 cm. Dark brown 7.5 YR), organic aggregates (220-1791 P, 3940-18,870 K, 8.2-13.3% OC).

Later Stone Age (Wilton/Wonderwerk)

WB-2B, 3. 10-20cm. Dark brown/reddish gray (5-7.5 YR), organic aggregates with residue of gritty to gravelly (subangular) sand in upper half (170-291 P, 255-4310 K, 0.5-2.6% OC). ^{14}C dates: 1210, 1890, 2910, 3060, 3990 B.P. During the sedimentation of WC-3, carbonate-saturated water from the cave roof locally seeped into lenses of WC-4 and cemented them next to the large stalagmite boss (^{14}C dates 2260, 2350, 2930 and 3360 B.P.).

WB-4A. 20-45cm. Reddish brown (5-7.5 YR), silty sand, organic debris, and subangular (corroded), medium-grade gravel (197-395 P, 335-5410 K, 0.5-2.4% OC). ^{14}C dates: 4240, 4890 B.P.

WB-4B. 45-65cm. Reddish brown (5 YR), silty sand with some subangular rubble and abundant organic aggregates (119-212 P, 278-1425 K, 0.5-2.6% OC). ^{14}C date: 5180 B.P.

WB-4C. 65-80cm. Reddish brown or brown (5-7.5 YR), organic aggregates with residue of gritty to gravelly (angular to subrounded), silty sand (40-456 P. 102-2600 K, 0.3-6.3% OC). ^{14}C dates: 5930, 5970, 7430 B.P.

Later Stone Age (Albany/Kuruman)

WB-4D. 80-95cm. Brown (7.5 YR), organic aggregates with residue of silty sand, forming matrix to minor roof spall horizon of subangular (corroded) gravel; abundant debitage (18-1184 P, 79-1063 K, 0.3-1.2% OC). ^{14}C dates: 9130, 10,000, 10,200 B.P.

WB-5A. 95-105cm. Reddish brown (5-7.5 YR), gritty sand (54-162 P, 90-2300 K, 0.1% OC). ^{14}C date: 9760 B.P.

WB-5B. 105-130cm. Dark brown (7.5 YR), silty sand as matrix to major roof spall horizon of medium-grade angular to subrounded gravel; abundant debitage (0-1700 P, 36-8075 K, 0-4.1% OC). ^{14}C date: 12,380 B.P.

WB-6. 130-145cm. Reddish brown to brown (5-7.5 YR) silty sand as matrix to major roof spall horizon (medium-grade, angular to subangular), directly overlying pinkish gray (7.5 YR) ash (5-4240 P, 0-8500 K, 0-1.4% OC).

Acheulian (Fauresmith)

WB-7. 145-185cm. Light to reddish brown (5-7.5 YR), silty sand with dispersed, medium-grade subangular gravel in middle and lower part, thin interbedded ash lenses, and local concentrations of debitage (1701-3024 P, 1720-7020 K, 0.1-0.3% OC).

WB-8. 188-195cm. Light brown (7.5 YR), partly calcified, ashy silty sand with some roof spall, ash lenses, and debitage (117-2153 P, 1470-7245 K, 0.1-0.2% OC).

(Contd.)

Table 22 (*Contd.*)

WB-9. 195-210cm. Brown to reddish brown (5-7.5 YR), partly calcified silty sand with some roof spall, ash lenses, and debitage (117-2153 P, 1470-7245 K, 0.1-0.2% OC).

WB-10. 210-218cm. Brown (7.5 YR), partly calcified silty sand and ash (57 P, 5445 K, 1.1% OC).

WB-11. 218-226cm. Pale to light brown or yellowish red (5-10 YR), silty sand and ash (444-639 P, 1455-6215 K, 0.1-0.5% OC).

WB-12. 226-242cm. Light to pinkish gray and brown (7.5-10 YR), partly calcified, silty sand and ash, with debitage (10-20 P, 1095-4720 K, 0-0.4% OC).

WB-13. 242-262cm. Pale to light or reddish brown (5-10 YR), partly calcified silty sand, ash, and fire-baked sediment (5 YR yellowish red, reddish yellow, pink) (214-307 P, 3115-5950 K, 0.1-0.2% OC).

WB-14. 262-312cm. Reddish brown (5-7.5 YR), silty sand, partly calcified in upper half, with white to pink ash lenses, traces of debitage, gypsum crystals, and much grit with subangular roof spall in lower middle part (406-5096 P, 285-2880 K, 0.1-0.5% OC).

WB-15. 322-361. Brown to very pale brown (5-10 YR), silty sand, with dispersed fine roof spall in upper part; with ash or firebaked lenses (5 YR reddish brown), and traces of debitage (268-2140 P, 475-7260 K, 0.1-0.3% OC).

WB-16. 361-422cm. Reddish brown (5 YR), silty sand, with zones of dispersed fine roof spall, and gritty at 370-390cm; extensive evidence of fire baking, with traces of debitage (734-4038 P, 1505-2235 K, 0.1-0.5% OC). Resting on dolomite floor.

---

sections recorded in 1974 and 1977 (Butzer, Stuckenrath & Vogel 1979) (Table 23).

^{14}C dates for the excavations by Beaumont, and those by A. and F. Thackeray in 1979, are given in Thackeray *et al.* (1981; Thackeray 1981), together with a plan of the excavated areas. More detailed analyses of the 1979 samples will be presented elsewhere by R.L. Burgess. The identification of the industry of WC-5 and 6 as Later Stone Age (Albany) is based on study of the 1944 excavation materials of B.D. Malan (Butzer, unpublished): these levels contain a high-debitage assemblage in cherty dolomite, mainly of small size, with amorphous cores and irregular scrapers. Ostrich eggshell from the Peabody excavations, unit WB-5B, supports this attribution with a date of 12,380 (Pta-2141). So far, no unequivocal Middle Stone Age deposits have been identified in the cave, despite the presence of some characteristic points reworked in younger deposits, e.g., in deposits excavated by Beaumont in 1980 adjacent to WA and equivalent to our units WA-3 and 4 (Table 23).

In terms of interrelating the two profiles, the youngest beds of Table 23 (WA) are 1.2m higher than those of Table 22 (WB), but the first 12,000 years of the column is missing from the interior sections. These form a quasi-horizontal deposit in WB, limited to the first 38m of the cave, and appear to rest on an erosional break that has removed the equivalent of WA-4 to at least below the incipient paleosol (WA-14A). With exception of a 10° westerly dip locally observed in WB-11 to 13 (P.B. Beaumont, pers. comm.), the individual strata are undulating, with some irregular contacts, but essentially horizontal. There is, then, only limited overlap between the two composite columns. One systematic difference is noteworthy: clays in the front sections average under 5% but this size fraction is over three times as great in the interior sections. This can be explained, in part, by slower accretion of external quartz sand farther from the entrance; this is supported by systematically finer and better sorted sands in column WA. A second explanation is that diverse organic colloids are twice as prominent in WA as in WB (mean 15.6% of non-carbonate residues, compared with 8.0%), in part reflecting the rapid increase in subrecent bat guano observed beyond 40m.

Table 23. Sediment stratigraphy of Wonderwerk Cave, Interior Sections (Units WA-1 to 9A at 59-62m from entrance, WA-9B to 14B at 41-42m).

WA-1. 0.10cm. Dark brown (7.5 YR), partly calcified, clayey silt sand, comprising organic hearths and interbedded ash lenses with debitage. ^{14}C date: 12,160 on carbon powder (SI-3561).

WA-2. 10-25cm. Very pale brown to pink (7.5-10 YR), partly calcified, silty clay sand and ash, with debitage and zones of medium-grade, angular gravel (3 P, 61,525 K, 0.5% OC).

WA-3. 25-40. Dark to very dark brown (10 YR), very organic silty clay sand, with stratified hearths, minor ash lenses, and dispersed medium-grade, angular to subangular gravel and roof slabs (626 P, 3510 K, 13.2% OC). ^{14}C dates on two organic fractions of carbon powder: 10,155 and 12,130 B.P. (SI-2043, 2032B).

WA-4. 40-55cm. Reddish brown (5 YR), partly calcified, silty clay sand with ash lenses, fire-baked laminae, zones of medium to coarse, subangular (corroded) gravel, and gypsum crystals (5 P, 3650 K, 0.4% OC).

WA-5. 55-60cm. White (2.5 Y), laminated ash, with residue of silty sand; horizon of very coarse, angular to subangular (corroded) gravel and some roof slabs (88 P, 7275 K, 0.2% OC). ^{14}C date on calcium carbonate powder of organic origin: 28,200 B.P. (SI-2031).

WA-6. 60-75cm. Brown to light brown (7.5 YR), partly calcified silty sand, with diffuse ash and fire-baked sediment, much debitage; horizon of medium to coarse, angular gravel (0 P, 3180 K, 0.2% OC).

WA-7. 75-90cm. Very pale brown (10 YR), partly calcified, clayey silt sand and ash, laminated (28 P, 6750 K, 0.6% OC); with 0-15cm lenticle of ash with residue of silty sand (193 P, 2185 K, 0.1% OC) in upper part, and abundant, very coarse, subangular (corroded) gravel in lower half.

WA-8. 90-120cm. Horizon of large slabs and roof blocks, to over a ton in mass, packed with and overlain by coarse angular gravel; average 15cm thickness at 43m beyond entrance, increasing to 40cm at 58m inside cave. Lime and gypsum efflorescence on blocks reflect roof weathering before detachment.

WA-9A. 120-135cm. Light brown (7.5 YR), clayey silt sand, interbedded with ash lenticles and zones of laminated carbon powder (carbonized grass matting?) (81 P, 3455 K, 0.3% OC). Forms matrix among lowermost roof blocks of WA-8. ^{14}C date on carbon powder: 38,600 ±1080 B.P. (SI-3562).

WA-9B. 135-140cm. Pinkish gray (7.5-10 YR), clayey silt sand, laminated organic structures (matting?), ashy and fire-baked (0 P, 3000 K, 0 OC).

WA-9C. 140-145cm. White, pink and reddish yellow (2.5 Y, 5 YR), silty sand, ash, and fire-baked laminae, thin-bedded; traces of debitage (1 P, 2925 K, trace OC).

WA-10. 145-165cm. Pink (7.5 YR), clayey silt sand, laminated, with diffuse ash and fire-baked zones; lateral lense of subangular gravel and slabs (254 P, 2975 K, 0.2% OC).

WA-11. 165-175cm. Brown (7.5 YR), silty sand, with diffuse ash, and capped by 5cm lense of laminated, very pale brown (10 YR) ash; dispersed coarse, subangular to subrounded gravel with some fresh fractures; trace of chert debitage and corroded/worn, lightly patinated biface on sharp lower contact (220 P, 2250 K, 0.4% OC).

WA-12. 175-185cm. Light brown (7.5 YR), partly calcified, clayey silt sand with diffuse ash; as matrix for coarse, subangular (corroded) gravel (1 P, 2650 K, 0.3% OC). Marks top of a calcified and reddish 'soil' zone extending down to 200cm.

WA-13. 185-195cm. Light reddish brown (5-7.5 YR), partly calcified, silty sand, with flakes and debitage, and horizons of very coarse, angular to subangular (corroded) gravel (0 P, 2075 K, 0.2% OC).

WA-14A. 195-200cm. Roof slabs and coarse, angular to subrounded gravel with matrix of light reddish brown (5 YR), partly calcified, silty sand and chert debitage (0 P, 1780 K, 0.6% OC).

WA-14B. 200-210cm (base not seen). Dark reddish brown (5 YR), organic silty sand, with gypsum crystals and dispersed, coarse subangular gravel (62 P, 3025 K, 4.0% OC).

The sedimentary components of the Wonderwerk columns can be identified to allow interpretation:

a) A wide range of organic materials is present, including plant residues, wood ash, and phosphates, as reflected in organic carbon, potassium and phosphorous values, as well as 7 times as much bulk loss of solubles in acid bath as compared with gasometric tests for calcium carbonate. These materials were introduced by people and animals.

b) A primary component of all the samples is the sand and grit fraction, consisting of quartz, with ancillary banded ironstone debris, dolomite, quartzite, jasper and chert in the coarser fractions. That these particles are overwhelmingly derived from the outside can be seen from the modal values for the Folk (1966) indices (0.37-6.35mm fraction) for the WA and WB columns:

	Meanϕ	Sorting	Skewness	Kurtosis
WA (Interior)	3.45	0.6	−0.35	+0.75
WB (Front)	3.20	1.2	−0.30	+2.00

Since wind velocities are rarely sufficient to transport more than dust-sized particles within a deep cave, most of this sand-grit fraction must have been introduced by running water, especially along the eastern side of the drip-line.

c) The stromatolitic bedrock averages 95% soluble, with a residue of sandy silt clay (23% sand), i.e. 3.85% of the bedrock consists of silt and clay. Slow bedrock solution could be expected to produce only a fraction of the 36% non-sand fraction in WB, or the 45% equivalent in WA. The bulk of this fraction must have been introduced by biological agents, with a subordinate eolian component probable but not quantifiable.

d) Some non-artifactual, ironstone pebbles in the 5-10mm grade are present in WB, but almost all of the gravel coarser than 5mm is dolomite. Some of this dolomite was edge-corroded ('subangular' or 'subrounded') prior to deposition, some was initially quite angular and subsequently either edge-corroded or unmodified. Two breccias just outside the drip-line were sampled: one contained no dolomite, the other only a fraction of corroded dolomite grit 2-5mm in size. Even these particles could not be washed about inside the cave on gradients of less than 0.3°. The gravelly horizons therefore represent detached roof spall, due collectively to pressure-release spalling, to cumulative corrosion along minor joints and bedding planes (at times of greater moisture), and to frost-fracturing (at times of greater cold). The cave is now absolutely dry, although stalagmites did form about 800-1900 years ago (WB-3) (carbonate dates estimated to be 1500 years too old, J.C. Vogel, pers. comm.). Detachment by corrosion would require a substantially moister climate, as would post-depositional corrosion, unless organic acids from intensive, long-term human occupation were mobilized. But the sediments give little evidence of ion dislocation and include 'desert rose' gypsum, so that deep wetting is unlikely. Further, pH values of the topmost organic strata (WB-3 to 4) range from 6.4-8.5. Corrosion is therefore a reasonable index of increased but limited moisture, whether before or after deposition. With mean free air temperatures of the coldest month now averaging near +10° C, frost would only be effective, deep within the cave, if temperatures were colder by about that amount. This may well apply to the angular gravel horizons. The great roof blocks, however, must be attributed to pressure release, since one 7 ton block fell on top of a branch dated 6000 B.P. (SI-2030) during warm, mid-Holocene times. Combinations of blocks and gravel are therefore more likely to be due to pressure release, whereas angular gravels mixed with occasional slabs are more suggestive of frost.

e) Finally, large-scale erosion of cave units, such as the removal of WA-4 to 14A in the front of the cave, cannot be attributed to prehistoric occupants. Water erosion, directed to a former sluice near the western side of the cave, is the only reasonable explanation, and requires repeated surges of runoff down the cliff face and into the cave, perhaps at a time when more rubble partially blocked the entrance than is the case today. In fact, an alternative explanation for the rubble horizon WB-5B is a gravel lag resulting from removal of the matrix from WA-4 to 14A by running water; considerable corrosion might be expected in the process.

These complex criteria can now be applied to interpret the WA and WB columns in environmental terms (Table 24). There nonetheless remains an element of subjectivity in evaluating the agencies for gravel accumulation, and the time of corrosion–during or after deposition. This interpretation should therefore be considered as provisional. The record of external grit is also incomplete, because the WA column is too far from the entrance. Nonetheless substantial environmental changes emerge, and there are many parallels with the Alex and Florisbad sequences. The indications of greater cold about 37,000-26,000 B.P., and of particularly wet conditions for much (but not all) of the time from 30,000-10,000 B.P. are noteworthy. The uneventful phases WW-2 and 7

Table 24. The environmental record of Wonderwerk Cave.

WW-17. Roof water and periodic runoff from outside, with transport of ironstone grit, corrosion of spall, and stalagmite formation (WB-2B, 3 upper). Much wetter. About 1900-800 B.P. Intensive Wonderwerk settlement.

WW-16. Little runoff or corrosion, no stalagmite growth (WB-3 lower). Dry as today. About 4000-1900 B.P. Intensive Wonderwerk settlement.

WW-15. Little runoff, some corrosion (WB-4A, 4B). A little wetter. About 5500-4000 B.P. Wonderwerk settlement.

WW-14. Some runoff and corrosion (WB-4C). Wetter. About 8000-5000 B.P. Intensive Kuruman settlement.

WW-13. Little runoff, much corrosion (WB-4D). Wetter, less seasonality. About 10,000-8000 B.P. Intensive Kuruman settlement.

WW-12. Some runoff, much corrosion in place probable (WB-5A). Wetter. About 11,800-10,000 B.P. Kuruman settlement.

WW-11. Little runoff, much corrosion probable, possibly assisted by frost-weathering (WB-5B, 6; WA-1 to 3). Moist, cool. About 13,000-11,800 B.P. Intensive Kuruman settlement.

(Major erosion by repeated, heavy runoff, between 24,000 and 13,000 B.P.; settlement lacuna, real or simulated).

WW-10. Much corrosion (WA-4). Moist. About 25,000 B.P. Settlement of uncertain affinity.

WW-9. Some corrosion, but major frost-weathering (WA-5 to 7). Moist and very cold. Perhaps 30,000-26,000 B.P. Intensive settlement of uncertain affinity.

WW-8. Catastrophic roof collapse, apparently followed by frost-weathering (WA-8). Probably very cold. About 35,000 B.P.

WW-7. Limited geomorphic change and dry, but intensive settlement of uncertain affinity (WA-9). Before 37,500 B.P.

WW-6. Some corrosion and some frost-weathering (WA-10, 11). Moist and cool. Early or mid-Upper Pleistocene. Settlement of uncertain affinity, possibly Middle Stone Age.

(Calcification of 25cm of sediment and gypsum at base, after corrosion of spall, suggesting short-term mobilization of carbonates, but drier than previous phase).

WW-5. Corrosion and strong frost-weathering (WA-12, 13, 14A). Moist and very cold. Some Acheulian (Fauresmith) settlement. Late Middle Pleistocene.

WW-4. Some corrosion (WA-14B). Relatively moist. Organic, but no evidence of human occupation.

(Probable gap in record)

WW-3. Little runoff, some corrosion (WB-7). Relatively moist. Intermittent Acheulian occupation, alternating with guano accumulation.

WW-2. Limited geomorphic change and dry, but intermittent, intensive, occupation (WB-8 to 14 upper).

WW-1. Episodes of runoff and some corrosion (WB-14 middle to 16 base). Relatively moist but fluctuating. Alternating, intensive Acheulian occupation and guano accumulation. Mid-Middle Pleistocene.

suggest quite dry conditions, and only one cold and moist phase stands out in the several hundred thousand years spanned by WW-1 to 7, much as does Kathu 2.

The moisture inferences of Table 24 for phases WW 13-17 match those derived by Van Zinderen Bakker (1982) on palynological grounds.

Rose Cottage Cave

Rose Cottage is a large sandstone overhang formed in Karoo sandstones, at 1660m near Ladybrand (29° 12′ 5″S, 27°28′ 19″E). It was the result of differential erosion along a major joint intersection, that allowed upland drainage to emerge intermittently from what is now the back of the cave. The breaking off of a huge roof block partly closed the overhang, and sediment began to accumulate. The filling consists of (a) material detached from the sandstone roof, in part through grain-by-grain weathering, in part by rock spalling; and (b) material washed in from the occasional spring at the back. Deposits generally dip down to the front block and then to a former sluice along its eastern side. Evidence for strong spring influx provides a good case for heavy and protracted rains, while angular roof rubble indicates rapid freeze-thaw weathering, as opposed to the subrounded spall due to gradual rock disintegration.

The sedimentary sequence, as established in 1977, has been presented earlier (Butzer & Vogel 1979) and is summarized in Table 25 together with the environmental interpretation. The latter takes into account my 1982 study of the external stream deposits and slope soils. Archeology is after Malan (1952) and Vogel (unpublished).

Table 25. Sedimentary sequence and environmental interpretation for Rose Cottage Cave, near Ladybrand.

RCC-13. Sorted, laminated and crossbedded, brown to dark brown (10 YR), silty sand (100 cm), filling small gully inside cave created by water spilling over drip-line. Lacks fissure spring components; due to strong spill-over and reduced groundcover on upland slope. Younger than Pottery Wilton occupation of RCC-12.

(Cutting of gully within cave [4m] parallel to dripline)

RCC-12. Dark grayish brown (10 YR), organic silty sand, poorly sorted with dispersed subangular roof spall (15-25cm). Intermittent water and sediment influx from spring, indicating high watertable; some slow roof spalling. Intensive Pottery Wilton occupation (1100 B.P., GrN-5298).

(Sedimentation break of at least 5000 years, with decalcification in cave)

RCC-11. Brown (7.5 YR) silty sand (50-100cm), poorly sorted, with multiple hearths and zones of subangular spall. Intermittent spring influx, seasonally high watertable; slow roof spalling. Intensive Wilton occupation (6850 and 8640 B.P., GrN-5299, Pta-2067).

(Sedimentation break, 5000-10,000 years, with decalcification).

RCC-10C. Brown (7.5 YR) stoney silty sand (40cm), with major subangular to angular spall, decalcified, and highly phosphatic. No spring influx and low watertable, much roof spalling with some frost-weathering. Early Later Stone Age occupations.

RCC-10B. Brown (7.5 YR) silty sand (30-60cm), with concentrations of subangular to angular spall, and distinct hearths. Some spring influx and higher watertable; some frost-weathering. Early Later Stone Age occupation (29,420 B.P., Pta-211).

RCC-10A. Yellowish brown (10 YR), sand (30-60cm), well sorted, with much coarse, angular spall; high phosphate levels, hearths and ash lenses. Some spring influx and relatively high watertable; quite cold. Early Later Stone Age occupations (inconsistent dates: 25,640, 23,380, 22,650 B.P. GrN-5300, Pta-1416, 1416B).

(Sedimentation break of several millennia; decalcification)

(Contd.)

Table 25 (*Contd.*)

RCC-9. Yellowish brown (10 YR) sand (20-45cm), well sorted, with dispersed subangular spall; decalcified, high phosphate levels, rodents, and snails. No spring influx and low watertable; little spalling. Only incidental human occupation, prior to 30,000 B.P.

(Long erosional interval, related to protracted spring influx with little sediment, suggesting very effective groundcover above site)

RCC-8. Light to pale brown (7.5-10 YR), stoney sand (40-85cm), poorly sorted but laminated, with subangular spall horizon near top. Strong and repeated spring influx, with high watertable and high-intensity rains; slow roof spalling. Occasional human occupation early and late in phase, prior to 40,000 B.P.

(Erosion dominant in cave)

RCC-7. Brown (7.5 YR), laminated sand (30-50cm), much subangular spall; with phosphates and hearths. High watertable and persistent spring influx; slow roof spalling. Late Middle Stone Age occupations (date of 'greater than 50,000 B.P.,' Pta-213).

RCC-6. Dark brown (7.5 YR) clayey silt sand (40-80cm), with much angular to subangular spall; major hearth complex, with abundant carbon and phosphates, ash laminae, and debitage. High watertable but little coarse-grained spring sedimentation, suggesting effective groundcover outside; efficient roof spalling and frost-weathering. Intensive Late Middle Stone Age occupation (with three infinite ^{14}C dates).

RCC-5. Yellowish brown (7.5-10 YR), silty clay sand (40-60cm), partly laminated; high phosphate levels but no hearths. High watertable but low-intensity rains and effective groundcover; no spalling. Incidental human occupation only.

RCC-4. Brown (7.5-10 YR), silty clay sand (40-45cm), partly laminated; high phosphate levels and minor hearths. High watertable, low-intensity rains, effective groundcover; no spalling. Some Middle Stone Age occupation.

RCC-3. Brown (7.5-10 YR), silty clay sand (5-50cm), bimodal sorting, strongly laminated; high phosphate and potassium levels, diffuse occupation evidence. Spring influx of variable intensity, high watertable, moderately effective groundcover. Some Middle Stone Age occupation.

(Erosion dominant, with partial reworking of RCC-2 as lag)

RCC-2. Brown (10 YR), clayey silt sand (10-50cm), partly laminated, grading into chaotic mass of coarse angular spall, with high artifact concentration. Strong spring influx, high watertable, moderately effective groundcover, very cold. Middle Stone Age (Howieson's Poort) occupations.

RCC-1 Reddish yellow to light brown (7.5 YR), clayey silt sand (20cm), thin-bedded; partly calcified, high phosphate and potassium levels, with diffuse hearths, resting on bedrock. Moderate-velocity spring waters, high watertable, low-intensity rains, and effective ground cover.

This record indicates a high watertable and effective groundcover, resulting in the trapping of fine suspended sediment during phases RCC-1 to 6. Climate was evidently humid and less prone to high-intensity rainfall. In addition, RCC-2 and 6 indicate quite cold conditions, before 50,000 or 60,000 B.P.

The watertable remained high, but a more torrential regime set in during RCC-7 and 8, before 40,000 B.P. By the time of RCC-9 deposition, before 30,000 B.P., the watertable was low, with conditions similar to those of today.

During the RCC-10 time span, the watertable was initially higher, then falling; it also was cold, later moderating, about 30,000-20,000 B.P. Subsequently, although relatively damp inside the cave, the watertable remained low and little roof disintegration took place until 10,000 B.P., when intermittent spring influx resumed for perhaps 4 millennia. A last phase of equally intermittent spring influx dates roughly 1500-1000 B.P. In recent times, devegetation above the cave, presumably by Sotho and Afrikaner grazing, has for the first time allowed spill-over water to dominate cave processes.

The Rose Cottage Cave sequence consequently appears to span the entire Upper Pleistocene and Holocene. The unusual nature of the equitable moist climate, with cold spells, over several tens of millennia before 60,000 B.P. is significant. There is nothing like it in the Wonderwerk or pan records; but at Border Cave (27° 1′S, 31° 59′E) a soil formed at the drip-line under similar conditions, and was followed by spall horizons (Butzer, Beaumont & Vogel 1978), while at Nelson Bay Cave there was intensive iron mobilization after deposition of the one major spall horizon, which contains a Howieson's Poort industry (Butzer 1973a, 1978b). On this key point, as well as on other details, Rose Cottage Cave parallels environmental trends in the southern coastal sector: the late Upper Pleistocene was cool and relatively humid, but the Holocene primarily on the dry side (summarized by Butzer, Stuckenrath *et al.* 1978).

Heuningneskrans Cave and Bushman Rock Shelter

The two major archeological caves of the eastern Transvaal are developed in a shaley Precambrian dolomite some 18km north of Ohrigstad.

Heuningneskrans (24° 35′07″S, 20° 39′21″E) is a west-facing overhang developed just above valley floor level, at 903m. A now defunct valley-side resurgence, i.e. a karst spring exit, directly behind the cavity, was instrumental in its formation. Soft, as well

Table 26. Sedimentary sequence of Heuningneskrans Cave, near Ohrigstad.

Iron Age

HNK-8. 0-20cm (Level 1A). Brown (7.5 YR), partly calcified, organic clayey silt sand (41-67 P, 1050-6545 K, 1.3-2.8% OC); with much debitage, and fire-baked zones. Top of mound.

Later Stone Age (Albany)

HNK-7. 20-85cm (Level 1B). Brown (7.5 YR), partly calcified, organic clayey silt sand (22-27 P, 2950-5630 K, 1.7-1.8% OC); with much debitage. ^{14}C date: 7200 B.P. (Pta-112). Part of mound.

HNK-6. 85-185cm (Levels 2A-3A). Gray to reddish brown (5-10 YR), partly calcified, clayey silt sand, gritty in lower third (1-46 P, 630-1570 K, 0.5-1.2% OC); with prominent ash horizons and abundant debitage. ^{14}C dates: 9780, 10,430 B.P. (Pta-099, 114). Base of mound.

Early Later Stone Age

HNK-5. 185-280cm (Levels 3B and 3C, upper). Brown (7.5 YR), partly calcified silty sand, coarser in lower half with some spall (0-8 P, 1170-1575 K, 0.7-1.7% OC). ^{14}C date: 13,100 B.P. (Pta-100). Coalesces with river terrace.

HNK-4. 280-360cm (Levels 3C, lower, to 3D upper). Brown to reddish brown (5-7.5 YR), silty sand, with lenticles dipping into overhang (3-40 P, 770-1650 K, 0.3-1.1% OC); with bone and debitage. Coalesces with river terrace.

HNK-3. 360-455cm (Levels 3D lower, to 3E). Brown to reddish brown (5-7.5 YR), silty sand, with lenses dipping toward valley; coarser in top third (0-39 P, 645-1415 K, 0.3-0.8% OC); with debitage and prominent ash horizons. ^{14}C date: 24,630 B.P. (Pta-101). Coalesces with fill coming out of karst pipe at back of cave.

HNK-2. 455-535cm (Levels 3F, 3G). Brown (7.5 YR), clayey silt sand, with lenses dipping to valley (0-22 P, 365-725 K, 0.2-0.7% OC); with prominent ash lenses, debitage, and bone. Coalesces with fill from karst pipe.

HNK-1. 535-630cm (Level 3H). Brown (7.5 YR), silty sand, with lenses dipping to river (2-11 P, 180-310 K, 0.2-0.6% OC); with prominent ash lenses, diffuse hearths, debitage, and bone. Coalesces with pipe fill. On dolomite bedrock.

as cemented silts, with proliferations of *Achatina*, indicate that this resurgence has been far more active in the past. The cave was originally excavated by P. Beaumont (1978; Beaumont & Vogel 1972a, 1972b), but samples were taken from a column recorded in 1974, as presented in Table 26 together with analytical data (Butzer, unpublished).

Bushman Rock Shelter (24°34′28″S, 30°37′31″E) is a large, open cave, facing south to a tributary of the Ohrigstad River, 2.2km northwest of Heuningneskrans. It is 15m

Table 27. Sedimentary sequence of Bushman Rock Shelter, near Ohrigstad.

Iron Age

BRS-7. 0-8cm (Level 1). Dark yellowish brown (10 YR), partly calcified, organic clayey sand (177 P, 38,920 K, 6.0% OC).

Later Stone Age (Albany)

BRS-6C. 8-60cm (Levels 2-5). Brown, light brownish gray and dark brown (7.5-10 YR), partly calcified, organic silty clay sand and silty sand (20-453 P, 12,415-14,175 K, 2.5-8.0% OC), with prominent zones of diffuse ash or hearths; debitage, seeds, land snails, and bone. Major dug hearth near base. ^{14}C dates: 9510, 9570, 9940 B.P. (GrN-4854, 5874 4813).

BRS-6B. 60-114cm (Levels 6-11). Brown (7.5 YR), organic silty sand and clayey silt sand, gritty near base (90-566 P, 10,060-53,705 K, 2.2-5.3% OC), with prominent hearths and lenticular bedding; bone and land snails.

BRS-6A. 114-130cm (Levels 12-14). Brown to light brownish gray (7.5-10 YR), clayey silt sand (1 P, 6555 K, 0.2% OC). ^{14}C dates: 12,090, 12,310, 27,400 B.P. (GrN-4814, Pta-1175, UCLA-1755B), with hearth and ash lenses.

Mixed Later and Middle Stone Age

BRS-5. 130-155 (Levels 15-18). Grayish to dark brown (7.5-10 YR), partly calcified, gritty and organic silty coarse sand, with corroded, subangular to subrounded spall increasing to base (5-313 P, 6245-9600 K, 3.4-4.5% OC); with diffuse hearths, debitage, bone and charcoal. ^{14}C dates: 12,160, 12,470, 12,510, 12,800, 12,950, and 31,900 B.P. (GrN-4816; Pta-1176, 1177, 1178; UCLA-1755C). Basal contact culturally disturbed.

(Moderate weathering horizon, with 20cm decalcification depth)

Middle Stone Age

BRS-4. 155-174cm (Levels 19-20). Dark brown (7.5 YR), clayey silt sand as matrix to coarse angular roof rubble, partly decalcified, resting on brown rubble-free lense as above, with sharp contact (1-19 P, 5015-6970 K, 1.6-3.4% OC); abundant debitage, with bone, shell and charcoal.

BRS-3. 174-294cm (Levels 21-30). Brown, dark brown, light yellowish brown (7.5-10 YR), partly calcified, silty sand and silty coarse sand, with dispersed, coarse angular to subangular spall and distinct rubble horizons at 190, 235 and 260cm (0-19 P, 3605-6970 K, 0.2-3.4% OC); some large, diffuse hearths, bone and debitage; gypsum crystals at 195 and 220cm. ^{14}C dates (Level 21): greater than 47,500, greater than 53,000 B.P. on charcoal (GrN-4855, 5116), and greater than 30,000 B.P. on bone (UCLA-1955-D).

(Major weathering horizon, with 100cm decalcification depth)

BRS-2. 294-350cm (Levels 31-37A). Dark to very dark brown (7.5-10 YR), sand and clayey silt sand, with three major rubble horizons of coarse angular to subangular, partly to entirely decalcified spall (177-1226 P, 3775-7570 K, 0.3-4.8% OC); with multiple hearths, some debitage, and occasional bone; gypsum crystals at 310cm.

BRS-1. 350 to below 480cm (Levels 38-71). Brown to dark reddish brown (5-7.5 YR), organic clayey sand, silty sand and clayey silt sand, lacking rubble, with calcified zone at 395-405cm (5-5920 P, 6100-20,015 K, 1.6-23.8% OC); with multiple hearth lines, abundant debitage, some bone and charcoal.

(Samples not available for levels 72-105, down to 690cm depth, over dolomite)

above the valley floor, at 928m, and developed through basal-sapping by a once higher-lying river. It has been primarily excavated by J.F. Eloff (unpublished; Plug 1981; also Beaumont & Vogel 1972a). The column recorded in 1974 was extended by further samples supplied by Eloff as the excavation was later deepened. This data is collated in Table 27 (Butzer, unpublished).

The HNK column (Table 26) consists of three components: HNK-6 to 8 form a 2m archeological mound within the overhang, and accumulated through human agency, with a little drip-line influx; only the basal 40cm show some grit probably introduced by flood waters from the resurgence. HNK-4 and 5 interdigitate with the +5m low terrace deposits of the Ohrigstad River, and interpolation of the ^{14}C dates suggests an age of 19,000-11,500 B.P. for this phase. By 11,220 B.P. the river had entrenched its bed and soil carbonates were being concentrated at 2.5m below the terrace surface (date on inorganic carbonate, Pta-233). The lowest units, HNK-1 to 3, include much material from the resurgence and, by extrapolation, should date about 30,000-19,000 B.P. The sediments relate to their changing landscape context, but provide no tangible record of an environmental shift, only in part because cultural agencies were fundamental in regulating sedimentation processes. Roof spall is almost absent, and variation in the statistical parameters of the sand and grit fraction suggest more relationship to the intensity of occupation than to flood energy. The mammalian fauna (R.G. Klein, unpublished) indicates more grassland forms before 11,500 B.P. possibly reflecting different edaphic conditons on the aggrading floodplain.

The BRS column (Table 27) presents an archeological and dating problem, firstly whether BRS-5 and 6A are 12,000-13,000 years old, or increase in age from 12,000 to 32,000 years, and secondly whether the industry from BRS-5 is Later or Middle Stone Age. The artifacts from BRS-5, as published by Plug (1981), suggest a mixture, a point substantiated in the excavation face. 'Scoops' of level 19 (BRS-4) were removed during the occupation of levels 16 and 17; reworked stringers of badly corroded pebbles tie these scoops directly into the younger strata, and the secondary, scoop filling of level 18a has the same texture as the matrix of level 19. As to the quasi-constant ^{14}C dates through levels 13-18, the sand statistics change too much to be consistent with a short span of rapid sedimentation. More probable is downward movement of charcoal through rodent burrows. It is therefore probable that BRS-5, 6 and 7 replicate the entire HNK sequence. No environmental change can be detected in the primary sediment of BRS-5 to 7, and the fauna (Plug 1981) is too sparse to allow conclusions.

The roof spall horizons, BRS-5 and 2, are classic frost-weathered rubbles and indicate two cold intervals against a background of an essentially modern micro-environment. The two decalcification episodes, after BRS-2 and again after BRS-4, are remarkable, and indicate somewhat damper conditions. These interpretations are presented in Table 28.

Comparing the Rose Cottage column and Table 28, it appears probable that the Middle Stone Age was broadly coeval at both sites, although the lithic assemblages are somewhat different and distinct raw materials were utilized. The paucity of more dramatic evidence for wetter climate at BRS and HNK may reflect in part on the drier regional climate; Ohrigstad has 492mm rainfall (85% in the summer half year) and Ladybrand 731mm (78% in summer). But the dry cave microenvironment is also responsible, because the watertable in nearby Echo Cave (2.7 km NW of BRS) was 20-30m higher 35,000-12,000 B.P. than it has been since (Brook 1982). Stream tufas in the

Table 28. Environmental interpretation of the Heuningneskrans and Bushman Rock columns.

HB-6C. Cultural residue, with undiagnostic mineral residue (BRS-7, HNK-8). Iron Age to modern.

(Long sedimentary and occupation break)

HB-6B. Cultural residue, with undiagnostic mineral residue (BRS-6C, 6B upper; HNK-7, 6 upper). Intensive Later Stone Age occupation, about 10,000-7000 B.P.

HB-6A. Cultural residue, with coarser mineral residues (BRS-6B lower; HNK-6 lower, 5), Ohrigstad River completes aggradation of 5m terrace, then incises, with 2.5m drop of watertable and soil decalcification 11,500 B.P. Intensive Later Stone Age occupation, about 15,500-10,000 B.P.

HB-5. Cultural residue, with zones of coarser mineral residue (BRS-6A, 5; HNK-1 to 4). Ohrigstad floodplain reaches floor level of HNK about 19,000 B.P. Intensive Later Stone Age occupation, about 33,000-15,500 B.P.

(Decalcification, of BRS-4; relatively humid climate)

HB-4. Major frost weathering with spall production, some roof fall near the BRS drip-line, and very cold (BRS-4). Intensive Middle Stone Age occupation, prior to 53,000 B.P.

HB-3. Cultural residue, interdigitated with minor roof spall levels near back wall, and cool to cold (BRS-3). Repeated Middle Stone Age occupation.

(Major decalcification, of BRS-2; long period of relatively humid climate)

HB-2. Cultural residue, interdigitated with three major roof spall horizons; cold (BRS-2). Intensive Middle Stone Age occupation.

HB-1. Cultural residue, with undiagnostic mineral residues (BRS-1). Intensive Middle Stone Age occupation (base not seen).

eastern Transvaal are dated 33,000-29000 B.P., 23,000-20,000 B.P., about 14,240 B.P., 3670 B.P., and 1400 B.P.; they suggest times of slightly greater moisture (Marker 1972; Brook 1982). Further, travertines in Wolkberg Cave, central Transvaal, indicate wetter conditions prior to 39,000 B.P., about 29,560 B.P., 19,800 B.P., 9070 B.P., 3390 B.P., and 1520 B.P. (Talma *et al.* 1974). These differences underscore the importance of geomorphic thresholds in determining sedimentation patterns. The primary interest of the BRS and HNK sediment columns may well lie in their archeostratigraphic implications.

INTERIM CONCLUSIONS

This paper has attempted (a) to provide the first comprehensive evaluation of the australopithecine breccias, based on new data, since the earlier synthesis of Brain (1958); (b) to continue to refine our understanding of the alluvial history of the lower Vaal drainage, buidling upon the systematic study of Helgren (1979a), who in turn had extended a tradition of work begun during the 1930s; (c) to generate long lake and spring histories, and so to delineate broader regional patterns of hydrological variation, and (d) to add informative cave sequences from the interior to those already established for the coastal regions of southern Africa.

These attempts have met with variable success. They have allowed the reconstruction of some regional sequences, spanning longer stretches of Quaternary time. Some of these sequences suggest fruitful links with data elsewhere or as derived from other geological contexts. But the results do not at the moment realistically lend themselves to chart and contrast key paleoclimatic trends for different areas. Such time-space comparisons appear possible for some parts of southern Africa during the last 20,000 years or so, but for the mid and early Pleistocene we have little more than a set of clues as to how a coarse stratigraphic framework might possibly look.

There are no sharp paleoclimatic demarcations when and where different sites or areas can indeed be compared at similar levels of detail. Instead, it appears that trends may retain a similar direction, with small shifts in their temporal spans, the more dramatic differences being in amplitude. But the climatic variables are too many to assess whether such differences in amplitude are indeed as small or as great as they seem to be. As a result there are many gradients, some in terms of geomorphologic sensitivity, i.e. proximity to thresholds for several variables, others in terms of real climatic patterns. Thus far these gradients are too subtle to allow a judgement as to whether or not there were different climatic provinces, e.g. with distinctive rainfall regimes, within the interior of the continent. But the picture of late Pleistocene and Holocene trends is indeed different from that of the coastal regions of the Cape Province, as I have attempted to show in earlier reviews (Butzer, Stuckenrath *et al.* 1978).

The most striking aspect of the climatic history of the interior is the rapidity of change, primarily in terms of phenomena that reflect effective moisture. It is the rapidity of such change, and the complexity of the evidence and the responsible variables, that makes synthesis so difficult. In this sense the paleoclimatic continuum resembles that of tropical Africa, with its wildly fluctuating lake levels and unpredictable spatial patterns, far more than it does that of Europe or North America, where few authors would have difficulty identifying a representative mode of, for example, *the* Holocene environment. Climatic changes had a wave length of only several hundred years to several thousand years, and longer-term periodicities, such as those with 20 or 100 millennia, are not apparent at the present stage of investigation. For this reason I continue to be skeptical of attempts to define *a* glacial mode of climate anywhere in Africa.

This analysis has dealt explicity with sediments and soils from contexts with archeological or fossil records and, consequently, with grounds and opportunities for greater chronometric precision. It requires no emphasis that the local or regional paleoclimatic histories emphasized here could be substituted for by applications to the understanding of a regional, environmental mosaic and its changes through time. This was in fact the primary goal of the research reported on, and it is in this perspective that the data are both more tangible and relevant. Each phase of human occupation can be delineated in its ecological context, and successive occupations can be viewed against temporal and spatial change in that context. The facies of an archeological sediment provide keys to broader ecological models that are normally expanded upon in geoarcheological research. Consequently the sequences outlined here, at some length, are nothing but a shorthand record that could be expanded upon by an order of magnitude. However the regional human ecology that is of interest to the prehistorian represents a different goal than the paleoenvironmental interpretation of a subcontinental area aspired to by the climatologist. In part, too, this is a matter of scale, because the big picture is best put together from many small pictures.

Finally, the geological evidence despite its merits, must be complemented by faunal and botanical work, as represented in this volume by the chapters of Klein and Scott. These other research methods and data also are imperfect, in that they have not marshalled sufficiently detailed or unequivocal information to reconstruct the big picture, either spatially or temporally. Unfortunate, even ironical, is that the best sites or study areas for sediments, for faunas, and for pollen rarely coincide. As a result, multidisciplinary integration, however desirable, remains rudimentary at the level of paleoenvironmental reconstruction for southern Africa.

ACKNOWLEDGEMENTS

During the course of my work in South Africa I have profited greatly from collaborative fieldwork with Louis Scott, David Helgren, Bob Brain, Leon Jacobson, and Peter Beaumont on various occasions, Robert Stuckenrath's continuing interest provided many of the radiocarbon dates cited. John Vogel, during several memorable discussions, over the years, sustained my enthusiasm at times when the work involved seemed interminable. Finally, Richard Klein has been that other Africanist at Chicago with whom I could share my thoughts. The research was supported in part by the National Science Foundation (Anthropology Program), the University of Chicago, the John Simon Guggenheim Foundation, the Wenner-Gren Foundation, and the Swiss Federal Institute of Technology (Zurich). This paper is warmly dedicated to Gerhard and Dora Fock, with whom I have collaborated for so many years.

L. SCOTT
Institute for Environmental Sciences, University of the OFS, Bloemfontein, South Africa

2

Palynological evidence for Quaternary paleoenvironments in southern Africa

INTRODUCTION

Pollen analysis can be a very useful instrument in the reconstruction of past climates. In order to assess the available evidence produced by this technique in Southern Africa the results from different sites will briefly be summarised and their significance outlined. From these findings it will be seen that the distribution pattern of the African vegetation types is constantly changing. The general environmental setting in the Quaternary already developed during the late Tertiary when more tropical vegetation types in the southern parts were probably exterminated by generally cooler conditions. This was concluded from palynological evidence which shows that subtropical to tropical palm vegetation existed in the south-western Cape during the Tertiary (Coetzee 1978a and b). Since it has been suggested that the preliminary palynological results from late Tertiary deposits in the Transvaal (Horowitz 1975) cannot be relied upon (Scott 1982d), there is no direct evidence for the Tertiary vegetation in the central parts of Southern Africa. It can, however, be assumed that more tropical conditions must also have existed here during these early times.

The method of Quaternary pollen analysis was first introduced in Southern Africa by E.M. Van Zinderen Bakker (Sr.). The valuable pioneering work which formed a basis for most later efforts, consisted of tasks like the monotoring of the modern pollen rain over the Orange Free State (Coetzee & Van Zinderen Bakker 1952), the preparation of pollen reference collection, the description of S. African pollen types (e.g. Van Zinderen Bakker 1953, 1956) and the study of pollen in peat deposits in Lesotho, the Orange Free State and various other parts of Africa (Van Zinderen Bakker 1955, 1957, 1963, 1964, 1969, etc.).

In comparison with the temperate regions of the Northern Hemisphere where numerous Quaternary pollen profiles have been produced, only very few were studied in Southern Africa. In general this is also the case with the rest of Africa. The only exception is perhaps East Africa where a number of long lake pollen profiles have been studied (Kendall 1966; Coetzee 1967; Livingstone 1967; Morrison 1968, etc.). As in most other relatively, dry parts of the world the small number of pollen profiles in Southern Africa is mainly due to the scarcity of rich pollen-bearing deposits like lake beds and peats. In general the mineral rich deposits of Quaternary age in this region do not preserve useful pollen assemblages. Therefore, some of the most productive sites up to now have been rare swampy areas like permanent springs with peat deposits.

There are several factors which complicate the palynological technique. Among these are the problem of accurate identification. Many pollen types can only be identified to family or genus level as for instance in the important grass group which is ubiquitous in Africa. Another problem is to distinguish between local swamp elements and the more distant regional pollen in peat sites. It is necessary to separate the latter as an indicator for the surrounding vegetation changes but this is complicated because members of certain palynomorph taxa are produced both locally and regionally. Grass pollen is also a good example of a group with this tendency. Apart from the fact that deposits older than about 30-40,000 yr cannot be effectively dated by the ^{14}C method a further problem is that the age determinations often provide anomalous values. This is especially the case with spring peat sites in the relatively dry central parts. It is believed that contamination of younger organic material such as deeply penetrating minute rootlets could be the main reason for this problem (Scott & Vogel, unpublished data). Another factor which complicates the palynological reconstruction of past environments is that different types of plants have very different pollen production and dispersal qualities. Since the palynological reconstruction of Quaternary environments in South Africa is mostly based on comparisons of the fossil with recent pollen spectra from different vegetation types and climatic regions, it is necessary to gain knowledge about the relationships between pollen spectra and various veld types. In this way pollen spectra of the modern vegetation can be consulted to provide a means of arbitrarily estimating the nature of past conditions. Therefore the first part of this paper will provide a short survey of the available modern pollen data. In general the information is too limited to allow firm or detailed reconstructions of the past vegetation types and climates and only isolated clues can be provided. The locality of the palynological sites mentioned below, is indicated in Fig. 1.

MODERN POLLEN SPECTRA

Very little information on the modern pollen rain produced by the different vegetation types for this vast region is available. Most of the recent spectra which were studied came from the few areas from where fossil spectra were also described. Most of these modern spectra contain high percentages of exotic pollen, especially that of *Pinus* (2% to 60%) which is derived from numerous introduced trees. In cases where these foreign elements are relatively prominent in the studied spectra mentioned below, the pollen percentages are calculated excluding *Pinus* numbers in the sum.

Modern pollen spectra in the southern Cape province have been described by Thiergart *et al.* (1962), Martin (1968) and Schalke (1973). In view of the small number of these spectra it is not possible to provide a clear picture of the pollen production of the region. The data show that the arboreal pollen is important in regions with extensive indigenous forests like at Knysna but in some cases the non-arboreal elements in forests are also abundant. *Podocarpus* pollen (12 to 52%) is the most prominent tree element but *Olea* pollen can be just as important (up to 50%) inside forests. It is curious that the *Podocarpus* pollen of 3,4% provided by Thiergart *et al.* (1962) from the Knysna forest near an old brown coal mine is considerably

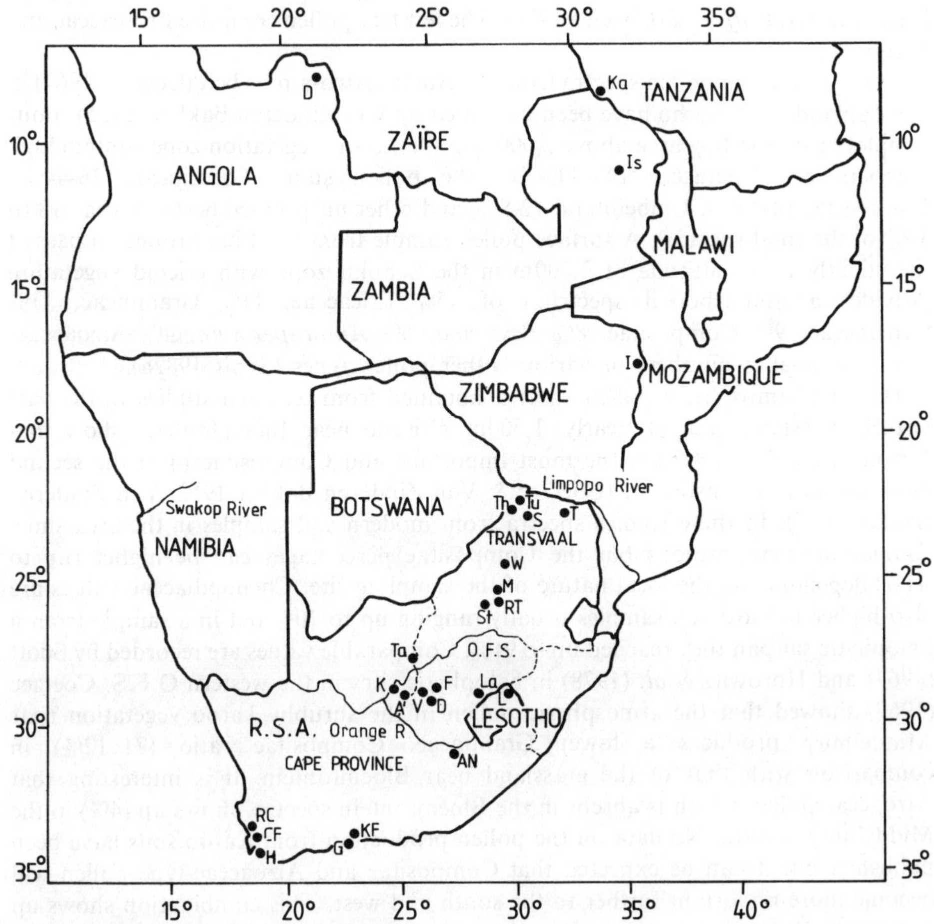

Figure 1. Locality map showing the positions of palynological sites mentioned in the text. H–Hangklip; CF–Cape Flats; RC–Rietvlei (Cape); G–Groenvlei; KF–Knysna forest; AN–Aliwal North, D–Deelpan; V–Voigtspost; A–Alexandersfontein; K–Klipfontein; F–Florisbad; La–Ladybrand; L–Lesotho Highlands; Ta–Taung; St–Sterkfontein; RT–Rietvlei (Transvaal); M–Moreletta River; W–Wonderkrater; S–Scot; Th–Theuniskloof; Tu–Tugela; T–Tate Vondo; I–Inyanga Mountains; Is–Ishiba Ngandu; Ka–Kalambo Falls; D–Dundo.

lower than those reported by Martin (1968). In contrast to the data from Knysna, Martin's spectra from the macchia on Table Mountain show no arboreal pollen although forest patches occur on the lower slopes. Tree pollen is probably obscured by the high production of local macchia elements like Restionaceae, Compositae and Ericaceae. High non-arboreal pollen values (87,5%) are recorded in a sample from the fynbos at Killarney on the plains 15km north-east of Cape Town (Schalke 1973). The 12,5% arboral pollen consists mainly of *Myrica* and Myrtaceae. Sediment samples from some river mouth samples in the southwestern Cape, passing through Cape macchia vegetation, also produce high non-arboreal pollen values consisting mainly of Compositae (including the *Elytropappus*-type), Restionaceae,

Ericaceae and *Cliffortia* (Coetzee 1976). The few tree pollen comprise Proteaceae and *Podocarpus*.

The first modern pollen spectra from the Austro Afro-alpine belt (Coetzee 1967) in the highlands of Lesotho have been described by Van Zinderen Bakker (1955). Four samples from the bog area above 3,000m in the Letsiri vegetation zone contain high numbers of Cyperaceae (43-50% of the pollen sum), Gramineae (36-44%), Compositae (4-13%), Umbelliferae (2-8%) and other unspecified herbs ('varia': 13 to 37% of the total spectra). A surface pollen sample from the Blue Mountain pass at the slightly lower altitude of 2,700m in the Seboku zone with ericoid vegetation provides a non-arboreal spectrum of 45% Ericaceae, 11% Gramineae, 11% Cyperaceae, 9% Compositae, 8% *Passerina*, 6% *Anthospermum*, 4% Aizoaceae-type and smaller numbers of various other pollen types (Scott 1982a).

The first atmospheric pollen spectra obtained from seasonal studies in the vast O.F.S. grassland area at nearly 1,500m altitude near Bloemfontein, show that Gramineae pollen (90%) is the most important and Compositae (4%) the second most prominent constituent (Coetzee & Van Zinderen Bakker 1952; Van Zinderen Bakker 1957). In these studies spectra from modern soil samples in the area show comparable compositions but the Compositae percentages can be higher (up to 41%) depending on the local nature of the sampling site. Chenopdiaceae values are also higher in most soil samples usually ranging up to 10% but in a sample from a halophytic saltpan they reached up to 91%. Comparable values are recorded by Scott (1967) and Horowitz *et al.* (1978) in halophytic sites in the western O.F.S. Coetzee (1967) showed that the atmospheric pollen in the shrubby karoo vegetation near Middelburg produces a lower Gramineae: Compositae ratio (71:12%) in comparison with that of the grassland near Bloemfontein. It is interesting that Aizoaceae pollen which is absent in the Bloemfontein spectra, shows up (4%) in the Middelburg spectra. No data on the pollen production from Karoo soils have been published but it can be expected that Compositae and Aizoaceae-type pollen will become more important further to the south and west. This combination shows up constantly in river mouth samples along the west coast from the Swakop River southwards (Coetzee 1976).

Arboreal pollen percentages are usually low in the O.F.S./Karoo region and *Rhus* pollen appears to be the most important constituent reaching up to 9% (Coetzee & Van Zinderen Bakker 1952) while very high values are recorded on wooded koppies (Scott 1982c). A mountain slope at 1,900m in the Cymbopogon-Themeda veld in the eastern O.F.S. (Ladybrand) also produced a high arboreal pollen content (Scott 1982a), consisting mainly of 10-40% *Celtis*, 4-30% *Rhus* and 2% *Clutia* pollen. The non-arboreal pollen here consists mainly of 15-21% Gramineae, 2,4-6% Compositae and 0,5-11% *Cliffortia*. As in the grasslands of the Orange Free State the Bankenveld of the northern parts of the highveld grassland plateau at Sterkfontein in Transvaal show a high Gramineae: Compositae ratio of 38:2% (Horowitz 1975). The woody vegetation of the rocky outcrops of the area produces relatively high arboreal pollen values consisting mainly of 34% *Rhus* and 5% *Olea*. A lower Gramineae: Compositae ratio of 7,5:8% is found in a disturbed vlei area in the Bankenveld at Rietvlei Dam where the spectrum is dominated by 66% Cyperaceae of local origin and the arboreal content is less than 1% (Scott 1982a).

Published information on the modern pollen spectra from the woodland areas of

Southern Africa is restricted to the Transvaal and N.W. Cape and mainly described by Scott (1982a). As can be expected the arboreal pollen taxa are important in these areas and their distinction can be useful in the characterization and diagnosis of various bushveld types. Firstly the pollen spectrum from the dense eastern montane forest area near Tzaneen at 1,700m altitude is not very illuminating (Scott 1982a) as it is overrepresented by local fern spores (70%) and pine pollen (18%) of the extensive exotic plantations in the region. The rest of the arboreal pollen which is poorly represented consists of *Podocarpus* (2%) Myrtaceae (3% and traces of *Myrica* and *Euclea*. Other studies as mentioned earlier (Martin 1968) recorded relatively high *Podocarpus* values in indigenous African forests and this is supported by Coetzee (1967) in a study on spectra from the East African Mountains.

In the open woodland types between 1,250 and 1,800m in the upland areas of the central plateau of the Transvaal Proteaceae pollen (2 to 22%) is quite important but pollen taxa of this family belonging to the relatively lower levels, like *Faurea saligna,* cannot be distinguished from *Protea* spp. of higher altitudes (Scott 1982a). The spectra from the lower-lying areas produced higher arboreal pollen values (up to more than 50%, including ±15% Combretaceae-type) than the higher-lying samples where only up to 17% arboreal pollen are recorded together with high non-arboreal values which include up to 30% Compositae. In the lower-lying sourish mixed bushveld, at 1,100-1,500m, the arboreal pollen consists mainly of Combretaceae-type (up to 45%) and *Euclea* (up to 15%) while the non-arboreal pollen consists mainly of Gramineae (9-80%). Tarchonantheae (up to 7%) and Capparaceae (up to 19%) are some of the most important elements of the arboreal pollen in samples of the relatively dry Kalahari thornveld. In the spectra of this vegetation type from a small outlying patch of this vegetation, in the Springbok Flats area north of Pretoria at 1,100m, the Capparaceae pollen is more prominent than at Buxton (near Taung) 480km towards the south-west in Bophuthatswana at 1,200m. About 85km south of Taung at Klipfontein no Capparaceae was found in three surface spectra from a spring site although the trees are growing in the vicinity (Butzer *et al.* 1979).

On the northern slope (at 1,100m) of the Soutpansberg high percentages of arboreal pollen occur including 19% Combretaceae-type but local swamp forest elements are prominent at the spring site, Scot (Scott 1982b). The arid sweet bushveld north of the Soutpansberg at 700 to 800m also produced high percentages of Combretaceae-type (up to nearly 50%) but other elements comprise *Burkea africana, Euclea, Acacia,* Capparaceae and others. Further north-west close to the Limpopo River comparable spectra are found at Tugela andTheuniskloof (Scott, unpublished data). It is surprising that small numbers of arboreal elements and no *Colophospermum mopane* pollen was recorded in a sample of the mopane veld in the Kruger National Park, which produced a high Gramineae percentage (95%) (Scott 1982a). Mopane pollen grains are, however, recorded among arboreal pollen like Combretaceae-type, *Croton* and *Acacia,* together with many grass pollen grains (75%) in grab samples at the mouths of rivers passing through Mopane veld in the northern parts of Namibia (Coetzee 1976).

FOSSIL POLLEN ASSEMBLAGES

The southern Cape Province

In the southwestern Cape the pollen sites in the Rietvlei basin, on the Cape Flats and at Cape Hangklip are all surrounded by Cape macchia vegetation (Schalke 1973). In view of its high arboreal pollen content the profile from the Cape Flats area can be viewed separately from the other two sites. A ^{14}C date of 41,500 (+2, 100, -1,800) yr B.P. was obatained from a paleosol level near the top of this polleniferous zone. The tree pollen content includes *Podocarpus* percentages ranging between 12 and 40%, suggesting that *Podocarpus* trees either surrounded the site or occurred fairly close to it on the Cape Flats. These values are comaparable to those in surface pollen spectra of the Knysna area (Martin 1968), which were mentioned earlier. Other prominent tree components are Proteaceae, Anacardiaceae, *Ilex mitis* and Myrtaceae. One of the Euphorbiaceae pollen types illustrated by Schalke (1973, Plate 4, Fig. 2) from the Cape Flats profile appears to be that of *Alchornea.* This subtropical and tropical type could indicate that the forest developed under relatively warm and moist conditions. *Alchornea* pollen was also recorded in the same area from Tertiary deposits which represent warm climates (Coetzee, 1978a).

Schalke attributes the pollen profile from the Cape Flats to zone T, also recognised in some profiles of the first sedimentation cycle in the Rietvlei basin. This correlation is based on ^{14}C determinations of more or less the same age, and a small peak of *Podocarpus* pollen percentage in the latter. *Podocarpus* values in the Cape Flats core are, however, considerably higher than in zone T of the Rietvlei basin where only up to nearly 5% is reached. It is therefore felt that the Cape Flats profile could quite possibly represent a different time period than the one covered by the Rietvlei pollen sections and that it is risky to rely on the relatively old, possibly 'inaccurate' ^{14}C dates to correlate the zone between the profiles.

The four pollen profiles from the Rietvlei basin comprise mainly Restionaceae, Ericaceae, Compositae, Gramineae, Cyperaceae, Proteaceae, Chenopodiaceae, *Myrica* and others and show constant oscillations in the percentages of these types. In the first sedimentation cycle, which is probably largely of middle Pleniglacial age, Schalke distinguishes six zones (s to x). They are not always very distinct and unlike the Cape Flats profile do not suggest climates which are very different from those of the present time. It is therefore not easy to see the rationale behind their correlation among the different profiles and in this respect the ten ^{14}C dates are not of much help mainly for the same reason as mentioned above. In some units like zone X (Rietvlei borehole 2) and zones T and V (borehole 1), the small percentages of *Podocarpus* appear relatively constantly in comparison with other zones in which they are absent. However, they never reach more than roughly 5%. These zones could possibly only represent a slight increase in distant mountain forests, but placing them into a chronological frame on the basis of the available data is not possible. It cannot be demonstrated with certainty that any part of the profiles at Rietvlei cover the upper Pleniglacial cold phase and there is a possibility that this phase is represented by some of the sandy layers without pollen. Therefore it remains unclear what the behavior of the macchia vegetation of the area was during the coldest phase of the Late Pleistocene.

Schalke assigns both the mentioned upper unit at Rietvlei as well as the well dated Holocene profile from Hangklip to zone Z. There are, however, no ^{14}C dates to confirm that the zone at Rietvlei belongs to the Holocene. The sites apear to represent different local environments and therefore it is difficult to see any similarities between them. At both sites, however, *Podocarpus* pollen grains are absent. At Rietvlei (boreholes 1 and 2) there is a marked increase in Compositae and Chenopodiaceae in the uppermost samples, while Restionaceae (borehole 1) and Gramineae (borehole 2) decline sharply. This is possibly the result of local developments and it cannot be determined if a change in climate played any role.

The peaty and sandy deposits from Hangklip provided three ^{14}C dates the oldest of which is about 7,280 yr B.P. The maximum date for the profile can be estimated to be roughly 7,500 B.P. The pollen profile is dominated by Restionaceae while Compositae, Ericaceae, Rosaceae, Gramineae, Aizoaceae and others are also prominent. It shows a break associated with sand deposits which roughly covers the period between 6,000 and 2,000 yr B.P. The arboreal pollen content is relatively low throughout consisting mainly of Proteaceae and *Myrica* while there are no signs of forest pollen. In general the profile seems to represent pure macchia vegetation with small changes in its composition. The significance of these oscillations is as yet difficult to estimate.

The Holocene palynological site, Groenvlei, is situated in macchia vegetation bordering on the extensive Knysna forests. In his study of a core from the fen at Groenvlei Martin (1968) provides a detailed pollen profile, based on large counts of palynomorphs. The work is supported by extensive background information on the diatom stratigraphy, ecology and history of the site (Martin 1959, 1960 and 1962). Despite all the information the pollen profile still remains a difficult one to interpret since its different responses to slight changes in past temperatures on the one hand and moisture conditions on the other are difficult to separate. This is due to the absence of good temperature sensitive indicators, the complex physiography of the area and the problem of facies change like the development of marine units in the profile. Martin's interpretation is mainly concerned with moisture changes. His cautious approach is justified but the actual nature and mechanism of the environmental changes remain somewhat unclear. In order to draw a broader insight from the data a wider hypothesis is proposed below together with a summary of Martin's results and interpretations.

The age of the deepest subzone, B1, is estimated to be between $>$ 7,000 and 8,000 yr B.P. Although it contains fair numbers of arboreal pollen, consisting of *Podocarpus, Olea* and *Rhus* etc., the arboreal content is the lowest for the profile suggesting the presence of somewhat restricted forests. Therefore fairly high non-arboreal pollen with especially Compositae and other macchia elements occur. Some freshwater aquatic plants are also recorded. The spectra of unit B1, compare best with the high-lying surface spectrum from Table Mountain. In view of this it is felt that speculations about the existence of past temperatures of about similar values than at present or slightly lower is not unreasonable. Relatively cool conditions could have helped to maintain somewhat restricted forests. The presence of various freshwater aquatic plants is probably solely a reflection of the fen's basin shape, decribed by Martin (1959), but if, as is suggested, dry conditions played a role the aquatic plants will be even easier to visualise if relatively cool conditions and low evaporation are also involved.

The next subzone, B2, represents a peat zone of 6,870 yr B.P. which shows a slight increase in arboreal pollen but with a lower species richness. Among the non-arboreal pollen *Artemisia* is more prominent. Martin tentatively suggests that this unit represents wetter conditions and possibly also the destruction of dune heath through the mobilization of sands and the consequent relative increase in distant forest pollen. A slight increase in temperature in comparison with the previous zone is, however, not an unreasonable alternative or complementary hypothesis for the explanation of the increase in forest and forest pollen with the simultaneous relative reduction of local heath pollen percentages. This could have been enhanced by more hostile conditions on the surrounding dunes.

In subzone B3 there is a change of facies from peat to sedimentation under marine conditions. It can be expected to lead to a different pollen-trap mechanism with modified pollen content. Without the control of recent pollen spectra from comparable waterlaid and other deposits it remains difficult to estimate what the effect of the facies change will be on the fossil pollen. Thus firm interpretations cannot be assured and it is therefore not certain if the spectra recorded in subzone B3 represent a different vegetation pattern than those of B2 or the same general type in which the pollen was trapped differently. Slightly lower arboreal pollen in subzone B3 leads Martin to suggest that the forests were restricted and that the non-arboreal elements remained important. Prominent Restionaceae values could indicate high watertables in the valley bottoms. Martin proposes two alternative interpretations. The first possibility implies drier, or hotter, conditions and the second that dune vegetation cover was restricted by sand movement of which there is independent evidence. It can, however, also be argued that the decrease in arboreal pollen percentages was possibly caused by the relative increase of heath elements like Restionaceae in the close surroundings, perhaps as result of the suggested favourably high watertables which could possibly be related either to high sea levels or wetter conditions. The lower arboreal pollen does therefore not necessarily imply a retreat of the forests. The change may have occurred without temperatures differing markedly from those of the previous zone. Relatively high temperatures, if they existed, could have resulted in strong evaporation and unstable dune ridges with more Chenopodiaceae around marshy places.

Diatoms in the top part of zone B3 (above 440cm) provide evidence that the ocean water cooled off slightly (Martin 1959). This coincides with only a very small change in the pollen diagram, viz: a slight increase in *Olea, Rhus* and *Myrica* pollen types and a gradual decrease of *Podocarpus* pollen. It could be a indication that scrub elements began to increase relative to forest elements. It is not certain if this is a climatic response but the development of spectra in the next zone, C1, could imply a continuation of such a process as *Podocarpus* pollen reaches its lowest level and certain other woody and scrub elements like *Euclea* and *Sideroxylon* become prominent. The effect of facies change here to gyttja is also uncertain.

In the next zone, C2, the arboreal pollen increases. Among the tree pollen grains those of *Podocarpus* reach very high levels. The size distribution of this pollen type suggests that *P. falcatus*, such as occurring in the drier nearby forest types, played an important role. Further certain other tree taxa like *Fagara, Clausena* and *Apodytes* begin to appear prominently indicating that forest, probably of the drier type was able to expand over the nearby dunes. In climatic terms this could possibly mean that more favourable moisture conditions developed on the surrounding dunes together with perhaps a more stable climate. If a cooling process occurred in the top of zone B3 and in C1, it could have been stopped in this zone.

The Lesotho highlands

The first preliminary pollen analyses of a high altitude peat deposit of 2m depth occurring at 3,000m in the Letsiri grassland zone of the Austro-afroalpine belt in Lesotho is provided by Van Zinderen Bakker (1955). It is believed that the peat bogs started growing at these high altitudes after the amelioration of the glacial climate. The oldest ^{14}C date obtained for a bog of this kind is 8,020 yr B.P. (Van Zinderen Bakker & Werger 1974). The pollen sum of the profile consists of Cyperaceae, Gramineae, Compositae and Umbelliferae and apart from it there are about 10-40% unspecified herbs. In general, the abundant Cyperaceae, which could mainly represent local moist conditions, show a gradual decline in percentage towards the top of the profile, and this is compensated for by a general increase in Gramineae and Compositae. It cannot be determined with certainty how climatic changes shaped the profile but the pollen fluctuations do not suggest any marked shifts.

The Karoo and grassland regions

The first fossil pollen analysis from the Orange Free State grassland region at Florisbad was done by Van Zinderen Bakker (1957) and the results are summarised by Van Zinderen Bakker and Butzer (1973). The age of the bottom of the Florisbad deposits (peat I) is beyond the range of ^{14}C determinations. The younger peat layers (II and III) provided ^{14}C dates of about 28,450 and 19,350 yr B.P. respectively but new evidence provided by Vogel (Rightmire 1978a) shows that these layers are much older. In view of the uncertainty with the age and the initiation of new palynological work at the site it is perhaps not wise to speculate much about the significance of the early pollen results. It was, however, noted that in the oldest deposit Gramineae and Compositae pollen percentages are roughly equal. When compared with surface pollen data this composition, together with increasing numbers of Chenopodiaceae, almost certainly points to a dry karoid type of vegetation. In peat II, somewhat higher up in the profile, the Gramineae/Compositae ratio is higher and it increases upwards towards peat III. This change and also the ultimate decline of Chenopodiaceae in the latter peat seem to indicate a general trend to wetter, less brackish, conditions. The degree of regional change is, however, not estimated, since it may be that local pollen does occur in the pollen sum. For instance it can be expected that certain pollen specimens like those derived from swamp grasses (e.g. *Phragmites*) could form a prominent part of the total Gramineae pollen. Due to the lack of sensitive indicator pollen types no conclusions could be drawn about past temperatures at Florisbad but certain possibilities have been put forward (e.g. Van Zinderen Bakker & Butzer 1973).

Pollen analyses of thermal spring deposits in the False Upper Karoo vegetation at Aliwal North was provided by Coetzee (1967). The vegetation is transitional to the

Karoo in the south and west and different grassland types to the north and east. The pollen data derived from four profiles cover a period estimated between an age of roughly 13,200 yr B.P. and 9,650 yr B.P. Six pollen zones (u to z) are recognised. They are based mainly on the oscillating ratio between Gramineae pollen on the one hand and those of Compositae (other than *Stoebe*) and Chenopodiaceae on the other. The pattern seems to indicate alternating changes in the moisture conditions at the swamp. These developments probably have regional implications for the rainfall pattern, but for the same reason as in the case of Florisbad it is difficult to estimate the degree of change. Apart from having a high Gramineae/Compositae ratio, the deepest zone U is distinct from the others in having high proportions of *Stoebe*, *Typha* and Cyperaceae. The latter two types indicate that it was locally much wetter and the *Stoebe* pollen suggests quite a different composition in the grassland vegetation. Coetzee assigns the *Stoebe* pollen to *S. plumosa* which could be an indicator for cooler conditions. These characteristic types in zone U are low in the rest of the sequence which started to form shortly after 12,600 B.P. and in general correspond more closely with the pollen spectra of the modern vegetation of the area. Zones V and W seem to indicate a slightly drier and wetter cycle with generally drier local conditions than in zone U but zone X suggests the onset of even drier conditions on the basis of low Gramineae and high Compositae pollen ratios with fair numbers of Chenopodiaceae pollen. According to interpolated estimations this zone could have lasted from roughly about 12,000 yr B.P. until after 11,000 yr B.P. Although some oscillations in the pollen profile follow during zones Y and Z, the general trend after zone X is to a slight improvement in the moisture conditions until about 9,650 B.P.

Preliminary palynological results from Holocene deposits immediately east of the O.F.S./northwestern Cape border is discussed by Scott (1976). The site is situated on the fringe of the large treeless Alexandersfontein basin and falls within the Kalahari thornveld vegetation invaded by Karoo. To the east it is bordering on the grassland (False Upper Karoo) of the Orange Free State. The pollen in two generations of spring deposit, provisionally dated to the Late Holocene younger than 5,000 yr B.P., is described. That in the disturbed spring eye suggests conditions not very different from the present. The deposits adjacent to the spring eye contain spectra which differ from the modern assemblage viz: high percentages of Gramineae, lower numbers of Chenopodiaceae and Aizoaceae, a different Compositae combination and virtually no pollen of the local tree species surrounding the site. This composition seems to point to a wetter, environment during some stage in the Late Holocene. The pollen in a nearby alluvial terrace (US-3) of at least 14,670 yr B.P. is possibly a modern mudcrack contaminant (Butzer, this volume).

A few pollen spectra from Holocene pan shore deposits in the False Upper Karoo vegetation to the east near Petrusburg have been described. From Voigtspost a spectrum from a reddish clayey marl, associated with a ^{14}C ostrich-eggshell date of 6,350 yr B.P., seems to indicate dry karoid conditions with a possible stronger influence of the Kalahari Thornveld which occurs towards the north-west (Horowitz *et al.* 1978). The anomalous combination of some forms in this single spectrum together with a pine pollen suggest some degree of contamination and therefore caution is necessary with such an interpretation. The spectrum of a younger dark clay appears to represent conditions which are not very different from the present, while the youngest spectrum from overlying sands associated with an ostrich-eggshell date of 1,220 yr B.P. points to karoid conditions. However, at Deelpan 40 km east of Voigtspost, in hyena coprolites from sands which in broad terms are stratigraphically comparable to the latter, about 90% grass pollen are recorded (Scott & Klein 1981). This differs markedly from the Voigtspost assemblage which is dominated by Chenopodiaceae and Compositae. The anomaly can possibly partly be attributed to the differing nature of the sediment types and coprolites as pollen traps.

The only pollen profile from the northernmost part of the highveld grassland was recorded from the Rietvlei Dam area 15 km southeast of Pretoria. The results which include nine ^{14}C dates are due to be published soon (Scott & Vogel, in prep.). The pollen spectra suggest that the general grassland vegetation and climate around 10,000 yr B.P. at the Rietvlei Dam site were not markedly different from those of modern times. Later, however, the climate became relatively drier before a temporary expansion of some bushveld elements occurred over this northern part of the highveld around 6,580 ±70 yr B.P. (Pta-3061). The migration could have been caused by relatively warm temperatures and favourable moisture conditions. The pollen spectra suggest that a virtually treeless grassland returned to the area during the Late Holocene.

The woodland areas south of the Limpopo River

The southernmost pollen spectra from the woodland area are those described by Butzer *et al.* (1979) from a rock engraving site at the spring, Klipfontein. The area is covered by fairly dense Kalahari thornveld. Two pollen spectra were found in marl deposits, one overlying a unit with a carbonate date of 6,820 yr B.P. and the other from a layer dated at 20,500 yr B.P. The spectra are more or less similar and in comparison with surface pollen spectra they could represent either reduced grass cover or expanded bush, consisting of nearly the same elements as recorded in the area at present. However, the scarcity of pollen, the remarkable similarity between the spectra and also to some extent between them and the surface spectra could suggest that the data are somewhat suspect.

In the woodland areas of the Transvaal detailed pollen diagrams have been described from two spring sites, viz: at Wonderkrater in the bushveld of the low central plateau (Scott 1982a) and at Scot in the bushveld at the northern foot of the Soutpansberg (Scott 1982b). Micrographs of the different fossil pollen types and discussions about their nature and significance are presented by Scott (1982c). The results of the palynological work together with some limited information about evidence from current work in the Transvaal have already been summarised (Scott 1982d and e).

The results from Wonderkrater especially suggest that extensive plant migrations and environmental changes took place. The 8m profile was dated by 12 ^{14}C dates and although some of them show anomalous values, it is suggested that the age of the profile is at least 35,000 yr B.P. The earliest phase at Wonderkrater (W1), is characterised by high numbers (up to 50%) of arboreal pollen, and the vegetation must have consisted of a more mesic bushveld with many *Olea* and Combretaceae trees on the plains. Relatively high proportions of *Podocarpus* (up to 34%) and *Kiggelaria* pollen also suggest a greater expansion of montane-type forests in the region. The next pollen zone (W2) shows a gradual decline in the arboreal pollen. The *Podocarpus* pollen percentages are reduced while the savanna elements are mainly represented by Capparaceae and Tarchonantheae, suggesting a drier climate. The following zones W3 and especially W4 provide evidence for a more open grassland vegetation with a macchia component consisting of Ericaceae, *Passerina, Cliffortia* and *Stoebe*. At present the nearest comparable types of vegetation occur in belts approximately 1,000m higher than the Wonderkrater site suggesting an average cooling of about 5 to 6° C. Conditions apparently also became wetter than in zone W2. ^{14}C-dates for zone W4 range between 14,180 and 24,340 yr B.P. According to the dates zone W5 of the Wonderkrater sequence probably corresponds with the late Glacial or early Holocene. The cool indicators disappeared and warmer, drier conditions developed. The vegetation then consisted of open dry grassland savanna. Further amelioration in the early Holocene allowed Kalahari Thornveld-type elements such as Capparaceae and Tarchonantheae to expand in the central plains around Wonderkrater (W6) between about 9,000 and 6,000 yr B.P. Zone W7 at Wonderkrater suggests that a warm, but more broadleafed type of bushveld with an increased Combretaceae component, was present between 6,000 and 4,000 yr B.P. The total AP declined after 4,000 yr B.P. but the arboreal pollen, especially the Compositae, increased. In comparison with some surface pollen data from upland woodlands these spectra can tentatively suggest slightly cooler conditions. Higher arboreal pollen percentages occurred again after 2,000 yr B.P. at Wonderkrater. Until 1,000 yr B.P. there are increased proportions of Proteaceae (probably *Faurea saligna*) pollen at Wonderkrater. This suggests that the climate was warm, but possibly still slightly cooler and/or wetter than at present. The modern broad-orthophyllous to microphyllous type of bushveld at Wonderkrater originated after about 1,000 yr B.P. and there is evidence of more *Burkea africana* trees near Wonderkrater during part of this period.

The peat deposit which developed at the swamp forest site, Scot, in the Soutpansberg, was dated by ^{14}C at about 5,000 yr B.P. (Scott 1982b). The pollen suggest that in general savanna resembling the present type, existed in the area throughout the period covered by the profile. However, between 3,530 yr B.P. and 2,460 yr B.P. the arboreal pollen is not as abundant and higher proportions of non-arboreal elements, especially Compositae occur. This seems to indicate that the savanna was more open like the upland woodland types. Therefore slightly cooler conditions could have existed but other alternatives concerning local developments are also considered.

Some preliminary palynological findings from four sites, in the Transvaal, which are currently being investigated are only briefly mentioned by Scott (1982d, e). These tentative findings consist of (1) evidence for more open vegetation at sites of presumably late Pleistocene or early Holocene age in Venda (Tate Vondo) and in Pretoria (Moreletta River), (2) relatively high non-arboreal pollen ratios for undated stream deposits near the Limpopo River (Theuniskloof), (3) increased *Burkea africana* pollen in relatively young spring deposits from Tugela near the Limpopo River, and (4) an increase in the tree pollen some time after 5,220 yr B.P. at Moreletta River. The spectrum from Tate Vondo has in the meantime been dated and it is indeed of terminal Pleistocene age.

Southern Africa north of the Limpopo River

Very few pollen profiles have been studied in the vast predominantly woodland area north of the Limpopo. Three Holocene pollen profiles from the grassland zone around the Inyanga highlands, Zimbabwe, have been described by Tomlinson (1974). The preliminary data show that Gramineae pollen, averaging between 60 and 90%, is the dominant element in the cores while Compositae (5%) is the second most prominent component. The tree pollen content is low in cores I and II, consisting mainly of *Brachystegia* but core V which is situated closer to the wooded eastern escarpment, has relatively high numbers of different arboreal elements while ericoid pollen is also fairly regular. From the oldest core (II), which has a ^{14}C date of 11,750 yr B.P. below the studied pollen levels, it can be estimated that grassland existed throughout most of the Holocene and the *Brachystegia* only became more prominent in the area during the late Holocene. However, this probably happened before 4,670 yr B.P. because core I shows that the genus was relatively prominent around that date and shortly afterwards. Core V of 2,4m which is not dated shows relatively high Compositae, gymnosperm, ericoid and proteoid pollen in the lower levels while Gramineae are very abundant towards the top. Cultivated grasses make a sudden prominent appearance at 50 cm. *Brachystegia* pollen is scarce or absent but seems to be more consistent in the lower levels.

A detailed pollen diagram, of about 5m of lake deposit, covering at least 22,000 yr B.P. is provided from the lake Ishiba Ngandu on the plateau of Zambia (Livingstone 1971). The vegetation around the site consists of miombo woodland with *Brachystegia* trees. There is a predominance of Gramineae pollen throughout the profile but is can nevertheless be subdivided into three units. The bottom part which is older than 21,600 B.P. coincides with a more sandy unit below 3,5m. Here the Gramineae pollen percentages reach their lowest levels (roughly 50%), while Compositae and *Cliffortia* pollen and also *Cyathea* spores occur relatively frequently. Livingstone proposed that the difference between this interval and the next one is only related to the change in sediment type and local environments but it is felt that this is not necessarily the only possibility. Whatever the case the spectra suggest a fairly open vegetation type with the arboreal pollen to some extent probably derived from long distance transport. Such a vegetation type could indicate relatively cool, dry conditions. In the next interval, between 3,5 and roughly 0,5 m. of which the base has been dated at 21,600 yr B.P., Compositae and *Cliffortia* decline gradually while *Podocarpus* and eventually *Olea, Myrica* and Ericaceae show an increase. Grass pollen percentages increase while different arboreal pollen types in general also appear to be only a little more persistent. It is not unlikely that the change in composition was caused by slightly different, perhaps wetter, climatic conditions. In the uppermost part of this interval, above the 1,5 m, level, however, the arboreal elements especially *Brachystegia, Acalypha*, Myrtaceae and Combretaceae (+Melastomataceae) seem to be even more persistent although in total they remain scarce. It is possible that this change reflects the beginning of slightly warmer conditions which culminated in the formation of the next unit, above 0,5m, where the mentioned arboreal elements become markedly prominent and others like *Olea, Podocarpus* as well as those of shrubs like *Myrica* and Ericaceae decline or disappear. It is therefore tempting to conclude that this upper type of assemblage represents warm Holocene conditions while the previous intervals belong to the cooler 'glacial' type of environment. If this is the case it is, however, difficult to draw a line between the Pleistocene and Holocene parts of the profile since the warming process already started before the 0,5m level. The shallowness of the upper section with more arboreal pollen could suggest that the sedimentation rate was relatively low or underbroken in comparison with the deeper levels. This is, however, contrary to what is implied by the recorded lowering of the pollen concentration in this section.

Further towards the north at Kalambo Falls near the south end of Lake Tanganyika, also in the miombo woodland, some fossil pollen spectra have been described by Clark and Van Zinderen Bakker (1964) and Van Zinderen Bakker (1969). Different spectra from two sites are grouped together in six pollen units (U to Z). Of these the second youngest one, Y, can be associated with ^{14}C age determinations, placing it between 27,000 and 30,000 yr B.P. This and also the older units suggest that the *Brachystegia*-type of woodland has been present in the area for at least a considerable part of the Pleistocene. One warmer, drier period (U) and two cooler phases (V and X) have been postulated during the period older than 30,000 yr B.P. but it is felt that these could not have been very marked due to the general similarity of the pollen spectra. In the upper unit, Z, younger than 27,000 yr B.P., *Brachystegia* pollen is still recorded but the high numbers of Ericaceae suggest a drop in the vegetation belts to a more open type. This lowering provides evidence for a distinct cold phase which probably coincides with the last glacial maximum period.

More than 1,000km to the west of Kalambo Falls at Dundo in the Luembe valley, north-eastern Angola, more fossil pollen spectra have been described (Van Zinderen Bakker & Clark 1962; Van Zinderen Bakker 1963). This area occurs at a relatively low altitude (700-800m) and it is surrounded by vegetation related to the *Brachystegia-Julbernardia* woodland. The oldest peaty clay sample which can be associated with ^{14}C dates of 38,000 and >34,000 yr B.P. provides evidence for the presence of woodland with *Brachystegia*. Samples of a younger age which are probably more than 14,503 years old, contain high percentages of *Syzygium* and *Podocarpus* together with smaller numbers of woodland elements such as Combretaceae, Proteaceae, *Uapaca, Peltophorum* and others. Van Zinderen Bakker suggests that such a combination could have been produced by a relatively cooler and wetter type of woodland. The two youngest spectra suggest that relatively warm woodland with *Brachystegia* and *Alchornea* existed in the area during the Holocene.

CONCLUSION

The palynological evidence for past vegetation types in Southern Africa is very limited for this vast region and at this stage the paleoenvironmental conclusions at the different sites can only be of a tentative nature. No overall reconstruction based on the pollen data alone can be attempted but the following short summary of the most important results can nevertheless be made: In the period before the last glacial maximum of the Northern Hemisphere it is suggested that *Brachystegia*-type woodlands were already widespread at some sites where they occur today like Kalambo Falls and N.E. Angola. Their composition was, however, probably not the same as at present. In the Transvaal at Wonderkrater there is evidence for a relatively mesic woodland with expanded forests at first and later for a relatively dry more open woodland before the last glacial maximum period. From the grassland in the central Orange Free State at Florsibad it it suggested that more dry karoid conditions existed at some undated stage in the Pleistocene. In the southern Cape there is evidence for relatively wet, warm times with expanded forests on the Cape Flats during a period which is difficult to date while conditions closer to the present occurred at Rietvlei during other times before the last maximum cold phase.

The pollen spectra which formed in southern Africa during the times of the last glacial maximum in the Northern Hemisphere suggest that woodlands to the north of the Limpopo River were in some areas replaced by more open vegetation like at Kalambo Falls and Ishiba Ngandu and that cooler, wetter woodlands occurred elsewhere like in north-eastern Angola. A general lowering of vegetation belts is implied in these regions and this is also evident in the Transvaal (Wonderkrater) where the bushveld vegetation was replaced by a cool ericaceous type of grassland. There is no palynological evidence that the precipitation was markedly different from at present but lower evaporation could have established more effective moisture. It is not clear what the vegetation consisted of further towards the south. At the end of the glacial period, however, the grassland in the vicinity of the upper Orange River (Aliwal North) seems to have had a different composition with more shrubs like *Stoebe*. It cannot be established with certainty how the Cape macchia vegetation reacted to this cold phase.

After the maximum cold phase, woodland types, close to but probably not the same as the modern ones, returned to their present habitats. In the Transvaal at Wonderkrater and Scot there is evidence that certain relatively small oscillations in temperature and moisture conditions occurred which regularly changed the composition of the vegetation during the Holocene. This includes evidence for relatively warm conditions at the end of the early Holocene as well as possible cooler, wetter conditions with more open vegetation in the Late Holocene. During this period the grassland areas of Zimbabwe and also the highveld region in the south and the Lesotho highlands remained essentially grassland with relatively small changes in composition. These changes are implied by relatively dry conditions during the terminal Pleistocene/Holocene around the Upper Orange River at Aliwal North, expanded woodland elements over the northern parts of the southern Transvaal highveld (Rietvlei) during relatively warm wet conditions at the end of the early Holocene, and wetter conditions at the western O.F.S./N.W. Cape border (Alexandersfontein) during part of the late Holocene. In the southern and western Cape at Groenvlei and Hangklip there is evidence for the persistence of macchia vegetation but small changes in the composition occurred during the Holocene. At Knysna the vegetation was relatively open in the early Holocene but forests became very prominent around 7,000 yr B.P. and only expanded further over coastal dunes in the late Holocene (2,000 yr B.P.). These developments could also be related to small temperature and/or moisture changes. Finally it seems that both from the Cape (Groenvlei) and the Transvaal (Wonderkrater) regions the present vegetation pattern only developed during the last one or two thousand years.

Palynological work on Quaternary deposits, especially those of the late

Quaternary, is continuing and a number of promising sites in the Transvaal, the Orange Free State, the western desert and Kalahari regions and the southern Cape Province are currently being investigated by different workers. In future this research will certainly help to expand knowledge about South African palaeoenvironments.

Q.B. HENDEY
South African Museum, Cape Town

3

Southern African late Tertiary vertebrates

INTRODUCTION

There is a widespread interest in the late Tertiary vertebrate fossil record of Africa, mainly because of this continent's role in the origin and early evolution of the Hominidae. Most attention has been focussed on east Africa (Tanzania, Uganda, Kenya, Ethiopia), where there are extensive and highly fossiliferous deposits of Miocene and Pliocene age (Figure 1). Significant fossil occurrences otherwise tend to be widely scattered in north African countries bordering the Mediterranean, and southern Africa (Namibia, South Africa). In the latter area, a few Late Pliocene fossil assemblages are recorded from localities on the inland plateau (Klein, this volume), but fossiliferous late Tertiary strata are best developed along the western coastal margin of the subcontinent. This paper deals mainly with them.

These strata range in age from Early Miocene to Late Pliocene, and most recorded fossil occurrences are in marine sediments. Terrestrial vertebrates occur most commonly in fluviatile sediments that have a more restricted distribution. Only a few of the recorded fossiliferous localities (Figure 2) have been investigated in any detail, and a considerably increased effort in this connection is clearly desirable. In addition, it is virtually certain that other vertebrate occurrences await discovery, but either the Tertiary deposits are buried beneath Quaternary sediments, with their exposure usually being dependent on activities related to the mining and construction industries, or they are located in areas where access is restricted for security reasons.

Deposition of the west coast late Tertiary sediments was largely controlled by the sea-level changes of this period. As a general rule, marine regressions were periods of erosion along the continental margin, and consequently there was comparatively little accumulation of sediment at such times. Conversely, marine transgressions sometimes resulted in substantial bodies of sediment being laid down, either on marine and river terraces, or in coastal barrier environments. These have been the most significant sources of late Tertiary fossils on the southern African west coast.

The proximity to the South Atlantic Ocean has made it possible to correlate the continental succession with those of the continental shelf and deep-sea. Since the latter successions tend to be more complete, and since they have recently been the subject of intensive study, they provide a wealth of information on age and past climatic conditions. This is particularly significant in a southern African context, where

Age in Myr	Epoch	Southern Africa	East Africa	North Africa
4	Early PLIOCENE		Kanapoi	
5		Langebaanweg (Varswater Fm)		Hamada Damous
6	Late MIOCENE		Sahabi	Lothagam 1
7			Mpesida	
8				
9				
10				
	Middle		Ngorora	Beglia Fm
			Fort Ternan	Beni Mellal
15		Arrisdrift	Maboko	
				Gebel Zelten
		? ↑ Langental, Plant 4, Grillental, Fiskus, Elisabethfeld ↓ ?	Moruorot	
			↕ Rusinga	
20	Early			

Figure 1. Some Miocene and Early Pliocene vertebrate localities in Africa. Adapted in part from H.B.S. Cooke (1978b) and Pickford (1981)

reliable dating information is often lacking in continental situations, and where the nature of palaeoenvironments is generally not well known.

The correlation of the continental and marine sequences is based on the cycles of sea-level change recognized by Vail *et al.* (1977) (see also Vail & Hardenbol 1979 and Kerr 1980). The effects of the sea-level changes were equally pronounced in coastal and continental shelf environments, and they provide a generally accepted global geochronological framework to which the west coast fossil occurrences can be related (Hendey 1981a, b).

The most significant of these occurrences are located in the southern Namib Desert, where they date from the earlier part of the Miocene, and in the vicinity of Langebaanweg in the Cape Province, where the Early Pliocene record is particularly good (Figure 2).

MIOCENE OCCURRENCES IN THE SOUTHERN NAMIB DESERT

During the Oligocene there was a major marine regression, during which local sea-level fell several hundred metres below that of the present (Siesser & Dingle 1981). There was little, if any, accumulation of sediment on the southern African west coast at this time. Sea-level began rising again towards the end of the Oligocene, and by the Early Miocene the present coastline was at, or near, sea-level. Although sediments of Early Miocene age were laid down over wide areas, it is only in the southern Namib Desert that significant fossil vertebrate occurrences of this age have been recorded.

The earliest of these occurrences are in fluviatile and lacustrine deposits near the coast between Lüderitz and Bogenfels. Stromer (1926 and earlier papers) described assemblages of fossils from Langental, Elisabethfeld and a borehole near the abandoned Plant 4 of the Kolonial Bergbau Gesellschaft. A little later, Hopwood (1928) described more material from the same area. The Namibian Mammal Age of southern Africa was defined on the basis of the material recorded by Stromer and Hopwood (Hendey 1974a, b). More recently, potentially important occurrences were recorded at Fiskus and Grillental, but to date little material has been recovered from these localities (Greenman 1966; Hamilton & Van Couvering 1977).

Although the recorded vertebrate material from this area is all believed to be of Early Miocene age, with a suggested age of 17 to 17.5 Myr (Pickford 1981), the localities from which it is derived are scattered over a wide area in differing geological contexts, and the assemblages are not necessarily contemporaneous. The individual assemblages are too small for precise faunal dating purposes, radiometric age determinations are not possible, and detailed information on the geological history of the deposits concerned is not available. There is, however, some faunal evidence which suggests that the Elisabethfeld occurrence is slightly older than that from Langental (see below).

The fossils, which are generally fragmentary, are apparently nowhere common, and those *in situ* are difficult to recover (Hamilton & Van Couvering 1977). Furthermore, access to the area is restricted because of the presence of diamondiferous deposits, a factor which compounds the difficulties of fieldwork in one of the most arid and desolate regions in the world. An appreciation of the fossils is further hampered by the fact that part of the original collection described by

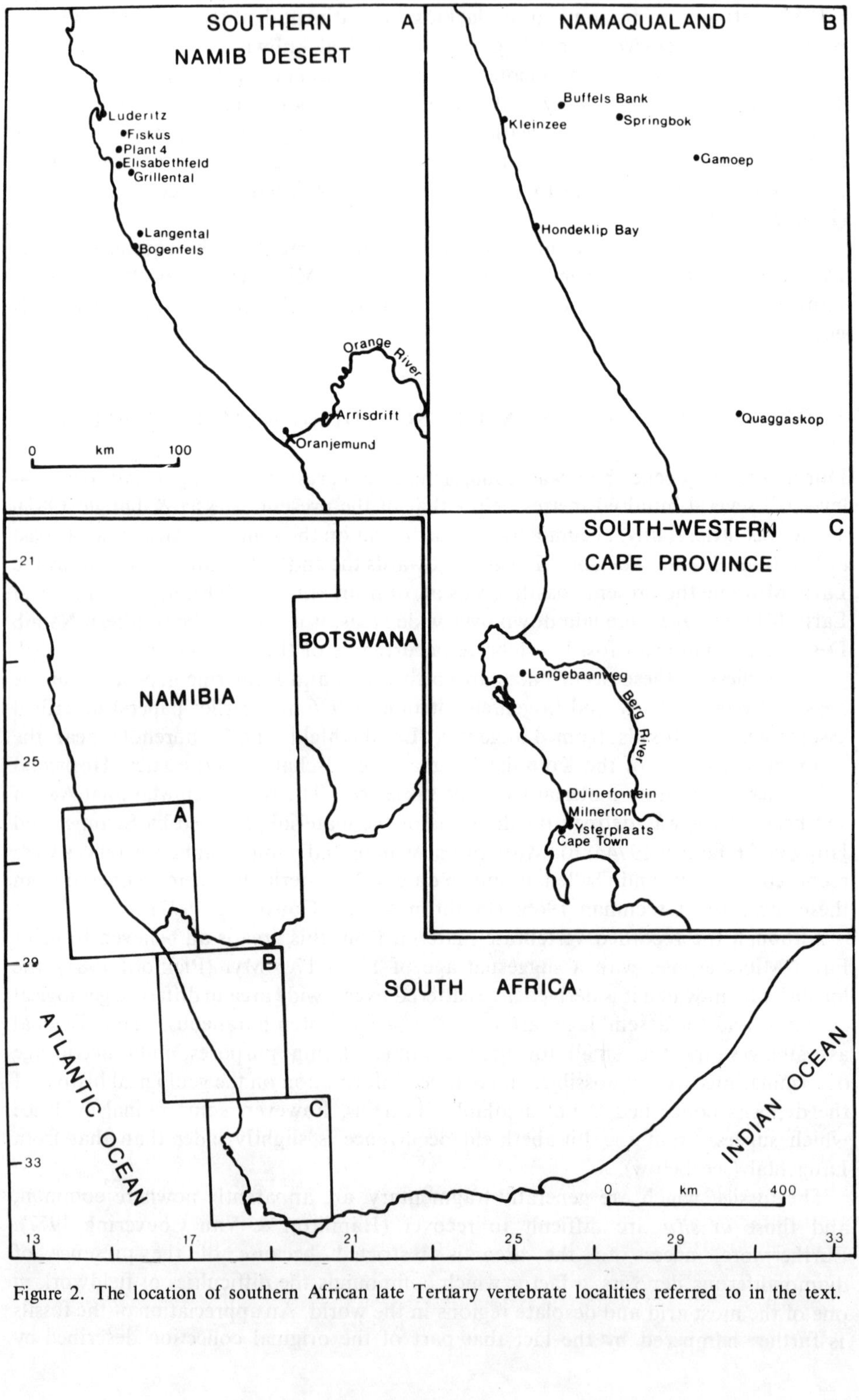

Figure 2. The location of southern African late Tertiary vertebrate localities referred to in the text.

Stromer (1926) was lost during the Second World War, while localities are not given for the material recorded by Hopwood (1928) and Hamilton and Van Couvering (1977). The amount of undescribed material available is limited.

In spite of these problems, the Lüderitz-Bogenfels occurrences are very important, since they provide the only record of the nature of southern Africa's terrestrial vertebrate fauna during the Early Miocene. In general, there are striking resemblances to the contemporary, but better known, faunas of east Africa, and some inferences may therefore be drawn from the latter. In addition, it is clear that the full potential of the Lüderitz-Bogenfels occurrences has yet to be realized.

Of slightly younger age (early Middle Miocene—*c.* 15 Myr), but still representing the Namibian Mammal Age, is the assemblage of fossils from Arrisdrift on the Orange River (Corvinus & Hendey 1978; Hendey 1978b). These fossils occur in river terrace deposits that are correlated with the +33m marine terrace on which the coastal mining town of Oranjemund is built (Hendey 1981a). Indications are that at the time that these terraces were developed, the global Early to Middle Miocene marine transgression had risen above present sea-level and had inundated the southern African western coastal margin. While the Arrisdrift fossiliferous river terrace and the +33m marine terrace were probably developed at the peak of this transgression, the actual elevation reached is not known, since the coast north of the Orange River is downwarped and the features in question are clearly well below their Miocene elevation. Marine deposits which abut the western escarpment in the more stable Namaqualand region at an elevation of 120-140m may be contemporaneous, but dating evidence is lacking and the possibility that they are of early Tertiary age cannot be excluded (Hendey 1981a).

The Arrisdrift fossils occur in a gravel lag deposit in a shallow channel incised in bedrock. This deposit is evidently extensive, but only about 180m^2 has been sampled. Nevertheless, several thousand specimens, representing at least 37 vertebrate taxa (Table 1), have been recovered, totals which exceed those of all the other Namib Miocene localities combined. Indeed, Arrisdrift ranks as one of the more prolific sources of Miocene vertebrates in Africa (see Van Couvering & Van Couvering 1976: Table 8). Unfortunately, the fossils tend to be fragmentary, and are often either poorly preserved or damaged by the recovery process. Almost all the material has yet to be described in detail. As with the more northerly of the Namib occurrences, there is clearly a potential for more productive collecting at Arrisdrift.

PLIOCENE OCCURRENCES IN THE LANGEBAANWEG AREA

The late Cenozoic succession in the vicinity of Langebaanweg is comprised of up to 120m of generally unconsolidated sediments ranging in age from Early Miocene to Holocene. The greater part of the succession is made up of the Elandsfontyn Formation that was laid down during the global Early to Middle Miocene transgression, and the Varswater Formation that was laid down during the Early Pliocene transgression. These two formations are separated by the Late Miocene 'Gravel Member', a lithostratigraphic unit formerly included in the Varswater Formation, but which is to be accommodated in the redefined 'Saldanha Formation' (J. Rogers, pers. comm.) (Figure 3).

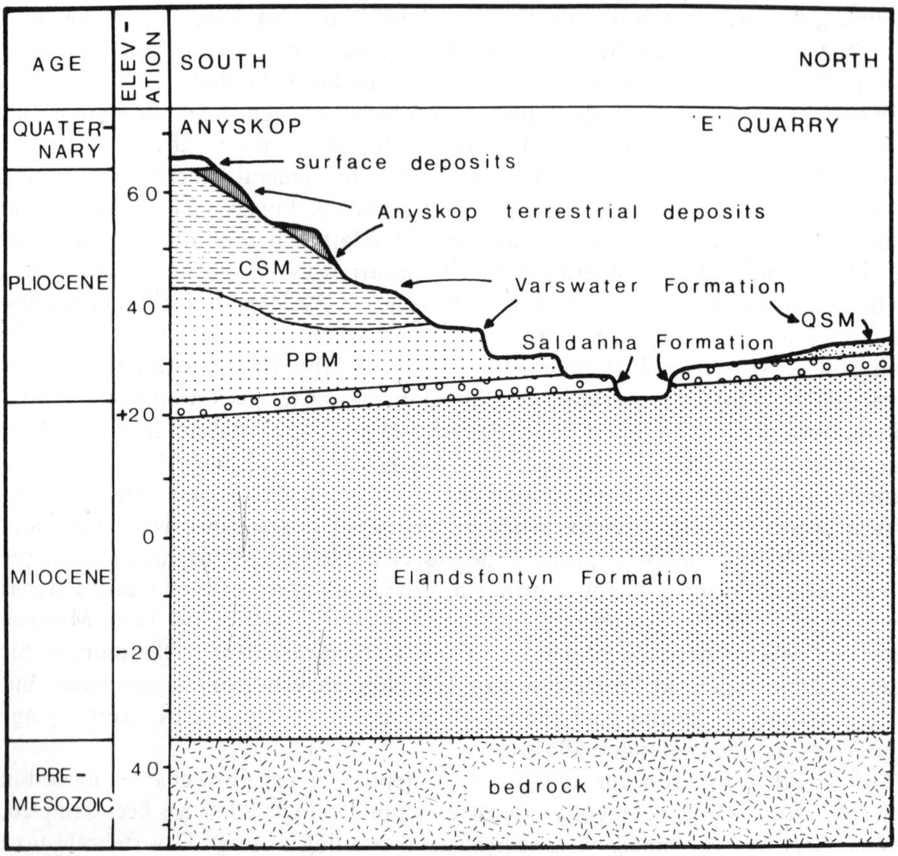

Figure 3. The geological succession at 'E' Quarry, Langebaanweg.

Fossils are recorded from all elements in the succession, but vertebrates are most common in the Varswater Formation. This formation is in fact the source of one of the largest and most diverse assemblages of late Tertiary vertebrates known anywhere in Africa. The best exposures of the Varswater Formation are in the phosphate mine, 'E' Quarry, and since 1965 hundreds of thousands of specimens, ranging from isolated teeth and bones to partial skeletons, have been collected at this site. Although much material has yet to be studied, nearly 200 invertebrate and vertebrate taxa, 80 of which are mammals, have already been recorded (Table 2; see also Hendey 1981a: 45-53). The 'E' Quarry deposits are so rich a source of information on the local fauna of the Early Pliocene, that they to some extent compensate for the paucity of the southern African record for the preceding 10 Ma or more. This fauna is the type for the Langebaanian Mammal Age of southern Africa (Hendey 1974a, b).

The pre-Varswater Formation deposits at Langebaanweg have been the source of few terrestrial vertebrate remains, but plant and marine fossils are common at certain levels. Post-Varswater Formation deposits comprise only a relatively small part of the succession as a whole, but they include two notable vertebrate fossil occurrences. They are the terrestrial sands overlying the Varswater Formation south and west of 'E'

Quarry ('Anyskop terrestrial deposits'), and the fluviatile deposits in the mined-out Baard's Quarry, 2km east of 'E' Quarry (Hendey 1978a, 1981a; Klein, this volume). Both these occurrences apparently date from the Late Pliocene.

The Varswater Formation is comprised of three main lithostratigraphic units, the Quartzose Sand, Pelletal Phosphorite and Calcareous Sand Members, which are abbreviated respectively as the QSM, PPM and CSM.

The QSM is a truncated complex of deposits laid down adjacent to an estuary, with a maximum development of about 2m. At the time that it was laid down, a precursor of the present Berg River passed south-westwards along the southern margin of 'E' Quarry, the QSM exposures having been laid down north of the river estuary and the sandbar which separated it from the sea. Three facies are represented, each with characteristic sediments and fossils assemblages. Much of this unit is a floodplain deposit (QSM I), comprised of a generally fine-grained quartzose sand, which in places is highly fossiliferous, with remains of terrestrial vertebrates predominating. Closer to the estuary is a salt marsh deposit (QSM II), comprised of carbonaceous sand and clay, which contain abundant fossil pollens and vertebrate remains. The third facies is a tidal flat deposit (QSM III), with fine-grained sediments incorporating a very rich, mainly invertebrate, fauna. A high proportion of the vertebrates from QSM I and II are believed to represent the remains of animals which lived in the immediate vicinity (Hendey 1974a, 1976a, 1981a).

As the Early Pliocene transgression progressed, the lower course of the proto-Berg River shifted northwards and the marine littoral environment encroached over the QSM deposits. The conditions in this environment were such that extensive deposition of a pelletal phosphorite occurred, the resultant lithostratigraphic unit being the PPM. River channel deposits associated with deposition of the PPM were exposed in two area of 'E' Quarry. These deposits are informally designated 'PPM 3aS' and 'PPM 3aN', with the latter postdating the former, and they are the source of extensive fossil assemblages in which terrestrial vertebrates predominate. Some of these fossils were reworked from the underlying QSM, some represent animals which lived in the immediate vicinity, but most represent animals washed into the area from upstream habitats.

It is evident that the QSM, PPM 3aS and PPM 3aN were deposited successively, and that there were intervals of time between their deposition. These intervals were of sufficient duration to show up evolutionary changes in taxa common to more than one level. Nevertheless, the total period which elapsed could not have been great, perhaps no more than 0,5 Myr, and an inferred age of *circa* 5 Myr applies to all three units.

The 'E' Quarry area was completely inundated by the sea during the latter stages of the Early Pliocene transgression, and the mouth of the proto-Berg River shifted away in an easterly direction. The local accumulation of terrestrial fossils and phosphatic sediment ceased. However, calcareous sediments (the CSM) were laid down over the PPM, apparently in the form of a coastal barrier or tombola linking granite outcrops north and south of 'E' Quarry. Vestiges of this feature survived to form the core of Anyskop, south of the quarry, and proved to be of great significance since it protected the underlying PPM and QSM from subsequent erosion, thus preserving an economically and palaeontologically important body of Early Pliocene sediment.

OTHER LATE TERTIARY OCCURRENCES

Since the late Tertiary vertebrate record of the west coast region is so dominated by the Miocene occurrences of the southern Namib and the Pliocene ones at Langebaanweg, other occurrences tend to be overlooked. However, they provide some valuable complementary data, and, more importantly, indicate a potential for future investigations which could appreciably increase the region's significance in African late Tertiary studies. The additional occurrences can conveniently be dealt with in three groups.

Namaqualand coastal occurrences

Stromer (1931a, b) described a small assemblage of Early Pliocene fossils from Kleinzee at the mouth of the Buffel's River. The condition of the specimens and their geological context are reminiscent of materials from the Varswater Formation (QSM, PPM) at Langebaanweg. In addition, although relatively few species were represented at Kleinzee, at least three are conspecific with Varswater Formation species. They are *Enhydriodon africanus, Hyaenictitherium namaquense* and a viverrid (Hendey 1974a, 1978a, d). The two occurrences are probably broadly contemporaneous. The Kleinzee fossiliferous deposits have long since been mined away, and part of the assemblage was lost during the Second World War. This occurrence is, therefore, of little more than historical interest.

A few vertebrate fossils were collected from the farms Koingnaas and Somnaas, near Hondeklip Bay, during a diamond prospecting operation a decade ago. They were from deposits associated with an old course of the Swartlintjies River. There are also physical and taxonomic similarities between these specimens and ones from the Varswater Formation, although the deposits from which they are derived may be younger (?Late Pliocene) (Hendey 1981a). It is virtually certain that more material awaits discovery, but there are practical difficulties which could prevent further collecting in this area.

The Langebaanweg, Kleinzee and Hondeklip Bay occurrences suggest that other areas on the west coast where ancient rivers met the sea may in future yield fossils of Pliocene age.

There are two little known vertebrate occurrences along the Namaqualand escarpment which are included under the present heading because they indicate the existence of an ancient shoreline in this position. They are at Buffel's Bank on the Kammagas Reserve west of Springbok, and on the farm Quaggaskop north of Vanrhynsdorp. They date from either the climax of the Early to Middle Miocene transgression, or an early Tertiary one. Both occurrences require investigation.

Namaqualand inland occurrences

The interior plateau of Namaqualand is a typical semi-arid environment of inselbergs with intervening sediment-filled basins. The latter sediments date back to the Mesozoic, but much is of Tertiary age, and there have been occasional reports of late Tertiary vertebrate fossils in this context. For example, Haughton (1932) recorded the partial dentition of an *Hipparion ('Notohipparion' namaquense)* from a well at

Areb, 65km east of Springbok. This specimen probably dates from the Pliocene.

An occurrence of greater potential is at Bosluispan near Gamoep, where the few specimens salvaged to date suggest deposits of Miocene age. The fossils occur in diamondiferous river channel deposits, the river in question being an extinct tributary of the Orange River, 100km further north (H. Jenner-Clark, pers. comm.).

Other potentially important deposits are the infillings of calderas in the Gamoep area, such as the one at Banke which has yielded plant and frog fossils (see Estes 1977). These fossils came from an excavation near the centre of the caldera. The apparently unexplored perimeters of the calderas are a likely source of terrestrial vertebrate remains. Although an early Tertiary age for the caldera infillings has been suggested, they may well be younger.

It is clear that Namaqualand, both inland and on the coast, is a potentially important source of Tertiary vertebrates, and that opportunities for collecting such materials have been lost.

South-western Cape occurrences

Apart from the occurrences at Langebaanweg, late Tertiary vertebrates have been recorded in the south-western Cape from Ysterplaats, Milnerton and Duinefontein. Marine vertebrates predominate, but some remains of terrestrial species are represented in existing collections.

The Ysterplaats occurrence was briefly exposed by building operations during 1973, and was described by Tankard (1975). The only published account of vertebrate fossils from this locality is that on the penguins by Simpson (1973).

The Milnerton fossils are known only out of their geological context, occurring as isolated and usually fragmented specimens washed up on the local beach. They are apparently derived from a north-westerly extension of the deposits encountered at Ysterplaats, and once again marine forms predominate. Only one apparently Tertiary terrestrial vertebrate (a gomphothere) is recorded. The Milnerton and Ysterplaats deposits are believed to be of Miocene age (Simpson 1973; Tankard 1975).

The Duinefontein Tertiary deposits were initially detected in boreholes, but were subsequently exposed during the construction of the Koeberg Nuclear Power Station. These deposits, like those at Milnerton and Ysterplaats, are of marine origin, and they are probably a northward extension of the latter. Apparently a vast number of vertebrate fossils, mainly shark's teeth and whale bones, from this occurrence have passed into private collections. Once again the only published account of the Tertiary vertebrates from this locality is one on the penguins by Simpson (1979).

THE MIOCENE FAUNA

The fauna from the Early Miocene localities in the southern Namib Desert (Stromer 1926; Hopwood 1929; Hamilton & Van Couvering 1977) is reminiscent of the much better known 'Rusinga-like' fauna of east Africa (see Van Couvering & Van Couvering 1976). The mammals include a variety of small forms (rodents, lagomorphs, insectivores), a creodont, a gomphothere, a hyrax, a rhinoceros, and several artiodactyls. Primates are noticeably absent. It remains to be seen if this absence is real rather than just a sampling deficiency, since the recorded fauna clearly represents only a part of the living fauna of that time. All the available material has been the subject of a recent, but as yet unpublished, study by the late W.R. Hamilton.

The early Middle Miocene fauna from Arrisdrift (Hendey 1978c) represents a period for which the African fossil vertebrate record is comparatively poor. Broadly contemporaneous faunas include those from Maboko in Kenya (Andrews *et al.* 1981), and Gebel Zelten in Libya (Savage & Hamilton 1973). As with the Namib Early Miocene assemblages, there are resemblances to the east African 'Rusinga-like' fauna, but it also includes elements which indicate a slightly younger age. The Arridrift fauna is, however, more primitive in character than that from Fort Ternan in Kenya, which is the best known of Africa's Middle Miocene faunas (see Shipman *et al.* 1981).

The Arrisdrift mammalian assemblage is similar to those from the other Namib localities in being comprised of several small mammals, a gomphothere, a hyrax, a rhinoceros, and artiodactyls. Primates are again conspicuously absent, and in this instance the absence is more likely to be real since the assemblage is fairly comprehensive (Table 1). The Arrisdrift assemblage is, however, significantly different in some respects. At least a few of these differences can be ascribed simply to the larger size of the Arrisdrift assemblage. For example, the aardvark and deinothere, recorded at Arrisdrift on the basis of only a few specimens, may well have had counterparts in the region during the Early Miocene. Of greater significance is the fact that some of the species represented at Arrisdrift are clearly more advanced than ones in the other Namib assemblages. This is perhaps most obvious in the case of the ruminant artiodactyls, but it is also striking that at Arrisdrift only non-creodont carnivores are recorded.

Although the Arrisdrift assemblage is comparatively large, it does have its disadvantages. Small mammals, which are the best known group from the other Namib localities, are clearly under-represented, evidently for taphonomic reasons. In addition, many specimens are fragmentary and/or poorly preserved, and there are only a few, notably the hyrax and crocodile, for which sample sizes are satisfactorily large.

The Namib Miocene assemblages are distinguished from contemporary ones elsewhere in Africa by the presence of some apparently endemic taxa. The endemism is, however, less marked than early studies had suggested. For example, *Prohyrax* was until recently known only from the Namib, but it is now also recorded in Kenya and Libya (Meyer 1978). In addition, a re-evaluation of African Miocene rodents has revealed that some Namib 'endemics' actually belong to more widespread taxa (Lavocat 1973, 1978).

One of the forms which apparently is endemic to the Namib is the large myohyracine, *Protypotheroides beetzi*. Although they belong to the order Insectivora, the myohyracines were herbivores, and have high-crowned lagomorph-like cheekteeth (Patterson 1965; Butler 1978). *P. beetzi* is the larger of the two recorded species, the other being *Myohyrax oswaldi*, which is more common and which occurs in both the Namib and east Africa. The myohyracines probably occupied an ecological niche similar to that of two other Namib endemics, the ochotonid lagomorph, *Austrolagomys inexpectatus*, and the pedetid rodent,

Table 1. The early Middle Miocene vertebrates from Arrisdrift (Prospect pit 2 of drill-line AD 8).

Taxon	
OSTEICHTHYES (fishes)	At least one species
AMPHIBIA (frogs)	At least one species
REPTILIA	
Squamata (snakes, lizards)	At least two species
Crocodilia (crocodiles)	*Crocodylus* sp.
Chelonia (tortoises, turtles)	At least one species
AVES (birds)	At least three species
MAMMALIA	
Insectivora	
Chrysochloridae (golden moles)	Gen. & sp. indet.
Macroscelididae (elephants shrews, myohyracines)	*Myohyrax* cf. *oswaldi*
Soricidae (shrews)	Gen. & sp. not det.
Insectivora not det.	At least two species
Tubulidentata (aardvarks)	*'Orycteropus'* cf. *minutus*
Carnivora	
Amphicyonidae (bear-dogs)	cf. *Amphicyon major*
	Amphicyon cf. *steinheimensis*
?Ursidae (bears)	?Hemicyoninae indet.
Mustelidae (weasels etc.)	?*Ischyrictis* sp.
Viverridae (mongooses etc.)	Gen. & sp. indet.
Carnivora not det.	At least two species
Hyracoidea (hyraxes or dassies)	*Prohyrax* sp.
Deinotherioidea	*Prodeinotherium hobleyi*
Proboscidea (elephants, mastodons etc.)	
Gomphotheriidae	Gen. & sp. not det.
Perissodactyla	
Rhinocerotidae (rhinos)	*Dicerorhinus* sp.
Artiodactyla	
Suidae (pigs)	Gen & sp. not det.
	Lopholistriodon moruoroti
Tragulidae (chevrotains)	*Dorcatherium* cf. *pigotti*
Climacoceridae (deer-like ruminants)	*Climacoceras* sp.
Bovidae (antelopes etc.)	Gen. & sp. not det.
Pecora	Gen. & sp. not det.
Lagomorpha	
Ochotonidae (pikas)	*Kenyalagomys* sp.
Rodentia	
Pedetidae (spring-hares)	*Parapedetes* sp.
Rodentia not det.	At least four species

Parapedetes namaquensis. The latter two species are from Elisabethfeld. An ochotonid and pedetid are also recorded from Arrisdrift, but in this instance the genera concerned are also known from further afield.

Parapedetes namaquensis is otherwise remarkable because the holotype is one of the most complete specimens of an African Miocene rodent yet described. The rodents in the Early Miocene assemblages from the Namib are usually represented by very fragmentary material. They are, however, important in having provided one of the earliest indications of the nature of an important phase in the evolution of

African rodents. Their significance has been re-affirmed by the recent studies of Lavocat (1973, 1978). Apart from *P. namaquensis*, the Namib rodent assemblages also include the endemic taxa, *Pomonomys dubius, Phiomyoides humilis* and *Paracryptomys mackennae*. The latter, together with the more widely distributed *Bathyergoides neotertiarius*, are early representatives of the fossorial Bathyergoidea, an endemic African group to which the living mole-rats belong. Another group of rodents which is comparatively well represented in the Namib Miocene faunas is the Thryonomyoidea. Like the Pedetidae, they are known as fossils in southern Eurasia, but they have an essentially African history. They were an extremely successful group during the Early Miocene (Lavocat 1978), but were subsequently largely superceded by other groups, and only the cane rates (*Thryonomys* spp.) are common today.

The hypsodont taxa amongst the small herbivorous mammals have been cited as an indication of the presence of savanna vegetation in the Namib during the Early Miocene (e.g. Hopwood 1928). It is, however, unlikely that true savannas did exist, since the larger herbivores are all brachyodont and probably woodland and forest forms. Although previously ubiquitous forests were probably beginning to open up at this time, it was not until the Late Miocene and Pliocene that larger mammals evolved to fill the niches offered by savannas and grasslands (see below).

Except for some carnivores which will be mentioned below, the remaining small mammals recorded from the Namib Miocene are all insectivores, and all are from Arrisdrift. Like the rodents from this locality, they are unstudied.

The next size category up from the small mammals includes representatives of the Hyracoidea, Suidae, Tragulidae and Bovidae. They are brachyodont herbivores of relatively small size that probably inhabited the undergrowth of forests adjacent to rivers and lakes, the preferred habitat of living tragulids.

With the exception of the Arrisdrift hyrax, these animals are poorly represented. The tragulids appear to be forms which were both common and widespread in Africa during the Early Miocene. The same applies in the case of one of the Arrisdrift pigs (*Lopholistriodon moruoroti*), and one from Langental (*Xenochoerus africanus*). Two other pigs from these localities have yet to be identified. The single bovid recorded is a poorly known form from Arrisdrift, and is one of the earliest known representatives of its family. Although the bovid radiation got underway during the Middle Miocene, they were only to reach their acme in Africa during the Pleistocene.

By the Early Miocene the hyracoids had already passed their period of greatest success, and in the Namib only single species are recorded from Langental and Arrisdrift. The Langental species is *Prohyrax tertiarius*, while that from Arrisdrift is an apparently new species of the same genus. *Prohyrax* is now known from elsewhere in Africa. A specimen from the Early Miocene of Loperot in Kenya is said to be more primitive than *P. tertiarius* (Meyer 1978), and it presumably could represent the stock from which the latter was derived. If such a relationship did indeed exist, then the Langental fauna may be younger than has hitherto been supposed. *Prohyrax* is otherwise recorded from Moruorot (*c.* 17-17.5 Myr) and Gebel Zelten (*c.* 16 Myr) (see Meyer 1978; Pickford 1981), and from Arrisdrift (*c.* 15 Myr). It is only at the latter locality that it is common.

The combination of these records with the fact that the most commonly occurring hyracoid in the 'Rusinga-like' faunas of east Africa is not a *Prohyrax*, but *Pachyhyrax championi*, suggests that although *Prohyrax* had its origins during the

Early Miocene, it was only during the early Middle Miocene that it became common, widespread and the dominant representative of its group.

The relationships of *Prohyrax* have yet to be settled. Meyer (1978) believes it is the earliest known member of the Procaviidae, the family to which living hyracoids belongs. The earliest undoubted procaviid is the Early Pliocene species from Langebaanweg (see below), and there is no obvious indication of a direct phylogenetic connection between it and the Arrisdrift *Prohyrax*, which is apparently the youngest representative of the genus. It is much more likely that the latter is an early representative of the subfamily Pliohyraciinae (Hendey 1978c), a group best known from large and specialized Late Miocene and Pliocene species, which were the most widely distributed of all hyracoids. Intermediate between *Prohyrax* and later pliohyracines is *Parapliohyrax*, a late Middle Miocene form from Beni Mellal in Morocco (Lavocat 1961) and Ngorora in Kenya (Bishop & Pickford 1975). According to this interpretation, *Prohyrax* and *Parapliohyrax* are the hyracoids characteristic of the African Middle Miocene. The origin of the Procaviidae is a matter which has still to be resolved.

Medium-sized ruminants form the next size category of herbivores in the Namib Miocene record. Stromer (1926) recorded two species amongst the material available to him, namely, *Propalaeoryx austroafricanus* from Elisabethfeld, and cf. *Strogulognathus* sp. from Langental. Hamilton & Van Couvering (1977) placed Stromer's material, and more collected by themselves, in a single species, *P. austroafricanus*. A recent re-examination of this material, together with that from Arrisdrift, has suggested that there may indeed be two lineages represented. The Elisabethfeld *P. austroafricanus* is here regarded as the most primitive form represented, with two species being represented at both Langental and Arrisdrift, the latter being the most advanced and probably descended from the Langental species. *P. austroafricanus* could be ancestral to either, or both, of the Langental-Arrisdrift lineages. The Langental cf. *Strogulognathus* sp. is tentatively referred to *Climacoceras*, a genus identified at Arrisdrift on the basis of its deer-like 'antlers' (ossicones). The nature of the frontal appendages, if any, of representatives of the second Langental-Arrisdrift lineage is not known, and the material is listed as an unidentified pecoran in Table 1.

Climacoceras was formerly included in the Palaeomerycidae, but Hamilton (1978) has placed it in a new family, the Climacoceridae. In his comprehensive review of African Miocene giraffoids, Hamilton (1978) excluded *Propalaeoryx* from consideration since the known material could not be interpreted using the cladistic methodology employed in the study. *Propalaeoryx* is here regarded as a primitive giraffoid, and, since it is possibly ancestral to *Climacoceras*, it is tentatively included in the Climacoceridae.

In view of the fragmentary nature of the Namib ruminant material, the small sample sizes, and the fact that the animals concerned represent primitive stages in the evolution of the groups to which they belong, it may not be possible to reach consensus on identification.

The large mammals in the Namib Miocene assemblages are all herbivores. The families represented are the Gomphotheriidae, Deinotheriidae, Rhinocerotidae and Anthracotheriidae. The deinothere from Arrisdrift, *Prodeinotherium hobleyi*, was described by Harris (1977), while the gomphothere and rhinoceros from this locality

Figure 4. Reconstructions of Miocene mammals from Arrisdrift in the southern Namib Desert. Clockwise from top left: Gomphotheriidae gen. & sp. not det.; *Climacoceras* sp.; cf. *Amphicyon major; Dorcatherium* cf. *pigotti; Prodeinotherium hobleyi*.

have received only superficial treatment, with the latter having been identified as *Dicerorhinus* sp. (Hendey 1978c). The rhinoceros from the more northerly localities represents a different species, *Brachypotherium heinzeilini* (Heissing 1971; Hamilton & Van Couvering 1977). The gomphothere and anthracothere from these localities are undescribed.

The only identified carnivore recorded from the Early Miocene localities is a small creodont from Elisabethfeld. By contrast, there are at least half a dozen carnivores represented at Arrisdrift. Although the available material is fragmentary and incompletely identified, it is clear that it belongs to members of the order Carnivora, rather than the more primitive Creodonta.

The only significant addition to the previously recorded carnivore material from Arrisdrift (Hendey 1978c) is a maxillary fragment of a small viverrid. Other remains of small carnivores may represent the same species, but as with much of the carnivore material from this locality it is likely to remain incompletely identified. The best specimens are incomplete hemimandibles belonging to a mustelid,? *Ischyristis* sp., and medium- and large-sized amphicyonids. The latter group (the bear-dogs) was widespread in Africa, Eurasia and North America during the Miocene, but were superceded by true bears (Ursidae) before the end of this epoch.

The only other mammal recorded from the Namib Miocene localities is an aardvark from Arrisdrift. It is a small species which may be conspecific with the east African Early Miocene *'Orycteropus' minutus* (Pickford 1975; Patterson 1978).

Non-mammalian vertebrates include birds, reptiles and amphibians, the only well represented species being a crocodile from Arrisdrift.

As indicated earlier, it is possible to draw inferences from the richer Miocene faunas of east Africa to fill out the rather sparse vertebrate record for this epoch in southern Africa. For example, it is certain that carnivores are grossly under-represented at the Namib Early Miocene localities. It is probable that, as in east Africa, there were several species of creodont of varying size occupying the available carnivore niches. The presence of amphicyonids and viverrids at east African Early Miocene localities and at Arrisdrift suggests that these groups may also have been present in southern Africa during the Early Miocene. It is also likely that there was a greater variety of small mammals (insectivores and rodents), and of small- to medium-sized herbivores in southern Africa than the existing Namib record indicates.

The apparent absence of primates in the Namib is an intriguing aspect of the Early to Middle Miocene record there. This very successful group of mammals is comparatively well represented in contemporary east African occurrences, and it would be truly remarkable if they were indeed not present in southern Africa. It was suggested elsewhere that a zoogeographic barrier between south and east Africa may have impeded the southward dispersal of primates during the Early Miocene (Hendey 1978c). This hypothesis seems improbable in view of the wide variety of other taxa which the two regions have in common. The problem may yet be resolved by further collecting at the localities in question.

The vertebrate fossil record in southern Africa is virtually non-existent for the period between 15 Ma and 5 Ma, that is, almost the entire Middle and Late Miocene. However, some indication of the pattern of vertebrate evolution during this period can once again be inferred from the record elsewhere in Africa. In addition, the exceptionally rich Early Pliocene assemblage from Langebaanweg (see below) provides an indication of the nature of the local fauna during the latter part of the Miocene.

The Middle to Late Miocene was a period when Africa's mammalian fauna underwent dramatic changes (see Maglio 1978). Some archaic groups declined in importance (e.g. Tragulidae), and others became extinct (e.g. Creodonta, Amphicyonidae, Anthracotheriidae, Climacoceridae). Some groups appeared for the first time, having either evolved locally or migrated from Eurasia (e.g. Hyaenidae, Equidae, Hippopotamidae). The latter, and other groups (e.g. Giraffidae, Bovidae), underwent radiations which changed the entire character of Africa's fauna.

The primates were as much involved with these changes as other groups of mammals. The Lorisidae and Pongidae declined in importance, the Hylobatidae became extinct in Africa, while the Cercopithecidae radiated (see Walker 1978; Simons & Delson 1978; Simons, Andrews & Pilbeam 1978). Most significantly though, this was the period when the Hominidae were differentiated (Simons & Pilbeam 1978), and it is thus the one to which man owes his origins.

The most important characteristic of the evolution of African mammals, and other vertebrates, was the rise of open country species at the expense of forest forms. This is a direct reflection of the changing environmental conditions of the epoch (see below).

THE EARLY PLIOCENE FAUNA

The fossils of the Varswater Formation at Langebaanweg provide an unparalleled record of the nature of Africa's vertebrate fauna during the Early Pliocene. This was a critical period of transition between the archaic world of the Tertiary and that of the Quaternary. It was the period when the Ethiopian Faunal Realm of Subsaharan Africa began to take on its distinctive character, and when the physical features, climate and vegetation of the continent were becoming established in the patterns which were to characterize the last 3-4 Myr of geological time. It was the last period during which the Hominidae were to have a status comparable to that of other mammals. The next phase in the faunal history of Africa saw the advent and rapid dispersal of bipedal, intelligent, tool-making man, whose profound impact on the world was thereafter to escalate continuously.

Resemblances between the Namib Miocene and Varswater Formation mammalian faunas are slight, and communality apparently exists only at family and higher taxonomic levels. The latter fauna had already largely taken on the composition of the faunas which were to characterize Africa during the Quaternary. Only one of the fourteen mammalian orders represented today in Africa, or the adjacent oceans, is not recorded from the Varswater Formation (i.e. the Sirenia). Communality is progressively reduced through lower taxonomic ranks, and it is only at the species level that the Varswater Formation fauna is entirely distinct. This fauna is unmistakeably 'African' in character (Table 2).

There are, however, 'non-African' elements. They include representatives of the families Ursidae, Phocidae and Tayassuidae, and the tribes Gulonini, Boselaphini and Ovibovini. No members of these groups are resident in Subsaharan Africa today. However, all had less restricted distributions during the late Tertiary, and their presence at Langebaanweg, which implies a wide distribution in Africa, is not necessarily unexpected. Some groups, such as the Agriotheriini, Gomphotheriidae and Sivatheriinae, are 'archaic', being unknown or relatively poorly represented in Quaternary contexts. They are, however, far outnumbered by groups which

Table 2. The Early Pliocene mammals of the Varswater Formation, Langebaanweg (Hendey 1981a, see this reference for lists of lower vertebrates and invertebrates).

	QSM	PPM 3aS	PPM 3aN
ORDER INSECTIVORA			
Family Chrysochloridae (golden moles)			
Chrysochloris sp.	×	×	×
Family Soricidae (shrews)			
Myosorex sp.	×		
Suncus sp.	×		
Soricidae gen. & sp(p) not det.	×	×	×
Family Macroscelididae (elephant shrews)			
Elephantulus sp.	×	×	×
ORDER CHIROPTERA (bats)			
Family Vespertilionidae			
Eptesicus sp.	×		
ORDER PRIMATES			
Family Cercopithecidae (monkeys etc.)			
Gen. & sp. indet.	×		
ORDER PHOLIDOTA (pangolins)			
Phataginus sp.	×		
ORDER TUBULIDENTATA (aardvarks)			
Gen. & sp. not det.	×		×
ORDER CARNIVORA			
Family Canidae (foxes, jackals etc.)			
Gen. & sp. not det. (?aff. '*Canis*' *brevirostris*)		×	?
Vulpes sp. (fox)		×	×
Family Ursidae (bears)			
Agriotherium africanum		×	×
Family Mustelidae (weasels, martins etc.)			
Plesiogulo monspessulanus (wolverine)	×	?	
Mellivora benfieldi (honey badger)		×	×
Enhydriodon africanus (otter)		×	×
Family Phocidae (seals)			
Homiphoca capensis	×	×	×
Family Viverridae (mongooses etc.)			
'*Viverra*' *leakeyi* (?aff. *Civettictis*) (civet)	×		×
Viverrinae gen. & sp. not det. (?aff. *Pseudocivetta*) (civet)	×	×	×
Genetta sp. (genet)	×		
Herpestes spp. A, B (mongooses)	×	×	
Herpestinae spp. C, D, E (mongooses)	×		
Herpestinae not det.		×	×
Family Hyaenidae (hyaenas)			
'*Adcrocuta*' *australis*	×	?	?
Ictitherium preforfex		×	×
Hyaena abronia (striped hyaena)	×	×	×
Hyaenictitherium namaquense	×		
Euryboas sp. (hunting hyaena)	×	×	×
Hyaenidae sp. E		×	
Hyaenidae not det.		×	×

(*Contd.*)

Table 2 (*Contd.*)

	QSM	PPM 3aS	PPM 3aN
Family Felidae (cats)			
'Machairodus' sp. (sabre-tooth)	×		
Homotherium sp. (sabre-tooth)	×	×	?
Felis sp. (wildcat)	×		
Felis aff. *issiodorensis* (lynx)	×	×	×
Felis obscura		×	
Dinofelis diastemata (false sabre-tooth)	×	×	×
Felidae not det.		×	×
CARNIVORA NOT DET.			
Gen. & sp. not det. (Canidae or Viverridae)	×		
Gen. & sp. not det. (?Procyonidae)	×		
Gen. & sp. not det. (?Lutrinae)	×		
ORDER PROBOSCIDEA (elephants, mastodons etc.)			
Family Gomphotheriidae			
Anancus sp.	×	×	
Family Elephantidae			
Mammuthus subplanifrons	×	?	×
ORDER HYRACOIDEA (hyraxes or dassies)			
Family Procaviidae			
Procavia cf. *antiqua*	×		?
ORDER PERISSODACTYLA			
Family Equidae (horses)			
Hipparion cf. *baardi*	×	×	×
Family Rhinocerotidae (rhinos)			
Ceratotherium praecox	×	×	
ORDER ARTIODACTYLA			
Family Tayassuidae (peccaries)			
Pecarichoerus (or *Barberahyus*) *africanus*		×	×
Family Suidae (pigs)			
Nyanzachoerus cf. *pattersoni* (or *kanamensis*)	×		
Nyanzachoerus cf. *jaegeri*		×	
Family Hippopotamidae (hippos)			
Gen. & sp. not det.		×	×
Family Giraffidae (giraffes, okapi etc.)			
Sivatherium hendeyi	×	×	×
Palaeotragus cf. *germaini*			×
Giraffa sp.	×	×	×
Family Bovidae (buffaloes, antelopes etc.)			
Tragelaphus sp. A (nyala-like)	×	×	×
Tragelaphus sp. B (nyala-like)			×
Mesembriportax (or *Miotragoceros*) *acrae* (kudu-like relative of the nilgai)	×	×	×
Simatherium demissum (buffalo)	×	×	×
Kobus subdolus (kob-like)		×	×
Kobus sp. B (kob-like)			×
Damalacra neanica (hartebeest-like)		×	×
Damalacra acalla (hartebeest-like)	×	×	×
Raphicerus paralius (steenbok)	×	×	×

(*Contd.*)

Table 2 (*Contd.*)

	QSM	PPM 3aS	PPM 3aN
Gazella sp. (gazelle)	×	×	×
Ovibovini gen. & at least 2 spp. not det.	×	×	×
ORDER LAGOMORPHA			
Family Leporidae (hares, rabbits)			
Pronolagus sp.	×	×	×
ORDER RODENTIA			
Family Bathyergidae (rodent moles)			
Bathyergus sp.	×	×	×
Cryptomys sp.	×		?
Family Hystricidae (porcupines)			
Gen. & sp. not det. A	×		
Gen. & sp. not det. B		×	×
Family Cricetidae (rats, mice, gerbils etc.)			
Mystromys sp. A	×		
Mystromys cf. *darti*	×		
Mystromys cf. *hausleitneri*	×		
Gerbillus or *Desmodillus* sp.	×		
Dendromus sp.	×		
Steatomys or *Malacothrix* sp.	×		
Family Muridae (rats, mice)			
Aethomys spp. A, B	×		
Mus spp. A, B	×		
Rhabdomys sp.	×		
Euryotomys pelomyoides	×		
Family Muscardinidae (dormice)			
Graphiurus sp.	×		
RODENTIA NOT DET.	×	×	×
ORDER CETACEA (whales, dolphins)			
Gen. & spp. not det.	×	×	×

flourished during the Quaternary, although their Varswater Formation representatives are 'primitive' in terms of their evolutionary state, and in the combinations of taxa.

The location of Langebaanweg near the southern extremity of Africa makes it possible that the Varswater Formation fauna was at least in some respects atypical of the Early Pliocene fauna of Africa as a whole. There is a potential in such situations for the presence of endemic taxa, late-surviving representatives of lineages which were already extinct elsewhere, and early representatives of lineages which still had only limited distributions. In addition, the south-western parts of southern Africa were already climatically distinct from the remainder of Subsaharan African by the Early Pliocene (see below), and it is likely that some taxa excluded this region from their areas of distribution. Nevertheless, the Varswater Formation mammalian fauna probably is a good reflection of that of the continent as a whole during the important period near the Miocene-Pliocene boundary.

One of the most remarkable aspects of the mammalian fauna is that it must represent something approaching the entire spectrum of terrestrial, freshwater and marine forms which lived in the area at that time. It is only the Chiroptera which are obviously under-represented. Although the fossils are often fragmentary, their preservation is generally good, and the sample sizes for individual taxa are usually adequate for the purposes of identification. Incomplete identifications on the accompanying faunal list (Table 2) may be taken as an indication that the relevant material is in need of study. Much the same applies in the case of the birds, the Varswater Formation being the source of the largest assemblage of pre-Quaternary bird bones in the world (Rich 1980). Other lower vertebrates are also common.

The small mammals of the Varswater Formation have been the subject of only a preliminary study (Pocock 1976), but it is evident that this element in the fauna had already taken on a character essentially similar to that of the present. Myohyracines and thryonomyoids, which were so characteristic of Africa's Early to Middle Miocene faunas, are absent. Although bathyergoids were still present, they appear to be no more than specifically distinct from those inhabiting the region today. Similarly, other rodents, as well as insectivores, a bat and a lagomorph, are closely related to forms still living in southern Africa today. The only small mammal which has so far been described in detail, *Euryotomys pelomyoides* (Pocock 1976), is of particular interest, since it is a primitive representative of a group of rodents, the vlei-rats (Otomyinae), which have since become a highly specialized and successful group in Africa.

The small herbivores form a less significant group in the fauna as a whole than their counterparts in Miocene assemblages, evidently because of a decline in the preferred habitats of such animals (e.g. the undergrowth of gallery forests). This group includes representatives of the Hyracoidea (Hendey 1976a), Tayassuidae (Hendey 1976b) and Bovidae (Gentry 1980). Although they are rodents, the Hystricidae are also included here. Tragulidae and small Suidae are not recorded, although the former did survive in forests elsewhere on the continent. The peccary is an ecological vicar of the small suids of the Miocene, and is an example of a late-survivor of a group which in Africa was characteristic of the Miocene. By contrast, the hyracoid, small bovids and porcupine are all forms which are similar to living representatives of their groups.

Medium-sized herbivores are represented in the Varswater Formation by Equidae (Hooijer 1976), Suidae and Bovidae (Gentry 1974, 1980). The equid, *Hipparion* cf. *baardi*, is a descendant of a late Middle Miocene immigrant to Africa, whose rapid dispersal through Eurasia and Africa at this time is an indication of the progressive development of grasslands at the expense of forests. *H.* cf. *baardi* belonged on a lineage which was endemic to the more southerly parts of Africa. Although *Hipparion* was to survive into the Pleistocene, it was eventually superceded by *Equus*, an immigrant from Eurasia which entered Africa towards the end of the Pliocene.

The two suids from the Varswater Formation were large members of their group, *Nyanzachoerus* cf. *pattersoni* and *N.* cf. *jaegeri*. *Nyanzachoerus* was a 'bushpig' from which a 'warthog', *Notochoerus*, was derived during the latter part of the Early Pliocene (see Harris & White 1979). This evolutionary development reflects the changing vegetation patterns in Africa at this time, with grasslands and savannas increasingly replacing woodlands and forests.

As with the small bovids referred to above, the larger bovids of the Varswater

Formation are mostly related to forms characteristic of younger faunas in Africa. However, one species, *Mesembriportax acrae* (Gentry 1974), is a late survivor of a group, the Boselaphini, which was most characteristic of the Middle to Late Miocene of Eurasia and Africa. Only two boselaphine species survive today, both occurring in India. The bovids also include two unidentified Ovibovini, a group best known from the northern continents, where they are still represented by the muskox and takin today. The bovids are a mixture of woodland and open country forms, with representatives of the latter group (e.g. the alcelaphines, *Damalacra* spp.) being much less specialized than their later relatives. This is a further indication of the environmental transition which was underway during the Early Pliocene.

The Varswater Formation fauna includes an impressive array of large herbivores. There were two proboscideans (Coppens *et al.* 1978), a rhinoceros (Hooijer 1972), a hippopotamus and three giraffids (Harris 1976). The proboscideans and giraffids were forest and woodland species, but the rhinoceros (*Ceratotherium praecox*) was an open-country grazer, and an early ancestor of the living white rhinoceros of Africa (*C. simum*). One of the proboscideans (*Mammuthus subplanifrons*) is an early representative of the true elephants (Elephantidae), but the other, a gomphothere, belongs to a group which was more characteristic of the Miocene, and which was superceded by the elephants before the end of the Pliocene.

The giraffids are of particular interest, with one representative of each of three subfamilies, namely, the Sivatheriinae, Palaeotraginae and Giraffidae. The Palaeotraginae were differentiated as a group during the Middle Miocene, and during the Late Miocene they underwent a radiation in Eurasia and Africa. By the end of this epoch their radiation was over and thereafter little is known of their history. The group has a single living survivor, the rare okapi of the central African forests. Coincident with the decline of the palaeotragines was the rise of the long-necked giraffes (Giraffidae). Apparently the decline of forests and woodlands into open woodlands and savannas reduced the preferred habitats of the palaeotragines and created new ones suited to the giraffes. The sivatheres have a long history, during which they were able to co-exist with both palaeotragines and giraffes, competing neither for the closed habitats of the former, nor for the canopy vegetation favoured by the latter.

The representation of the three giraffids at Langebaanweg is unequal. The sivathere (*Sivatherium hendeyi*) is the most common, followed in turn by the giraffe (*Giraffa* sp.) and the as yet undescribed palaeotragine (*Palaeotragus* cf. *germaini*). This suggests that the preferred habitat of the latter was the least extensive, the vegetation being more open and suited to giraffes and sivatheres. A pathological condition observed in the sivathere (Hendey 1981a), and a massive mortality amongst these animals (Klein 1981b), suggests that the local population was under stress, probably because of prevailing environmental conditions.

One of the most striking features of the Varswater Formation fauna is the diversity of carnivores represented, with a third of the recorded species of mammals belonging to this group (Hendey 1974a; and other references cited below). They ranged in size from a very small mongoose to a bear of gigantic proportions. This diversity testifies to a high biomass amongst prey species. The present situation is in sharp contrast to that which existed during the Early Pliocene, with the local sclerophyllous fynbos vegetation being capable of supporting only a relatively impoverished mammalian fauna. The high diversity of carnivores is also at least partly due to the fact that material derived from a variety of habitats is represented. These are the local marine and freshwater environments, the local terrestrial environment, and others upstream. Nevertheless, there is probably no region of comparable extent anywhere in the world today where a similar array of carnivores would be encountered. The high proportion of carnivores in the fauna is presumably an archaic pattern. The critical factor in the subsequent change in this pattern may have been the rise of the Hominidae (see Klein, this volume).

The small carnivores are almost all members of the family Viverridae (a genet and several mongooses). Although this group was already diverse, individual species show only the beginnings of the specializations which characterize Quaternary viverrids (Hendey 1974b). Other small species are a honey badger and a cat. Descendants of the former were appreciably larger and adapted to an omnivorous diet. The small mustelids of the modern African fauna have no relatives in the Varswater Formation fauna.

Medium- to large-sized carnivores are the most numerous and most diverse. There are two foxes, a wolverine, two civets, at least five hyaenas, and five members of the cat family in addition to the one referred to above. Also included in this size category are aquatic forms, namely, an otter (Hendey 1978d) and a seal (De Muizon & Hendey 1980). With the exception of the sabre-toothed cats and some of the hyaenas, these carnivores still have extant relatives.

The foxes and civets have dentitions adapted to a largely carnivorous way of life, whereas at least some later relatives have insectivorous or omnivorous specializations. The civets were apparently ecological vicars of the jackals that only appeared on the African scene later and which were to become the most ubiquitous and successful of the continent's carnivores. The living African civet is by contrast a much rarer animal with a more limited distribution. It has removed itself from competition with the jackals by becoming a solitary, nocturnal omnivore.

The relatively unspecialized character of the carnivores is well illustrated by the hyaenas, which are distinguished from their Late Miocene ancestors and from one another partly on the basis of size, with morphological differences being relatively slight. Their diversity is more characteristic of the Late Miocene than later times (Hendey 1978a). They include an early ancestor of the hunting hyaenas, which were ecological vicars of the cheetah and hunting dog during the Late Pliocene and Early Pleistocene. An ancestor of the living striped hyaena, as well as a possible ancestor of the brown hyaena, are also represented. The insectivorous aardwolf has no counterpart in the Varswater Formation fauna.

Apart from the hunting hyaena, the larger primary predators were the true and false sabre-toothed cats (Machairodontinae and Felinae), and the wolverine (*Plesiogulo*). The latter was a widely distributed in Africa and the northern continents during the Late Miocene and Early Pliocene, but the living wolverine is confined to the Arctic and Subarctic. The sabre-toothed cats were later replaced by the lion and leopard.

The bear (*Agriotherium*) is by far the largest of the Varswater Formation carnivores (Hendey 1980a). Its nearest living relative is the giant panda (Hendey 1980b), of which only a small number survive in a natural state in west-central China. *Agriotherium* was probably an ecological vicar of the large bear-dog recorded from

Figure 5. Reconstructions of Pliocene mammals from Langebaanweg in the south-western Cape Province. Clockwise from top left: *Sivatherium hendeyi; Agriotherium africanum; Homotherium* sp.; *Simatherium demissum; Mesembriportax* (or *Miotragoceras*) *acrae*.

Arrisdrift, and it is the most obviously 'archaic' form amongst the Langebaanweg carnivores. *Agriotherium* had a distribution similar to that of *Plesiogulo*, but unlike the latter it has no living decendant.

Rare elements in the mammalian fauna are an aardvark, a pangolin and a cercopithecoid primate. The latter is the most poorly represented of all the mammals. The situation in respect of primates in the Varswater Formation therefore contrasts with that at the Namib Miocene localities in that the group is represented, although it is an exceedingly rare element in the fauna. This cannot be ascribed to sampling deficiencies, and is probably due to the prevailing temperate climate having made the environment unsuitable for primates (Hendey 1981a). This factor does not apply in the case of the Namib early in the Miocene, where there still existed a tropical environment favoured by primates.

The lower vertebrates of the Varswater Formation are largely unstudied, but they were apparently already established in the associations which characterized the region during the Quaternary. Tropical forms such as crocodiles are absent, although at least one essentially tropical group, the parrots, were still represented locally (Rich 1980).

LATE TERTIARY PALAEOENVIRONMENTS

The principal factors influencing the evolution of Africa's late Tertiary vertebrates were, firstly, the faunal interchange which took place with Eurasia, and, secondly, the environmental conditions on the continent itself.

During much of the early Tertiary Africa was effectively isolated from other continents in a zoogeographic sense, but early in the Miocene the northward moving Afro-Arabian plate came into contact with the Eurasian plate and a land link via the Middle East was established (Berggren & Van Couvering 1974). For the first time in tens of millions of years a 'corridor route' of faunal interchange was opened, with Africa receiving Eurasian immigrants and *vice versa*. The vertebrate faunas of both continents underwent profound changes. This land connection was maintained for the remainder of the late Tertiary, but there were evidently periods when faunal interchange was restricted, and others when appreciable interchange took place. For example, the formation of a land connection between the Iberian Peninsula and north-west Africa during the Terminal Miocene, and the consequent dessication of the Mediterranean Basin (Hsü *et al.* 1977), allowed for a significant faunal interchange between Africa and Europe (e.g. see Azzaroli 1975). Some of the effects on the African fauna of the interchange during the late Tertiary are discussed by Maglio (1978).

The Early Miocene was also the period when development of the East African Rift System was intensified as a result of considerable uplift of the continent (H.B.S. Cooke 1978b). The rifting, mountain building, vulcanism, lake formation and related events created a great variety of habitats out of a previously more homogeneous environment, and this led the resident vertebrate fauna to diversify.

Although the events in east Africa and the intercontinental faunal interchange were geographically far removed from southern Africa, their effects were also felt in this region, since Eurasian immigrants and taxa evolved elsewhere on the continent spread their ranges southwards. In addition, the southern parts of the continent experienced environmental changes which influenced the evolution of the local fauna. These environmental changes were due largely to the changing climatic conditions of the period.

The oceans of the world have a considerable influence on climates and in the south-western parts of Africa it is conditions in the South Atlantic and Southern Oceans which to a large extent determine the local climate. During the late Tertiary very dramatic changes took place in the Southern Ocean. According to Kennett (1980: 123), 'The climatic history of the Southern Ocean is, in general, one of decreasing temperatures, including the development of increased Antarctic glaciation and later ice-sheet formation, a climatic regime which itself had a profound effect on environmental evolution, and thus the biogeography at high southern latitudes.' Although less extreme, the effects in southern mid-latitudes, including the South Atlantic and south-western Africa, were no less significant. The most important events in Antarctica and the Southern Ocean during this period were summed up by Kennett (1980: 130-131) as follows:

1. Early Miocene (*c.* 22 Ma): First unrestricted Circum-Antarctic oceanic circulation following the separation of Antarctica from South America. This resulted in the development of the Antarctic Convergence and the thermal isolation of Antarctica.
2. Middle Miocene (14-11 Ma): Expansion of the Antarctic ice-cap.
3. Late Miocene: Although the Antarctic ice-cap has remained in existence since the Middle Miocene, there have been changes in its volume, the most important of which occurred towards the end of the Miocene (*c.* 6.6-5.2 Ma), when the ice volume increased markedly beyond that of the present. Related events included a global lowering of temperatures and a sharp eustatic drop in sea-level.

The Late Miocene was the period during which the Benguela Current System off the south-west African coast was developed (Siesser 1980), and its cold water is a major factor contributing to the aridity of this region (Tankard & Rogers 1978). The lowering of temperatures in southern mid-latitudes was a direct result of the glaciation of Antarctica, and it caused the climate in the extreme south of Africa to change from tropical to temperate during the latter half of the Miocene. The wind systems of the southern oceanic regions also changed, and coupled with the lowering of rainfall along the west coast was a change in the orographic rain pattern with a summer dry/winter wet regime being established south of the Namib Desert.

Vegetation altered accordingly, with open types progressively replacing forests and woodlands. In the south-western Cape a sclerophyllous fynbos vegetation was established, grading northwards into 'Karoo' scrubland and finally desert. It is only in the south-western Cape that some of the changes are documented in the fossil record (Coetzee 1978a), with the Langebaanweg succession providing both direct and indirect evidence of the nature of these changes (Hendey 1981a).

Vertebrates characteristic of tropical environments were reduced in the south-western Cape, and by the Early Pliocene relatively few were left. Similarly, open country species made their appearence, and became a progressively more significant element in the fauna. Forest and woodland species were under stress by the Early Pliocene, and were thereafter never to recover their supremacy in south-western Cape. At the same time environmental changes were having an effect on faunas elsewhere on the subcontinent. For example, the fauna further north along the west coast adapted to the semi-desert and desert conditions which developed in this region. The nature of these changes are unrecorded. Also undocumented are the changes which occurred along the southern and eastern coastal margins, and on the interior plateau. In the latter regions, the climate is more influenced by conditions in the tropics and the warm south-western Indian Ocean, and the environmental changes were probably not as pronounced as those on the west coast. Nevertheless, they must still have had a significant effect on local faunas.

Late Tertiary environments were also influenced by physiographic changes. Along the west coast the most important of such changes were caused by fluctuations in sea-level. The importance of sea-level fluctuations in respect of deposition in this region has already been mentioned, but they were also significant in that the alternate exposure of part of the continental shelf and inundation of the coastal margin changed the local geography appreciably. This must have had an effect on microclimates and the local biota.

The relevant sea-level changes as recorded by Vail & Hardenbol (1979), and by local observations (Hendey 1981a, b; Siesser & Dingle 1981), were as follows:

1. Early to Middle Miocene (22-13 Myr): A period when a series of transgressions brought the sea from below to above its present level, perhaps up to an elevation of 120 to 140m. The Namib Miocene faunas date from this period.
2. Middle to Late Miocene (13-6.6 Myr): Two regressive cycles lowered the sea to near its present level.
3. Terminal Miocene (6.6-5.2 Myr): A further period of regression when the sea dropped to an undetermined level below that of the present.
4. Early Pliocene (5.2-4.2 Myr): A transgressive period when the coastal margin was inundated up to an elevation of 90m. The Varswater Formation was laid down during this transgression.

The environmental changes in southern Africa during the Miocene and Early Pliocene were increasingly rapid and dramatic. They irrevocably altered the biology of the subcontinent, and set the scene for events during the Late Pliocene and Pleistocene which culminated in the situation as we know it today.

ACKNOWLEDGEMENTS

I am indebted to the many persons whose research and technical assistance made the compilation of this paper possible. Special thanks are due to Miss Lindsay Scott who prepared Figs 3, 4 and 5. The support for the Langebaanweg and Arrisdrift Research Projects by the following organizations is gratefully acknowledged: CHEMFOS Ltd, Consolidated Diamond Mines of South West Africa (Pty) Ltd, SHELL (South Africa) (Pty) Ltd, South African Air Force, South African Council for Scientific and Industrial Research, South African Museum, and Wenner-Gren Foundation for Anthropological Research.

RICHARD G. KLEIN
Department of Anthropology, University of Chicago, Ill., USA

4

The large mammals of southern Africa: Late Pliocene to Recent

AIMS AND BASIC DEFINITIONS

This paper aims to summarize the known history of large mammals in southern Africa, from the late Pliocene, roughly 3.5 million years (Myr) ago, to the recent past. A 'large' mammal is defined as one in which adults weigh at least 0.75 kg. Partial justification for restricting the discussion to 'large' mammals is that they are the ones with which people most often directly interact and they are thus of paramount importance in any attempt to reconstruct prehistoric human ecology (see the papers by Deacon, Volman, and Maggs, this volume). Additionally, in southern Africa, as in many other parts of the world, 'large' mammals are the ones which are best known from both the paleontological and modern behavioral/ecological points-of-view.

The emphasis on 'large' mammals here is not meant to deny the potential of 'small' ones (small rodents, insectivores, and bats) in paleoenvironmental research and especially in biostratigraphy and 'dating.' In a southern African context, D.M. Avery (1982a) in particular has shown how studies of small mammals can both complement and supplement information on late Pleistocene and Holocene paleoenvironments gained from large mammals and other sources. There are numerous paleoenvironmental studies involving small mammals found at sites in the Northern Hemisphere, where they have also proven exceptionally useful for biostratigraphy and 'dating' in the late Pliocene to Recent interval (see, for example, Janossy 1975 or Repenning 1980.)

Preliminary identifications of small mammals from australopithecine sites and like-aged localities in South Africa [by De Graaff (1960) and by Davis and Meester as summarized by H.B.S. Cooke (1978a)] suggest that turnover (first and last appearances) in the 'small' mammal fauna was much less dramatic than in the Northern Hemisphere in the late Pliocene to Recent interval. Faunal turnover also seems to have been less marked in East African 'small' mammals (Jaeger & Wesselman 1976). Still, it can be assumed that in the future, the potential of small mammals for biostratigraphy and their proven utility for documenting environmental change will make them an object of more intensive research in southern (and East) African paleontology.

The last 3.5 million years or so forms a natural unit for discussion here, since, unlike earlier intervals in southern Africa (as discussed by Hendey, this volume), it is characterized from the very beginning by a mammal fauna that is distinctly 'African.'

Moreover, this interval includes hominids more or less throughout (as discussed by Rightmire, this volume). Southern Africa is defined here broadly to include the Zambesian, Transvaalian, Kalaharian, Basutolian, Karoo-Namaqualian, and Cape ecozones of Devred, as presented by De Vos (1975) (Figure 1 here). This is Africa roughly between 8 and 35 degrees south latitude. The modern countries included are southern Zaire, southern Tanzania, Angola, Zambia, Malawi, Mozambique, Zimbabwe, Botswana, Namibia (South West Africa), Lesotho, Swaziland, and South Africa.

THE HISTORIC LARGE MAMMAL FAUNA OF SOUTHERN AFRICA

The species of large mammals that occurred in southern Africa historically are presented in Table 1, which also lists some of their outstanding behavioral/ecological characteristics and the ecozones in which they occurred. The ecozones are abstractions, within which there is considerable environmental and ecological variability and between which there are rarely sharp lines. However, they are based on broad correspondences between major climatic, topographic, phytogeographic, and zoogeographic variables, and they provide a very useful first basis for understanding the historic distributions of large mammals and thus for evaluating past distributions and more general faunal history, as revealed in the fossil record. Their geographic configuration is shown in Figure 1, and they may be briefly characterized as follows [based largely on information in Brown (1965), De Vos (1975), and Werger (1978)]:

1. The *Zambesian Ecozone* is a vast stretch of wooded grassland extending across southern Zaire, southern Tanzania, Angola, Zambia, Malawi, northern Mozambique, and northern Zimbabwe. Average rainfall is generally more than 500 mm/a, and surface water is available in many places throughout the year. Vegetational communities typically comprise numerous trees – mainly *Brachystegia* and *Julbernardia* in the north and *Colophospermum mopane* in the south – scattered through a more or less continuous mat of grasses.

2. The *Transvaalian Ecozone* is a relatively narrow belt of mainly semi-arid country immediately to the south of the Zambesian Zone. Rainfall generally averages less than 500mm/a and perennial surface water is less common than in the Zambesian Zone. It is difficult to characterize the vegetation succinctly, and from a vegetational point-of-view, the zone is perhaps best divided into three parts: (i) a western portion, comprising the northern Transvaal, southern Zimbabwe, northern Botswana, northern Namibia, and southern Angola, in which grassland with shrubby acacias predominates. This part of the zone is really not so much a separate zone as a transitional area between the Zambesian Zone and the Kalaharian one; (ii) an eastern portion comprising the eastern Transvaal, adjacent southwestern Mozambique, and eastern Swaziland, in which bush and tree cover tends to be relatively dense (true 'bushveld'); and (iii) a southern extension comprising the Natal and southeastern Cape coasts and their hinterlands, in which subtropical thornbush and scrub forest dominate.

Table 1. The species of large mammals recorded historically in the six southern African ecozones of Devred. '(x)' indicates that a species was only marginally or sparsely represented in a particular zone, insofar as I could determine this from the publications I consulted. I obtained information on species distributions mainly from Ansell (1971a, 1971b, 1971c), du P. Bothma (1971), Coetzee (1971), Ellerman *et al.* (1953), Joubert & Mostert (1975), Osman Hill & Meester (1971), Petter (1971), Roberts (1951), Shortridge (1934), Sidney (1965), Smithers (1966, 1971a, 1971b), and Smithers & Labão Tello (1976). I supplemented information on habitat preferences in authoritative summaries by Astley Maberley (1967), Bigalke (1978), Dorst & Dandelot (1970), and Smithers (1966, 1971a) with information in numerous technical articles (appearing especially in *The East African Wildlife Journal* and *The South African Journal of Wildlife Research*).

	Zambesian	Transvaalian	Kalaharian	Basutolian	Karoo-Namaqualian	Cape	Habitat and dietary preferences
INSECTIVORA							
Erinaceus frontalis South African hedgehog	(×)	(×)		×	×		Grassland & woodland; terrestrial; omnivorous; nocturnal
Potamogale velox otter shrew	(×)						Rivers, swamps in forest; diet – aquatic vertebrates and invertebrates
PRIMATES							
Galago senegalensis Senegal bushbaby	×	×	(×)				Dense woodland; arboreal; mainly insectivorous nocturnal; solitary or in pairs
Galago crassicaudatus thick-tailed bushbaby	×	×					Dense woodland; arboreal; mainly insectivorous; nocturnal; solitary or in pairs
Colobus polykomos black and white colobus	×						Forest; arboreal; diet mainly leaves; diurnal; gregarious
Cercopithecus aethiops vervet monkey	×	×	×		(×)	×	Woodland & riverine forest; mainly arboreal; diet mainly leaves, fruits, other vegetal items; diurnal; gregarious
Cercopithecus mitis samango monkey	×	×					Forest & dense woodland; arboreal; diet mainly fruits, leaves, other vegetal items; diurnal; gregarious
Cercopithecus nictitans white-nosed monkey	×						Forest & dense woodland; arboreal; diet mainly fruits, leaves, other vegetal items; diurnal; gregarious
Papio cynocephalus yellow baboon	×						Woodland; mainly terrestrial; omnivorous, but mainly vegetarian; diurnal; gregarious

Table 1 (*Contd.*)

	Zambesian	Transvaalian	Kalaharian	Basutolian	Karoo-Namaqualian	Cape	Habitat and dietary preferences
Papio ursinus chacma baboon	×	×	(×)	×	×	×	Subdesert to open woodland; mainly terrestrial; omnivorous, but mainly vegetarian; diurnal
Pan troglodytes chimpanzee	(×)						Woodland & forest; arboreal & terrestrial; mainly herbivorous; diurnal; gregarious
Homo sapiens people	×	×	×	×	×	×	Grassland & woodland; terrestrial; omnivorous; diurnal; gregarious
PHOLIDOTA							
Phataginus temmincki Cape pangolin	×	×	×	×	(×)		Woodland; terrestrial; insectivorous; nocturnal; solitary
Phataginus tricuspis tree pangolin	(×)						Forest; arboreal; insectivorous; nocturnal; solitary
CARNIVORA							
Otocyon megalotis bat-eared fox	(×)	×	×	×	×	×	Semi-desert, grassland, open woodland; terrestrial; mainly insectivorous; mainly nocturnal; gregarious
Vulpes chama Cape fox		(×)	(×)	×	×	×	Semi-desert, grassland, open woodland; terrestrial; omnivorous; mainly nocturnal; solitarv or in pairs
Canis mesomelas black-backed jackal	(×)	×	×	×	×	×	Semi-desert, grassland, woodland; terrestrial; omnivorous; mainly nocturnal; solitary or in pairs
Canis adustus side-striped jackal	×	×					Grassland & woodland; terrestrial; omnivorous; mainly nocturnal; solitary or in pairs
Lycaon pictus hunting dog	×	×	×	×	×	×	Grassland & open woodland; terrestrial; carnivorous; diurnal; gregarious
Ictonyx striatus striped polecat	×	×	×	×	×	×	Desert to well-watered woodland; terrestrial; carnivorous & insectivorous; nocturnal; solitary
Poecigale albinucha white-naped weasel	×	×	×	×	(×)	×	Scrubland, grassland, woodland; mostly terrestrial; carnivorous; nocturnal; solitary

Mellivora capensis honey badger	×	×	×	×	×	×	Semi-desert, grassland, woodland, forest; terrestrial; omnivorous; mainly nocturnal; solitary & in pairs
Lutra maculicollis spotted-necked otter	×	×		×	(×)	(×)	Rivers, swamps; diet mainly aquatic vertebrates and invertebrates; nocturnal; mainly solitary
Aonyx capensis clawless otter	×	×		×	(×)	×	Rivers, swamps; diet – crabs, small vertebrates; nocturnal & diurnal; solitary or in pairs
Nandinia binotata palm civet	×	(×)					Forest & dense woodland; arboreal; omnivorous; nocturnal; solitary
Civettictis civetta civet cat	×	×					Well-watered woodland; terrestrial; omnivorous; nocturnal; solitary
Genetta genetta small-spotted genet	×	×	×	×	×	×	Scrubland, grassland, woodland; terrestrial & arboreal; carnivorous & insectivorous; nocturnal; solitary or in pairs
Genetta tigrina large-spotted genet	×	×		×			Well-watered woodland; terrestrial & arboreal; carnivorous & insectivorous; nocturnal; mainly solitary
Herpestes ichneumon Egyptian mongoose	×	×					Well-watered woodland; terrestrial; carnivorous & insectivorous; diurnal; solitary, in pairs, sometimes small packs
Herpestes sanguineus slender mongoose	×	×	×	×	×	×	Mainly, but not exclusively well-watered woodland; terrestrial; carnivorous & insectivorous; diurnal; mainly solitary
Herpestes pulverulentus Cape grey mongoose				×	×	×	Scrubland, grassland, woodland; terrestrial; carnivorous & insectivorous; diurnal; solitary or in pairs
Atilax paludinosus water mongoose	×	×		(×)	(×)	×	Woodland & forest near water; aquatic & terrestrial; omnivorous; nocturnal; solitary or in pairs
Helogale parvula dwarf mongoose	×	×					Woodland & scrub; terrestrial; mainly insectivorous; diurnal; gregarious
Mungos mungo banded mongoose	×	×	(×)				Woodland & scrub; terrestrial; mainly insectivorous; diurnal; gregarious
Crossarchus obscurus dark mongoose	(×)						Dense woodland & forest; terrestrial; mainly insectivorous; diurnal; gregarious

Table 1 (*Contd.*)

	Zambesian	Transvaalian	Kalaharian	Basutolian	Karoo-Namaqualian	Cape	Habitat and dietary preferences
Ichneumia albicauda white-tailed mongoose	×	×					Well-watered woodland & bush; terrestrial; carnivorous & insectivorous; nocturnal; solitary or in pairs
Rhynchogale melleri Meller's mongoose	×	×					Woodland; terrestrial; omnivorous; nocturnal; solitary
Bdeogale crassicauda bushy-tailed mongoose	×						Woodland; terrestrial; ?omnivorous; nocturnal; solitary
Paracynictis selousi Selous' meerkat	×	×					Open woodland and scrub; terrestrial; insectivorous & carnivorous; nocturnal; solitary or in pairs
Cynictis penicillata yellow mongoose		×	×	×	×		Open woodland & scrub; terrestrial; insectivorous & carnivorous; mainly diurnal; gregarious, but forage singly or in pairs
Suricata suricatta grey meerkat		(×)	×	×	×		Grassland, open woodland; terrestrial; mainly insectivorous; diurnal; gregarious
Proteles cristatus aardwolf	(×)	×	×	×	×	×	Semi-desert, grassland, open woodland; terrestrial; insectivorous; nocturnal; solitary or in pairs
Hyaena brunnea brown hyena		×	×	×	×	×	Desert, scrub, grassland, woodland; primarily scavenges, also hunts; mainly nocturnal; mainly solitary
Crocuta crocuta spotted hyena	×	×	×	×	×	×	Arid scrub, grassland, woodland; hunts & scavenges; mainly nocturnal; mainly solitary
Felis libyca wildcat	×	×	×	×	×	×	Semi-desert, grassland, woodland; active predator; mainly nocturnal; terrestrial; solitary
Felis nigripes black-footed cat			×	×	×		Scrub, grassland; active predator; mainly nocturnal; terrestrial; solitary
Felis serval serval cat	×	×		×		×	Well-watered grassland & woodland; active predator; nocturnal; terrestrial; solitary or in pairs
Felis caracal caracal	×	×	×	×	×	×	Semi-desert, grassland, woodland; active predator; mainly nocturnal; mainly terrestrial; solitary or in pairs
Panthera pardus leopard	×	×	×	×	×	×	Semi-desert, scrub, grassland, but especially woodland; active predator; mainly terrestrial; mainly nocturnal

Panthera leo lion	×	×	×	×	×	×	Subdesert, scrub, but especially grassland and woodland; predator and scavenger, terrestrial; diurnal & nocturnal; solitary, in pairs & in prides
Acinonyx jubatus cheetah	×	×	×	×	×	×	Semi-desert, grassland, open woodland; active predator; terrestrial; mainly diurnal; mainly in pairs
PINNIPEDIA							
Arctocephalus pusillus Cape fur seal					×	×	Cape & Namibian coasts; diet – fish and other marine creatures; gregarious
TUBULIDENTATA							
Orycteropus afer aardvark	×	×	×	×	×		Scrub, grassland, woodland; insectivorous; terrestrial; nocturnal; solitary
PROBOSCIDEA							
Loxodonta africana elephant	×	×	(×)	(×)	(×)	×	Woodland; browse & graze; mainly gregarious
HYRACOIDEA							
Procavia capensis rock dassie	×	×	(×)	×	×	×	Grassland & woodland with rocky outcrops; graze & browse; mainly terrestrial; diurnal; gregarious
Heterohyrax brucel yellow-spotted dassie	×	×					Grassland & woodland with rocky outcrops; browse; terrestrial; diurnal; gregarious
Dendrohyrax arboreus tree dessie	×	×					Dense woodland & forest; arboreal; browse; nocturnal; gregarious
PERISSODACTYLA							
Diceros bicornis black rhinoceros	×	×	×	(×)	×	×	Scrub & woodland; browse; mainly solitary
Ceratotherium simum white rhinoceros	×	×	(×)	×	(×)		Grassland, open woodland; graze; variably solitary and gregarious
Equus zebra mountain zebra					×	×	Sparsely vegetated hilly country; graze; gregarious
Equus burchelli Burchell's zebra	×	×	(×)	×	(×)		Grassland & open woodland; graze; gregarious
Equus guagga guagga				×	×		Scrub & grassland; graze; gregarious

Table 1 (*Contd.*)

	Zambesian	Transvaalian	Kalaharian	Basutolian	Karoo-Namaqualian	Cape	Habitat and dietary preferences
ARTIODACTYLA							
Potamochoerus porcus bushpig	×	×				×	Well-watered bush and dense woodland; omnivorous, but diet mainly subterranean parts of plants; mainly nocturnal; in pairs or groups ('sounders')
Phacochoerus aethiopicus warthog	×	×		(×)	×		Grassland & open woodland; mainly graze; diurnal; in pairs or family parties
Hippopotamus amhibius hippopotamus	×	×		(×)	×		Grassland & open woodland with perennial streams or pools; graze; mainly nocturnal; gregarious
Giraffa camelopardalis giraffe	×	×	×		(×)		Woodland; browse; gregarious
Cephalophus natalensis red duiker	(×)	(×)					Dense woodland & forest; browse; mainly nocturnal; solitary or in pairs
Cephalophus silvicultor yellow-backed duiker	(×)						Dense woodland & forest; browse; mainly nocturnal; solitary or in pairs
Cephalophus monticola blue duiker	×	×				(×)	Dense woodland & forest; browse; mainly nocturnal; solitary or in pairs
Sylvicapra grimmia grey duiker	×	×	×	(×)	×	×	Scrub & woodland; browse; mainly nocturnal; solitary or in pairs
Raphicerus campestris steenbok	(×)	×	×	×	×	×	Scrub, grassland, open woodland; graze & browse; mainly diurnal; solitary or in pairs
Raphicerus melanotis Cape grysbok		(×)			(×)	×	Scrub & forest; mainly browse; mainly nocturnal; solitary or in pairs
Raphicerus sharpei Sharpe's grysbok	×	×					Bush & woodland; mainly browse; mainly nocturnal; solitary or in pairs
Ourebia ourebi ourebi	×	×		×		(×)	Grassland & open woodland near water; mainly graze; mainly diurnal; in pairs or small groups
Nesotragus moschatus suni	×	×					Dense bush & thickets; browse; mainly diurnal; solitary or in pairs
Oreotragus oreotragus klipspringer	×	×	(×)	×	×	×	Rocky hills; mainly browse; in pairs or small parties
Madoqua kirkii dikdik		(×)			(×)		Dry woodland or thicket; browse; solitary or in pairs

Pelea capreolus vaalribook		(×)	(×)	×	(×)	×	Grassy slopes in hilly country; mainly graze; gregarious
Redunca fulvorufula mountain reedbuck		(×)	(×)	×		×	Grassland or woodland in hilly country; mainly graze; gregarious
Redunca arundinum southern reedbuck	×	×			(×)		Grassland or open woodland near water; graze, solitary, in pairs, or small herds
Kobus ellipsiprymnus waterbuck	×	×					Grassland & open woodland near water; mainly graze; gregarious
Kobus vardoni puku	×	(×)					Open woodland & grassland along swamps rivers; graze; gregarious
Kobus leche lechwe	×	×					Grassland on fringes of streams or swamps; graze; gregarious
Aepyceros melampus impala	×	×		(×)			Open woodland & scrub; browse & graze; gregarious
Antidorcas marsupialis springbok		(×)	×	×	×		Grassland, open scrub, light woodland; graze & browse; gregarious
Oryx gazella gemsbok		(×)	×		×		Grassland, light woodland; mainly graze; gregarious
Hippotragus equinus roan antelope	×	×	×			(×)	Open woodland; graze; gregarious
Hippotragus niger sable antelope	×	×					Open woodland; mainly graze; gregarious
Hippotragus leucophaeus blue antelope						×	Open woodland; graze; gregarious
Damaliscus lunatus tsessebe	×	×	×				Grassland & open woodland; graze; gregarious
Damaliscus dorcas bontebok/blesbok				×		(×)	Grassland; graze; gregarious
Alcelaphus buselaphus red hartebeest		(×)	×	×	×	×	Grassland & open woodland; graze; gregarious
Alcelaphus lichtensteini Lichtenstein's hartebeest	×						Grassland & open woodland; graze; gregarious
Connochaetes gnou black wildebeest				×	×		Grassland & scrub; graze; gregarious
Connochaetes taurinus blue wildebeest	×	×	×	(×)			Grassland & open woodland; graze; gregarious
Tragelaphus scriptus bushbuck	×	×				×	Thick bush & forest; mainly browse; solitary or in pairs
Tragelaphus spekei sitatunga	(×)	(×)					Swamps; graze; solitary; in pairs or small herds

Table 1 (*Contd.*)

	Zambesian	Transvaalian	Kalaharian	Basutolian	Karoo-Namaqualian	Cape	Habitat and dietary preferences
Tragelaphus angasi nyala	(×)	×					Bush & dense woodland near water; mainly browse; gregarious
Tragelaphus strepsiceros greater kudu	×	×	×	(×)	(×)		Woodland & thickets; mainly browse; gregarious
Taurotragus oryx eland	×	×	×	×	×	×	Semi-desert, grassland, and especially open woodland; browse & graze; gregarious
Syncerus caffer Cape buffalo	×	×	(×)				Woodland near water; graze; gregarious
LAGOMORPHA							
Lepus capensis Cape hare	(×)	(×)	(×)	×	×	×	Grassland, grassy scrubland; graze; nocturnal; usually solitary
Lepus crawshayi Crawshay's hare	×	×	(×)	×			Woodland, grassland; graze; nocturnal; usually solitary
Lepus saxatilis scrub hare			×		×	×	Scrub & open woodland; graze; nocturnal; usually solitary
Pronolagus crassicaudatus red rock hare	(×)	(×)		×	×	×	Rocky hills with some grass & shrub cover; graze; nocturnal; solitary or in pairs
Bunolagus monticularis Bushman hare					×		Scrub; graze; nocturnal; solitary
RODENTIA							
Bathyergus suillus Cape mole rat					(×)	×	Various vegetational settings in sandy areas; fossorial; diet–roots, tubers, bulbs; gregarious
Hystrix africaeaustralis porcupine	×	×	×	×	×	×	Various vegetational settings; herbivorous, especially subterranean parts of plants; terrestrial; nocturnal; solitary
Thrynomys swinderianus greater cane rat	×	×					Grassy areas and reed beds near water; terrestrial; mainly nocturnal; herbivorous; gregarious
Thryonomys gregorianus lesser cane rat	×						Moist woodland; terrestrial; mainly nocturnal; herbivorous; gregarious
Pedetes capensis springhare	(×)	×	×	×	×		Grassland, woodland on sandy soils; terrestrial; graze; nocturnal; in pairs
Cricetomys gambianus giant rat	×	×		×			Grassland, woodland near water; terrestrial; mainly herbivorous; nocturnal; gregarious

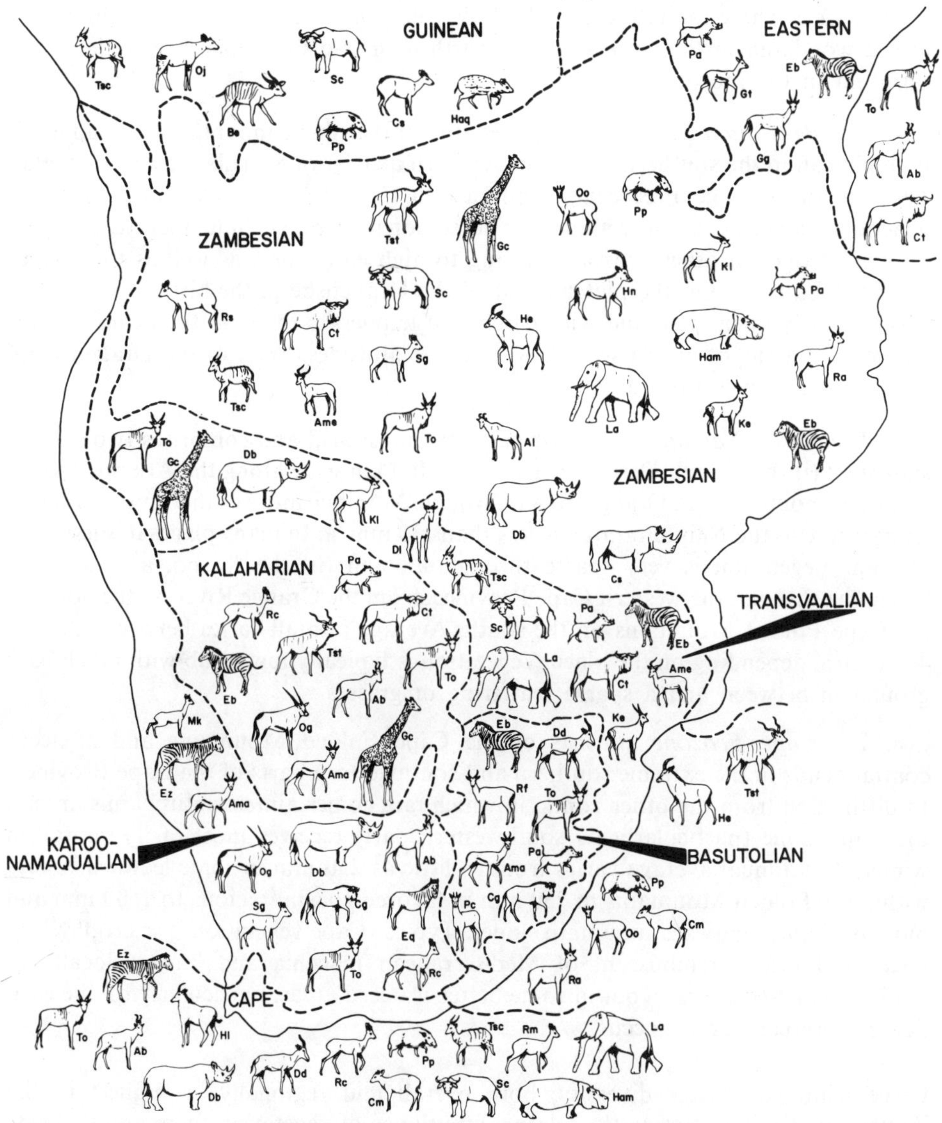

Figure 1. The Southern African ecozones of Devred, with some of the most common ungulates found in each. Next to each ungulate are the first initials of its technical generic and species names. The technical names may be found in Table 1, except for *Okapia johnstoni* (Oj) (okapi), *Boocercus euryceros* (Be) (bongo), *Hyemoschus aquaticus* (Haq) (water chevrotain), *Gazella thomsoni* (Gt) (Thomson's gazelle), and *Gazella granti* (Gg) (Grant's gazelle).

3. The *Kalaharian Ecozone* corresponds broadly to that part of the southern African interior plateau often known as the 'Kalahari Desert.' It covers eastern Namibia, the western two-thirds of Botswana, and a large portion of the adjacent (northern) Cape Province of South Africa. Rainfall decreases from an average of nearly 500mm/a in the northwest to as little as 200mm/a in the southwest. Highly porous sandy soils soak up rainfall rapidly, so that surface water is relatively rare,

even in areas where rainfall is fairly high. Vegetation varies from relatively luxuriant acacia woodland in the better-watered north to sparse shrub acacia woodland and bushveld in the more arid south.

4. The *Basutolian Ecozone* comprises the Drakensberg mountains and adjacent high plateau in the southern Transvaal, western quarter of Swaziland, western Natal, Lesotho, the Orange Free State, and eastern Cape Province. Average rainfall is generally between 620 and 750mm/a, and seasonal contrasts in temperature tend to be greater than in other ecozones, owing to high elevations, as well as to distance from the Equator. On the plateau east of the Drakensberg, the historic vegetation was primarily open grassland with patches of temperate forest at the heads of river valleys and areas of acacia woodland at lower altitudes. West of the Drakensberg, pure grassland was the rule.

5. The *Karoo-Namaqualian Ecozone* is the most arid and comprises two distinct subzones: (i) the Namib Desert, a strip up to 160 km wide along the Atlantic Coast, from the mouth of the Orange River through Namibia into southernmost Angola. Everywhere in the Namib rainfall is less than 130 mm/a. In many places it is less than 25 mm. Vegetation is very sparse throughout; and (ii) the Karoo, a great plain extending through the western Cape Province, from the Orange River on the north to the Cape Folded Mountains on the south. Average rainfall varies between 130 and 400 mm/a, depending on the place. Vegetation is typically low scrub with much bare ground in between and a sparse scattering of grasses.

6. The *Cape Ecozone* consists of the Cape Folded Mountains and adjacent coastal plains in the extreme southern and southwestern parts of the Cape Province. In distinction from the other zones, in which rain comes almost entirely in summer, the Cape Zone (particularly its southwestern part) receives most of its rainfall in winter. The annual average varies from as little as 250 mm in some leeward valleys within the Folded Mountains to 750 mm in favored coastal regions to 1500 mm and more on some windward mountain slopes. Typical Cape vegetation is sclerophyllous scrub and heath, reminiscent of Mediterranean macchia and known locally as 'fynbos.' Grasses are rare, but in some better-watered areas, particularly on the east, dense evergreen forests occur.

Large mammal species diversity, both overall and regionally, is highest in the Zambesian Zone, reflecting the relative luxuriance of vegetation there and the wide variety of opportunities the zone offers both grazers and browsers. The Transvaalian Zone, which is the one most similar to the Zambesian Zone in climate and vegetation, has a comparable species diversity overall, but diversity at the regional level is generally lower. In the remaining four zones, greater vegetational homogeneity (especially in the Basutolian Zone), a less complete vegetation mat (especially in the Kalaharian and Karoo-Namaqualian Zones), vegetation less suitable as food for large mammals (especially in the Cape Zone), or some combination of these factors has led to lower species diversity, both overall and locally. However, at least historically, in spite of their relatively low species diversity, the grasslands of the Basutolian Zone probably supported a greater biomass of large mammals than any of the other zones. Next in average biomass was probably the Zambesian Zone, followed in order by the Transvaalian, the Kalaharian, the Cape, and the Karoo-Namaqualian.

Southern Africa shares most of its large mammal species with other parts of Africa, particularly with the adjacent Eastern Ecozone to the northeast (Figure 1), which includes a variety of grassland and woodland habitats very similar in many important respects to those in various parts of southern Africa. At present, ecological barriers between or within southern African ecozones inhibit the spread of some species among them and between them and the Eastern Zone, but in the relatively recent past, both the configuration of the ecozones and the nature of internal mosaics were different, promoting different species distributions (see below).

Of the relatively small number of species that are essentially endemic to southern Africa (Cape fox, Cape grey mongoose, Selous' meerkat, yellow mongoose, grey meerkat, brown hyena, black-footed cat, mountain zebra, quagga, Cape grysbok, vaalribbok, puku, lechwe, springbok, blue antelope, bontebok/blesbok, black wildebeest, nyala, scrub hare, Bushman hare, and Cape mole rat), some occurred outside southern Africa in the recent geologic past (springbok, ?nyala), while others have very close relatives to the east and north from which they may only recently have differentiated (?Cape fox, ?puku, ?lechwe). The remaining ('true') endemics are mainly restricted to the three southernmost ecozones (Basutolian, Karoo-Namaqualian, and Cape), where they probably evolved as a result of relative geographic isolation and a comparatively unique set of vegetational/climatic settings. The temperate grasslands of the Basutolian Zone and the immediately adjacent grasslands/scrublands of the Karoo-Namaqualian Zone were probably particularly important as a center of speciation, fostering, among other species, the black wildebeest, blesbok, and quagga, which ranged there in great numbers into the last century.

Table 1 shows that within southern Africa many species were widely distributed, occurring at least marginally in most, if not all ecozones. This is particularly true of the large carnivores, reflecting their tendency to feed on whatever prey species occur locally, as well as their common ability to obtain their moisture requirements largely or entirely from meat. Some herbivores also had very broad distributions, and the eland is perhaps the most notable in this respect. However, most herbivores were somewhat more restricted, and broadly speaking, two great geographic groups may be recognized. The first prominently includes the elephant, giraffe, blue wildebeest tsessebe, roan and sable antelopes, kudu, waterbuck, lechwe, impala, buffalo, and Burchell's zebra, which were limited to or most numerous in the wooded grasslands of the Zambesian and Transvaalian Zones. The second includes especially the black wildebeest, red hartebeest, blesbok, springbok, gemsbok, vaalribbok, and quagga, which were most at home in the open grasslands or sparse, shrubby grasslands of the Basutolian and Karoo-Namaqualian Zones.

Overall, the herbivore communitiesof the Kalaharian Zone were more similar to those of the Zambesian and Transvaalian Zones, reflecting the greater vegetational similarity, but two prominent Karoo-Namaqualian species – the springbok and the gemsbok – were also numerous in the Kalaharian Zone, probably due to their ability to thrive without surface water. The herbivore communities of the remaining zone – the Cape – may be characterized as impoverished versions of Transvaalian communities on the east and of Karoo-Namaqualian ones on the west. In keeping with its unique climate and flora, the Cape Zone was also unique in being the only zone in which browsers probably outnumbered grazers in both diversity and overall biomass.

TAXONOMIC CHANGE IN THE LARGE MAMMAL FAUNA: LATE PLIOCENE TO RECENT

On the basis of taxonomic composition, H.B.S. Cooke (1963; Ewer & Cooke 1964) suggested that southern African fossil faunas dating within what is now clearly the late Pliocene to Recent time interval could be organized broadly into four successive 'stages' (from older to younger): (1) the Sterkfontein Stage; (2) the Swartkrans Stage, (3) the Vaal-Cornelia Stage, and (4) the Florisbad-Vlakkraal Stage. Wells (1962, 1967) published a similar scheme, referring to the four units first as 'stages' and later as 'faunal assemblages,' while Cooke later (1967, also 1978b) suggested that they be called 'faunal spans.' More recently, Hendey (1974a, b) proposed that sufficient information had accumulated to replace the four 'faunal spans' with four 'Land Mammal Ages,' patterned after the Land Mammal Ages of North America and Europe and labelled the 'Early Makapanian,' the 'Late Makapanian,' the 'Cornelian,' and 'the Florisian' respectively. Hendey also defined two geologically earlier ages (the Namibian and the Langebaanian), which are discussed in his paper in this volume.

I have adopted Hendey's scheme here (Figure 2), and the principal sites providing faunas assigned to each stage are listed in Table 2. The approximate locations of the sites are shown in Figure 3. The most serious weakness in the scheme is that the boundaries between the Ages are not firmly fixed in time, except for the upper boundary of the 'Florisian' at 12,000-10,000 B.P. The main difficulty with establishing the other boundaries is that they all lie well beyond 40,000 years ago, the lower limit of the radiocarbon method as practiced by most laboratories, while radiocarbon is the only absolute dating technique that is widely applicable in southern Africa.

Taxonomic comparisons to (mainly potassium-argon) dated faunas in East Africa, in combination with paleomagnetic determinations at Makapansgat (Brock *et al.* 1977; McFadden *et al.* 1979), suggest that the Early/Late Makapanian boundary probably lies around 2.0-1.8 Myr, while geomorphic/sedimentological analyses of the contexts of faunas from Duinefontein 1 and 2, Swartklip 1, Klasies River Mouth and other Florisian sities (references mainly by Butzer in Table 2) indicate that the Cornelian/Florisian boundary lies in the Middle Pleistocene, prior to 130,000 years ago. Firmer estimates for this boundary and also for the Late Makapanian/Cornelian one may prove particularly difficult to obtain, since the first and certainly the second falls between 1,000,000 and 130,000 years B.P., an interval in which radiometric dating methods have proven difficult to apply, even in East Africa. The point to be emphasized is that the absolute boundaries between Ages which have been suggested here (in Figure 2 and Table 2) are highly provisional.

Table 3 lists the historic and extinct species of large mammals that are known from the various Land Mammal Ages. It is immediately obvious that there are many differences among Ages in species composition. The reasons are varied. Some differences are probably due to the presence of species in earlier Ages that are

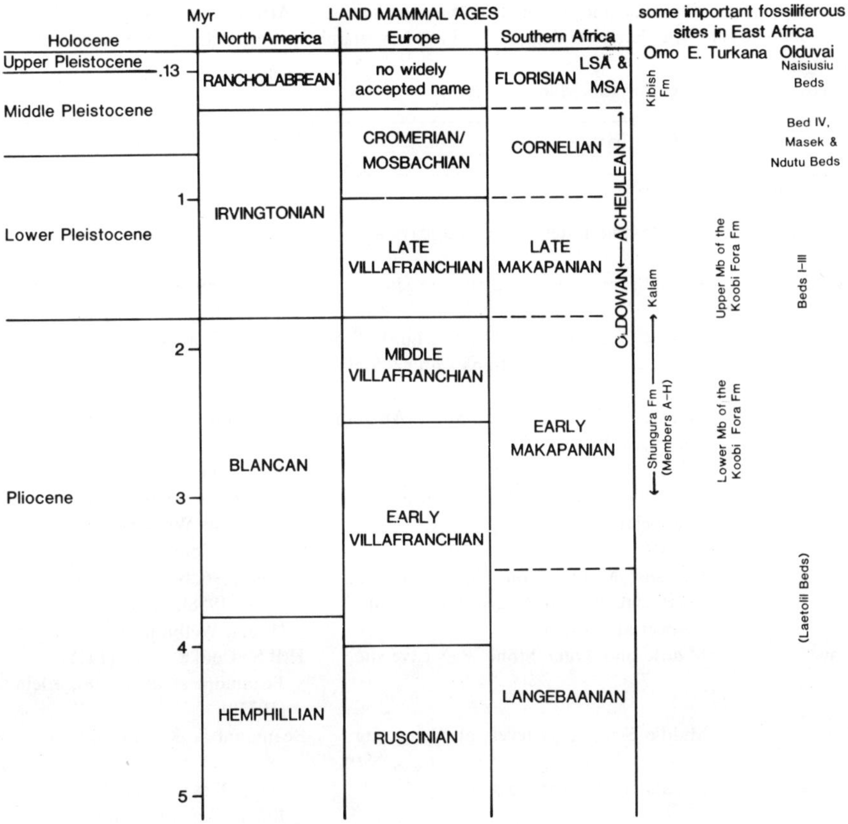

Figure 2. Land Mammal Ages of southern Africa in relation to Land Mammal Ages of North America and Europe. In each case, with the exception of the terminal boundary for the most recent Age, the absolute dates for the boundaries between Ages are subject to revision, particularly in southern Africa. The estimated ages for the southern African boundaries are based in part on radiometric dates from important East African fossiliferous localities, some of which are listed in the Table. General sources for the East African localities are: Howell (1978b) (Omo); M.G. Leakey & R.E.F. Leakey (1978) (East Turkana); and M.D. Leakey (1979) (Olduvai).

ancestral to species in later ones, as indicated in the table. Other differences, particularly in the abundance of relatively small 'large' mammals (small carnivores, hares, hyraxes, large rodents, and small primates) are probably due to collection bias in which smaller bones were simply overlooked or not systematically collected, particularly at some important Cornelian and Makapanian (*s.l.*) sites. Finally, some of the differences may reflect great differences between Ages in the geographic concentration of important sites (Figure 3) or in the agents primarily responsible for bone accumulation (Table 2). Only for the Florisian is there a reasonable scatter of sites throughout the historic ecozones (compare Figures 1 and 3), combined with a reasonable variety of depositional agencies at different sites. Even for the Florisian, however, there are few or no sites that probably sample dense woodland or forest.

Table 2. The principal Pliocene and Pleistocene sites in southern Africa that have provided fossil assemblages assigned to the Early Makapanian, Late Makapanian, Cornelian, and Florisian Land Mammal Ages.
Asterisks indicate name sites for the Mammal Ages.

Site	General nature	References
Florisian (?400,000 – 10,000 B.P.)		
Kalemba Rock Shelter	Middle and Later Stone Age cave site	Phillipson (1976)
Leopard's Hill Cave	Later Stone Age and carnivore site	Miller (1969, 1971); Klein (unpub.)
Twin Rivers Kopje	Breccia ?formed in sinkhole; Middle Stone Age artifact associations; ?archeological site	Clark (1971a); Oakley (1954)
Redcliff Cave	Middle Stone Age, Later Stone Age, and carnivore site	C.K. Cooke (1978); Klein (1978a)
Pomongwe Cave	Middle and Later Stone Age cave site	C.K. Cooke (1963); Brain (1981)
Chelmer	Fluviatile deposit without cultural associations	Bond & Summers (1951); H.B.S. Cooke & Wells (1951); H.B.S. Cooke (1963)
Kalkbank	Pan site; Middle Stone Age cultural associations; bone accumulation of uncertain origin	Mason (1962b, 1967b); Mason *et al.* (1958); H.B.S. Cooke (1962); Welbourne (1971)
Border Cave	Middle and Later Stone Age cave site	H.B.S. Cooke *et al.* (1945); Beaumont *et al.* (1978); Klein (1977b)
Kathu Pan (in part)	Middle Stone Age levels at spring site	Beaumont & Klein (unpub.)
Witkrans Cave	Middle Stone Age Cave site	Clark (1971a); Peabody (1954); Butzer, Stuckenrath *et al.* (1978); H.B.S. Cooke (n.d.)
Equus Cave	Carnivore cave site	Butzer, Stuckenrath *et al.* (1978); Beaumont & Shackley (unpub.); Klein (n.d.)
*Florisbad	Middle Stone Age spring site	Dreyer (1938); Hoffman (1955); Meiring (1956); Butzer (1971); H.B.S. Cooke (1963); Gentry & Gentry (1978)
Vlakkraal	Middle Stone Age spring site	Wells *et al.* (1942); Butzer (1971); H.B.S. Cooke (1963)
Elands Bay Cave (levels 18-22)	Later Stone Age Cave site	Parkington (1972, 1979, 1981); Klein (unpub.)
Boomplaas Cave A (BRL7Y – LOH)	Middle and Later Stone Age cave site	H.J. Deacon (1979); Klein (1978b and unpub.)
Klasies River Mouth Cave 1 (layer 14 and below)	Middle Stone Age Cave site	Wymer & Singer (1972); Butzer (1978b); Klein (1976a); Volman (1981); Singer & Wyner (1982).
Nelson Bay Cave (BSL – YGL)	Later Stone Age Cave site	J. Deacon (1978); Klein (1972b); Butzer (1973a)
Byneskranskop 1	Later Stone Age Cave site	Schweitzer & Wilson (1978 & 1982); Klein (1981a)
Die Kelders 1	Middle Stone Age Cave site	Tankard & Schweitzer (1974, 1976); Tankard (1976); Klein (1975a); Volman (1981)

Site	General nature	References
Swartklip 1	Carnivore site in ancient gully fill within eolianite	Hendey & Hendey (1968); Klein (1975b)
Duinefontein 1 and 2	Carnivore site ('1') and Middle Stone Age 'buffalo kill site' on margin of ancient marsh ('2')	J. Deacon (1976a); Klein (1976b)
Sea Harvest	Carnivore accumulation(s)	Hendey (1974a); Northey (1979); Klein (unpub.)
Cornelian (?1,000,000–?400,000 B.P.)		
Namib 4	Acheulean Pan site	Shackley (1980); Klein (unpub.)
Kathu Pan (in part)	Acheulean spring site	Beaumont & Klein (unpub.)
Vaal River 'Younger Gravels' (in large part)	Acheulean cultural associations, but nature of bone accumulations probably mixed fluviatile and archeological	H.B.S. Cooke (1949); Wells (1964); Helgren (1977)
Swartkrans Member 2 (in part)	?Acheulean archeological site; ?carnivore site	Brain (1976, 1981); Butzer (1976); Vrba (1975, 1976)
Sterkfontein Member 6 (?)	?carnivore site; ?archeological site	Partridge (1978); Vrba (1974, 1975, 1976)
*Cornelia	?fluviatile accumulation	Butzer *et al.* (1974)
Elandsfontein (most)	Archeological and ?carnivore site near waterhole(s)	Singer & Wymer (1968); Gentry & Gentry (1978); Hendey (1974a); Klein (1978c); Butzer (1973b)
Late Makapanian (?1,800,000–?1,000,000 B.P.)		
Kromdraai B	?carnivore accumulation	Brain (1975); Vrba (1981)
Kromdraai A	?carnivore accumulation	Vrba (1975, 1976, 1981)
Swartkrans Member 1	?carnivore accumulation	Brain (1976, 1981); Butzer (1976); Vrba (1975, 1976)
Sterkfontein Member 5	?carnivore site; ?archeological site	Partridge (1978); Vrba (1974, 1975, 1976)
Bolt's Farm (in part)	?carnivore accumulation	H.B.S. Cooke (1963, 1978a)
Baard's Quarry, Langebaanweg (?in part)	?fluviatile accumulation	Hendey (1978b)
Early Makapanian (?3,500,000–?1,800,000 B.P.)		
Chiwondo Beds (in part)	Fluviatile or lacustrine accumulations	Clark *et al.* (1970); Kaufulu *et al.* (1981)
Bolt's Farm	?carnivore accumulation	H.B.S. Cooke (1963, 1978a)
Sterkfontein Member 4	?carnivore accumulation	Partridge (1978); Vrba (1974, 1975, 1976)
*Makapansgat	?carnivore accumulation	Brock *et al.* (1977); Brickhill (1978); Collings *et al.* (1975); Maguire (1978); H.B.S. Cooke (1978a); Wells & Cooke (1956); Vrba (1978 & in prep.); Gentry & Gentry (1978); McFadden *et al.* (1979); Partridge (1979)
Taung	?carnivore accumulation	Peabody (1954); Butzer (1974a, 1980a); H.B.S. Cooke (1978a)
Baard's Quaary, Langebaanweg (?in part)	?fluviatile accumulation	Hendey (1978b)

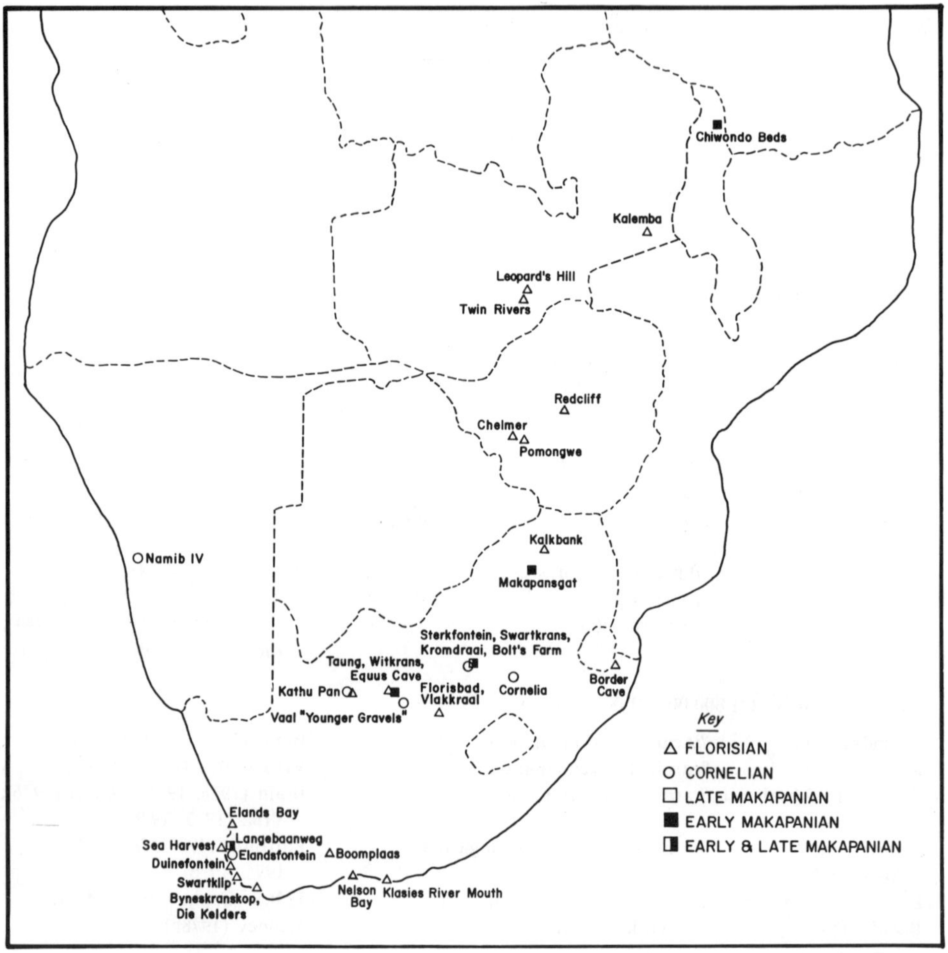

Figure 3. Map showing the approximate locations of the southern African fossil sites mentioned in the text.

The near absence of such sites probably accounts for the generally poor representation of species with a preference for closed habitats (e.g. bushpig or duikers of the genus *Cephalophus.*)

Bearing in mind the sampling problems outlined in the last paragraph, there are still many differences in species composition among Ages that probably reflect the real presence or absence of taxa, or more precisely their appearance or disappearance. Particularly secure 'first appearances' are ones that involve taxa which are almost certainly immigrants to southern Africa, as judged from their paleontological record elsewhere. Particularly secure 'last appearances' involve taxa which do not occur in large samples from the same region(s) where they were well-documented earlier. Such taxa are presumed to have become extinct, at least locally.

Taking into account East and North African data (as reported in many of the same references listed in the caption to Table 3), the most certain and prominent immigrants to southern Africa during the late Pliocene to Recent interval were large cats of the genus *Panthera*, appearing during the Early or Late Makapanian, and one-toed horses of the genus *Equus*, appearing at or near the beginning of the Late Makapanian. It is also possible that *Homo* sp. immigrated to southern Africa at or near the beginning of the Late Makapanian, but it remains possible that it evolved locally.

Notable and reasonably secure last appearances include *Chasmaporthetes ('Euryboas') nitidula, Hyaena bellax, Hyaenictis ('Leecyaena') forfex, Machairodus transvaalensis, Dinofelis piveteaui, Makapania broomi, Notochoerus capensis, Parapapio* spp., *Dinopithecus ingens,* and *Gorgopithecus major*, all probably during the Late Makapanian; *Megantereon gracile, Hippotragus gigas, Parmularius* sp(p), *Elephas* sp(p), *Kolpocherus ('Mesochoerus') limnetes, Hippopotamus gorgops, Sivatherium maurusium, Hipparion* sp(p), and *Theropithecus (Simopithecus) oswaldi,* all probably during the Cornelian; and *Damaliscus niro, Megalotragus priscus, Antidorcas bondi, Antidorcas australis, Pelorovis antiquus, Metridiochoerus compactus,* and *Equus capensis*, all during the Florisian. Some of the extinct forms are illustrated (reconstructed) in Figures 4 and 5.

The loss of *Makapania broomi*, an ovibovine bovid ('muskox'), in the Late Makapanian, and of *Mesembriportax acrae*, a boselaphine which may have persisted nearly as long, removed the last major 'Eurasiatic' faunal elements which had been part of the southern African fauna in the earlier Pliocene (see the paper in this volume by Hendey). From the Late Makapanian on, the composition of the fauna was distinctly and overwhelmingly African in the historic sense.

Except in the case of some of the very latest extinctions – those of *Megalotragus priscus, Pelorovis antiquus, Antidorcas australis*, and *Equus capensis* – occurring at or very nearly the end of the Florisian, the absolute dates of final disappearances remain unknown, and it is not even clear whether the disappearances occurred gradually or in clusters (Klein in press b). Causation also remains undetermined. It is possible that early hominids played a role, and I have argued that the particularly competent hunter-gatherers who existed at the end of the Florisian may have been directly responsible for the extinctions that occurred then, in combination with environmental change limiting the numbers and ranges of the species that became extinct (Klein 1979a, 1980, in press b).

A more pervasive 'cause' of extinction was probably long-term environmental change, especially the dramatic alternation between drier and wetter climates that occurred in southern Africa in complex relation to the alternation of 'glacials' and 'interglacials' at higher latitudes (see the paper by Butzer, this volume). 'Glacial'/'interglacial' alternation profound enough to have a significant impact even at lower latitudes began in the late Pliocene, approximately 3.2 Myr, prior to which there was a lengthy period of relatively stable 'interglacial' climate, as documented in the deep-sea record (Shackleton & Opdyke 1977). It seems possible, perhaps even likely that initiation of this set of glacial/interglacial cycles was partly responsible for the disappearance of many Langebaanian (early Pliocene) taxa in southern Africa. At about 2.5 Myr, the glacials may have become longer, more intense, or both, a trend which perhaps accelerated in the Pleistocene, especially after 700,000 B.P.

Bonifay (1980) has shown how change within Villafranchian (late Pliocene and early Pleistocene) faunas and between Villafranchian and Cromerian/Mosbachian (early and mid-Pleistocene) ones in Europe clearly relates to climatic change, with

Table 3. Large mammal species known from the late Pliocene to Recent 'Land Mammal Ages' of southern Africa. A double line indicates the certain or nearly certain presence of a species; a single line indicates its probable or very possible presence; a '?' indicates possible presence. Uncertainties are introduced by problems of dating or identification, particularly of specimens that are often very fragmentary. Langebaanian (early Pliocene) species are listed only if they bear on the ancestry of later species. (For sources on the Langebaanian, see the paper by Hendey, this volume). At least three species that occurred in southern Africa in the late Pliocene or Pleistocene may have extant descendants that only occur elsewhere. The extant descendants (*Hyaena hyaena*, the striped hyena; *Gazella granti*, Grant's gazelle; and *Gazella thomsoni*, Thomson's gazelle) are listed in parentheses at the appropriate point in the table.

Sources for the table include those listed in the caption to Table 2, as well as the following for particular taxa: Cercopithecidae - Simons, & Delson (1978); Carnivora - Ewer (1954, 1955a, 1955b, 1955c, 1956a, 1956b, 1956c, 1956d) and Hendey (1973, 1974a, 1974b, 1974c, 1978a); Proboscidea - Coppens *et al.* (1978); Hyracoidea - Meyer (1978); Rhinocerotidae - Hooijer (1978), Equidae - Churcher & Richardson (1978); Suidae - H.B.S. Cooke & Wilkinson (1978) and Harris & White (1979); Hippopotamidae - Coryndon (1978); Giraffidae - Churcher (1978); Bovidae - Gentry (1978, 1980) Gentry & Gentry (1978), and Vrba (1975, 1976, 1978, 1979); and Hystricidae - Maguire (1978).

Myr Langebaanian	3.5 Early Makapanian	1.8 Late Makapanian	1.0 Cornelian	0.4 Florisian	0.01 Recent	
						INSECTIVORA
						Erinaceidae (hedgehogs)
				========	×	*Erinaceus frontalis*
						Potamogalidae (otter-shrews)
					×	*Potamogale velox*
						PRIMATES
						Galagidae (bushbabies)
					×	*Galago senegalensis*
					×	*Galago crassicaudatus*
						Cercopithecidae (monkeys)
				========	×	*Cercopithecus aethiops*
					×	*Cercopithecus mitis*
					×	*Cercopithecus nictitans*
	========	========				*Parapapio* spp.
		======== - -				*Papio robinsoni*
			? - - - - - -	========	×	*Papio ursinus/cynocephalus*
	========	========				*Papio* spp.
		========				*Dinopithecus ingens*

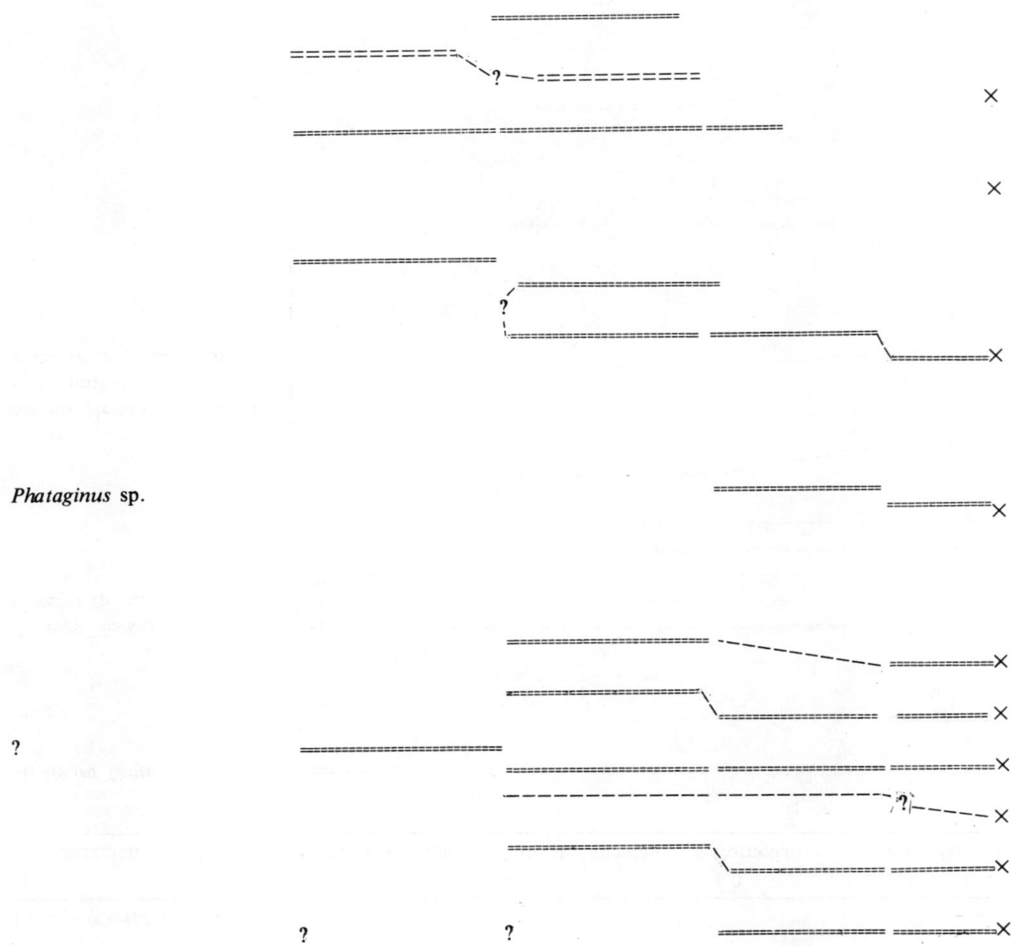

Gorgopithecus major
Theropithecus darti
Theropithecus oswaldi
Colobus polykomos
Cercopithecoides williamsi

Pongidae (apes)
Pan troglodytes

Hominidae (people)
Australopithecus africanus
Australopithecus robustus
Homo habilis
Homo erectus
Homo sapiens

PHOLIDOTA

Manidae (pangolins)
Phataginus sp.
Phataginus temmincki

CARNIVORA

Canidae (foxes, jackals, etc.)
Otocyon recki
Otocyon megalotis
Vulpes pulcher
Vulpes chama
'Canis' brevirostris
Canis mesomelas
Canis terblanchei
Canis adustus
Lycaon atrox
Lycaon pictus

Mustelidae (weasels, otters, etc.)
Ictonyx striatus

Table 3 (*Contd.*)

Myr Langebaanian	3.5 Early Makapanian	1.8 Late Makapanian	1.0 Cornelian	0.4 Florisian	0.01 Recent	
					×	*Poecigale albinucha*
Mellivora benfieldi	- - -	- - -	===	===	×	*Mellivora capensis*
					×	*Lutra maculicollis*
	?	?		===	×	*Aonyx capensis*
						Viverridae (mongooses, etc.)
					×	*Nandinia binotata*
'Viverra' leakeyi	- - -	- - -	===	===	×	*Civettictis civetta*
Genetta sp.	? - - -	- - -	- - -	?	×	*Genetta genetta*
	? - - -	- - -	- - -	===	×	*Genetta tigrina*
	- - -	- - -	===	===	×	*Herpestes ichneumon*
		- - -	- - -	===	×	*Herpestes sanguineus*
				===	×	*Herpestes pulverulentus*
		===	- - -			*Atilax mesotes*
				===	×	*Atilax paludinosus*
					×	*Helogale parvula*
various Herpestinae, at least some probably ancestral to later forms					×	*Mungos mungo*
		===	- - -			*Crossarchus transvaalensis*
				- - -	×	*Crossarchus obscurus*
				===	×	*Ichneumia albicauda*
					×	*Rhynchogale melleri*
					×	*Bdeogale crassicauda*
					×	*Paracynictis selousi*
	- - -	===	===	===	×	*Cynictis penicillata*
			===			*Suricata major*
				===	×	*Suricata suricatta*
						Protelidae (aardwolves)
		===	- - -			*Proteles transvaalensis*
				===	×	*Proteles cristatus*
						Hyaenidae (hyenas)
		===				*Chasmaporthetes nitidula*

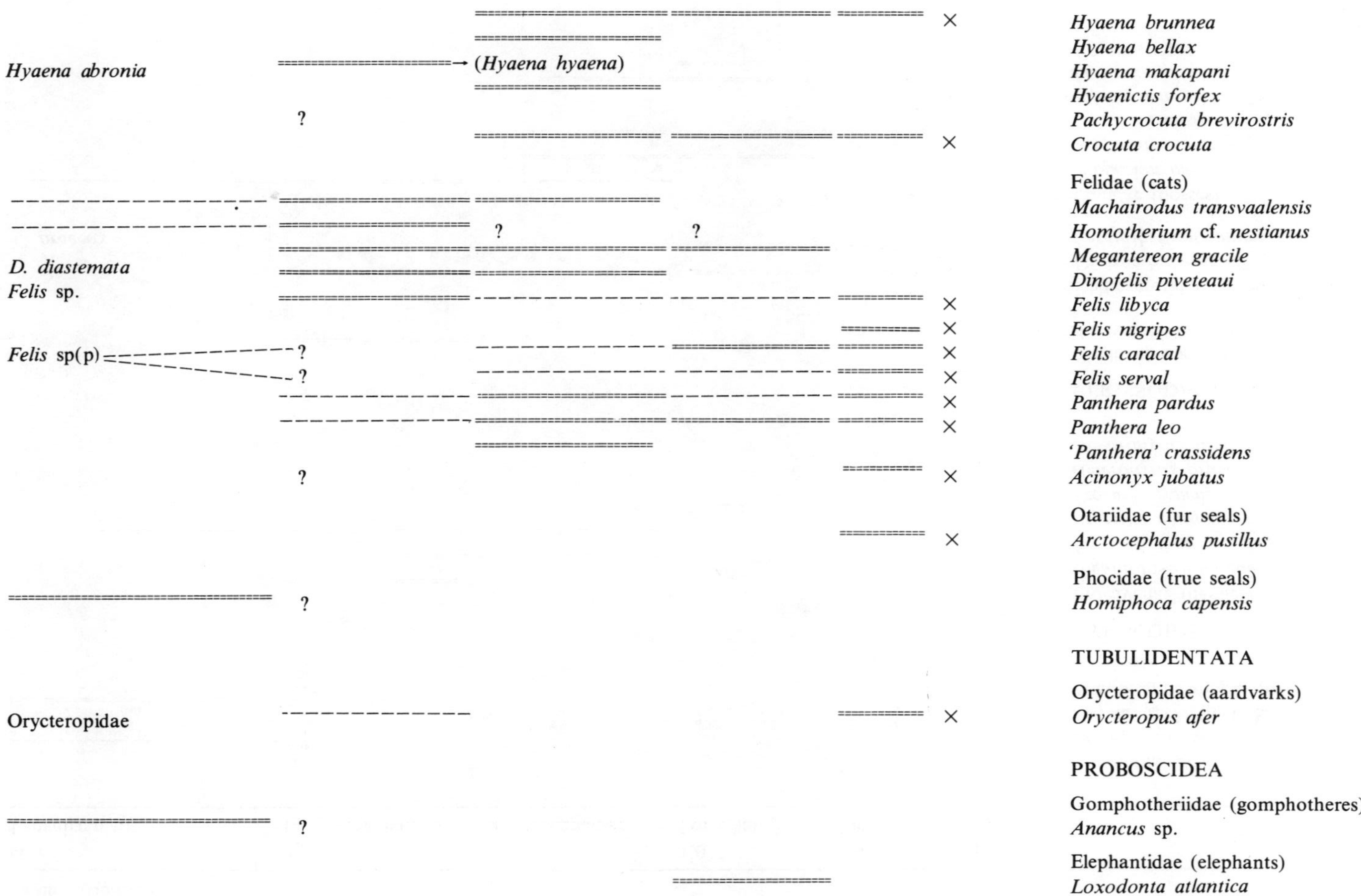
Hyaena abronia
(Hyaena hyaena)
D. diastemata
Felis sp.
Felis sp(p)
Orycteropidae
?
×
Hyaena brunnea
Hyaena bellax
Hyaena makapani
Hyaenictis forfex
Pachycrocuta brevirostris
Crocuta crocuta
Felidae (cats)
Machairodus transvaalensis
Homotherium cf. nestianus
Megantereon gracile
Dinofelis piveteaui
Felis libyca
Felis nigripes
Felis caracal
Felis serval
Panthera pardus
Panthera leo
'Panthera' crassidens
Acinonyx jubatus
Otariidae (fur seals)
Arctocephalus pusillus
Phocidae (true seals)
Homiphoca capensis
TUBULIDENTATA
Orycteropidae (aardvarks)
Orycteropus afer
PROBOSCIDEA
Gomphotheriidae (gomphotheres)
Anancus sp.
Elephantidae (elephants)
Loxodonta atlantica

Table 3 (*Contd.*)

Myr Langebaanian	3.5 Early Makapanian	1.8 Late Makapanian	1.0 Cornelian	0.4 Florisian	0.01 Recent	
				====	×	*Loxodonta africana*
----	?	?				*Elephas ekorensis*
			?	====		*Elephas recki* (incl. *E. iolensis*)
====	?					*Mammuthus subplanifrons*
						HYRACOIDEA
						Procaviidae (dassies)
	====					*Gigantohyrax maguirei*
	====	====				*Procavia transvaalensis*
----	====	====				*Procavia antiqua*
			?----	====	×	*Procavia capensis*
				====	×	*Heterohyrax brucei*
				====	×	*Dendrohyrax araboreus*
						PERISSODACTYLA
						Chalicotheriidae (chalicotheres)
	====					*Ancyclotherium hennigi*
						Rhinocerotidae (rhinoceroses)
	----	====	====	====	×	*Diceros bicornis*
C. praecox	----	----	====	====	×	*Ceratotherium simum*
						Equidae (horses)
----	====	----				*Hipparion baardi*
		====	====			*Hipparion libycum*
		====	====	====		*Equus capensis*
				====	×	*Equus zebra*
		====	====	====	×	*Equus burchelli*
			====	====	×	*Equus quagga*

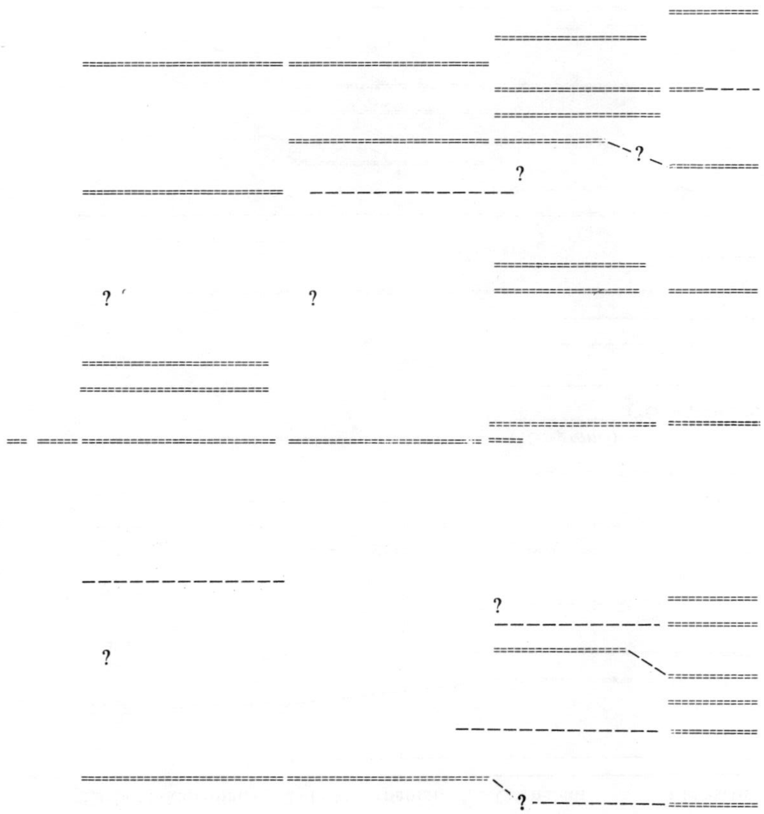

		ARTIODACTYLA
		Suidae (pigs)
	×	*Potamochoerus porcus*
		Kolpochoerus limnetes
		Metridiochoerus andrewsi
	×	*Metridiochoerus compactus*
		Metridiochoerus hopwoodi
		Phacochoerus antiquus
	×	*Phacochoerus aethiopicus*
Nyanzachoerus sp(p)		*Notochoerus capensis*
		Hippopotamidae (hippopotamuses)
		Hippopotamus gorgops
Hippopotamus sp.	×	*Hippopotamus amphibius*
		Giraffidae (giraffes)
Giraffa sp.		*Giraffa jumae*
		Giraffa gracillis
	×	*Giraffa camelopardalis*
		Sivatherium maurusium
		Bovidae (antelopes, buffaloes)
	×	*Cephalophus natalensis*
	×	*Cephalophus silvicultor*
	×	*Cephalophus monticola*
	×	*Sylvicapra grimmia*
	×	*Raphicerus campestris*
Raphicerus paralius		*Raphicerus* sp.
	×	*Raphicerus melanotis*
	×	*Raphicerus sharpei*
	×	*Ourebia ourebi*
	×	*Nesotragus moschatus*
		Oreotragus major
	×	*Oreotragus oreotragus*
	×	*Madoqua kirkii*

Table 3 (*Contd.*)

Myr Langebaanian	3.5 Early Makapanian	1.8 Late Makapanian	1.0 Cornelian	0.4 Florisian	0.01 Recent	
		---	---	===	×	*Pelea capreolus*
				===	×	*Redunca fulvorufula*
	===	--- \				*Redunca darti*
			===	===	×	*Redunca arundinum*
Kobus sp(p)			---	===	×	*Kobus ellipsiprymnus*
					×	*Kobus vardoni*
			===	===	×	*Kobus leche*
				===	×	*Aepyceros melampus*
		===	===	===		*Antidorcas bondi*
	---	===	=== \			*Antidorcas recki*
				===	×	*Antidorcas marsupialis*
			===	===		*Antidorcas australis*
Gazella sp.	===	===	→(?*G. granti*)			*Gazella vanhoepeni*
			=== (?*G. thomsoni*)			*Gazella* sp.
			---	---	×	*Oryx gazella*
				===	×	*Hippotragus equinus*
			---	===	×	*Hippotragus niger*
			---	===	×	*Hippotragus leucophaeus*
	===	===	===			*Hippotragus gigas*
				---	×	*Damaliscus lunatus*
	===					*Damaliscus gentryi*
	---	---	===	===		*Damaliscus niro*
			===	===	×	*Damaliscus dorcas*
		===				*Parmularius angusticornis*
	===					*Parmularius braini*
		===	?			*Parmularius parvus*
Damalacra sp(p)		===	=== \			*Rabaticeras arambourgi*
				===	×	*Alcelaphus buselaphus*
			===	---		*Alcelaphus lichtensteini*
	---	---	===	===		*Megalotragus priscus*
			===	===	×	*Connochaetes gnou*
	---	---	===	===	×	*Connochaetes taurinus*

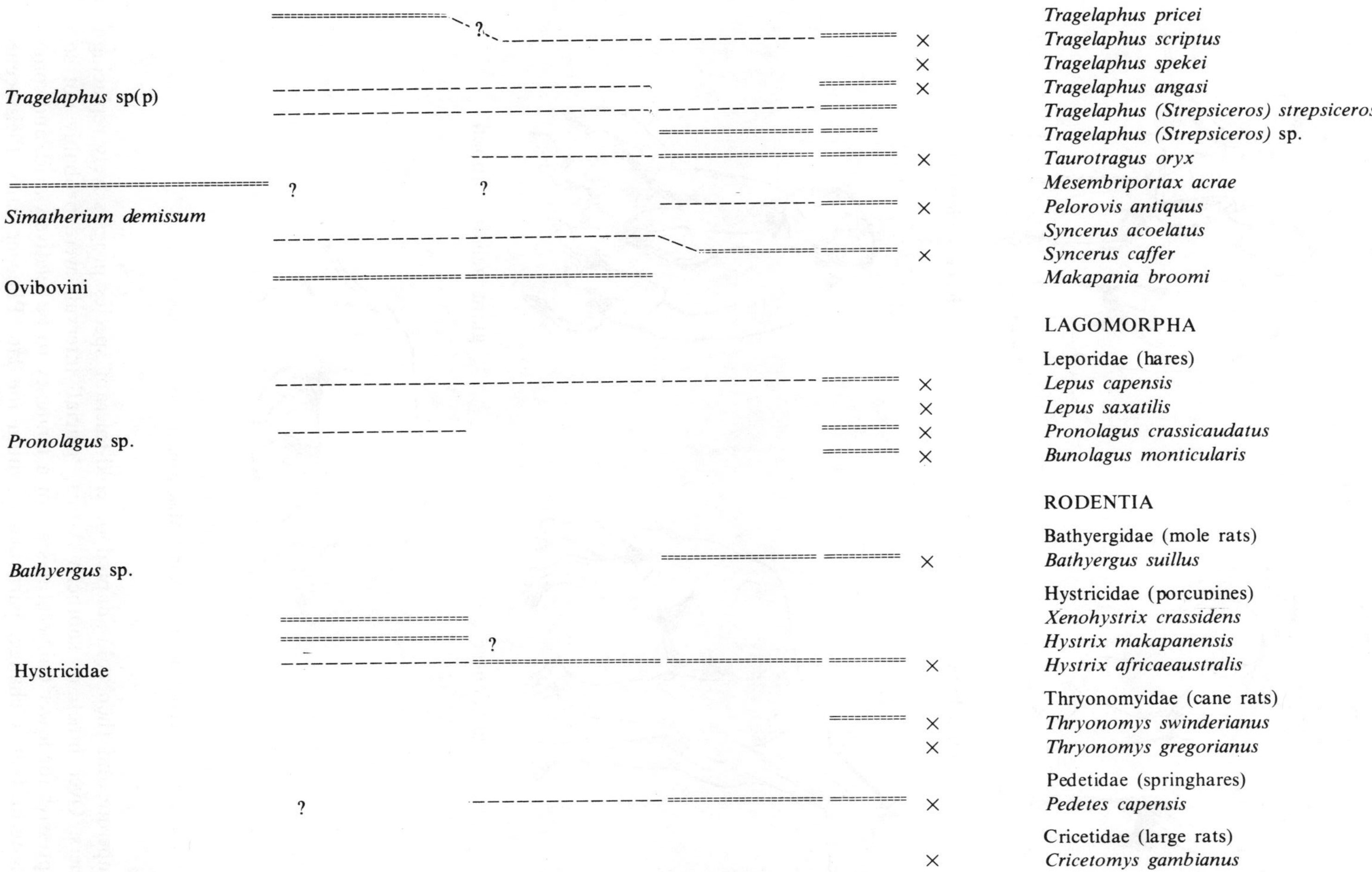
Tragelaphus sp(p)
Simatherium demissum
Ovibovini
Pronolagus sp.
Bathyergus sp.
Hystricidae
?
Tragelaphus pricei
Tragelaphus scriptus
Tragelaphus spekei
Tragelaphus angasi
Tragelaphus (Strepsiceros) strepsiceros
Tragelaphus (Strepsiceros) sp.
Taurotragus oryx
Mesembriportax acrae
Pelorovis antiquus
Syncerus acoelatus
Syncerus caffer
Makapania broomi
LAGOMORPHA
Leporidae (hares)
Lepus capensis
Lepus saxatilis
Pronolagus crassicaudatus
Bunolagus monticularis
RODENTIA
Bathyergidae (mole rats)
Bathyergus suillus
Hystricidae (porcupines)
Xenohystrix crassidens
Hystrix makapanensis
Hystrix africaeaustralis
Thryonomyidae (cane rats)
Thryonomys swinderianus
Thryonomys gregorianus
Pedetidae (springhares)
Pedetes capensis
Cricetidae (large rats)
Cricetomys gambianus

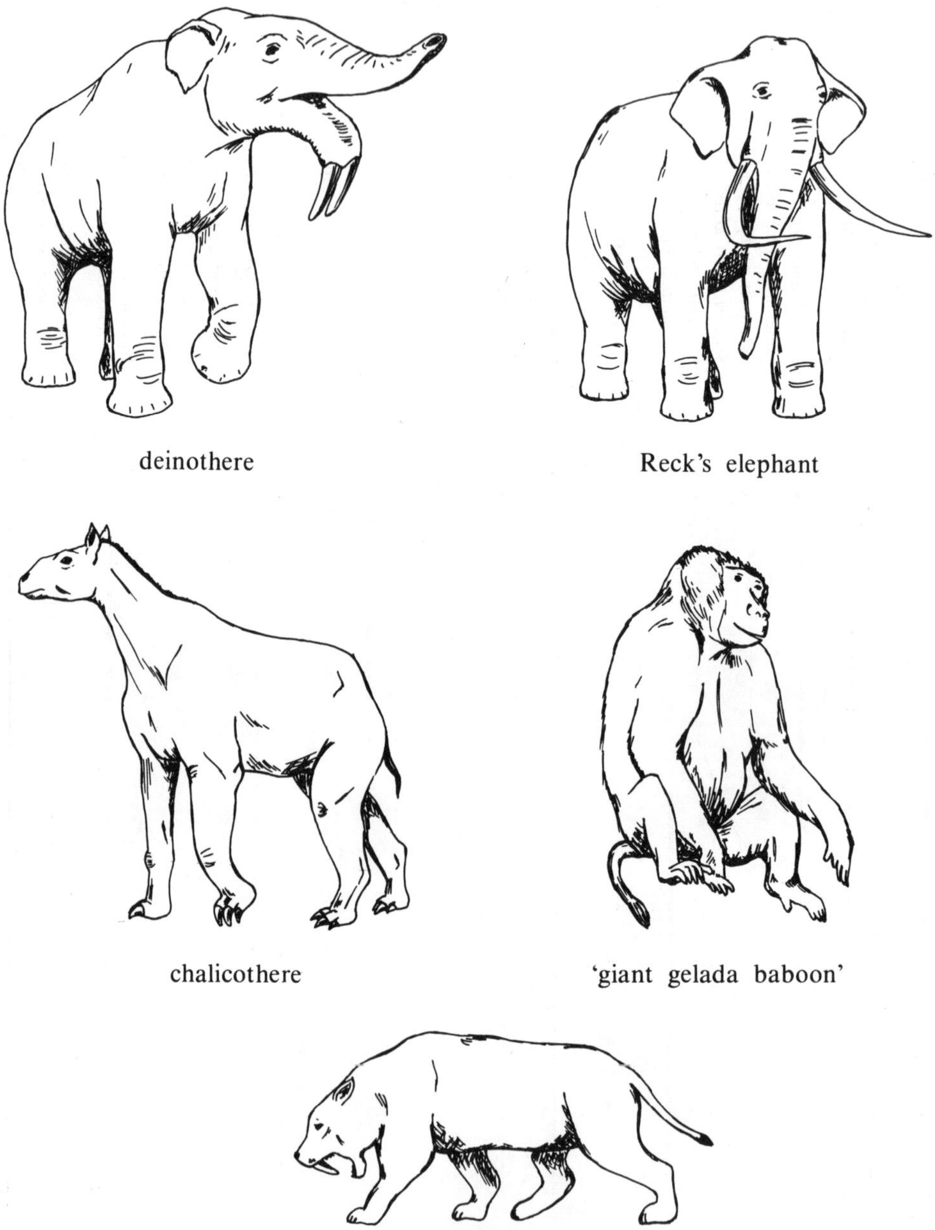

Figure 4. Reconstructions of some extinct Makapanian and Cornelian species.

temperate and then cold-adapted or cold-tolerant species progressively replacing others. Over much of southern Africa, 'glacial' intervals were probably not so impressive for lower temperatures as for a tendency to be relatively arid, and some long-term faunal change, particularly within the late Pliocene/early Pleistocene

Figure 5. Reconstructions of some extinct Florisian species.

Makapanian complex is probably a result of community adjustment to a climatic regime in which recurrent periods of relative aridity became longer, more intense or both. Such climatic change was probably particularly responsible for possible or probable Makapanian losses in suids, giraffids, and proboscideans.

A further 'cause' of faunal change, probably interwoven with climatic change, was almost certainly the dynamic interplay among species. It is not hard to visualize a connection between the decline and disappearance of the sabertooths (*Machairodus, Homotherium,* and *Megantereon*) and false sabertooths (*Dinofelis*) in southern Africa and the broadly contemporaneous arrival and rise to prominence of large cats of the genus *Panthera*, with whom the sabertooths and false sabertooths certainly competed for prey. Additionally, from the nature of their dentition, it may be inferred that the sabertooths and false sabertooths left more for scavengers than do *Panthera* spp., so that the demise of the sabertooths could be directly responsible for the broadly contemporaneous reduction in hyaenid taxa (Ewer 1973). At least one of the extinct hyaenids – *Chasmaporthetes nitidula*, which was probably a highly predaceous, cursorial form (Hendey 1978a) – may also have been adversely affected by the appearance (?in the Late Makapanian) and subsequent success of hunting dogs [*Lycaon* sp(p)] and of the cheetah (*Acinonyx*), whose niches probably overlapped that of the hyaenid substantially. Finally, it is possible that the overall reduction in large carnivore taxa which occurred from the Early Makapanian to the Cornelian at least partly reflects the entry and subsequent expansion of hominids into a meat-eating niche.

The increasing sucess of the hominids, especially of *Homo*, as terrestrial omnivores in grassland and open woodland settings may also have been largely responsible for the reduction in cercopithecoid primate diversity during the Late Makapanian and Cornelian. Virtually all the cercopithecoids that became extinct were large terrestrial forms (Simons & Delson 1978), probably broadly similar in character to living baboons. The like-aged reduction in suid taxa probably reflects their continuing failure (beginning earlier on in the Pliocene or late Miocene) to compete for grazing and browsing niches with burgeoning bovid lineages. It is possible that the decline of the suids is also partly responsbile for the disappearance of some of the sabertooths, if it may be assumed that large, thick-skinned, relatively slow suids were among the favored prey of some sabertooths (Hendey 1974a). Finally, the disappearance of *Hipparion* sp(p) in the Late Makapanian and Cornelian probably culminates a decline that was initiated by the appearance (immigration) of *Equus*.

Two particularly intractable puzzles in any attempt to 'explain' extinctions via climatic change or unsuccessful competition are the disappearance of *Gazella* and of *Elephas* sp(p) from southern Africa. Modern field studies of *Gazella* and *Antidorcas*, together with their close phylogenetic relationship and their completely non-overlapping distributions, clearly suggest that they occupy very similar niches and that the presence of *Antidorcas* essentially excludes *Gazella* from southern Africa (while *Gazella* excludes *Antidorcas* elsewhere). Yet, in the Early Makapanian, Late Makapanian, and Cornelian, the two taxa overlapped extensively in southern Africa and even co-existed at its southwestern extremity. There is similar, like-aged evidence for their coexistence (sympatry) in East and even in North Africa (Gentry 1978). The circumstances that led to their total historic separation remain unclear, but perhaps they were separated at a time (?in the Florisian) when East Africa was relatively moist while much of southern Africa was relatively dry. Butzer (1978a) discusses the former occurrence of this circumstance. A return to sympatry may have become impossible due to the creation of those conditions in the Zambesian Zone that presently make it virtually impenetrable to either *Antidorcas* or *Gazella* (see the next section).

The fate of African *Elephas* poses a similar mystery. *Elephas* sp(p) appear to have been the most common elephants in southern (and eastern) Africa as late as the Cornelian. This is in contrast to the Florisian and the Recent from which only the widespread genus *Loxodonta* is known. It seems unlikely that *Loxodonta* is filling the niche that *Elephas* occupied previously. Partly this is suggested by significant differences in dentition (*Elephas*, particularly advanced *Elephas*, is much more hypsodont, with, thin, tightly-folded enamel, and relatively closely packed molar plates), pointing to probable differences in diet (?significantly more grazing in *Elephas*). Partly it is suggested by the fact that the two taxa rarely occur in the same site (so far not at all in southern Africa), while at least in southern Africa, fossils of *Elephas* have been found in arid regions (at Kathu Pan and especially Namib 4; see Figure 3) where *Loxodonta* was rare or absent historically and is unknown as a fossil. Additionally, in the multilevel site of Kathu Pan, a presumably mid-Pleistocene (Cornelian) fauna in which advanced *Elephas recki* is the most abundant element is replaced upwards in the sequence by a presumed later Pleistocene (Florisian) fauna which is broadly similar in overall composition, except that elephants of any kind are absent and their numerical position is now occupied by bovines (*Pelorovis* and or *Syncerus*), which were relatively rare or absent in the *Elephas*-dominated layers (Beaumont & Klein, unpub.) This raises the possibility that *Elephas* occupied a niche broadly similar to that of the indigenous African bovines, without unfortunately shedding further light on why *Elephas* became extinct in Africa.

As a final point here, it is interesting to note that Table 3 is richer in probable disappearances than in appearances, which is to say that the Recent fauna of large mammals in southern Africa is impoverished vs. that of earlier times. Impoverishment of Recent vs. late Pliocene and Pleistocene faunas is also characteristic elsewhere in the world and is particularly well-documented in Europe (Kurtén 1968; Maglio 1975; Bonifay 1980) and North America (Kurtén & Anderson 1980). Everywhere, it probably reflects long-term global climatic deterioration, beginning in the late Miocene, as much as ten million years ago.

Given sampling and dating problems (which are perhaps particularly acute for southern Africa), it may be premature to compare the extent and rate of faunal turnover among continents. However, North America clearly has the largest number of both appearances and extinctions, with Europe a poor second and southern (and eastern) Africa third. Similarly, North America is also the undeniable leader in the rate of appearances and extinctions, again with Europe a distant second and southern Africa third. The large number of first appearances and extinctions on both Northern Hemisphere continents almost certainly reflects the fact that ecological change during 'glacial'/'interglacial' alternations was more radical there than in southern (or eastern) Africa. This point is reinforced by the observation that taxa which disappeared from temperate latitudes in North America often managed to survive near the Equator (Webb 1977).

The very high number of first appearances and extinctions in North America vs. either Europe or southern Africa reflects the remarkable extent to which North America exchanged species with Asia in the late Pliocene to Recent interval, with North America acting mainly as a recipient (Webb 1969, 1977; Repenning 1969, also 1980). By definition, this would lead to more 'first appearances,' and the systemic instability that resulted from so much immigration was probably also responsible for the remarkable number of extinctions that took place in North America. It is particularly from a North American perspective of so much loss that it seems reasonable to refer to recent Africa as "the living Pleistocene.'

CHANGES IN LARGE MAMMAL DISTRIBUTIONS: LATE PLIOCENE TO RECENT

Several prominent southern African mammals have strikingly discontinuous distributions, as summarized succinctly by Meester (1965). To begin with, there are 'forest species'-bushpig, red duiker, blue duiker, tree dassie, samango (or blue) monkey, and palm civet-whose heartland is in the lowland forests of the Guinean Ecozone (Figure 1). From there they occur discontinuously in patches of forest or dense woodland through the Zambesian Zone into the eastern part of the Transvaalian Zone, and in the case of the blue duiker and bushpig, to the eastern part of the Cape Zone.

Even more notable are several species more at home in arid or semi-arid grassland and woodland-dikdik, steenbok, gemsbok, tsessebe (topi), red hartebeest (kongoni), mountain reedbuck, giraffe, aardwolf, bat-eared fox, black-backed jackal, and springhare-which are distributed very discontinuously within the Zambesian Zone or absent from it altogether, though they are relatively abundant in zones both to the south and northeast of it. To these open-country species may be added antilopine antelopes (the 'gazelles' of which the springbok is a southern African representative), which are totally absent in the Zambesian Zone, though one or more antilopines are common in all grassland and woodland zones north and south of it (Gentry 1971). The implication of these disjunct distributions is that the historic ecozones once had a different configuration.

Taking into account similarly disjunct distributions in insects and lower vertebrates, as well as mammals, Balinsky (1962) has suggested that rainforest must once have existed more or less continuously from northwest to southeast across (what is here termed) the Zambesian Zone. At other times, the zone was crossed from northeast to southwest by an arid 'drought corridor' even now detectable in a continuous belt of country in which rainfall averages less than 10 mm/month for more than three consecutive months each year. It is interesting, however, that species which indicate more or less continuous forest are fewer in number, and in general, probably more ancient than species pointing to more widespread aridity. Additionally, at least with respect to the large mammals, the two most widespread 'forest species' (blue duiker and especially bushpig) could probably have achieved their present distributions with only slightly moister conditions, while the samango monkey, which may have required much moister ones, is characterized by well-differentiated subspecies in the different forest islands it occupies. This suggests it achieved its present distribution a very long time ago. More generally, it may be argued that the much moister conditions suggested by the 'forest species' occurred much less recently than the more arid ones suggested by the open-country group. This is in accord with paleoclimatological (mainly geomorphic) evidence that significantly greater aridity was probably common during the Pleistocene in Africa, while climate moist enough to support continuous forest through the Zambesian Zone has probably not existed since the late Tertiary (see Butzer 1978a for a general summary of Plio-Pleistocene climatic change in Africa)

Important and often dramatic changes in precipitation in southern Africa have been especially well-documented for the late Pleistocene and Recent (Holocene), beginning approximately 130,000 years ago, when they were clearly related to the same changes in global temperatures and atmospheric circulation patterns that resulted in 'ice-ages' in the Northern Hemisphere. Although the precise pattern of late Pleistocene and Holocene climatic change has yet to be worked out for any southern African ecozone, it is clear that there was no simple correlation between temperature and precipitation (see Butzer, this volume). Cooler/drier, cooler/wetter, warmer/drier, and perhaps warmer/wetter conditions are all recorded in one or another ecozone. In most places, 'glacial' maxima probably witnessed drier conditions, but this was not universally true, and more generally, it is clear that there were times when one area was drier, while others were wetter than at present. In most places, relative to the present, periods of greater aridity were probably more extreme than periods of greater moisture.

I have recently reviewed the evidence for late Pleistocene and Holocene large mammal distributions in southern Africa (Klein 1980) and found that most fossil faunas, whether of 'glacial' or 'interglacial' age, are dominated by the same species that predominated historically. The principal exceptions are in the Cape Zone. Here (and in adjacent parts of the Karoo-Namaqualian and Transvaalian Zones), faunas dominated by historic species (mainly browsers) alternated with ones dominated by grassland species – black wildebeest, springbok, warthog, quagga, white rhinoceros, and others – which were absent from the Cape historically or occurred only on its peripheries. Sedimentological studies suggest that these species entered the Cape Zone primarily when climate was cooler and wetter, presumably from the neighboring Basutolian or Karoo-Namaqualian ecozones. The fact that faunal response to climatic change was apparently much greater in the Cape than in other ecozones almost certainly reflects its greater distance from the Equator, combined with its unique climate and vegetation versus that of all other zones.

Although late Pleistocene faunal change was more subtle in other zones, it did occur. Thus, at one or more times during the late Pleistocene, blesbok, springbok, and mountain reedbuck lived in the southern part of the Zambesian Zone, suggesting that the local vegetation may have acquired some characteristics of Kalaharian or even Basutolian communities. It remains unclear whether northwards extension of the springbok and other species occurred under wetter conditions, drier conditions, or perhaps both. The occurrence of springbok in part fills the 'gazelle-free' gap mentioned previously, as does the occurrence of Thomson's gazelle at late Pleistocene and Holocene sites in the northern part of the Zambesian Zone.

To the south, lechwe has been recorded in late Pleistocene sites in the southern part of the Kalaharian Zone and northern part of the Basutolian Zone. This is 700-800 km south of its nearest historic record (in the swampy regions of northern Botswana in the Transvaalian Zone.) The precise date is unclear, but sedimentological and palynological evidence point to one or more intervals within the 'Last Interglacial,' when conditions were probably periodically moister than they were historically.

Possibly wetter and warmer conditions, promoting the spread of trees or bush, are also reflected in the occurrence of giraffe, Cape buffalo, roan antelope, and bushbuck at various imperfectly dated late Pleistocene sites within the Basutolian Zone. It is important to reiterate, however, that these species are very much second in importance to alcelaphine antelopes (wildebeest, hartebeest, etc.), equids, and other species similar or identical to the ones that dominated the Basutolian grasslands historically, so that the extent of environmental change that is implied is not great.

Overall, in fact, with the exception of the Cape Zone, the kinds of changes suggested by the late Pleistocene and Holocene large mammal data are adjustments in ecozone boundaries or shifts in internal mosaics, rather than wholesale reformulations. This accords broadly with other paleoenvironmental (sedimentological/palynological) data (as discussed by Butzer, this volume).

Sampling and dating problems make it difficult to assess possible relationships between yet earlier (late Pliocene to mid-Pleistocene) mammal distributions and general climatic/environmental change, but there is at least a strong hint of a major change in fauna and environment within the Makapanian complex (*s.l.*), followed by relative stability thereafter, except perhaps in the Cape Zone. Thus, the Early Makapanian (late Pliocene) faunas of Makapansgat and Sterkfontein are characterized by lower proportions of hypsodont to hyperhypsodont alcelaphine and antilopine antelopes than the Late Makapanian and Cornelian (early and mid-Pleistocene) faunas of Swartkrans and Kromdraai A (Wells & Cooke 1956; Vrba 1975, 1976, 1980).

Since Makapansgat is located in the (modern) Transvaalian Zone, while Swartkrans and Kromdraai are in the Basutolian one, the difference between them could perhaps be ascribed to geography, but that does not explain the difference between Sterkfontein and Swartkrans/Kromdraai, which are all located within sighting distance of one another. It seems at least possible, perhaps likely that the difference between Makapansgat and Sterkfontein on the one hand and Swartkrans and Kromdraai on the other reflects a long-term trend, beginning in the late Pliocene, for periods of greater aridity to become relatively longer, more intense, or both, as discussed in the previous section. An interpretation linking the difference between Early Makapanian and later faunas to global climatic change is supported by like-aged trends for proportional increases in open-country species in East Africa (Gentry & Gentry 1978) and also in Europe (Bonifay 1980) and North America (Webb 1977).

In the Cape Zone, the Early and Late Makapanian are poorly known (?at Baard's Quarry, Langebaanweg), but there is an exceptionally rich Cornelian fauna from the site of Elandsfontein (= Hopefield = Saldanha). This fauna is overwhelmingly dominated by grazers – especially alcelaphine antelopes and equids – which could not have existed in such numbers or diversity in a vegetational setting similar to the historic fynbos. The relative and absolute abundance of large grazers suggests one or more mid-Pleistocene climatic episodes in which cooler, wetter conditions prevailed, similar to those which periodically promoted the dominance of large grazers during the late Pleistocene in the Cape Zone.

Curiously, however, the Elandsfontein fauna also contains a sivathere, implying the presence of trees, and a civet cat, most at home in dense woodland or thicket, though entirely terrestrial. Suitable habitat for either species was absent historically, nor is there any evidence it existed nearby in the late Pleistocene. It is possible that giraffe, found in fossil form at Bloembos near Elandsfontein, and also indicative of trees, dates from the same interval as the sivathere and civet. Unfortunately, most fossils at Elandsfontein were unsystematically collected from the surface, and it is impossible to determine whether the sivathere and civet were contemporaneous with the alcelaphines, equids, etc. However, they still imply that at least a part of the mid-Pleistocene was characterized by environmental change that was qualitatively different from late Pleistocene and Holocene change. This may also be inferred from geomorphic evidence (Butzer 1973b) and from study of the mean size of individuals in some Elandsfontein species, as reported in the next section.

MAMMAL SIZE VARIATION AND CLIMATIC CHANGE

Mean individual size has been shown to vary with climate in a number of Northern Hemisphere mammal species, particularly in carnivores and micromammals (small rodents and insectivores). Kurtén (1959, 1965) has presented some examples and discussed the reasons for size variation, including Bergmann's Rule and so-called population density factors. Bergmann's Rule states that, all other things equal, individuals of a species will tend to be larger in colder climates. This is because as an animal grows, its volume, which produces heat, increases more rapidly than its skin area, which dissipates it. Bergmann's Rule has been shown particularly to characterize carnivores, though there are many species that do not obey it. The reason is apparently that 'other things' are not always equal. Probably most important is the fact that individuals must be able to find sufficient food to maintain a larger body. It is in this context that 'population-density factors' become important. In particular, increased body size will probably be selected against, if it reduces the number of individuals the environment can support below the critical level necessary for successful reproduction. Through its effect on food supply, a change in climate, even greater cold, might thus actually induce a reduction in mean individual body size.

Avery (1982a) has shown that mean size in several southern African micromammal species is probably related to variation in temperature or precipitation, but otherwise very little research of this kind has been done in southern Africa. In my own work, designed to explore the possible linkages between climate and size in somewhat larger mammals, I have focussed on species which are particularly common at late Pleistocene and Holocene sites in the Cape Zone. It will be remembered that this is the zone in which late Pleistocene changes in large mammal distributions and relative species abundance were most dramatic, indicating that local climatic/environmental fluctuations were certainly profound enough to affect local mammals. Additionally, it is the zone in which I have done most of my own research, and it has an especially rich late Pleistocene and Holocene faunal record.

So far, I have concentrated especially on the black-backed jackal, rock hyrax, Cape mole rat, and Cape grysbok. I have found that mean size in the jackal increases with decreasing temperature, while mean size in the other three species appears to be related to precipitation, at least within the Cape Zone.More specifically, as rainfall increases, the mole rat and grysbok appear to increase in size, while the hyrax decreases. The inverse relationship between grysbok and hyrax probably reflects the fact that as rainfall increases, preferred grysbok food (browse) becomes proportionately more abundant relative to preferred hyrax food (grass). I have been unable to locate detailed information on the preferred food of the mole rat, but in any case, it may be responding not so much to a change in the proportion of preferred food as to a general increase in vegetation density as rainfall increases in that part of the Cape Zone it inhabits.

Some of my data on the mole rat, based on the medio-lateral diameter ('breadth') of the distal humerus are presented here in Figure 6. The right-hand side of the figure

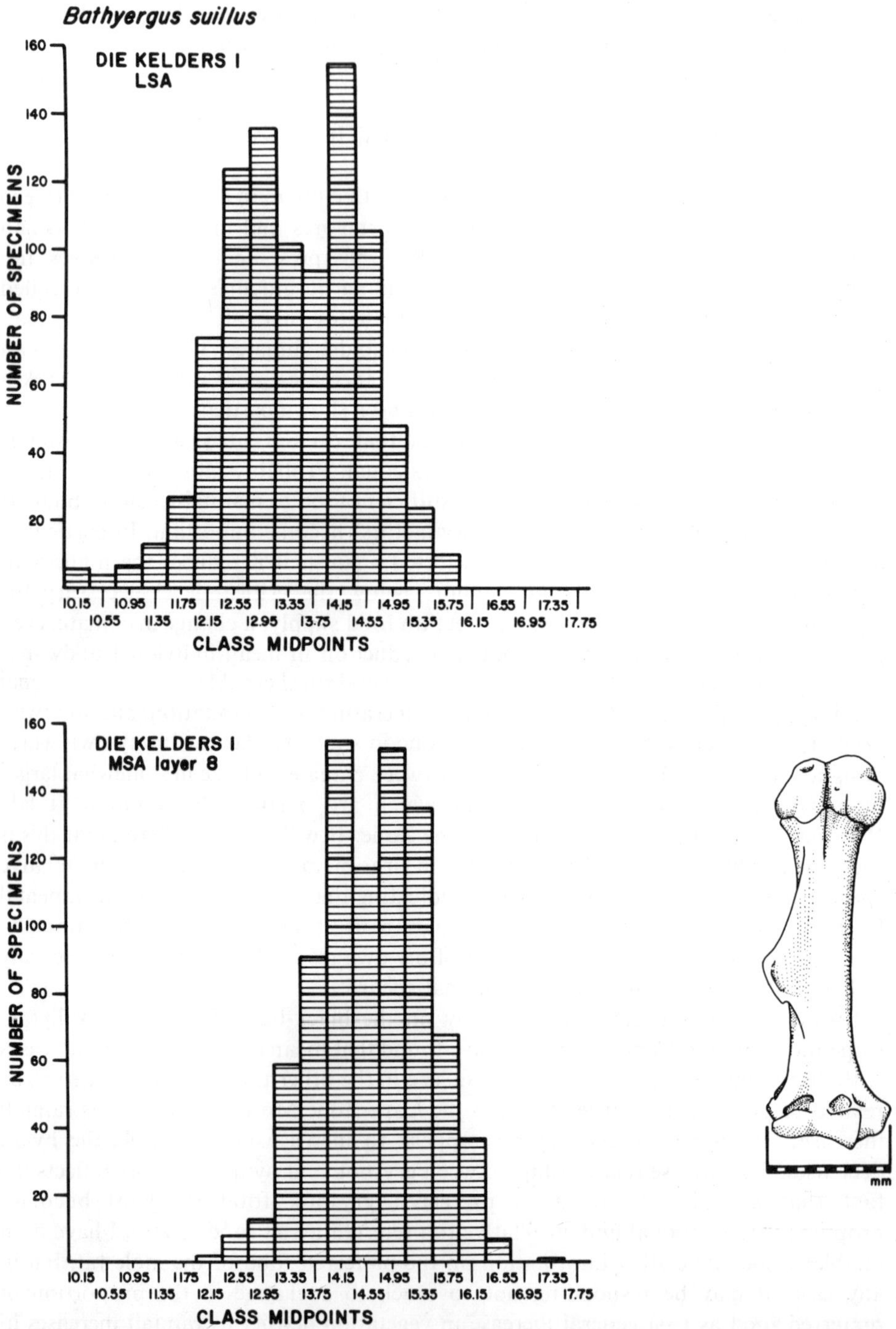

Figure 6. Left: Frequency distributions of Cape Mole Rat distal humerus 'breadths' in the Later Stone Age (LSA) levels and in Middle Stone Age (MSA) level 8 at Die Keiders. Measurements which were originally made to the nearest tenth of a millimeter have been grouped into 0.4 mm classes.

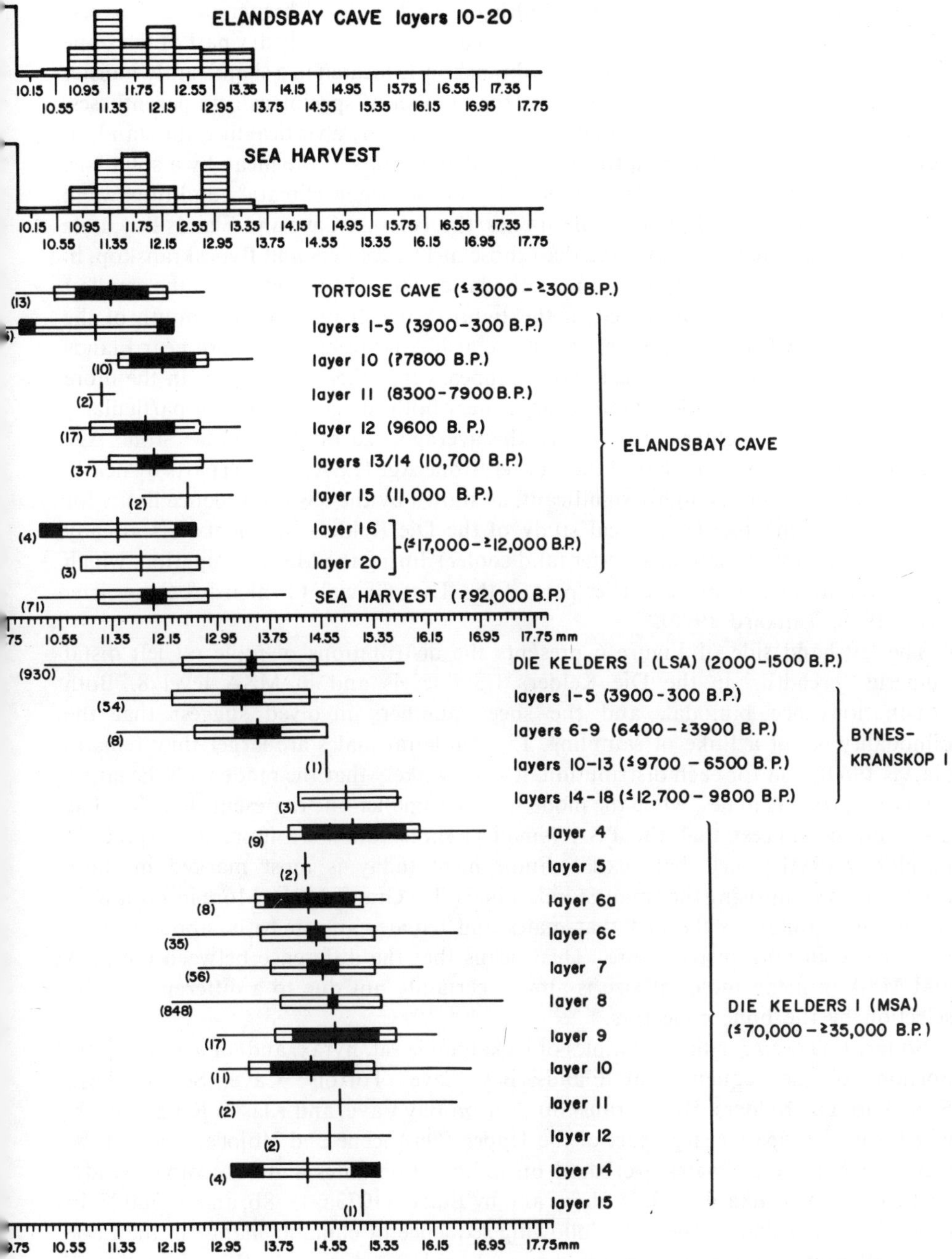

Figure 6 (continued. Upper: Frequency distributions of mole rat distal humerus 'breadths' in layers 10-20 of Elands Bay Cave and at Sea Harvest.
Lower: the mean (vertical bar), range (horizontal bar), standard deviation (open bar), and 95% confidence limits of the mean (solid bar) for mole rat distal humerus 'breadths' from Tortoise Cave, Elands Bay Cave, Sea Harvest, Die, Kelders, and Byneskranskop. For discussion see the text.

presents basic parameters for the breadth of the left distal humeri from Tortoise Cave, Elands Bay Cave, and Sea Harvest, located in a relatively dry part of the Cape Zone, and from Die Kelders 1 and Byneskranskop 1, located in a significantly moister area. For each site or level within a site, the number of specimens is in parentheses, the range is represented by a horizontal line, the mean by a vertical line, the standard deviation by an open bar, and the 95% confidence limits of the mean by a solid bar.

The figure shows that during the late Holocene, when climatic conditions were almost certainly very similar to modern ones, the Tortoise Cave and Elands Bay mole rats were much smaller on average than those at Die Kelders and Byneskranskop, in keeping with greater precipitation near the latter sites. (In recent times, the smallest mole rats of all, not represented in the figure, came from near the mouth of the Orange River, where they live in even more arid conditions than those near Elands Bay and Tortoise Cave.) Interestingly, it appears that there were times in the more distant past when the mole rats were larger near both site groups. This is particularly clear for the Die Kelders area, where the average size of the Middle Stone Age (MSA) mole rats exceeds that of their Later Stone Age (LSA) counterparts by nearly 10%. (The difference is highly significant, as shown by the 95% confidence limits for the respective means.) Geological study of the Die Kelders sediments has already indicated that the climate was wetter (and cooler) during the MSA occupation, which is believed to date from the earlier part of the 'Last Glacial' (Tankard & Schweitzer 1974, 1976; Tankard 1976).

The left-hand side of Figure 6 presents the distributions of mole rat left distal humerus 'breadths' in the Die Kelders LSA levels and in MSA level 8. Both distributions are bimodal, and the sheer numbers involved suggest that the bimodality is not a fluke of sampling. Live mole rat males are larger than females (Jarvis 1969), and for each distribution, it is very likely that the mode with the larger humeri represents males, while the mode with the smaller one represents females. The distributions suggest that the LSA (smaller) mole rats were more dimorphic, in keeping with the fact that sexual dimorphism today is most marked in those populations comprising the smallest individuals (De Graaff 1971). More important in the present context is the fact that males and females appear to be about equally represented in both distributions. This means that the difference between the LSA and MSA in mean mole rat size is almost certainly not due to a difference in size selection by the bone collectors.

So far, I have size data on samples of jackal, mole rat, hyrax, and/or grysbok from portions of the sequences at Elands Bay Cave, Tortoise Cave, Sea Harvest, Swartklip, Die Kelders, Byneskranskop, Nelson Bay Cave, and Klasies River Mouth, which together span a large part of the Upper Pleistocene and Holocene within the Cape Zone. [Age estimates are based on radiocarbon dates and on geologic study, particularly by Tankard as cited above and by Butzer (1973a, 1978b, and unpub.)] In general, the size data suggest the following sequence of climatic change in the Cape Zone: alternating warm/wet, warm/dry, and cold/wet during the 'Last Interglacial' (ca. 130,000 to 70,000 years ago), followed by cold/wet in the earlier part of the 'Last Glacial,' followed probably by cold/very dry during the later part (there are few fossil samples from this interval – the best ones, from Boomplaas Cave, have yet to be measured), followed by warm/dry in the early Holocene, followed finally by warm/wet (like present) in the later Holocene. This scheme agrees well with ones based on sedimentological/geomorphic observations made by Butzer (Butzer & Helgren 1972; Butzer, Stuckenrath *et al.* 1978, and unpub.), and in terms of temperature, is consistent with the global marine record (Shackleton & Opdyke 1973, 1976). Particularly noteworthy in this context are the presence of pronounced cold intervals during the 'Last Interglacial.'

One interesting aspect of the study so far concerns the mid-Pleistocene fauna from Elandsfontein, discussed in the last section. The Elandsfontein jackal is quite small, while the grysbok is enormous, suggesting conditions that were at least as warm and probably wetter than any that occurred during the Upper Pleistocene and Holocene. As noted in the last section, a mid-Pleistocene climatic interval(s) qualitatively different from any known Upper Pleistocene/Holocene one may also be inferred from the species composition of the Elandsfontein fauna and from the geology of the site.

Although the research has only just begun, it seems likely that mammal size variation will prove to be a powerful tool for inferring past environments and climates in southern Africa. This is particularly important since pollen is often absent or poorly preserved in local fossil sites, and paleoenvironmental reconstruction must often rest on fauna and sediments, or on fauna alone. A firmer grasp of environmental/climatic change is important not only in its own right, but also if we are to achieve a firmer grasp of the selective forces behind the changing nature of the southern African fauna between the late Pliocene and the Recent.

SUMMARY AND CONCLUSION

The large mammal fauna of southern Africa has had a distinctively African character during the entire late Pliocene to Recent interval, covering approximately the last 3.5 million years. However, it was far from static in composition. Losses are particularly apparent in the suids, proboscideans, and carnivores, while gains are most obvious in the bovids. Although change was somewhat less dramatic than in Europe and much less dramatic than in North America, the recent fauna is still impoverished versus that of earlier intervals and particularly versus that of the late Pliocene.

The reasons for faunal change are complex, and probably involve climatic change, especially long-term fluctuation in available moisture, and the dynamic interplay among species. Particularly important in the letter context has been the burgeoning of the bovids, the rise to prominence of large cats of the genus *Panthera,* and perhaps very importantly, the continuing evolution of the hominids, especially the emergence of the genus *Homo* and the species *Homo sapiens.* It seems quite possible that this species was responsible for the extinction of a series of others at or near the end of the Pleistocene, 12-10,000 years ago.

In addition to taxonomic change, the fauna has also been subject to repeated redistribution, probably mainly as a result of climatic/environmental change. Climatic/environmental change has also been responsible for long-term fluctuations in the average size of individuals in several species. Taken together, changes in species distributions and in average individual size in fact constitute important sources of information about the course of past environmental change.

There is still much to be learned about the nature and causes of long-term faunal change in southern Africa, especially prior to the Upper Pleistocene, 130,000 years ago and before. Future progress will require the excavation of large samples from as wide a range of contexts as possible, accompanied by the careful collection of sedimentological, palynological, and other contextual data, and followed by detailed laboratory analyses of sample composition.

ACKNOWLEDGEMENTS

I thank K.W. Butzer, C.S. Churcher, H.B.S. Cooke, A.W. Gentry, D.P. Gifford, and T.P. Volman for comments on a preliminary draft and the National Science Foundation for research support. K. Cruz-Uribe helped to draft the figures.

G.P. RIGHTMIRE
Department of Anthropology, State University of New York, Binghamton, N.Y., USA

5

The fossil evidence for hominid evolution in southern Africa

Hominids of the genus *Australopithecus* are now thought to have emerged in Africa, probably 4 million years ago or earlier, and that continent has continued throughout the Pleistocene to be of central importance to human evolution. Many significant discoveries have been made in eastern and in southern Africa since the beginning of the present century. In this review, the fossil evidence from southern Africa is emphasized, but it is not really feasible to consider one body of material in isolation from the other, and I have referred from time to time to major East African remains as well. This is important, for example, in any discussion of the australopithecines. Quantities of these hominids are known from both regions, and the best indications of the age of the Transvaal limestone cave deposits still stem from comparisons of faunal assemblages with East African mammal collections that have been dated radiometrically. Parallel studies are also required of later, mid-Pleistocene hominids, which have been found on the Cape coast, in the interior of southern Africa, and at eastern African localities. This work, and examinations of Upper Pleistocene skeletons, have proceeded to the point where it is reasonable at least to outline a story of human evolution. Fossils are still all too scarce, however, particularly at mid-Pleistocene sites in southern Africa where few hominids have come to light in association with Acheulian tools. We are frequently left with more questions than answers, but perhaps the questions can now be framed somewhat more carefully and explicitly than was possible a decade or so ago.

PLIO-PLEISTOCENE HOMINIDS OF THE TRANSVAAL

While many different chapters in human evolutionary history are recorded at sites unearthed in southern Africa, none is more amply documented than that of the australopithecines. The story of the discovery of these small brained hominids is well known, and the significance of Dart's (1925) description of the Taung child is now widely recognized. The Taung skull was followed later by the first adult *Australopithecus,* found by Broom at Sterkfontein in 1936, and by additional finds at Kromdraai in 1938. Subsequent discoveries of hominids and other extinct animals were made at these localities and also at Makapansgat and Swartkrans during the next decade. By the early 1950's, after the flow of fossils had slowed, it was clear that the (South African) australopithecines could reasonably be sorted into two groups, one comprising the smaller individuals from Taung, Sterkfontein and Makapansgat (now collectively referred to as 'gracile'), the other the more heavily built specimens from Kromdraai and Swartkrans (now called 'robust'). Early classifications (Robinson 1954a) recognized this distinction at a generic level, and enough evidence had been gathered by 1956 to permit a detailed accounting of the major dental and cranial differences between the two forms (Robinson 1954b, 1956). Following a gap in fieldwork in South Africa, and a period of spectacular discoveries of hominid remains at Olduvai Gorge in East Africa, excavations were reopened at Swartkrans in 1965 and at Sterkfontein in 1966 (Brain 1967; Tobias & Hughes 1969). Most recently, work has begun again at Kromdraai (Vrba 1981). The last 15 years have seen the accumulation of some important new fossils, and these together with the wealth of hominid material now known from East African localities have stimulated fresh thinking about the role of the australopithecines in human evolution.

Sites, dating and paleoecology

With the exception of Taung, in the northern Cape, the sites yielding australopithecine fossils are located in the Transvaal. Sterkfontein, Swartkrans and Kromdraai are in the Sterkfontein Valley, not far from Johannesburg, while Makapansgat lies to the north, near Potgietersrus. The hominid fossils, together with other bones and sometimes crude stone artifacts as well, are found in dolomitic limestone cave deposits. These ancient caverns have lengthy histories, differing somewhat from site to site, but in each case a cave system formed by solution in the dolomite and subsequently enlarged by further solution and roof collapse was eventually opened to the surface. This initial opening may have been narrow and nearly vertical as at Swartkrans for example, and the cave at this stage would not be suitable for occupation. Sediments, rock, bone and other debris could accumulate slowly in the cavity, however, and this fill was typically cemented by deposition of calcium carbonate into a hard breccia. Later events might change considerably the form of the cavern and its entrance(s), and so initiate new periods of sediment collection. Successive breccias may therefore be of quite different age and composition, and the lithostratigraphy of these sites is frequently complex.

Another problem is dating. It is most unfortunate that the radiometric dating methods used so successfully on volcanic materials in the East African sites have so far proved inapplicable to the australopithecine cave breccias and travertines. Cores for determination of paleomagnetism have recently been obtained from several of the Transvaal localities, and results especially for Makapansgat are encouraging. However, the most reliable estimates of the ages of the South African hominids continue to be based on faunal analysis and comparison with the radiometrically dated assemblages from East Africa. Cooke's (1970) work on fossil suids and elephants is useful in this regard, and Vrba's (1975) studies of bovid remains help to assess the relative ages of different site units within the Sterkfontein Valley sequence. White and Harris (1977) in a comprehensive review of suid evolution and phylogeny, have also employed suid faunas to refine stratigraphic correlations at a number of South and East African localities.

Evidence which is increasingly convincing now suggests that the Makapansgat Limeworks may be the oldest of the Transvaal sites. Two suid species are represented,

Table 1. Relative ages of important hominid-bearing members at South African australopithecine caves. The sequence is based primarily on faunal studies, and absolute dates are not established with much certainty. For example, Sterkfontein Member 4 may be as old as Makapansgat Member 3 or may be younger by half a million years or more. Taung (not listed) may be relatively recent (younger than Swartkrans Member 1), but the Taung fauna contains few bovids and is difficult to date with any accuracy.

Age (my)	Site/member	Hominids represented
1.0	? Swartkrans Member 2	*Homo* cf. *erectus*
	? Kromdraai B	*A. robustus*
1.5	Swartkrans Member 1	*A. robustus* and *Homo* sp.
	Sterkfontein Member 5	*Homo* sp.
2.0		
2.5	Sterkfontein Member 4	*A. Africanus*
3.0	Makapansgat Member 3	*A. Africanus*

and these are best matched in assemblages from Members B and C of the Shungura Formation in the Lower Omo Basin, Ethiopia. This long Omo sequence of fluvial and lacustrine deposits contains numerous volcanic tuffs suitable for potassium-argon dating, and Members B and C are apparently between 3.0 and 2.5 million years in age (Brown & Nash, 1976; see also Brown, Howell & Eck, 1978). A date of this magnitude is in keeping with paleomagnetic results reported for Makapansgat by Brock, McFadden & Partridge (1977), and there are indications that the Makapansgat Member 3 deposits, from which most of the hominids are derived, are in fact closer to 3.0 million years old (McFadden, Brock & Partridge 1979.)

Suid remains from Sterkfontein are scarce and unfortunately not very useful for correlation studies. Bovid crania and teeth are more plentiful, and Vrba's (1974, 1975) work shows that these fossils can be divided into three groups, most probably accumulated at different times and under varying environmental conditions. The first of these bovid phases is represented by specimens excavated from the Sterkfontein type locality, now know as Member 4 of the Sterkfontein Formation (Partridge 1978). This fauna is similar to that recovered at Makapansgat, so Member 4, in which the Sterkfontein australopithecines are found, must be approximately contemporary with the Limeworks. An age of 3.0 to perhaps 2.5 or 2.0 million years is now widely accepted. Other Sterkfontein bovids are clearly younger, and the material associated with Member 5, present in the Extension Site or West Pit, is probably less than 2.0 million years old. Both suid and antelope data suggest that Swartkrans is also younger than Sterkfontein Member 4 and Makapansgat. The fauna grouped as SKa, from (the earlier) Member 1 of the Swartkrans Formation, can be compared with assemblages from Members H to J of the Shungura sequence and from lower middle Bed II at Olduvai Gorge. A date of 2.0-1.5 million years may be about right, and the SKa robust australopithecines seem therefore to have succeeded the gracile species in the Sterkfontein area. The robust hominids from site B at Kromdraai may be still younger, although faunal information from this site is so far less complete. Satisfactory dates for Kromdraai and for Taung are not presently available.

Other current questions concerning the australopithecine sites relate to mechanisms whereby the fossils may have been concentrated in the breccias during some periods in a cavern's history. Earlier investigators thought it likely that hominids used the caves as shelters, but Brain's (1970, 1976, 1981) work has made it clear that Swartkrans and probably other localities were inaccessible both to humans and to other animals for long periods. Systematic collection or modification of bones by australopithecine hunters is unlikely, and skeletal remains must generally have reached the deeper parts of underground cavern systems by other means. In some cases, leopards may have played an important role. Brain (1980) has argued that the SKa assemblage reflects leopard feeding activities in the vicinity of the Swartkrans entrance. Trees present in the sheltered hollow near the cavern mouth would provide convenient retreats for leopards attempting to keep their kills away from other predators, and bones could then easily be deposited within the catchment area of the cave opening. The remains of leopards and other large carnivores are present at Swartkrans and also in Member 4 at Sterkfontein, where sabre-toothed cats may have been the principal bone accumulators (Brain 1981). At Makapansgat, the fossils are heavily concentrated in Member 3 deposits, located far from the entrance and close to the cave roof (Partridge 1976). Predators and scavengers probably did not live in this area but could well have left bones near the entrance, where they would then be washed into the fill during episodes of flooding. Klein (1977a) has noted that the Makapansgat assemblage resembles that from Swartklip in the Cape, in that the ratio of postcranial to cranial remains increases with the body size of the species represented; also, the post-cranial parts of larger animals are mostly adult, while crania are frequently juvenile. The Swartklip bones are said to be hyena-accumulated, and it is quite possible that hyenas were responsible for concentrating the fossils at Makapansgat as well. There is little or no evidence from this early time period that the australopithecines themselves played any active role as hunters or scavengers, and only in some later assemblages, such as Sterkfontein Member 5 containing stone tools, is there any indication that hominids were bringing meat to the vicinity of the caverns.

Advances in functional anatomy

Rather large samples of hominid skeletal remains have been extracted from the Transvaal cave breccias, and these together with australopithecines from East Africa continue to be extensively studied. We now have quite a lot of information about endocranial volume, dental morphology, and shape of the skull, although more comparative work is certainly needed. Other recent investigations have been designed to answer important questions about the function of anatomical complexes such as the teeth, jaw apparatus, and parts of the postcranial skeleton. Biomechanical models have been employed to analyze the forces operating on the mandible during mastication, for example, and White (1977) has confirmed that both gracile and robust australopithecines, with high, vertical ascending rami and relatively small braincases, are funetionally equipped to maximize occulusal forces along the entire postcanine tooth row. Especially in the robust forms, the molars and premolars present large flat crown surfaces, and crushing or grinding of large quantities of food is implied. In these creatures, the temporal muscle fibres pass medially as well as upward from the lower jaw to the side of the cranium, and these muscles tend to squeeze the two halves of the mandible together during chewing. White follows Walker (1977) in suggesting that increased buttressing, in the form of superior and inferior transverse tori on the inner aspect of the symphysis, is needed to withstand the resulting torsional and compressive stresses acting on australopithecine jaws.

Studies of dental wear in relation to diet are also proceeding. Walker has used a scanning election microscope to examine wear patterns on teeth from a variety of mammals, including grazing and browsing forms, carnivores and living primates. In some cases, characteristic microwear can be associated with the type of food preferred, so there is clear potential for determining diet from fossil dental remains (Walker 1980). Preliminary examination of teeth from (East African) robust australopithecines suggests similarities to modern chimpanzees, and fruit is likely to have been an important food item. Whether other australopithecines from South Africa show the same sort of dental microwear has still to be determined.

Work on the postcranial skeleton has stepped up tremendously during the last decade. Because trunk and lower limb parts are relatively well represented in the assemblages, and these skeletal elements are particularly relevant to posture and locomotion, many investigations have emphasized comparative and biomechanical aspects of the *Australopithecus* femur and pelvic girdle. Foot bones and other fossils are also available, mostly from East African localities. There is still incomplete agreement concerning the extent or significance of differences between the australopithecines and more advanced *Homo*, but most workers now believe that the gracile hominids at least were capable bipeds. The ilia from Sterkfontein and Makapansgat are low and broad, and similar to those of modern humans in general proportions. A sciatic notch is developed, and in adult specimens a strong anterior inferior iliac spine marks the attachment for rectus femoris and the illiofemoral ligament. A zone of thickening of the outer table establishes the presence of a distinct iliac pillar, although this feature may occupy a more anterior position than is usually the case in *Homo sapiens.* Positioning of the greater and lesser trochanters may be slightly different in australopithecine femora, but there is a lot of variation, and these changes are not likely to have much mechanical significance. More important features are the apparently long femoral neck and small head, well documented for robust australopithecines but not adequately preserved in the one crushed proximal femur (Sts 14) from Sterkfontein. The distal femur is very much like that of modern humans.

The most extensive studies of South African postcranial remains have been carried out by Robinson (1972; see also Robinson, Freedman & Sigmon 1972). His analysis has demonstrated that both the Sterkfontein and Swartkrans hominids have the short expanded iliac blade associated with a human type of lateral balance control, and Robinson has strongly maintained that gracile *Australopithecus* displays a complex of vertebral, pelvic and lower limb characteristics which are consistent with the 'habitual and effective' use of erectly bipedal posture and locomotion. In Robinson's view, only some slight improvements are needed to make this hominid as efficient a strider as *Homo sapiens.* The robust australopithecine is said to possess a proportionately longer ischium, however, and this difference is important. A longer ischium, and consequent displacement of the hamstrings from the hip joint, would make these muscles less effective in extending the femur during walking in a fully upright position This orientation of the hip extensors coupled with what Robinson feels may be a shorter lower limb and relatively mobile foot suggests a less efficient adaptation to bipedal locomotion. Instead, the Swartkrans hominid may have been more ape-like in locomotor behavior and was perhaps structurally suited to climbing in trees.

This interpretation of robust australopithecine anatomy has been questioned by Lovejoy and his colleagues, who have examined East African as well as South African postcranial material. Lovejoy (1975) has claimed that the Swartkrans innominate (SK 50) does not have a short ischium, when length is measured in a biomechanically significant way and normalized for body size. Functional ischium length for the Swartkrans hominid is said to be comparable to that of Sts 14 and modern humans. There is no basis either here or in the foot material for assuming that robust australopithecines were less capable than the smaller forms of using a bipedal gait. Lovejoy has also argued that the *Australopithecus* hip complex is well adapted to the load bearing demands of upright posture, despite differences in femoral neck length and head size. In his opinion, the adjustments in pelvic and femoral anatomy that have occurred in the evolution from *Australopithecus* to *Homo sapiens* are more likely to be responses to a requirement for increasing birth canal size than refinements related to locomotion. Increases in the coronal diameter of the pelvic opening to accommodate large fetal crania would result in enlargement of the femoral head for biomechanical reasons, as a means of dissipating the higher joint pressures generated by a decrease in length of the abductor lever arm. Some doubt about these calculations has been expressed by McHenry and other recent workers (see McHenry & Temerin 1979 for a comprehensive review), but Lovejoy's argument is compelling and should be developed further.

A challenge to both Robinson's and Lovejoy's views has been issued by Oxnard (1975a, 1975b). With measurements taken on the shoulder, pelvis, limbs and foot bones of living primates, Oxnard has used canonical analysis and generalized distance to explore the similarities of fossil bones to recent genera. His results are certainly provocative, as the skeletal fragments of early hominids are shown in most cases to be unlike the corresponding parts of modern humans and also to be statistically distant from larger African apes. Functional resemblance to the Asiatic orang is considered most likely, and Oxnard's principal conclusion is that species of *Australopithecus* practiced an imperfect form of bipedal locomotion quite different from that of *Homo sapiens.*

This finding has aroused considerable controversy, centering especially around multivariate studies of the innominate bone and talus. The hip from Sterkfontein (Sts 14) is said to be dissimilar to that of humans and other extant primates when canonical analysis is based on nine measurements relating to muscle attachment and joint position, and this result is interpreted by Oxnard to mean that the australopithecine pelvis is functionally unique. Contradictory conclusions have been reached by McHenry and Corruccini (1975) and by Steudel (1978), however. These authors have dealt with the more fragmentary innominates from Swartkrans and Makapansgat as well as Sts 14, and the 16 iliac dimensions of McHenry and Corruccini are converted to 'shape variables' in an attempt to correct for size differences among the hominoid genera used for comparison. Distances computed from these shape variables indicate that the hominid fossils are closer to modern humans than to any of the apes. In this case, the australopithecine remains appear to share a number of distinctly human characteristics, and it may not be fair simply to categorize them as unique on the evidence available.

The Kromdraai talus, claimed by Oxnard to resemble the foot bone of an orang, has also been examined by other workers. Principal coordinates analysis performed by Corruccini (1975) on distances computed from transformed measurements suggests ties with *Homo sapiens*, while Lovejoy (1978) has emphasized that small differences exhibited by available fossil hominid tali may not be functionally significant. Given these doubts (see also Howell, Washburn & Ciochon 1978), it is not clear that australopithecine postcranial parts are as unlike those of modern humans as Oxnard states. Certainly there are differences, as brought out by his analyses and others, and how to interpret this morphological variation is an important question, to which we still have no entirely satisfactory answer. That at least some australopithecines were bipedal is no longer questioned, in light of the discovery at Laetoli in Tanzania of hominid footprints 3.8 to 3.6 million years old (M.D. Leakey & Hay 1979; Day & Wickens 1980; White 1980). That all species were fully adapted to an upright striding gait has not been firmly established, however.

Postcranial remains can also be employed to estimate body weights of australopithecine individuals, and differences in body size between species are important in allometric considerations. Determinations of weight have usually been obtained by measurement of reconstructed long bones or vertebrae, and none can be very reliable. Uncertainties stem from the incomplete nature of the fossil material and from the necessary assumption that australopithecines had body proportions similar to those of living hominoids, which is not the case. For the gracile species, weights of less than 20 kg to nearly 40 kg have been reported, and certainly some individuals such as Sts 14 were quite lightly built. However, Steudel (1980) has recently employed cranial measurements to arrive at an average of about 36 kg for hominids at Sterkfontein. Robust australopithecines are larger, although there is apparently a lot of variation. Postcranial parts from Swartkrans generally suggest weights of 35 kg to 55 kg, but Robinson (1972) has argued that figures as high as 80 kg are reasonable. Steudel's (1980) average of 59 kg for robust hominids is based on both cranial and postcranial dimensions.

If the robust australopithecines are consistently larger than the gracile forms, it is appropriate to consider size as a factor accounting for some of the dental, cranial and (perhaps) postcranial differences distinguishing the two species. That the shape of the heavy facial skeleton and crested braincase in the robust hominids may be influenced by allometry has been noted a number of times, and Pilbeam and Gould (1974) have presented important evidence to support this point of view. In their paper, the relationships to body size of cranial capacity (brain volume) and posterior tooth surface area are emphasized. Brain volume in South and East African *Australopithecus* species is found to increase (slightly) in robust forms at the rate expected for a series of related animals differing only in size, and there is no indication that one species shows a greater degree of cephalization than the others. At the same time, australopithecines exhibit positive allometry for cheek tooth surface area, as relatively larger teeth occur in the robust and hyper-robust East African individuals. Pilbeam and Gould argue that this is the situation to be expected among species which differ in body size but not in basic design, and they conclude that in these features, australopithecines are simply scaled versions of the 'same' animal displayed over a range of body weights. In this metric evidence, they find no support for the idea that gracile and robust species are differently adapted, and they specifically reject the dietary hypothesis (long favored by Robinson) which holds that many of the cranial and dental distinctions of the two forms are related to feeding preferences.

Some recent studies, such as that of Corruccini and Ciochon (1979) on facial dimensions, have tended to confirm this view, while others suggest that allometry may not satisfactorily explain all australopithecine variation. Wood and Stack (1980) have examined canine base area and molar crown surface areas in relation to cranial length (a 'reference variable' for size) in a series of extant hominoids and Old World monkeys, and they note that in animals of increasing size, canine base area shows positive allometry. Canine size in fact increases at a rate faster than that recorded for molar crown area. Robust australopithecines stand in marked contrast to this trend, however, as there is a decrease in canine base area in these taxa, in which larger cranial lengths are recorded. Relatively small anterior teeth are coupled with massive postcanine dentitions, and in this respect the robust species are quite different from the gracile forms. This change in dental proportions has been emphasized before by Robinson (Steudel & Robinson 1976, and earlier publications), and it apparently cannot be dismissed as a natural consequence of increasing body weight. Some dietary influence is likely, and results of continuing investigations of dental microwear should eventually help to resolve the question of whether australopithecines differed significantly in feeding behavior.

Classification and phylogeny

Anatomical studies of the now quite impressive fossil collections from southern Africa make it clear that australopithecines were relatively small brained creatures with pelvic and lower limb parts functionally suited to providing lateral balance control during upright locomotion. While the postcranial skeleton exhibits some structural differences from that of modern humans, there is ample evidence indicating that some or even all species were capable of striding. Facial skeletons are large in comparison to braincase size, and the heavy lower jaw is buttressed internally to withstand the stresses generated by powerful chewing muscles. Teeth are also large, especially in the more robust forms where the premolars and molars present extensive crown surface areas. This communality of form suggests that all australopithecines share a broadly similar adaptation to a bipedal, probably predominantly vegetarian type of existance. Dental and mandibular morphology is generally what would be expected in a primate adapted to processing bulk amounts of vegetable food, which might include small, hard items such as seeds and roots, and perhaps some fruits as well.

While there is still disagreement, most workers now feel that this information is consistent with a classification in which all of the South (and East) African material is referred to one genus.* It is also clear that, within this genus *Australopithecus*, at

*The principal alternative classification, favored by Robinson, groups the robust australopithecines in the genus *Paranthropus*, while individuals from Taung, Sterkfontein and Makapansgat are referred instead to the genus *Homo*. Robinson (1967, 1972) has thus elected to sink *Australopithecus* into *Homo*, which is in keeping with his opinion that the gracile hominids are members of the lineage leading to more advanced humans. Most other students choose to retain *Australopithecus*, but some support for a generic distinction of robust from gracile australopithecines has recently come from R.E. Leakey (1980) who recognizes *Paranthropus, Australopithecus* and *Homo*.

least two and probably three species can be recognized. One species, named *Australopithecus africanus* by Dart in 1925, is known from Taung, Sterkfontein and Makapansgat. These gracile hominids are characterized by small body size, rounded crania lacking crests (except perhaps in some of the largest individuals), a prognathic lower face, and anterior teeth which are large relative to the premolars and molars. Remains from Swartkrans and Kromdraai are usually referred to a second species, *Australopithecus robustus,* which exhibits larger body size, a sagittal crest located anteriorly on the braincase, a flatter facial skeleton, and small incisors and canines associated with large cheek teeth. Howell (1978a), following an earlier classification proposed by Broom in 1950, feels that further subdivision of the South African material is justified and has proposed that the fossils from Swartkrans be assigned to a separate species, *Australopithecus crassidens.* In his view, the Swartkrans and Kromdraai assemblages differ (perhaps) in endocranial volume, in molar size, and robusticity of the limb skeleton. Such differences are difficult to document, however, as there are still only a few, rather fragmentary finds from Kromdraai. There are considerably firmer grounds for setting the hyper-robust hominids of East Africa apart as *Australopithecus boisei.* Crania and mandibles, especially well known from Olduvai and the East Turkana localities, are somewhat larger than the South African remains, and the cheek teeth are huge, both absolutely and in comparison to the anterior dentition. At these eastern sites, associations of cranial with postcranial parts are usually uncertain, and it may be impossible to assign isolated limb bones to any given taxon. But there are suggestions that *Australopithecus boisei* may exhibit distinctive features of the forelimb skeleton as well. Still another, older and apparently more primitive, form of *Australopithecus* has recently been identified at Laetoli and Hadar in East Africa, and the implications of these discoveries are discussed briefly below.

The question of how these several Plio-Pleistocene species are related phylogenetically and which if any may be directly ancestral to *Homo* is still hotly debated. If there is currently a consensus opinion, it is probably that *Australopithecus africanus* is best qualified chronologically and anatomically to be antecedent both to the robust lineage and to an early species of the genus *Homo.* Remains of the gracile hominid found at Makapansgat and at Sterkfontein are rather clearly older than *Australopithecus robustus* at Swartkrans and Kromdraai and probably predate the robust forms known from East Africa as well. *Australopithecus africanus* is also older than most (or all) of the South and East African fossils that have been assigned to early *Homo* sp. A principal proponent of this view has been Tobias (1967, 1978a, b, 1980a), who suggests that the *Australopithecus africanus* lineage split approximately 2.5 million years ago. In one branch leading toward the more robust species, body size increased while the chewing apparatus and posterior teeth show successively greater specialization for a crushing and grinding form of mastication. In the other, trends toward reduction of tooth size (narrowing of the premolar and molar crowns) and increase in cranial volume were established, although hominids with significantly larger brains do not appear in the record until after about 2.0 million years ago. Presumably some adjustments to the postcranial skeleton were also occurring in the *Homo* lineage during this period, but direct evidence is mostly lacking.

Such an ancestral role for *Australopithecus africanus* is not contradicted by the

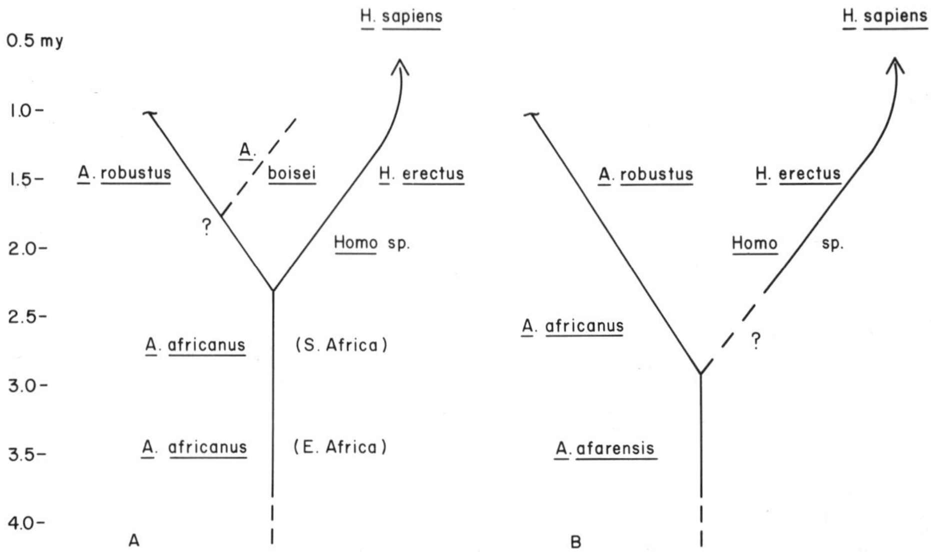

Figure 1. Tree diagrams representing alternate views of hominid phylogeny during the last 4.0 million years. The relationships shown in (A) have been widely accepted. Here it is assumed that the hominids from Hadar and Laetoli are conspecific with those from Sterkfontein, Makapansgat and probably Taung. *Australopithecus africanus* is ancestral to two later lineages. One is comprised of the robust australopithecines, although there is incomplete agreement about the status of *A. boisei* in East Africa. The two robust forms have been viewed either as subspecies or as closely related species, and the split marked with a (?) may not have occurred. The other lineage is made up of *Homo* species, and many workers refer the earliest *Homo* specimens to *Homo habilis*. This species is considered to be ancestral to *Homo erectus*, which in turn gave rise to modern humans.

Tree (B) reflects the opinion that the fossils from Hadar and Laetoli represent a new species, *Australopithecus afarensis*. This taxon exhibits several primitive dental and mandibular characteristics which suggest that it may be the common ancestor of all later australopithecines and early *Homo*. *Australopithecus africanus* as known from South Africa is considered to lie in a descendant lineage and to be the direct antecedent of *Australopithecus robustus*. A separate species of hyper-robust East African australopithecines is not recognized. In this scheme, the *Homo* lineage must be derived from *Australopithecus afarensis* at an early date, and the (?) indicates uncertainty about the ancestry of early *Homo*. Fossils presently assigned to *Homo* sp. are not older than about 2.0 million years. Relationships within the later segment of the *Homo* lineage are assumed to be the same as those in (A).

allometric investigations of Pilbeam and Gould (1974), and several workers undertaking a cladistic analysis of hominid taxa have concluded that the more gracile australopithecine is indeed conservative in most features, when compared with the robust forms or early *Homo* (Eldrege & Tattersall 1975; Delson, Eldredge & Tattersall 1977). On the basis of his multivariate statistical studies. Oxnard has disputed this view, claiming that australopithecines are in many ways unique in pelvic and limb structure. According to Oxnard, these differences are cause to regard *Australopithecus* as unsuitable as an ancestor to later humans, These findings have proved to be quite controversial, however, and other statistical and biomechanical evidence does not provide consistent support, as noted previously. A more recent challenge to accepted australopithecine phylogeny has been issued by Johanson and White (1979). These authors points to the fossils discovered at Laetoli and Hadar as representative of a new species, *Australopithecus afarensis*, known to be of mid-

Pliocene antiquity (Johanson, White & Coppens 1978). This species exhibits a suite of primitive cranial and dental characteristics, which suggests that *Australopithecus afarensis* may be ancestral to all later australopithecines and also to *Homo*. Compared to the Laetoli and Hadar specimens, the other gracile and robust forms appear to be more specialized in such features as their heavily molarized premolar teeth, relatively larger cheek dentition, and strongly buttressed mandibular symphyses. These shared derived characters may necessitate placing both *Australopithecus africanus* and later *Australopithecus robustus/Australopithecus boisei* in one lineage, so that all australopithecines save *Australopithecus afarensis* are removed from the stream leading to *Homo*. There is no agreement about this, and both the new name and the systematic position of the older East African fossils are presently questioned (Day, M.D. Leakey & Olson, 1980; R.E. Leakey & Walker, 1980; Tobias 1980a). Issues relating to the extent of morphological variation within the Hadar assemblage and the significance of features said to distinguish this material from *Australopithecus africanus* will probably not be settled for some time.

EARLY REPRESENTATIVES OF THE GENUS *HOMO* IN SOUTHERN AFRICA

The origins of our own genus *Homo* are obscure, but fossils which are different in certain respects from australopithecines have now been recovered at several localities, particularly in eastern Africa. Remains from Bed I and the lower parts of Bed II at Olduvai Gorge were referred to *Homo habilis* by L.S.B. Leakey, Tobias and Napier in 1964, and more material has since been reported from the Omo and from East of Lake Turkana (Wood 1976; Boaz & Howell 1977; Howell 1978a; Coppens, 1980). There is a great deal of variation in these assemblages, and fossils which are assigned to *Homo habilis* or early *Homo* sp. by some workers are still viewed by others as better attributed to *Australopithecus*. However, several of the crania do suggest that brain volume had increased beyond the level expected in australopithecines, and the teeth are generally small in relation to the size of the skull and mandible. A few of the Turkana fossils have no firm geological provenience but material from Olduvai and the Omo Shungura sequence is quite well dated. Available radiometric determinations indicate that the earliest of these East African remains are approximately 2.0 million years old, while the younger specimens may be 1.6-1.5 million years in age.

In South Africa, fossils which may represent an ancient species of *Homo* have come from two of the sites in the Sterkfontein Valley. A pair of broken mandibles, a maxillary fragment, and other fossils found during earlier excavations at Swartkrans were orginally assigned to a new genus, called *'Telanthropus'* by Broom and Robinson. At the time, there was some doubt about the date of the breccia containing these remains, although Robinson subsequently claimed that all of the *'Telanthropus'* specimens were as old as the Swartkrans australopithecines. This genus name has since been sunk into *Homo*, and more recent fieldwork at the site has made it clear that some (not all) of the material is contemporary with the Member 1 australopithecine assemblage. The broken maxilla (SK 80) has now been joined to other parts of a braincase, to produce an incomplete composite cranium known as

SK 847 (Clarke, Howell & Brain 1970). This cranium together with the SK 45 mandible and the remains of (possibly) one or two additional individuals provide reasonable evidence for the presence at Swartkrans of a second type of hominid, said to be anatomically distinct from *Australopithecus.*

There is enough of the frontal of the reconstructed skull to show that the supraorbital torus is rounded and thicker than expected for a robust australopithecine of comparable size. Behind the torus, there is a definite depression or sulcus, and the forehead then appears to rise rather steeply upward. Unfortunately, most of the rest of the frontal squama and temporal fossa are missing, though it is evident that post-orbital constriction is only moderately developed. Other features of the face and cranial base are also cited by Clarke and Howell (1972) to support their referral of this material to the genus *Homo*, and Olson (1978) has recently dealt with nasal, palatal and mastoid characteristics in considerable detail. There are definite resemblances to several of the more complete crania from Olduvai and Koobi Fora. However, the Swartkrans hominid is badly damaged, and much of the vault is lacking. Endocranial volume cannot be determined, though it has been argued that the high SK 847 frontal profile is in keeping with a brain expanded slightly beyond australopithecine standards. The skull is not that of a robust *Australopithecus,* but without more compelling evidence, it is probably best to leave the question of (species) indentification open, as Clarke and Howell (1972) have recommended.

Additional finds are known from Sterkfontein, where another broken cranium was recovered from Member 5 in 1976 (Hughes & Tobias 1977). Stw 53 consists of fragments of a vault, base and facial skeleton, and many of the upper cheek teeth are preserved. So far, it has been determined that the new fossil represents a small individual, but, as with SK 847, brain volume has not been measured. Post-orbital constriction, or narrowing of the braincase behind the brows, is again less marked than usual for an australopithecine, and the nasal region exhibits some of the same resemblances to *Homo* as have been noted for the Swartkrans specimen. However, the two fossils are different in other respects, and the Sterkfontein supraorbital torus is substantially thinner and more delicate (Tobias 1978c). In its frontal features, Stw 53 is similar to O.H. 24, a Bed I Olduvai individual assigned to *Homo habilis* by a number of workers but viewed as *Australopithecus*-like by others.

The Sterkfontein teeth are heavily abraded. A pronounced helicoidal pattern of attrition is present, in that the lingual aspects of the crowns of the premolars and first two molars are most worn, while in the posterior part of the tooth row, attrition is most marked on the buccal side of the crown surface. The occlusal plane is thus twisted rather than horizontal, apparently as a consequence of the relative widths of the upper and lower jaws. Tobias (1980b) has called attention to this helicoidal wear pattern as characteristic of the dentitions of early *Homo* and has observed that it does not occur in australopithecines.

The hominids from Member 5 at Sterkfontein are not very securely dated, but Vrba's bovid studies indicate that they are younger than the australopithecine fossils found in Member 4. It is possible that Member 5 is approximately the same age as Member 1 at Swartkrans, thought to be 2.0-1.5 million years old. It is also apparent that the fossils from these deposits can be described as more 'advanced' in some features than gracile or robust australopithecines. But the material is very fragmentary, and a list of characters shared by the South African forms and *Homo* in East Africa is still difficult to compile in satisfactory detail. How these southern populations may be related to other early *Homo* groups or to later species of humans is very much an issue, and the speciational event that gave rise to *Homo erectus* is only poorly understood.

We do know that *Homo erectus* was present in eastern Africa before 1.5 million years ago, as there are good crania, lower jaws, and other remains of about this age in the Upper Member of the Koobi Fora Formation, exposed to the East of Lake Turkana. Other fossils have been recovered from Bed II and from the higher levels at Olduvai Gorge. Further to the north, at sites in Algeria and Morocco, *Homo erectus* is documented from later, Middle Pleistocene deposits, and there is much evidence to show that by this time the species had spread over most of the Old World. In South Africa, however, the mid-Pleistocene record is very sparse indeed, and there is little with which to chart the course of human evolution after the deposition of Member 1 at Swartkrans. One mandible, considered at first to be part of the *'Telanthropus'* assemblage, may in fact belong to *Homo erectus.* Along with one or two teeth and a radius, the SK 15 jaw is now thought to be derived from Member 2 at Swartkrans, which is likely to be substantially younger than Member 1. Faunal comparisons suggest an age broadly equivalent to that of Bed IV at Olduvai, so the hominids are not as old as *Homo erectus* at Koobi Fora and Ileret. Unfortunately, the mandible is not well preserved, and it is therefore not particularly informative.

Homo erectus populations continued to flourish in Africa and elsewhere over a long period of time. This species is characterized by cranial capacities considerably larger than those of earlier *Homo*, and the skull itself is thicker, with heavier brows, a low frontal profile, and usually a well defined torus on an angulated occiput. Dentitions are not easily distinguishable from those of some specimens assigned to archaic *Homo* sp. Postcranial remains have been recovered most recently at East African localities, and these along with limb bones known from China suggest that there are some differences from modern humans. While the fossil record is still quite incomplete, there are at least indications that *Homo erectus* was a stable species, exhibiting little significant morphological change during a span of well over a million years.

THE EMERGENCE OF *HOMO SAPIENS*

With the likely exception of the Swartkrans mandible, fossils representing *Homo erectus* have not been uncovered in the southern part of Africa. Sites containing Acheulian artifacts do occur in some abundance, however, and it is probable that at least the older tools were made by members of this species. Unfortunately the sealed sites which have been excavated are still few in comparison to the much more common but less useful surface scatters of stone artifacts, and dating is frequently uncertain. As a result, we have relatively little information about the behavior of these early Acheulian populations. Upper Acheulian assemblages are more numerous, and at several of these later Middle Pleistocene localities important human remains have come to light. These fossils provide some clues concerning the first appearance in southern Africa of *Homo sapiens.*

Later mid-Pleistocene fossils

One collection of bones and tools was recovered at Broken Hill in 1921, three years before the first australopithecine was picked up at Taung in 1924. The Broken Hill cranium is quite complete, and like the Taung child, it has engendered a substantial amount of controversy over the course of human evolution. In addition to the adult cranium, a broken maxilla, a parietal bone and postcranial remains representing several individuals were found in cave deposits at the site (now Kabwe), north of Lusaka in Zambia. The fossils are known to have come from different sections of the cave fill, and analysis of the chemical content of the bones has never satisfactorily settled the question of whether the postcranial parts are contemporary with the skull. Dating of the hominids is also a problem. Radiometric methods cannot be used at Broken Hill, but faunal remains collected at the site suggest considerable antiquity. Of at least 23 larger mammal species identified, five are extinct, and this is indicative of a later Middle Pleistocene or earliest Upper Pleistocene age (Klein 1973). Amino acid racemization results obtained by Bada *et al.* (1974) support this conclusion. One of the femoral fragments gives a racemization age of approximately 110,000 years, while other hominid bone samples have produced aspartic acid ratios consistent with even greater antiquity. However, the cranium itself has not been subjected to racemization analysis, and dates determined by this method must in any case be treated with extreme caution. Before a D/L aspartic acid ratio can be converted to an age in years, the racemization reaction must be calibrated, using samples dated by other means. For Broken Hill, the reaction has been calibrated with radiocarbon dated bone from Mumbwa Cave, some 150 km away, and there is no certainty that the temperature histories of the two sites are comparable. Water distribution in the environment may also affect the extent of racemization in bone, and this adds further complications (Hare 1980).

Evidence for a Middle Pleistocene age and association with Acheulian artifacts is somewhat stronger at the farm Elandsfontein, near Hopefield in South Africa's Cape Province. Another cranium, less complete but still resembling that from Broken Hill, was discovered at this locality in 1953. Systematic excavations were conducted later, in 1965-66, and this work (Singer & Wymer 1968) together with Butzer's (1973b) restudy of the geology has clarified the association of much of the Elandsfontein fauna with Acheulian tools on or within a calcareous crust related to a former land surface. The hominid fossils, including the cranium and a mandibular fragment, should probably be grouped with the older animal remains and aritfacts, although there is still a possibility that they are derived from younger stratigraphic levels at the site. As a whole, the fauna is similar to that recovered from the Middle Pleistocene Cornelia Beds in the Orange Free State (Butzer, Clarke & Cooke 1974), and a number of extinct species are present. Most of these extinct forms are unknown in earlier Upper Pleistocene assemblages from the southern Cape (Klein 1980). On faunal grounds, therefore, it is likely that the Elandsfontein material is of Middle Pleistocene antiquity and perhaps significantly older than that from Broken Hill.

Two other localities have yielded human skeletal remains which are usually referred to early or archaic *Homo sapiens.* One is the Cave of Hearths in the Transvaal, where a broken mandible was discovered in company with Upper Acheulian tools in 1947. Fauna from the lower deposits in the cave, where the

Table 2. Southern African localities which have yielded human skeletal remains presumed to be of later Middle Pleistocene antiquity. Actual dates are uncertain, but Elandsfontein and probably also Broken Hill can be assigned a Middle Pleistocene age on faunal grounds. Evidence at Florisbad and the Cave of Hearths is less satisfactory.

Site	Type of deposit	Hominid remains and artifacts
Cave of Hearths, Transvaal	Cave fill	Broken juvenile mandible, associated with Acheulian tools. A proximal radius from the cave is of uncertain stratigraphic provenience.
Florisbad, Free State	Spring vent	Fragmentary cranium. Crude artifacts found at about the same level are said to be undiagnostic.
Broken Hill (Kabwe), Zambia	Cave fill	Cranial and postcranial remains representing several individuals. 'Sangoan' artifacts may better be classified as belonging to the Acheulian Industrial Complex.
Elandsfontein, Cape	Former land surface	Fragmentary cranium, unassociated lower jaw. Acheulian tools.

mandibular fragment was located, is sparse and not particularly helpful. However, the fossils may be quite ancient, if current thinking regarding the age of final Acheulian industries is correct. It is now apparent that these industries were replaced by the flake tools of the Middle Stone Age sometime late in the Middle Pleistocene, well in excess of 130,000 years ago (Beaumont, De Villiers & Vogel 1978; Szabo & Butzer 1979). Another important site is Florisbad, which consists of a series of sand and peat deposits above a number of hot spring vents. The Florisbad cranium was found in a small spring eye which had penetrated an early peat layer (Peat I) before being sealed by sand underlying the next peat level in the sequence. This second peat horizon (Peat II) has been dated as >42,600 years. Peat I and probably also the hominid remains can be shown to be a whole landscape cycle earlier than Peat II, so a late Middle Pleistocene age is not unlikely. Five extinct species of large mammals are found in the Florisbad fauna, but these forms do occur in Upper Pleistocene contexts in southern Africa and so do not point unequivocably to an age of more than 130,000 years (Klein 1980).

These hominids from Broken Hill and the South African sites are neither plentiful nor well enough preserved to tell us a great deal about later Middle Pleistocene populations. But there are a few other fossils from eastern Africa which seem to be similar, and all are decidedly archaic in anatomical appearance. The cranium, of which the most complete is still the first to be discovered, at Broken Hill, is low in outline, with massive brows. The supraorbital torus of Broken Hill is actually thicker than that of even such large *Homo erectus* individuals as Hominid 9 from Olduvai Gorge. Greatest breadth of the braincase falls at the level of the mastoid processes rather than higher on the parietal bones, which are flattened laterally and not rounded or bossed as in later *Homo sapiens.* Because of these features, both Broken Hill and the Elandsfontein specimen have sometimes been referred to *Homo erectus*. But cranial capacity is expanded well beyond the *Homo erectus* average, and the vault bones are reduced in thickness. The occiput is less strongly curved, and its upper scale is nearly vertical above an apparently well defined torus. The nuchal area is flattened so as to lie fully in the horizontal plane, and the mastoids themselves are not so laterally prominent. Studies of the Broken Hill cranial base, which exhibits a short basioccipital and steep clivus, indicate that this is essentially modern in its sagittal dimensions (Laitman, Heimbuch & Crelin 1979). Details of sphenotemporal articulation and of mandibular fossa construction are also like those of modern humans, although there are some peculiar features of the tympanic plate. The postcranial remains which have been examined appear to possess few of the characteristics associated with *Homo erectus.* However, it should be kept in mind that at Broken Hill there is no certain association of the limb and pelvic parts with the adult skull, and fossils from the cave may be of quite different ages.

A view which is now widely accepted is that this material represents a distinct subspecies of *Homo sapiens*, properly labelled *Homo sapiens rhodesiensis.* It has been argued that such subspecific status is justified on anatomical, geographic and chronological grounds, as the sub-Saharan fossils differ in a number of features from the Neanderthals of Europe as well as from *Homo sapiens sapiens* (Rightmire 1976). They are also older than the European remains, if a later Middle Pleistocene date can be established at most of the African localities. This designation of subspecies also implies that the Broken Hill people are tied into the ancestry of more recent *Homo sapiens*, or at least that the African populations share with other *Homo sapiens* a common ancestor that had evolved earlier from *Homo erectus* (Howells 1980). Neither of these suppositions can be documented with present evidence, as is discussed further in the next section. Given these uncertainties regarding the phylogenetic significance of the archaic African fossils, it may be best to place less emphasis on formal subspecies names, at least for the moment.

Problems in the ancestry of Upper Pleistocene humans

Relationships of the archaic South and East African populations to other early *Homo sapiens* groups and to anatomically modern humans are obscure. It is reasonably clear that Broken Hill and the other fossils exhibit cranial and (perhaps) postcranial features which set them apart from *Homo erectus.* It is possible that these hominids are part of a lineage derived locally (and directly) from earlier mid-Pleistocene Africans, as has been suggested before (Rightmire 1976). If this is the case, there may be few links between the sub-Saharan people and other roughly contemporary 'preneanderthal' populations in Europe. Here we are on shaky ground, as there is as yet incomplete understanding of similarities (and differences) between Broken Hill and the Bodo material from Ethiopia on one hand, and such fossils as Steinheim, Arago and Petralona, all of which are *Homo sapiens* according to one recent view (Stringer, Howell & Melentis 1979). Some of the latter, particularly Steinheim and Arago, are likely to be ancestral to later Neanderthals, however, and the 'classic' remains of Upper Pleistocene Europe are more definitley distinct from the East and South African material. That Neanderthals possess a suite of derived characters which are not shared by other hominids has been emphasized recently by several workers (Hublin 1978; Santa Luca 1978), and probably none of the sub-Saharan fossils can meaningfully be termed Neanderthal-like. Large brow ridges and low cranial vaults alone do not prove that a 'Neanderthaloid' phase or grade extended into southern Africa.

At the same time, it is difficult to relate any of the mid-Pleistocene hominids to others from later, Upper Pleistocene African contexts. Stringer (1974, 1978), Beaumont, De Villiers & Vogel (1978), and others have pointed especially to Florisbad and to the Omo (Kibish Formation) skeletons from Ethiopia as probable ancestors of recent *Homo sapiens*, but the impression that the Florisbad cranium is modern anatomically is not supported by analysis of measurements. A general resemblance to Broken Hill in frontal form and torus development is suggested by multivariate statistics, and surely there are no special ties with Bushman (San) or other living African populations (Rightmire 1978a). In fact, there is no solid evidence for continuity of archaic South or East African people with more modern *Homo sapiens* of the same geographic regions. This seems to be true not only in sub-Saharan Africa but also in other parts of the Old World where comparable skeletal assemblages have been uncovered (Howells 1976). Accordingly, the crucial question of where and how *Homo sapiens sapiens* originated cannot readily be answered. Modern humans may have evolved gradually in several regions, including southern Africa where skeletons of apparent early date have been found at Border Cave and Klasies River Mouth. Or the first modern people may have appeared in only one center, perhaps removed from Africa entirely, and then migrated rapidly into other regions. Some version of the first hypothesis has probably attracted most support, but no consensus can be reached without better fossils, more securely dated.

EARLY UPPER PLEISTOCENE POPULATIONS

Both chronology and cultural associations are somewhat more reliably established at several additional sites in southern Africa, where there is good evidence for human occupation in deposits of earlier Upper Pleistocene age. These assemblages contain Middle Stone Age (MSA) industries which lack cleavers and handaxes and are characterized instead by a predominance of flake tools such as points and scrapers (Volman, this volume). Variability in tool types seems to be greater than for the preceeding Acheulian, and backed blades and segments represent an important lithic innovation which occurs in MSA sequences over much of southern Africa at about the same time. Fire was used at several of the MSA sites, and on the Cape coast there are indications that marine resources were systematically exploited (Klein 1977a). Unfortunately, human skeletal remains from MSA contexts are generally scarce and fragmentary.

One site which has yielded human material in definite association with MSA artifacts is Klasies River Mouth, located on the Cape coast some 130 km to the west of Port Elizabeth. The Klasies caves were excavated between 1966 and 1968 (Wymer & Singer 1972; Singer & Wymer 1982). At cave 1, the MSA occupation sequence starts at or near the beginning of the Last Interglacial, approximately 125,000 to 130,000 years ago. Butzer's (1978b) analysis of the sedimentary stratigraphy suggests that MSA people were present at the site until perhaps 60,000 years ago, and the culture-stratigraphic sequence thus records more than 50,000 years of nearly continuous habitation. The Klasies MSA fauna have been studied by Klein (1976a), who notes that only a small number of mammal species are extinct. This fauna is therefore quite distinct from that at Elandsfontein, where a whole series of archaic and extinct forms are found.

Human remains have been recovered mainly from earlier Klasies levels, where the stone industries are termed MSA I and MSA II. All of the cranial material is badly broken, and it is difficult to characterize the MSA population in more than very general fashion. However, most of the specimens, including at least some of the mandibular remains, appear to be anatomically modern. A zygomatic bone is large in comparison to that of recent Africans and is rather flat in facial aspect, but its morphology is not noticeably archaic. Also interesting is a piece of frontal which preserves the right supraorbital margin and the nasal root. The superciliary eminence is not especially prominent, although glabella is moderately projecting. The nasal bones are broad and flat, as in many modern individuals. This Klasies frontal is very different from that of Florisbad, in which glabella is more prominent and the supraorbital torus is quite thick (Rightmire 1978a). Such evidence in not conclusive but certainly suggests that the Klasies people were less archaic in appearance than those represented by the Free State fossil.

Little additional information can be gained from Die Kelders, although the MSA sequence at this Cape coastal cave is characterized by excellent faunal preservation. Unfortunately the sample of human material recovered by Schweitzer (1979) is quite small, consisting only of isolated teeth. No signs of deliberate burials have turned up either in the earlier Die Kelders levels or at Klasies River Mouth, and at other coastal sites such as Nelson Bay Cave and Elands Bay, preservation of all bone in the MSA layers is relatively poor. Elsewhere, useful hominid remains in plausible association with MSA tools are known only from Border Cave, far to the north.

Border Cave, situated in the Lebombo Range on the border of Swaziland and Kwa-Zulu, has yielded lithic and skeletal material of considerable interest. Human bones were first recovered during the course of Horton's digging in 1940, but controlled excavations were not conducted until 1941 and 1942. Further work at Border Cave was initiated by Beaumont in 1970. The history of these excavations and new findings at the site have received a good deal of attention in the recent literature, and an extensive review is not required here. Lithostratigraphic and sedimentological studies, radiocarbon dates, microanalytical data for bone samples, and details of the cultural sequence are discussed by Beaumont, De Villiers & Vogel (1978) and by Butzer, Beaumont & Vogel (1978), while the fauna has been dealt with by Klein (1977b). Some of these results are controversial, but the evidence does indicate that MSA occupation began before the start of the Last Interglacial and continued until approximately 50,000 BP. The MSA inhabitants of Border Cave are represented by cranial and mandibular remains of three adult individuals, an infant skeleton, and perhaps by postcranial fragments of uncertain provenience. One of the mandibles was recovered in 1974 from deposits unquestionably containing MSA artifacts and dated by radiocarbon as >49,000 BP. The infant skeleton was excavated *in situ* by Cooke, Malan & Wells (1945) from a shallow burial cut into level 10 of Butzer, Beaumont & Vogel (1978) and is not likely to be younger than the overlying level 9. Pieces of an adult cranium were dug out of the cave by Horton, and the original position of this individual in the deposits was not directly verified. However, Cooke, Malan & Wells hae argued that soil found attached to the cranium is best matched by that of level 10, identified by Butzer with deep-sea oxygen isotope stage 5d. If this is correct, then the cranium and infant burial should be approximately 115,000 years old, while the 1974 mandible is somewhat younger.

All of the Border Cave hominids are anatomically modern in morphology (De Villiers 1973; 1976a), but there is some difference of opinion regarding the extent to which one or more of the specimens may resemble living African poulations. Because of its rugged appearance and well developed superciliary eminences, the adult cranium has been compared to such other skulls as Florisbad and Springbok Flats, while ties with Negroes, Bushmen (San) or Hottentots (Khoi) have been viewed as fairly remote. Analysis of cranial measurements has led De Villiers (1973; also Beaumont, De Villiers & Vogel 1978) to emphasize similarities of Border Cave to Springbok Flats and to assign both skulls the role of 'undifferentiated' ancestors from which recent populations ultimately arose. My own multivariate statistical treatment of Border Cave measurements suggests a rather different conclusion. When cranial dimensions are used in discriminant analysis, Border Cave approaches several modern populations and in fact lies closest to Hottentot (Khoi) and Bushman (San) groups (Rightmire 1979).

If the adult cranium is actually associated with a MSA tradition of early Last Interglacial age, these statistical results have important implications. The cranium and the two mandibles argue for the presence of fully modern *Homo sapiens sapiens* in southern Africa at an unexpectedly early date, as noted by Beaumont, De Villiers & Vogel (1978). This implies that Hottentots (Khoi) or large Bushman (San)-like people have inhabited southern Africa for a long time, as would be consistent with a phyletic view of later human evolution. Although very limited, the evidence surely does not point toward any abrupt change in local populations during the Upper Pleistocene, and it is unlikely that this part of Africa witnessed any large scale replacement of one people by another shortly before 35,000 BP.

However, the Border Cave and Klasies materials do not to my mind prove that modern humans evolved first in sub-Saharan Africa. Problems with the stratigraphic provenience of some of the Border Cave hominids have still to be unequivocably resolved, although this case for antiquity is certainly more convincing than any built around the undated and problematical remains from such sites as Boskop and Springbok Flats. The Florisbad fossil, which has also been advanced as a candidate for *Homo sapiens sapiens*, may well not qualify for reasons discussed previously. Even if all of the Border Cave individuals are accepted as ancient, the evidence does not permit us to disallow an equally early appearance of modern humans in other regions of the Old World. Some 16 skeletons have now been recovered from Mousterian levels at Qafza in Israel, for example, and some or all of this material is anatomically modern rather than Neanderthal-like. There are no good radiocarbon dates for Qafza, so the age for these individuals, which are presumed to be quite old, has not been reliably determined.

SKELETONS FROM THE LATE UPPER PLEISTOCENE AND HOLOCENE

The few useful fossils known or suspected to date from Last Interglacial or early Last Glacial times are followed by a long gap in the southern African record. Sites containing stratified MSA sequences seem to have been abandoned for long intervals before the onset of typical Later Stone Age (LSA) traditions, some 20,000 to 15,000 years ago (Klein 1977a; J. Deacon 1978). Early post-MSA occupations are known for only a handful of localities, such as Boomplaas in the Cape (H.J. Deacon & Brooker 1976), and no discoveries of significant human remains have been reported for this period. Only with the coming of LSA industries in the late Upper Pleistocene and Holocene is there any substantial increase in the inventory of human skeletons.

In Zambia, late Upper Pleistocene occupation has been documented at several rockshelters, including Mumbwa, Leopard's Hill Cave and Kalemba. The Mumbwa caves were excavated more than 40 years ago, and there has been considerable confusion about the number of skeletons recovered and their cultural associations. Some of the material has apparently been lost, and the skeletons now housed in the University of the Witwatersrand are in rather fragmentary condition. The cranium of Mumbwa 2 lacks all of its base and facial parts, while other sections of the vault are broken and slightly distorted. Number 3 has been heavily reconstructed with plaster, and Mumbwa 4 is also missing the facial skeleton. Given the state of these remains and other problems, it is difficult to extract from the assemblage much information about the Mumbwa population. The Leopard's Hill bones are also extremely fragmentary, but at Kalemba several individuals are associated with a long LSA succession. Most of the human material is not more than about 8,000 years old, but does display features which fall broadly within the range of variation of recent Negroes (De Villiers 1976b).

This evidence relating to the evolution of LSA people in the very late Pleistocene and early Holocene is extremely sketchy, and finds from this period in the interior of South Africa do not help a great deal. At Bushman Rock Shelter in the eastern Transvaal, a broken mandible was recovered in 1969 from levels apparently containing LSA artifacts. The jaw has been described by Protsch and De Villiers (1974) as that of a human infant between six and eight months of age, and all of the right half of the bone is preserved. The left ramus is missing and on this side the body has been broken just behind the socket for the second deciduous molar. Measurements do not distinguish the Bushman Rock Shelter mandible from those of South African Negroes of comparable age, while the specimen is said to share few anatomical features with the jaws of Bushmen (San). However, the number of infant mandibles available to Protsch and De Villiers (1974) for comparative purposes was small, and some caution in identifying the Transvaal individual as an early Negro is advised.

Fortunately many more skeletons have been exhumed in the southern Cape Province, particularly in caves located along the Cape coast. Extensive new excavations at several of these sites have been conducted during the last decade. Much of this work has been and continues to be multidisciplinary, and sedimentological studies, plant and faunal analyses in addition to reports on artifacts are increasingly important. Information concerning hunting strategies and food preferences is filling out our knowledge of the ecology of these coastal Stone Age people, and in some instances it is possible to relate major changes in artifact typology to shifts in climate and vegetation as inferred from faunal remains (J. Deacon 1978). Human burials associated with LSA occupation at a number of caves and shelters are also informative, although the larger collections of skeletons have come mostly from earlier excavations, where the stratigraphic provenience of the material is frequently in doubt.

Such is the case, for example, with Peers Cave at Fish Hoek. This site was first

Table 3. Principal sites in South Africa where human skeletal remains have been uncovered in association with MSA and LSA artifacts. All of the localities listed are mentioned in the text. Human burials have also been found at other sites, particularly in caves and shelters situated along the Cape coast, but this material is generally fragmentary and little of it has been studied systematically.

Site	Human skeletal remains	Cultural associations
Border Cave	Cranium, 2 mandibles, infant burial, other remains	MSA, but some remains not *in situ*
Die Kelders (earlier levels)	Isolated teeth	MSA
Klasies River Mouth	Fragments of several individuals	MSA
Bushmen Rock Shelter	Infant mandible	LSA (?)
Die Kelders (later levels)	Juvenile burial	LSA
Elands Bay	3 adults, 2 juveniles	LSA
Matjes River Shelter	>80 individuals	Albany/ Wilton
Nelson Bay Cave	Fragments of 2 adults, 3 juveniles	Wilton
Oakhurst	14 adults, 17 juveniles (?)	Albany/ Wilton
Fish Hoek (Peers Cave)	8 or 9 individuals, several juvenile	LSA (?), although 1 individual said to be associated with MSA tools

visited in the 1920's, and as many as 11 skeletons were excavated by members of the Peers family. Most of these individuals were apparently found in the upper levels at the cave, in company with LSA tools and ornaments. However, one skeleton (No. 4) from a lower layer has been viewed as ancient and thought to be associated with a MSA ('Stillbay') industry. On the basis of radiocarbon dating of *Equus* bone presumed to have been found at the site, Protsch (1974) has claimed that the assemblage including the human skeleton may be more than 35,000 years old. This assertion rests on contemporaneity of the faunal sample with the hominid, and there is reason to doubt that the *Equus* bones can be associated with any of the human material. If they cannot, then the dates are of little value, and claims for great antiquity cannot be substantiated. A more likely conclusion is that Peers No. 4 was buried into the MSA deposits, probably late in the Pleistocene or early in the Holocene (Rightmire 1978b).

When several of the Fish Hoek crania are measured, ties with living African populations emerge quite clearly. The controversial No. 4 is rather heavily constructed, with brows that are well developed in comparison to those of other recent individuals. But the face is short, and the cranium can hardly be regarded as other than fully modern in its morphology. In discriminant anaylsis, this specimen is aligned with Khoi or San populations, as has been shown both by Howells (1973) and by myself (Rightmire 1981). Discriminant functions can also be used to demonstrate that three other Peers Cave individuals are similar to Khoi or San, and they are certainly more akin to these southern African peoples than to Negroes.

This question of resemblance to recent Africans is important. The fragmentary Border Cave skull hints that large Bushman-like populations may have inhabited southern Africa for a long time, although the evidence is not really very solid, and it is not clear when the earliest Bushmen may have evolved. Tobias (1972, 1978d) has noted that the San and Negroes are probably derived from a common stock, and the time of splitting of these major groups is placed between 15,000 and 25,000 years ago, largely on archaeological grounds. If Border Cave is as old as claimed and if its significance has been read correctly, then this estimate is conservative, and the split must have occurred much earlier. Skeletons which settle this issue are not available, but Bushman-like individuals can certainly be identified in a late Upper Pleistocene or Holocene setting, not only at Fish Hoek but also at other shelters.

Material from LSA levels at Elands Bay and Nelson Bay Cave is of limited utility, as the skeletons are fragmentary and frequently subadult. Human remains from the famous shelter at Matjes River are better preserved in some cases, but the site has been worked over a span of roughly 30 years by different investigators, and this has led to problems in establishing the provenience of individual burials. It is difficult to determine even how many skeletons were finally excavated. Some are associated with a Wilton industry, while material from the earlier part of the sequence can probably be linked to an Albany assemblage, characterized by the presence of large scrapers with few other formal tools. It is too bad that all of the 'Wilton' skulls have suffered damage and that facial parts have been heavily reconstructed with plaster. Several individuals can still be measured, and braincases are a little larger than those of modern San. However, differences of this sort have surely been overemphasized in the past, and it would be surprising if the skulls actually lie outside of a Bushman or Hottentot range of variation. At Oakhurst, where some 30 burials were unearthed by Goodwin, the affinities of the more complete individuals can be described in similar fashion (Rightmire 1970).

These remains from coastal shelters seem to confirm the presence of Bushman-like people in South Africa during late Upper Pleistocene times. There is little doubt of their association with a Wilton way of life, while some of the earliest burials may belong with Albany tools, 12,000 to 8,000 years old. More than this is hard to say, although skeletons from Holocene coastal midden deposits and some inland regions are under study by Hausman (1983), and this work should shed light on associations between some Stone Age people and a pastoralist economy. Additional skeletons from carefully excavated contexts are urgently needed. Holocene developments should be explored further, and it would be especially useful to have more complete fossil remains linked with earlier Upper Pleistocene MSA and pre-Wilton LSA assemblages. Only when such material becomes available will it be possible to extend or modify hypotheses suggested by Border Cave and Klasies River Mouth.

ACKNOWLEDGMENTS

For advice and assistance as well as access to skeletal material, I am indebted to C.K. Brain, Q.B. Hendey, E.N. Keen, J.J. Oberholzer, R. Singer, C.B. Stringer and P.V. Tobias. My research in southern and eastern Africa has been supported by National Science Foundation grants BNS 75-02679 and BNS 80-04852, and by the National Geographic Society.

THOMAS P. VOLMAN
Cornell University Intercollege Program in Archaeology, Ithaca N.Y., USA

6

Early prehistory of southern Africa

INTRODUCTION

Aims and background

The purpose of this paper is to summarize human cultural developments in southern Africa, from the first traces of an archaeological record to the appearance of 'modern' material culture and human behaviors beginning about 40,000 years ago. In terms of the conventional subdivision of the sub-Saharan archaeological record into an Earlier (or Early), Middle and Later (or Late) Stone Age (ESA, MSA & LSA), I am concerned here with the ESA and MSA, and especially with the results of recent research. Particular attention is paid to the nature of the archaeological record, culture-stratigraphy, technology and subsistence, and to a much lesser extent to demography and non-subsistence behaviors.

From the first collection and excavation of Stone Age materials in southern Africa more than a century ago, research on the long archaeological record in this portion of the Old World has contributed in many important ways to our knowledge of human evolution. Modern approaches to prehistoric research began in the 1920s. It was during this decade that Raymond Dart described the Taung infant, the first australopthecine fossil, and defended its hominid status (as discussed by Rightmire, this volume). At the same time, A.J.H. Goodwin (1926, 1928; Goodwin & Van Riet Lowe 1929) was laying the foundations for systematic archaeological investigation with his studies of stone artifact collections and his development of a local terminology, including his recognition and characterization of the ESA, MSA and LSA.

In recent years, Stone Age research in southern Africa has tended to focus on the behavioral and ecological information available from later time ranges. Radiocarbon dating and a variety of sites with good to excellent preservation of organic materials have allowed archaeologists to identify temporal and geographical patterning, reconstruct paleoenvironmental conditions, and use ethnographic and historic information to good effect (see J. Deacon, this volume). At the other extreme, an inherent fascination with very early hominids has continued to attract attention to the interpretation of very early sites, though recent work in southern Africa has been overshadowed by the spectacular discoveries of Plio-Pleistocene fossils and archaeological sites in East Africa. The intervening time ranges have been relatively neglected. This is especially true for most of the Lower and Middle Pleistocene, for which chronological controls are difficult to establish, informative occurrences are rare, and most sites merely establish a human presence in an area at some time in the past. However, especially during the last two decades, great advances have been made in research on the Upper Pleistocene, and the archaeological record of the past 200,000 years is becoming increasingly well known.

Terminology

As noted above, this paper covers the archaeological record commonly assigned to the ESA and MSA in Goodwin's scheme. This terminology was based on Goodwin's study of the technological and typological characteristics of hundreds of South African archaeological collections. Other archaeologists subsequently applied the scheme throughout sub-Saharan Africa. Though officially abandoned today, because the terms have both cultural and chronological connotations, ESA, MSA and LSA remain convenient labels at a higher order of abstraction than is provided by the current hierarchical scheme, which recognizes occurrences, aggregates (or assemblages), phases, industries, and industrial complexes (Bishop & Clark 1967:892-897; Clark *et al.* 1966). In practice, this hierarchical scheme has been applied very differently by various researchers. It seems to have greatest utility for LSA materials younger than 20,000 years B.P. (Before Present). A more informal approach is adopted here for the ESA and MSA archaeological record.

For purposes of this review, ESA refers to the archaeological record from the first appearance of stone artifacts until the disappearance of large cutting tools such as handaxes and cleavers. The following MSA assemblages are usually characterized by a variety of prepared cores and retouched flakes, many of which have no counterparts in ESA assemblages. Beginning about 40,000 years ago in some areas of sub-Saharan Africa, microlithic assemblages appear in which characteristic MSA retouched tools, flakes and cores are absent. The formal properties of these assemblages, and their association with new items of material culture, make it convenient to distinguish these as early LSA (as discussed by J. Deacon, this volume).

Dating

Most of the archaeological record considered here is older than the practical limits of radiocarbon dating as practiced by most laboratories. In the past decade or so, it has become increasingly clear that many finite radiocarbon dates for MSA occurrences in southern Africa, and all finite radiocarbon dates for ESA sites, can only be regarded as minimum estimates of the true ages of the materials. Cultural developments in southern Africa can no longer be viewed as lagging behind those in other parts of the Old World (Clark 1975). There is, moreover, a growing body of geomorphic/sedimentologic and paleontological evidence, supported by infinite radiocarbon dates and other absolute dating methods, which allows some occurrences to be correlated with the world-wide sequence of oxygen isotope stages defined in deep-sea cores. The chronological framework used here, based mainly on

Table 1. Chronological framework.

Approximate years B.P.	Oxygen isotope stage	Inferred world-wide climate		Other terms used here
0				
	1	very warm		Holocene
12,000				
	2	very cold		
32,000				
	3	cold with warm oscillations	Last Glacial	
64,000				
	4	very cold		Upper
75,000				Pleistocene
c. 82,000	5a	warm		
	5b	cold	Last	
c. 105,000	5c	warm	Interglacial	
	5d	cold		
	5e	very warm		
128,000				
	6	cold with warm oscillations		
195,000				Middle
	7	warm with cold oscillations		Pleistocene
251,000				
730,000	19			
				Lower Pleistocene
1,700,000				
				Pliocene

Shackleton and Opdyke (1973, 1976) and Mankinen and Dalrymple (1979), is presented in Table 1.

Geographical coverage

As used here, the term southern Africa refers loosely to that part of the continent south of 8 degrees south latitude, including Angola, southern Zaire, southern Tanzania, Zambia, Malawi, Mozambique, Zimbabwe, Botswana, Namibia (South West Africa), Swaziland, Lesotho and South Africa.

At present, information about the prehistoric record in various regions is very uneven, a result of such factors as the geographic distribution of early peoples, archaeological research interests, and differences in site preservation and accessibility. Over large areas the Stone Age record is known mainly from selected collections made at disturbed open-air sites revealed by erosion or commercial activities, especially mining and quarrying. Much of the northwestern portion of southern Africa is underlain by redistributed Kalahari sands. Recent investigations by Cahen and Moeyersons (1977, 1979; Cahen 1978a, 1978b; Moeyersons 1978) raise many questions about the reliability of earlier reconstructions of paleoenvironmental conditions and culture-stratigraphic sequences in this vast area, and it will be largely ignored here. The distribution of sites in Figure 1 provides a fair indication of our knowledge of the early prehistoric record in different parts of southern Africa.

THE EARLY STONE AGE

East African background

The early archaeological record in southern Africa is best considered against the background of more extensive Plio-Pleistocene finds in East Africa, especially because a major source of chronological controls is provided by faunal comparisons with radiometrically dated East African sites.

Recent discoveries of footprint trails 3.6-3.8 million years (Myr) old at Laetoli in northern Tanzania (Leakey & Hay 1979; Day & Wickens 1980; T. D. White 1980; Leakey 1981) leave no doubt that by this time some hominids were upright bipeds with their hands habitually free to carry and manipulate objects. Although it is generally assumed that hominid adaptations from at least 4 Myr ago included a variety of carrying and manipulative behaviors, involving both organic and inorganic materials, the first unequivocal evidence for such activities is provided by stone artifacts which occur outside their source areas and show signs of deliberate manufacture. At Kada Gona in the Hadar region of Ethiopia, 18 stone artifacts excavated *in situ* plus scattered surface finds are believed to be only slightly younger than an underlying volcanic tuff, dated at 2.9 Myr and 2.7 Myr, respectively, by potassium-argon and fission-track techniques (Roche & Tercelin 1977; Johanson *et al.* 1978; Taieb *et al.* 1978; Lewin 1981a; Walter & Aronson 1982). Other early stone artifact assemblages from the Eastern Rift Valley System (Shungura, KBS and Oldowan/Developed Oldowan A industries) date to 2.1-1.6 Myr ago and have been found in a variety of open-air contexts, including sealed, barely disturbed, occupation surfaces with associated faunal remains.

A variety of primary forms occurs in the early East African stone artifact assemblages, including pebbles and cobbles, slabs and blocks, and intentionally produced flakes and angular fragments. Overall, trimming seems informal, with an emphasis on edge modification rather than shaping whole pieces. The typological system most commonly used to describe these assemblages (Leakey 1971) recognizes several heavy- and light-duty tools, though many of the heavy-duty forms are considered cores by other workers. There is considerable variability within the types recognized by Leakey and, instead of types being characterized by clear-cut boundaries, groups of types intergrade in series of more or less continuous variability.

With the possible exception of the Kada Gona artifacts, these early assemblages coincide broadly with the appearance of *Homo habilis* and *Australopithecus boisei* in the East African fossil record. It seems certain that *H. habilis* made stone artifacts and abandoned them in sufficiently restricted areas to create recognizable archaeological sites. Whether any australopithecines also made stone artifacts is an open question. Although evidence is lacking, some australopithecines probably were at least as proficient in and dependent on tool use and manufacture as chimpanzees, which show 'the greatest frequency and diversity of tool behavior of any nonhuman animal' (Beck 1980:79). Wild chimpanzees modify a variety of plant parts, and some populations occasionally use unmodified stone cobbles as hammers or anvils for breaking open tough fruits and nuts. Digging sticks, containers and hammerstones are often mentioned as tools that early hominids might have found especially useful. Of these, only hammerstones are preserved in very early assemblages, where they are far outnumbered by sharp-edged stone artifacts. This suggests that pounding with stones was far less common than activities involving cuting, chopping and the like. However, it is not known what proportion of the unretouched flakes and angular fragments in any assemblage are unutilized waste, nor how frequently various artifacts were discarded or replaced by fresh ones.

Although chimpanzees and baboons sometimes kill and eat small mammals, the use of bone tools is not recorded among living wild primates. Perhaps a greater dependence on meat and more frequent involvement with animal remains (especially those of larger mammals) is a necessary precondition for the use and manufacture of bone tools. Although well-fashioned unquestionable bone artifacts are found throughout the Old World in the late Upper Pleistocene (e.g. J. Deacon, this volume), the recognition of bone tools during earlier time ranges is a controversial topic (R. White 1982). Some pieces of bone, horn, teeth and ivory from Lower Pleistocene sites in East Africa, such as Melka Kunturé and Olduvai Gorge, show signs of flaking and abrasion that are interpreted as evidence for utilization, and a few pieces from Olduvai Gorge are even thought to be deliberately modified (Chavaillon *et al.* 1978; Leakey 1971). If these interpretations are correct, the implication is that over most of the Pleistocene the use of faunal remains as tools was rather casual, and deliberate modification did not lead to forms which are strikingly different from those produced on some stone artifacts.

Recent studies of mammal bones older than 1.5 Myr from Olduvai Gorge and Koobi Fora have found cutmarks produced by stone flakes used to process carcasses, as well as fractures from hammerstones used to smash bones, presumably for their marrow (Bunn 1981; Potts & Shipman 1981). Bones at some sites where stone artifacts are absent also show cutmarks. Several sites provide evidence for both hominid and carnivore activities, including instances where cutmarks either overlie or underlie marks made by carnivore teeth (Potts & Shipman 1981; Lewin 1981b). Bunn emphasizes the evidence for meat-eating by the early makers of stone tools, but notes that the relative importance of hunting versus scavenging is uncertain. Potts and Shipman found that cutmarks are relatively more common than tooth marks on non-meat-bearing bones; they suggest a possible hominid interest in ligaments and bones in addition to meat. Alternatively, Binford (1981) suggests that most cutmarks represent a stage in processing scavenged bone to obtain marrow. More recently, M.C. Wilson (1982) has suggested that it is non-meaty limb bones that are most likely to receive slicing marks during skinning operations.

Microscopic analyses of relatively rare, cryptocrystalline stone artifacts from sites older than 1.5 Myr at Koobi Fora also have yielded important information on early hominid tool use. Keeley and Toth (1981) found microwear polishes on 9 of the 54 artifacts they examined. Their replication experiments indicate that stone artifacts were used as meat-cutting knives, to cut grass stems or reeds, and to scrape and saw wood, probably to fashion wooden implements.

From about 1.6-1.5 Myr ago, East African assemblages (Acheulean, Developed Oldowan B and Karari) often contain an increased variety of retouched stone artifact forms, including true bifaces, such as handaxes and cleavers, whose finished shape sometimes has been determined entirely by trimming. This 'quantum jump' in internal differentiation of artifact assemblages (Isaac 1969, 1975) coincides with the replacement of *H. habilis* by *H. erectus* in the East African fossil record, and with the appearance of sites with higher densities of flakes and flaking debris than had occurred previously. Artifact types are sometimes more standardized than before, but intertype boundaries still are hard to define in many assemblages.

Faunal remains, including those of very large and dangerous mammals, are associated with the ESA assemblages at some East African sites. However, hunting capabilities and other aspects of Plio-Pleistocene hominid adaptations are not easily determined from the archaeological record. Earlier interpretations which stressed the importance of hunting have been increasingly replaced by models which emphasize the sharing of a variety of foodstuffs and propose that the earliest toolmakers did not kill animals larger than about 30 kg (Clark 1970; Isaac 1978). *H. erectus* may well have been a more capable hunter than *H. habilis*, but this has not been demonstrated with the available archaeological evidence. It is possible, perhaps even likely, that scavenging and opportunistic killing of disabled animals, not active hunting, was responsible for most of the remains of larger mammals at ESA sites long after the appearance of *H. erectus*.

The first traces of human behavior in southern Africa

At present there is little evidence in southern Africa for the early stage of artifact manufacturing recorded in East Africa before 1.6 Myr ago. Except for the isolated Taung specimen, the relevant sites are all former deep, subterranean, dolomite caverns in the Transvaal (see Rightmire & Klein, this volume; lithostratigraphic units are those of Partridge 1978, 1979; Butzer 1976; and Vrba 1981). These caves, and the deposits within them, have complex histories. All contain fossil-bearing sediments, derived from internal and external sources, which have been cemented by calcareous solutions to form cave breccias. Infilling and breccia formation were not continuous, however, and the deposits were affected by decalcification and erosion at various times in the past. Great quantities of fossiliferous breccia were destroyed or left in dumps in the course of mining operations which relied on blasting to uncover pure travertine. Recent scientific excavations have employed drills and wedges for the controlled removal of hard breccias. More conventional archaeological techniques are suitable for the unconsolidated deposits and have been particularly successful for excavating the extensive decalcified breccias at Swartkrans.

No artifacts were associated with the gracile australopithecine at Taung or with the *A. africanus* fossils in Member 4 at the Sterkfontein Type Locality. The evidence from Makapansgat for an 'osteodontokeratic' (bone, tooth and horn) industry in Member 3 (Dart 1957; see Brain 1981 for an excellent summary of the development of Dart's osteodontokeratic hypothesis) and for stone artifacts in Members 3 and 4 (Maguire 1980) is not compelling. Although hominid activity cannot be discounted entirely, it appears that the *A. africanus* and associated larger mammal remains at both Makapansgat and Sterkfontein represent carnivore prey (?primarily hyenas at Makapansgat and various large cats at Sterkfontein), and that the presumed artifacts from Makapansgat Members 3 and 4 could all result from natural agencies and processes (Brain 1981, with references; Klein 1975b, 1977a; Vrba 1974, 1975, 1976, 1980; Maguire *et al.* 1980; Mason 1965).

Some cobbles, angular fragments and rare flakes of quartzite from Member 5 and the overburden at Makapansgat, and perhaps some pieces of dolomite and chert, may be very old artifacts (Brain *et al.* 1955; Maguire 1980). Alternatively, except for some Acheulean and younger artifacts in the unconsolidated deposits, all of the fractured stone from Makapansgat may have resulted from geological processes (Mason 1965). There are also differences of opinion about the origin of the stones in Member 5. Partridge (1979) suggests that they were washed into the cave, but Maguire (1980) proposes an intracavernous origin for the dolomite and chert, and a hominid introduction of the quartzite.

The possible artifacts from Member 5 come from well above the *A. africanus* fossils in Member 3 and the single individual recorded from Member 4. Paleomagnetic determinations for Members 1, 2 and 4 (McFadden *et al.* 1979) suggest that the Member 3 australopithecines are either about 3.0 Myr or about 2.5 Myr old. Faunal comparisons with East African sites support the more recent date (White *et al.* 1981). The top of Member 4 may be considerably younger than 2.4 Myr whichever correlation is made between the Makapansgat paleomagnetic results and the geomagnetic polarity time scale (Mankinen & Dalrymple 1979). Member 5 is clearly unconformable with Member 4 and its age is uncertain. Unfortunately, the Member 5 faunal remains (mainly bovids) are scanty, badly fragmented and not yet analyzed. In sum, the presence of an early stone industry at Makapansgat, predating the Acheulean, remains to be established.

The three Transvaal caves containing robust australopithecines or early *Homo* fossils, or both, in possible or certain association with artifacts, pose many problems of interpretation. Faunal correlations suggest some temporal overlap between Sterkfontein Extension Member 5 and Swartkrans Member 1, and ages of roughly 2.0-1.5 Myr seem likely. At Kromdraai B East it is now recognized that the materials recovered from unconsolidated deposits by Brain (1958, 1975) derive from more than one of the five breccia members at the site (Vrba 1981). Vrba found remains of robust australopithecines only in Member 3 of the new lithostratigraphy. The implication is that all australopithecine fossils from the site probably come from this member. Associated faunal remains are not very informative for dating purposes, but Vrba's paleoenvironmental reconstruction suggests that the Member 3 deposits accumulated before a shift to drier conditions reflected in Sterkfontein Member 5 and Swartkrans Member 1. In addition, Grine's (1981) studies of trends in australopithecine dentition suggest that the Kromdraai B East fossils are intermediate between *A. africanus* ones and those of robust australopithecines at Swartkrans.

Vrba (1981: 17) proposes that during Member 3 times the Kromdraai B East cave 'probably served as an occasional sheltering place for primates, perhaps including the apeman, a death trap for some animals which fell in, and a source for opportunistic feeding by carnivores.' Vrba attributes the considerable fragmentation of faunal remains, formerly interpreted as a result of hominid food processing, to the effects of water-logging, trampling and pressure from overlying deposits. Vrba found no stone or other artifacts during her excavations of intact breccia. Earlier, Brain (1958, 1975) recovered a few chert flakes, a broken quartzite pebble and some quartzite blocks and pebbles at Kromdraai B. Further work at the site may provide some indication of the age and associations of these possible artifacts.

Swartkrans Member 1 has yielded abundant robust australopithecine fossils and remains of as many as three individuals attributed to early *Homo*, with affinities to both *H. habilis* and *H. erectus* (Tobias 1978c; Rightmire, this volume). Brain (1981, with references) has developed a convincing case that the bulk of the Swartkrans Member 1 larger fauna, including the robust australopithecine (?and early *Homo*)

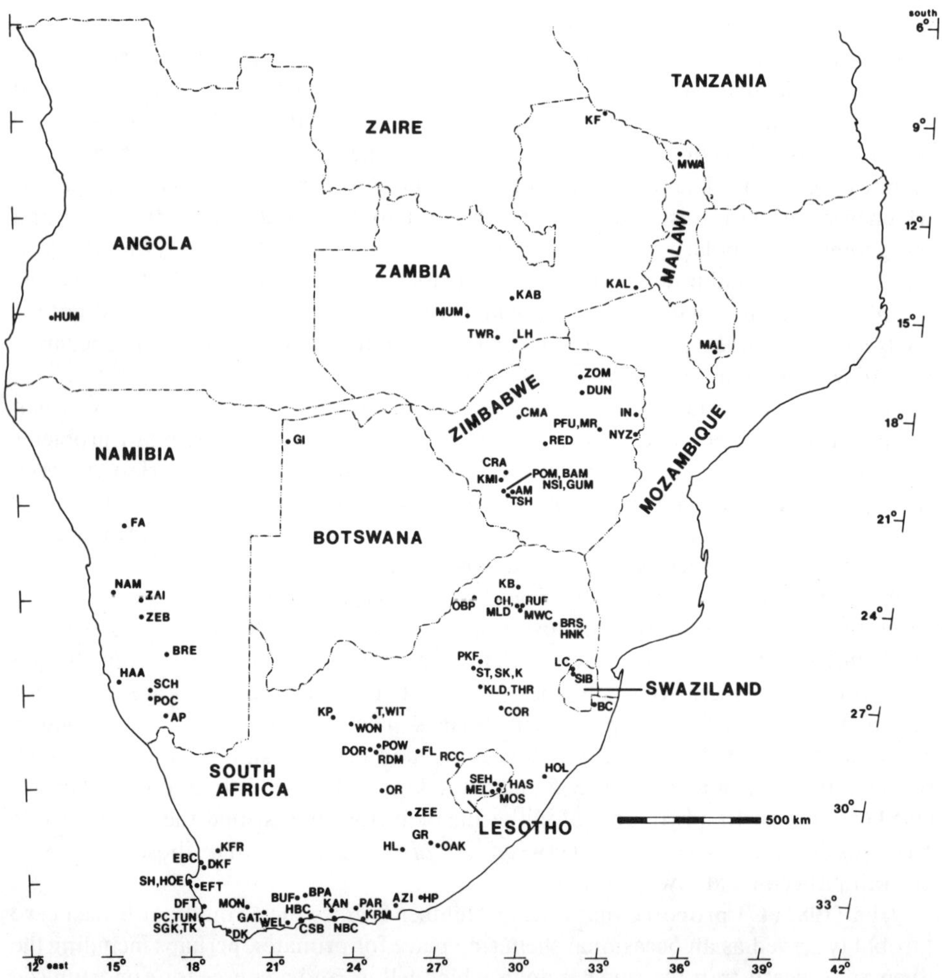

Figure 1. Approximate locations of southern African sites mentioned in the text.

Name of site	Abbreviation	Country
Amadzimba	AM	Zimbabwe
Amanzi	AZI	South Africa
Apollo 11	AP	Namibia
Bambata	BAM	Zimbabwe
Boomplaas	BPA	South Africa
Border Cave	BC	South Africa
Bremen 1C	BRE	Namibia
Buffelskloof	BUF	South Africa
Bushman Rockshelter	BRS	South Africa
Cape St. Blaize Cave	CSB	South Africa
Cave of Hearths	CH	South Africa
Charama Plateau sites	CMA	Zimbabwe
Cornelia	COR	South Africa
Cranmore Farm	CRA	Zimbabwe
Die Kelders 1	DK	South Africa
Diepkloof	DKF	South Africa
Doornlaagte	DOR	South Africa
Duinefontein 2	DFT	South Africa
Duncombe Farm	DUN	Zimbabwe
Elands Bay Cave	EBC	South Africa
Elandsfontein	EFT	South Africa
Fackelträger	FA	Namibia
Florisbad	FL	South Africa
Gatboskloof	GAT	South Africa
≠Gi	GI	Botswana
Grass Ridge	GR	South Africa
Gumani	GUM	Zimbabwe
Haalenberg	HAA	Namibia
Ha Soloja	HAS	Lesotho
Herolds Bay Cave	HBC	South Africa
Heuningsneskrans	HNK	South Africa
Highlands	HL	South Africa
Hoedjies Punt	HOE	South Africa
Holley Shelter	HOL	South Africa
Howieson's Poort	HP	South Africa
Humpata	HUM	Angola
Inyanga sites	IN	Zimbabwe
Kabwe (Broken Hill)	KAB	Zambia
Kalambo Falls	KF	Zambia
Kalemba	KAL	Zambia
Kalkbank	KB	South Africa
Kangkara	KAN	South Africa
Kathu Pan	KP	South Africa
Khami Waterworks	KMI	Zimbabwe
Klasies River Mouth	KRM	South Africa
Klipfonteinrand	KFR	South Africa

Name of site	Abbreviation	Country
Klipplaatdrif	KLD	South Africa
Kromdraai	K	South Africa
Leopard's Hill	LH	Zambia
Lion Cavern	LC	Swaziland
Makapansgat	MLD	South Africa
Malowa	MAL	Malawi
Melikane	MEL	Lesotho
Montagu Cave	MON	South Africa
Moshebi's Shelter	MOS	Lesotho
Mtemwa Rocks	MR	Zimbabwe
Mumbwa	MUM	Zambia
Mwanganda	MWA	Malawi
Mwulu's Cave	MWC	South Africa
Namib IV	NAM	Namibia
Nelson Bay Cave	NBC	South Africa
Nswatugi	NSI	Zimbabwe
Nyazongo	NYZ	Zimbabwe
Oakleigh Farm	OAK	South Africa
Olieboompoort	OBP	South Africa
Orangia 1	OR	South Africa
Paardeberg	PAR	South Africa
Peers Cave	PC	South Africa
Pfupi	PFU	Zimbabwe
Pietkloof	PKF	South Africa
Pockenbank 1	POC	Namibia
Pomongwe	POM	Zimbabwe
Power's Site	POW	South Africa
Redcliff	RED	Zimbabwe
Rooidam	RDM	South Africa
Rose Cottage Cave	RCC	South Africa
Rufus	RUF	South Africa
Schlangengrotte	SCH	Namibia
Sea Harvest	SH	South Africa
Sehonghong	SEH	Lesotho
Sibebe	SIB	Swaziland
Skildergatkop	SGK	South Africa
Sterkfontein	ST	South Africa
Swartkrans	SK	South Africa
Taung	T	South Africa
Three Rivers	THR	South Africa
Trappieskop	TK	South Africa
Tshangula	TSH	Zimbabwe
Tunnel Cave	TUN	South Africa
Twin Rivers	TWR	Zambia
Welgevonden	WEL	South Africa
Witkrans	WIT	South Africa
Wonderwerk	WON	South Africa
Zais	ZAI	Namibia
Zebrarivier	ZEB	Namibia
Zeekoegat 27	ZEE	South Africa
Zombepata	ZOM	Zimbabwe

Figure 1 (*Contd.*)

remains, represents a carnivore accumulation, with leopards and perhaps saber-toothed cats and hyenas the major contributors. No damage attributable to hominids has been identified on any of the faunal remains, though this may change when detailed studies are undertaken on bones from Brain's recent excavations. Moreover, for the sample he analyzed, Brain (1981:239) reports that 'details of damage most of the [non-hominid] bones may have suffered are still obscured by enclosing breccia.'

The small collection of stone artifacts recovered from breccia dumps at Swartkrans (Brain 1958, 1970; Leakey 1970, 1971) now is thought to come mainly from the Member 2 breccia and younger channel fills, which are Middle Pleistocene or younger in age (Brain 1976, 1981). After examining the matrix still adhering to some of the 35 pieces in the Transvaal Museum collection, Brain (1976) concluded that a quartzite, heavy-duty scraper, and perhaps 6 other pieces, came from Member 1. The 6 possible Member 1 pieces are also made of quartzite; they are all trimmed cobbles with the exception of a small, broken flake. The presence of a cleaver, a pick and 3 bifaces among the remaining artifacts may indicate brief occupations in or near the site by later makers of Acheulean tools, perhaps represented by the probable *Homo erectus* fossils from Member 2 (discussed by Rightmire, this volume). Many of these artifacts are abraded, perhaps through stream transport, which Brain (1981:227) suggests as a possible cause of general abrasion on the Member 1 scraper.

Brain's excavations of unconsolidated Member 1 breccia have provided more definite evidence for a stone artifact industry in Member 1, as well as a blesbok horncore and some 8-10 bone fragments which show signs of utilization (Brain, pers. comm.). In contrast to the collection from breccia dumps, the material from Brain's controlled excavations consists almost entirely of small flakes and angular fragments, primarily of quartz and chert, but also including some pieces of quartzite. The utilized bones are particularly interesting finds and complement Robinson's earlier discovery of a similar piece at the Sterkfontein Extension site (see below). The bone tools are relatively thin fragments, with polish and faceting extending roughly 30 mm from one tip, which is blunted in some cases. Ongoing experiments by Brain (pers. comm.) suggest one possible use for the bones. He found that thin bone fragments are useful for extracting bulbs from rocky soils when used with stone hammers and a digging stick. Experimental replicas showed wear similar to that on the Swartkrans pieces after less than a half hour of digging.

The presence of stone artifacts and bone tools in Member 1 at Swartkrans, along with hominid fossils, suggests that the artifact makers, whether early *Homo*, robust australopithecines, or both, were occasionally killed or died of natural causes while sheltering in or near the cave opening. The stone artifacts may represent manufacturing debris for the most part, or perhaps some pieces provided sharp cutting and slicing edges. If the utilized horncore and bone fragments were used for digging up bulbs at some distance from the cave, their presence in the deposits implies that some digging implements were conserved for future foraging trips, rather than being discarded soon after use.

The archaeological situation at the Sterkfontein Extension site is particularly difficult to interpret at present. Although some stone artifacts and faunal remains have been excavated from *in situ* Member 5 breccia, most come from superficial colluvium, miners' rubble dumps and unconsolidated weathering residues, which include solution cavities that extend through the Member 5 breccia. LSA, MSA and

ESA artifacts occur in the overburden. The Member 6 breccia and weathering residue contain probable Acheulean artifacts, perhaps some MSA artifacts, and a fauna which, overall, seems referable to the Middle Pleistocene with perhaps some younger elements (Mason 1976; Vrba 1974, 1975, 1976; Klein, this volume).

Vrba (1974, 1975, 1976, 1980) and Brain (1981) have recently reported on the faunal remains from Sterkfontein Extension Member 5. The best collection comes from Robinson's excavations, but it should be noted that even this sample may include remains from older and younger deposits (Vrba, pers. comm.). Vrba, who analyzed the relatively abundant bovid remains, interprets their high degree of fragmentation as the result of hominid food processing. She further suggests that the relatively high proportion of adults in all weight classes indicates hominid scavenging as opposed to active hunting.

Vrba's interpretations may well be correct, but neither of her criteria points exclusively to hominid involvement in accumulating or processing the faunal sample. Bone fragmentation, as was the case at Kromdraai, may reflect postdepositional damage rather than hominid butchery and marrow extraction. Mortality profiles also are affected by postdepositional disturbances, and by carnivore feeding as well (Klein 1982a, 1982a,b). Brain found damage from carnivore chewing on 37 of the 1202 macrovertebrate remains he analyzed from Robinson's excavations. Only one piece, 'the shaft of a small bovid humerus bearing three clear chop marks that were presumably caused by the edge of a stone tool' (Brain 1981:216), provides evidence for hominid butchery. Large, well-provenienced samples providing information on species composition, age profiles, body part representation and damage patterns are needed to better establish the contribution of hominid activities to the faunal assemblages in Sterkfontein Extension Member 5.

Artifacts of foreign stone from Member 5 are mostly quartz and quartzite, with rare diabase and a few pieces of hematite. Quartz flakes and angular fragments are relatively common, but flakes of quartzite and diabase are very infrequent. The few hematite pieces are all small flakes and fragments with no signs, such as abrasion, for use as pigment. Some of the quartzite and diabase pieces are unmodified, waterworn cobbles. Others are battered, broken or have a few flake removals, consistent with utilization, attempts at flake production, postdepositional damage, or a combination of these effects. Still others show evidence of intentional flaking, either to produce flakes, a retouched edge, or both. Some of the abundant pieces of chert and much rarer vein quartz, derived from the dolomite in which the cave was formed, probably are artifacts, but most are unmodified or naturally fractured. Many of the foreign stones in the breccia are unweathered or preserve original cortex. However, some pieces are slightly to heavily abraded, probably as a result of running water.

Member 5 was first excavated by J.T. Robinson (1962; Robinson & Mason 1957), who recovered 286 pieces of worked or foreign stone from his West Pit, as well as a utilized longbone fragment with polish and fine striations at one end (J.T. Robinson 1959). Among the faunal remains excavated by Robinson, Brain (1981:216) has also found a possibly utilized bone flake and what 'appears to be a piece of a bovid horn-core, three surfaces of which have been worn to a tapered point.' The single definite handaxe from this excavation (Figure 2) is slightly waterworn and came from a solution cavity, so that its derivation from Member 5 is uncertain. Robinson's collection has been called Developed Oldowan B by Leakey (1970, 1971) and

Acheulean by Mason (1962a, 1976). The main difference between their classifications is that most of the pieces called cores by Mason are classified as choppers and polyhedrons by Leakey.

Both Mason (1976) and Stiles (Stiles & Partridge 1979) have published classifications of much larger series of probable ESA artifacts, including material recovered from dumps and unconsolidated deposits during initial stages of renewed work at the site directed by Tobias and Hughes (1969). Stiles notes similarities to both Developed Oldowan B and Acheulean assemblages from Bed II at Olduvai and suggests that all of these assemblages should be placed in the Acheulean Industrial Complex. Whatever the final resolution of the nomenclature controversy, most workers who have studied the Member 5 and other probable ESA artifacts from Sterkfontein are agreed in grouping these with East African assemblages presumably manufactured by *Homo erectus* after 1.6 Myr ago. It is therefore somewhat surprising to find that the major hominid fossil from Member 5, the Stw-53 cranium, is considered most similar to fossils classified as *Homo habilis* in East Africa (Hughes & Tobias 1977; Tobias 1978c; Rightmire, this volume). This Sterkfontein hominid discovery raises questions about the nature of the Member 5 artifacts, and whether more than one archaeological entity is represented among the ESA materials from the site.

The data presently available suggest that the artifacts from Member 5 may be comparable to Oldowan/Developed Oldowan A assemblages from East Africa, not Acheulean/Developed Oldowan B. So far as I can tell, all of the true bifaces, and perhaps all of the artifact types which appear after about 1.6 Myr ago in East Africa, were found only in unconsolidated deposits at Sterkfontein, or in circumstances where their derivation from Member 5 is not certain. Thus, Member 5 may contain one ESA entity, and the later deposits another. It is also possible that more than one archaeological entity is represented in Member 5. Large scale excavations of *in situ* Member 5 breccia, recently initiated by Hughes, should provide the stratigraphic controls, artifact samples and faunal remains necessary to resolve this and other issues concerning hominid activities at the site.

Acheulean chronology and cultural stratigraphy

In southern Africa, as in East Africa, only broad divisions can be suggested for the more than one million year time span of the Acheulean and related industries, and even these are poorly distinguished. In East Africa, recognition of a high degree of interassemblage variability among roughly like-aged assemblages at such sites as Olduvai Gorge, Isimila and Olorgesailie has led to the widespread adoption of an 'activity pattern' model of explanation for interassemblage differences (Isaac 1972). In southern Africa, developmental schemes of progressive refinement within the Acheulean have been abandoned (Sampson 1974; H.J. Deacon 1975), but poor chronological controls and the few primary or semi-primary sites, usually lacking organic remains, do not allow for assessment of interassemblage variability over short time intervals. Only a Lower and Upper Acheulean are distinguished here for sub-Saharan Africa (see also H.J. Deacon 1975). In southern Africa, the Upper Acheulean includes assemblages assigned by various researchers to the Middle and Later Acheulean (also called Final Acheulean and Fauresmith).

The only unselected artifact samples from southern Africa that have been classified

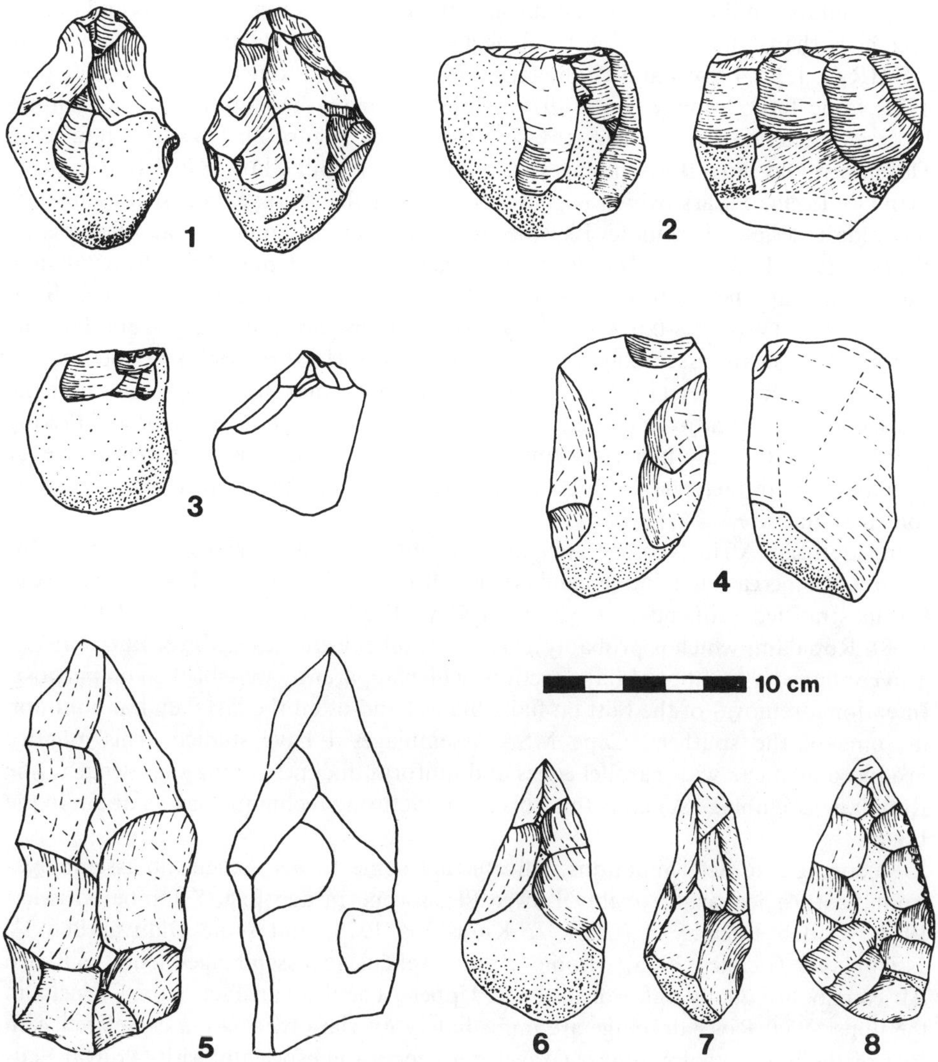

Figure 2. Lower Acheulean artifacts from Sterkfontein Extension Member 5 (1–3) and Three Rivers (4–8) (after Mason 1962b). Types: 1, 5–8: handaxes; 2: core; 3: chopper; 4: cleaver. Handaxe 1 came from a solution cavity and may be younger than artifacts from the Member 5 breccia.

as Lower Acheulean, aside from the Sterkfontein material discussed above, come from the 10 m terrace of the Klip River in the southern Transvaal. No fauna was preserved with these Three Rivers and Klipplaatdrif assemblages, which are assigned to the Lower Acheulean on typological grounds (Mason 1962b). Correlations of the Klip River gravels with alluvial sequences elsewhere in the Vaal River Basin are uncertain.

However, clear-cut typological differences between the Lower and Upper Acheulean have not been established on any firm basis, and perhaps a chronological boundary at about 1 Myr may prove more satisfactory. In both eastern and southern

Africa, bifaces in the oldest Acheulean assemblages have fewer flake scars and deeper scar beds than bifaces in some younger assemblages. Handaxes from Olduvai site EF-HR (c. 1.3-1.4 Myr) and Peninj (c. 1.4 Myr) average about 10 flake scars (Leakey 1971; Isaac 1967; Isaac & Curtis 1974). Handaxes from the Klip River series seldom have more than 10 flake scars (Mason 1962b; see Figure 2). However, bifaces in Developed Oldowan B assemblages (c. 1.5-1.2 Myr) from Olduvai Bed II average as many as 18 flake scars, which is about the same as the average flake scar counts for the oldest series of Middle Pleistocene Acheulean assemblages at Olorgesailie (Leakey 1971; Isaac 1977). Moreover, while many African Upper Acheulean handaxe samples average from 20 to 50 or more flake scars, several handaxe samples from Olduvai Bed IV (c. 0.8-0.6 Myr) (Hay 1976) average no more flake scars than the Lower Acheulean assemblages mentioned above (Leakey 1975). Experimental studies by P.R. Jones (1979:836) suggest that both the number of scars and the thickness/breadth ratio of handaxes may be controlled by the size, shape and flaking properties of the raw materials employed, so that 'some bifaces termed crude or primitive are, in fact, the products of sophisticated and efficient techniques of stone tool manufacture.'

In southern Africa, Upper Acheulean assemblages provide some evidence for progressive increase in relative handaxe breadth in the Transvaal (Mason 1962b), and within stratified sequences at Montagu Cave (Keller 1973) and Rooidam (Fock 1968). Rooidam, which is probably a very late Acheulean site (Szabo & Butzer 1979), also contained a number of flake sections which apparently were broken on purpose. Intentional removal of the butt on flake-blades, and often the distal end, is common in some of the southern Cape MSA assemblages I have studied. This practice produced artifacts with parallel edges and uniform thickness along their length. The Rooidam assemblage suggests that this manufacturing technique has its origin in the ESA.

Paleomagnetic determinations indicate that some Upper Acheulean assemblages from East Africa are probably Lower Pleistocene in age [e.g. Kilombe (Gowlett 1978), Gadeb Locality 8 (Clark & Kurashina 1979) and some Olduvai Bed IV assemblages (Leakey 1975)]. Some Upper Acheulean assemblages from southern Africa may be equally old. For instance, Upper Acheulean artifacts from Cornelia in the upper Vaal River drainage are immediately overlain by a fauna correlated with part of the lower Vaal Younger Gravels sequence, and especially with Olduvai Beds III/IV (Butzer *et al.* 1974). The Younger Gravels of the Lower Vaal Basin may span a large portion of Lower and Middle Pleistocene time (Helgren 1979a). Apparently even the oldest of these gravels (Van Riet Lowe 1945, 1952; Helgren 1978) provided Upper Acheulean collections with evidence of Levallois techniques, including Victoria West and perhaps some Levallois flake and blade cores (Figure 3).[1] With the possible exception of late ESA assemblages from Zimbabwe and Zambia, Levallois techniques seem to be restricted to the South African interior throughout the ESA in southern Africa (Sampson 1974; C.K. Cooke 1966a).

Unfortunately, the integrity of many of the lower Vaal collections is open to question. Some sites, such as the impressively rich Power's Site where the Upper Acheulean artifacts are probably of Middle Pleistocene age (Power 1955; Helgren 1978), are channel surface or channel margin deposits that have been more or less disturbed by water action. Such localities may have been repeatedly occupied over

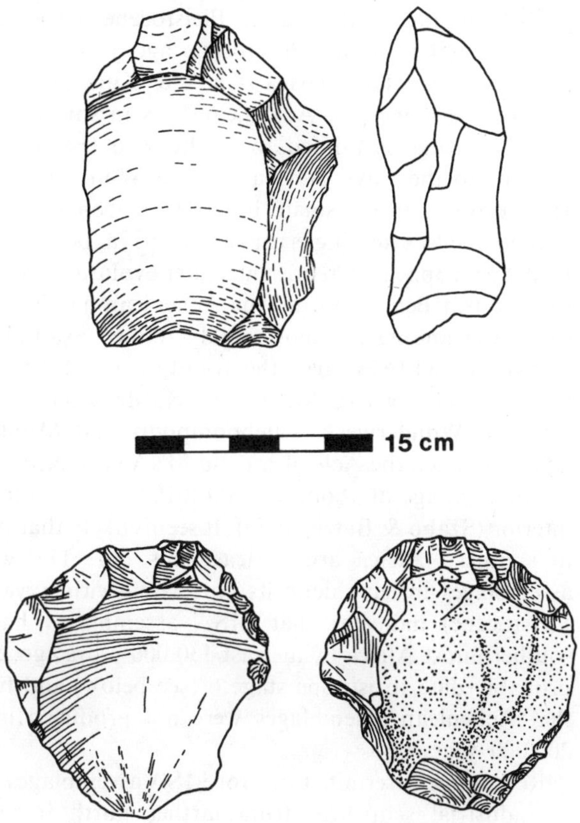

Figure 3. Victoria West prepared cores from the South African sites of Blaaubank (above; after Mason 1962b) and Canteen Kop (below; after Goodwin 1934).

long time periods. Other sites may contain materials washed down from higher ground; the artifacts could be mixed, and some pieces might be considerably older, or even younger, than the river gravels with which they are associated.[2] How far any of the Acheulean artifacts have been transported in the river beds is also an open question, though some answers might be provided by experimental studies. It is possible that even the most abraded pieces still recognizable as artifacts have traveled only short distances, and that long-distance transport could have obliterated any signs of manufacture.

Although further research is needed, the present Acheulean evidence suggests that fairly refined flake production techniques and artifact forms may date back beyond the beginning of the Middle Pleistocene in southern Africa. For many years the Acheulean artifacts from Elandsfontein in the southern Cape were considered very young ('Coastal Fauresmith') because the handaxe collections included numerous small, finely-worked forms less than 100 mm in length (Singer & Crawford 1958). It now appears that the fauna from Elandsfontein may span much of the Middle Pleistocene and perhaps some of the Lower Pleistocene as well (Klein 1978c, this volume, pers. comm.). The bulk of the Acheulean artifacts from Elandsfontein may date to the late Middle Pleistocene, but some may be considerably older. Occasional handaxes, similar in size and refinement to the smaller Elandsfontein ones and perhaps dating to the Lower Pleistocene, have been found in the Vaal Younger Gravels, and a Lower Pleistocene age also seems likely for possibly comparable handaxes from East Africa (Clark & Kurashina 1979).

Faunal evidence for Acheulean sites in southern Africa that postdate Bed IV at Olduvai Gorge is limited to Kathu Pan (Beaumont & Klein, pers. comm.), and perhaps to the Cave of Hearths and Wonderwerk Cave, although the absence of archaic forms in the small, highly fragmented samples from the last two sites may well be due to sampling error (Klein 1977a, pers. comm.). *Elephas recki* molars from Kathu Pan appear to represent a later evolutionary stage of this elephant species than the Olduvai Bed IV form. The Acheulean handaxes from Kathu Pan are relatively small and among the most highly refined examples from southern Africa.

It is difficult to estimate the age of the latest Acheulean in southern Africa. At the few cave sites where MSA artifacts definitely overlie Acheulean ones (Cave of Hearths, Wonderwerk, Olieboompoort and Montagu), there is probably a long hiatus between the Acheulean and MSA occupations. Recent determinations suggest a minimum age of about 200,000 B.P. for the latest Acheulean in the South African interior (Szabo & Butzer 1979). It seems likely that Acheulean assemblages elsewhere in southern Africa are at least this old. The widespread occurrence of MSA assemblages in cave deposits correlated with oxygen isotope stage 5 (early Upper Pleistocene) indicates that MSA assemblages had replaced Acheulean ones in southernmost Africa by at least 130,000 years ago. Several MSA occurrences which probably date to isotope stage 6 (see below) add further support to the suggestion that Acheulean assemblages were not produced in southern Africa after roughly 200,000 years ago.

Recent age determinations for MSA assemblages in East Africa and reevaluations of industrial sequences from farther north indicate that the southern African evidence for a late Middle Pleistocene replacement of Acheulean by MSA is neither unique nor unusual. Potassium-argon dates from central Ethiopia place the earliest MSA assemblages in this region at more than 180,000 years ago (Wendorf *et al.* 1975) and MSA artifacts from Laetoli in Tanzania also may be Middle Pleistocene in age (Day *et al.* 1980). Mousterian assemblages from North Africa and Europe date from the Last Interglacial and are preceded by late Middle Pleistocene assemblages which often are indistinguishable from Mousterian assemblages of Last Glacial age (Tuffreau 1979; Freeman 1975). As will be discussed more fully below, it would appear that throughout the western portion of the Old World, ESA/Lower Paleolithic industries were replaced by MSA/Middle Paleolithic ones in the late Middle Pleistocene.

Some ESA assemblages from the more forested portions of equatorial and southcentral to southeastern Africa have been assigned to a Sangoan Industrial Complex, separate from the Acheulean. Although some isolated artifact occurrences in southeasternmost Africa have been called Sangoan, the distribution of Sangoan assemblages essentially is restricted to regions north of the Limpopo. Most Sangoan samples are highly selected and there are no unselected collections from undisturbed contexts, with the possible exception of artifacts from the lowest levels at Pomongwe Cave and perhaps from Mwanganda and Kalambo Falls.[3] Handaxes and cleavers are relatively rare in Sangoan samples, which according to Clark (1970:111-112) 'are composed of a number of small scraping forms and crude pick-like tools,' termed 'core axes.' Clark further points out that there is a high frequency of denticulate and concave edges on the light duty tools.

In the few cases where Sangoan assemblages have been found in stratigraphic relationship to Acheulean ones, the Sangoan assemblages are younger. The time span covered by Sangoan assemblages is unknown, though, and they may have been produced over a long period of Middle Pleistocene time.

For Zimbabwe, C.K. Cooke (1966a) suggests that collections formerly assigned to an early, 'Proto-Stillbay' phase of the MSA are, in fact, a late Sangoan which he calls Charama. Cooke assigns the artifacts from the lowest levels at Bambata and Pomongwe caves to the Charama, but his large samples come from open-air sites, especially the sealed Khami Waterworks site and surface exposures at Cranmore Farm and on the Charama Plateau (C.K. Cooke 1957, 1966a, 1966a,b). Because they come from disturbed contexts, the reliability of these open-air collections is uncertain. Cooke reports that Charama samples from open-air sites contain fewer handaxes or picks, but more retouched points, than earlier Sangoan collections from the same or comparable contexts.

Collections which are perhaps similar to Zimbabwean Charama ones have been excavated at two cave sites and one localized open-air occurrence in Zambia. A small, selected assortment of stone artifacts that may have been associated with the Broken Hill skull at Kabwe (see Rightmire, this volume) was originally called 'Proto-Stillbay' and later classified as Sangoan (Clark *et al.* 1950; Clark 1970). There are no handaxes, picks or other large retouched pieces in this collection, however, and affinities with the Zimbabwean Charama material are not obvious. At Mumbwa the Italian Scientific Expedition (Dart & Del Grande 1931) found artifacts in their lowest levels in Cave 2 which may be similar to Zimbabwean Charama ones. Unfortunately, the few pieces described or illustrated are not very informative, and this level did not extend into the portion of Cave 2 that was excavated later by Clark (1942).

From the breccia atop Twin Rivers Kopje, Clark (1971a) made a collection of 929 stone artifacts which includes a variety of small flake tools, unifacial points and bifacial points including lanceolates, as well as pieces classified as picks, core-axes, core-scrapers, choppers, polyhedrons, spheroids and grindstones. This collection is perhaps the closest Zambian equivalent to the materials classified as Charama in Zimbabwe.

Finally, the brief, preliminary descriptions of some assemblages from deposits younger than those containing Sangoan artifacts at Kalambo Falls suggest that these 'Nakisasa Industry' artifacts may be technically and typologically comparable to the Zimbabwean Charama material (Clark 1974).

A few radiocarbon dates and analyses of some small faunal samples suggest that Charama and similar assemblages are beyond the range of conventional radiocarbon dating, but do not preclude an Upper Pleistocene age. The Kabwe assemblage may be considerably older and certainly seems to differ typologically from this apparently younger group. Some of the fauna from Kabwe can be assigned to the Cornelian 'Land Mammal Age' (Klein 1973, this volume), but the situation at the site could have been very complex and the Kabwe assemblages may be a mixture of ESA and MSA. Redcliff and Nswatugi caves, where the oldest assemblages are MSA, suggest that MSA assemblages had replaced Charama ones in Zimbabwe by the beginning of the Last Glacial, if not before. Faunal and sedimentological studies at Redcliff (Klein

1978b, 1980; Brain 1969b) suggest that MSA occupation began by the early Last Glacial. An infinite radiocarbon date places the entire MSA sequence at Nswatugi at more than 42,000 years ago (Walker 1978, 1980) and if the sequence spans the same time range as typologically comparable ones from south of the Limpopo, then perhaps Charama assemblages are more than 115,000 years old.

Adaptations

Unfortunately, data pertinent to an understanding of ESA adaptations are rather limited and seldom permit firm conclusions. The widespread southern African Acheulean assemblages demonstrate a high degree of variability in the morphology and relative frequency of artifact types, but there is little evidence for increasing differentiation through time. If later ESA populations could more effectively exploit African environments than earlier peoples – possibly indicated by greater numbers of sites, reflecting higher population densities, and the expansion of populations into the drier western areas of southern Africa (H.J. Deacon 1975) – this is not apparent in their stone artifact assemblages. Rather, the technological conservatism of ESA assemblages over more than a million years suggests that, whatever the improvements in ESA adaptations, the improvements required little innovation or technological refinement in stone tools.

Except for a bone point and two 'gouges' from possible ESA context at Kabwe (Clark *et al.* 1950; Clark 1977b), there is no evidence for intentionally shaped bone tools at any southern African ESA site, and only Kalambo Falls and perhaps Amanzi Springs provide evidence of woodworking. Several pieces of wood from Upper Acheulean levels at Kalambo Falls show signs of deliberate modification; among the items that have been illustrated, 'a wooden club and a smoothed and pointed implement' look particularly convincing (Clark 1970:286). At Amanzi a piece of wood from the Acheulean deposits may have cutmarks at one end (H.J. Deacon 1970b).

At present, there are no specific data to support Clark's suggestion (1964, 1970) that the latest Acheulean assemblages reflect a continuation of Acheulean adaptations to grassland or savanna regions, while the apparently crude Sangoan assemblages, with their heavy-duty equipment, reflect a new level of adaptation to more wooded areas based on extensive woodworking. There is no direct evidence that any Sangoan artifacts were used to fashion wood tools or equipment, nor are wood artifacts preserved at any Sangoan site. Nonetheless, the geographical distribution of Sangoan sites strongly suggests some connection between technology and environmental factors.

Only seven sites contain ESA artifacts and bones in primary or near primary context. These are the Cave of Hearths, Wonderwerk, Pomongwe, and the open-air sites of Elandsfontein, Mwanganda, Namib IV, and Kathu Pan (see below, Table 2, and Klein, this volume for references).[3] Interpretations of associated artifacts and bones in terms of human activities have been offered only for Mwanganda Area 1 (Clark & Haynes 1970; Clark *et al.* 1970) and Elandsfontein 'Cutting 10' (Klein 1978c), both of which are considered butchery sites.

The Mwanganda Area 1 site provided 314 stone artifacts in association with numerous remains of a single adult elephant plus fragmentary remains of hippopotamus and giraffe. Most of the retouched artifacts are light-duty forms which Clark and Haynes suggest were used to process the elephant. On the basis of the Mwanganda occurrence and comparisons with sites in Africa and Europe, Clark and Haynes (1970:409) suggest that 'Palaeolithic butchering and meat processing equipment consisted predominantly of small numbers of Light Duty tools and cutting flakes and small scraping tools with a few only of larger elements.'

This conclusion may be questioned in the light of the Elandsfontein 'Cutting 10' evidence. Klein's analysis of the 'Cutting 10' fauna indicates that carnivore predation and human and carnivore scavenging are reflected at the site in addition to human hunting and butchery. The ratio of bones to stone artifacts at 'Cutting 10' falls well within the range for other Acheulean 'kill/butchery' sites, suggesting that some of these, which have been assumed to represent only human predation, may result from a complex combination of human and carnivore hunting and scavenging, natural mortality, and geological processes.[4]

I studied the 'Cutting 10' artifacts at the South African Museum and counted 260 pieces from within the grid excavated by Singer and Wymer (1968). These included 37 handaxes, 2 cleavers, a thin chopping tool or bifacial knife, a bifacial knife, and only 14 definitely retouched or utilized light-duty pieces. Thus, at 'Cutting 10' the large cutting and scraping tools comprise 16% of the total artifacts and 75% of the modified pieces. As Klein has pointed out, the proportion of handaxes at 'Cutting 10' probably is higher than in any undistorted sample from any other well-excavated Acheulean site. Klein's conclusion that the 'Cutting 10' artifacts and bones are in meaningful association and not brought together by deflation is supported by the low coefficients of variation and standard deviations for measurements taken on the handaxe sample (Volman, in prep.).

Recent butchering experiments by P.R. Jones (1980:63) suggest 'the superior efficiency of large bifacial cutting tools over small retouched or unretouched flakes for almost all purposes of heavy-duty butchery,' especially when larger mammals and long work periods are involved. If Jones's experimental inferences are correct, the high proportion of handaxes at Elandsfontein 'Cutting 10' is consistent with Klein's explanation for differences in the minimum numbers and in the body part representation of different sized animals in the faunal sample. In particular, Klein argues that elephants and larger bovids were butchered on the spot, while small bovids either were removed intact for processing at camps located elsewhere, or were completely consumed by carnivores.

Kalambo Falls and Amanzi, mentioned previously in connection with woodworking, are the only ESA sites to preserve plant remains other than charcoal. At both sites the waterlogged conditions that favored plant preservation have resulted in the destruction of any faunal remains that may have been present. The plant remains at Amanzi (H.J. Deacon 1970b; Wells 1970) come from plants which grew in the vicinity of this spring site; none are sources of food and, with the possible exception of the piece of wood which may have cutmarks, none appear to have been introduced into the deposits by human activity. The most likely plant foods in the area would have been geophytes, both the edible and waste portions of which are less resistant to humification than the plant parts found in the site (H.J. Deacon 1975). At Kalambo Falls the macrobotanical materials include fruits and seeds of edible plants

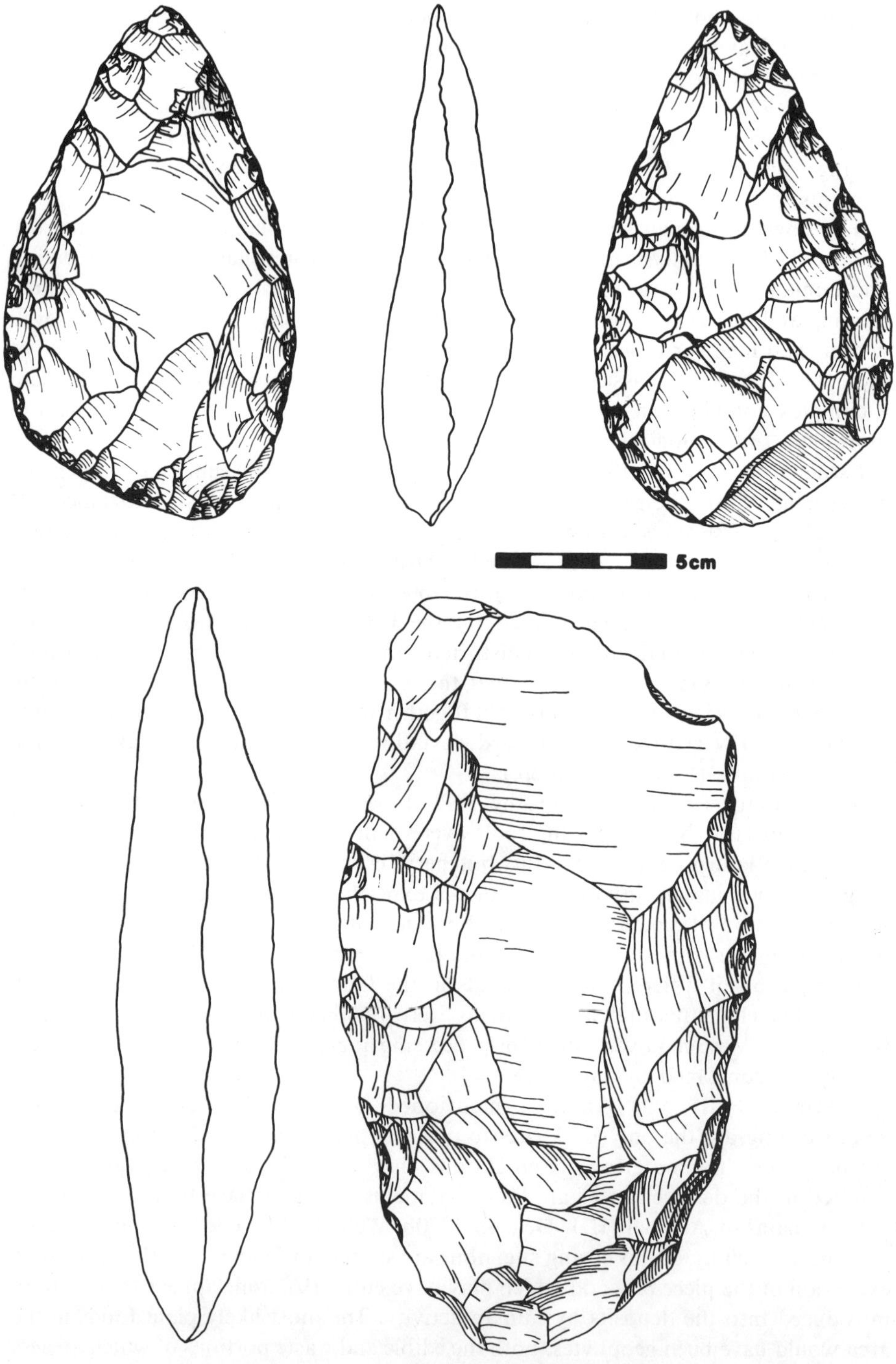

Figure 4. Upper Acheulean artifacts from Elandsfontein 'Cutting 10.' Types: above, handaxe; below, cleaver. Raw material: silcrete.

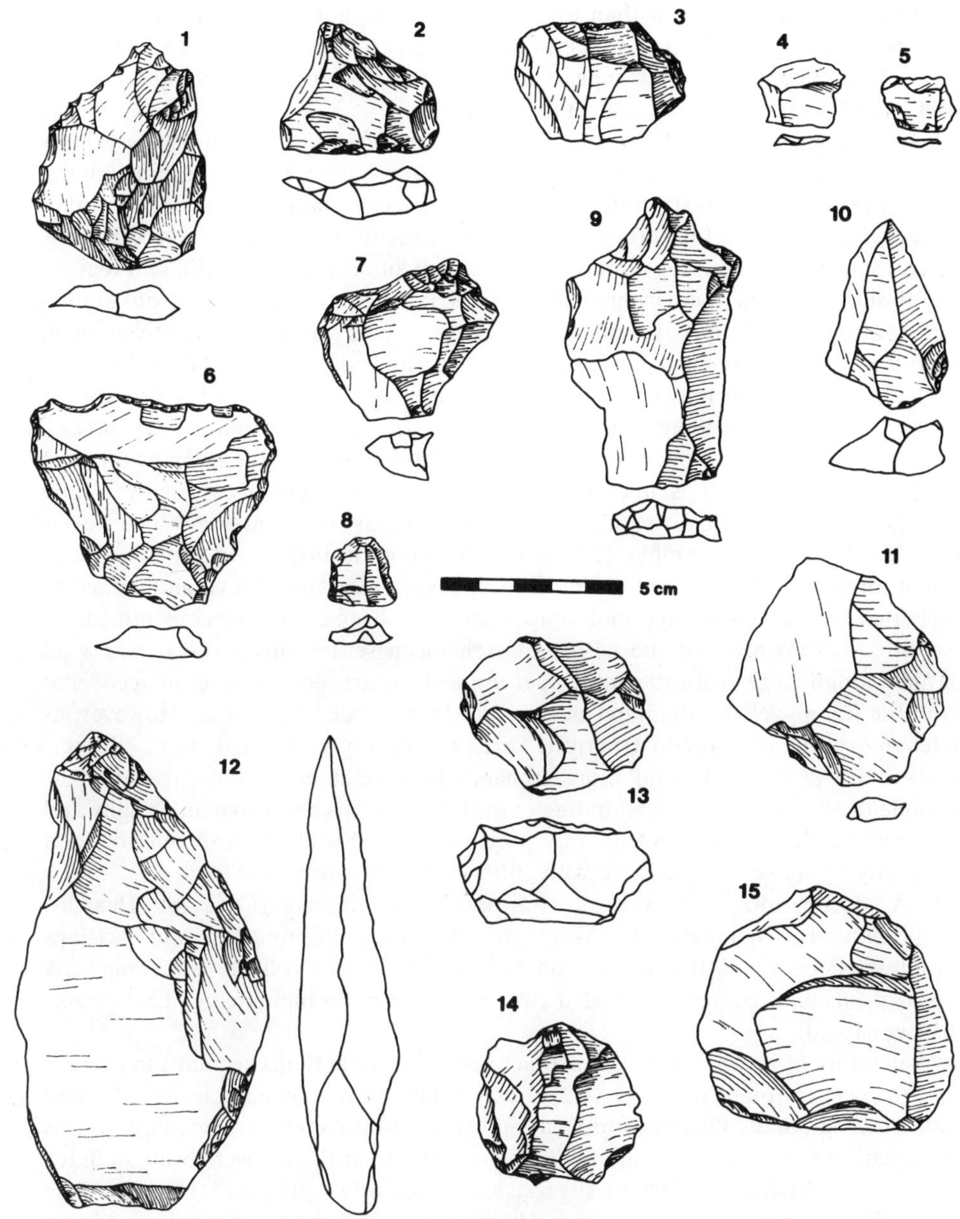

Figure 5. Upper Acheulean artifacts from Elandsfontein 'Cutting 10.' Types: 1: denticulate; 2: flake with continuous retouch; 3: scraper; 4–5: biface trimming flakes; 6–8: utilized/retouched flakes; 9–11: flakes; 12: bifacial knife; 13–15: cores. Raw materials: 1–10, 12–13, 15: silcrete; 11, 14: quartz.

which grow in the region today and which have been interpreted as the remains of human meals (Clark 1969).

ESA site locations are quite varied, apparently excluding only the extremes of dry desert and dense forest (H.J. Deacon 1975). The association of sites with water sources seems more marked than for the most recent hunter-gatherers in the subcontinent, perhaps indicating a lack of devices to carry water. Evidence for ostrich eggshell water containers only appears in the southern African archaeological record after 15,000 B.P. (see J. Deacon, this volume).

The vast majority of ESA sites in southern Africa are disturbed, open-air occurrences without organic remains. Evidence for temporal relationships among sites, even within restricted regions, is usually lacking, as is detailed paleoenvironmental information. Thus, it is difficult to provide even the most general statements about the duration and intensity of occupation in different regions and about the effects of environmental changes, much less make comparisons among regions. Before the Upper Acheulean, occupation seems restricted to relatively open interior areas. Upper Acheulean sites are much more widely distributed, and suggest population expansion into more arid and more wooded regions. Sites also appear near coasts, although there is no evidence for exploitation of marine resources.

Large collections of Upper Acheulean artifacts have been made in many areas of southern Africa, but it is unclear whether these areas were occupied over long periods, or intermittently, or only during especially favorable intervals. The Lower Vaal Basin is a good example. If LSA peoples found it difficult or unprofitable to inhabit this region during part of the Holocene (see J. Deacon, this volume), it seems likely that Upper Acheulean populations would have been even more constrained. As yet, little is known about the nature of Acheulean settlement in the Lower Vaal Basin, though huge quantities of Upper Acheulean artifacts have been recovered from the dumps left by diamond miners and from eroded exposures. However, as paleoenvironmental reconstructions are increasingly refined (e.g. Butzer, Stuckenrath *et al.* 1978) and more archaeological research is done, it should be possible to gain a better understanding of the relationships between demography and environment during the ESA in this portion of the South African interior, and eventually to make comparisons with other areas in southern Africa.

ESA occupations of cave sites appear very limited (Cave of Hearths, Olieboompoort, Wonderwerk, Montagu, Bambata, Pomongwe, and perhaps Mumbwa, Peers Cave, Bremen 1C and Kabwe), although it is likely that some ESA cave deposits have completely eroded away and some caves inhabited by ESA people have collapsed.

Availability of raw materials for artifact manufacture perhaps accounts in part for repeated occupation of some open-air sites associated with alluvial gravels, indurated shale outcrops, and colluvial deposits at pan sites. Doornlaagte is an example of this last situation and if the large, unmodified stone blocks at the site were brought in for use as seats, as Mason (1967b) suggests, or some other purpose, they would be practically the only evidence for site modification by ESA people in southern Africa. However, it seems more likely that the unmodified stones at Doornlaagte accumulated naturally (Butzer 1974b; H.J. Deacon 1975). Indeed, unlike the situation at some early sites in East Africa, there is no evidence during the ESA in southern Africa for the long-distance transport of either raw materials or finished artifacts. Perhaps the best evidence for an ESA structure comes from 'Acheulean Occupation Floor V' in Site B 2 at Kalambo Falls (Clark 1969: 172; 1970: 143). Here people apparently carried artifacts and natural stones to a temporary camp at a riverside sandbank. Some of the larger, unmodified stones form a rough semicircle, demarking an area about 2-1/2 m by 1 m, and might have served as the base for a windbreak.

Localized concentrations of charcoal from the Charama levels at Pomongwe were interpreted as hearths (C.K. Cooke 1963) and similar evidence for hearths is described from the lowest levels at Bambata (Armstrong 1931). If hearths were present in Charama levels at these caves, they would be important additions to the limited evidence for the control of fire at southern African ESA sites. Some Acheulean levels at the Cave of Hearths probably contain hearths (Mason 1962b), but the published information is not conclusive. Bed 1 is composed almost entirely of burnt bat guano, which may have been ignited naturally, or accidently by possible hearths at the top of Bed 1. The widespread ash in Bed 3 may represent a large hearth, but could also result from the natural burning of vegetation, such as plant food debris and bedding, brought in and discarded by the Acheulean inhabitants. High concentrations of mineral ash in the Acheulean levels at Montagu Cave may have a similar origin (Butzer 1973e). Intentional burning may account for some of the carbonized wood and other vegetation in ESA sites at Kalambo Falls. However, natural ageing processes are responsible for much of the partially carbonized or apparently charred wood (Clark 1969: 160). ESA control of fire at Kalambo Falls may be indicated by the possible use of fire to make wooden artifacts and by the presence of a reddened area of burnt clay, about 1 m in diameter, that might be the base of a hearth. Unequivocal hearths are common in cave sites in Upper Pleistocene MSA levels. Among the sites with MSA assemblages possibly dating to isotope stage 6, there are good hearths at Cave of Hearths Bed 4, Border Cave and Peers Cave. It would appear, then, that human control of fire was widespread in southern Africa by 130,000 years ago, and perhaps much earlier.

The end of the ESA

Recent discussions of the replacement of the ESA by the MSA (Clark 1970; H.J. Deacon 1975) have emphasized the 'interplay between demographic, ecological, technological, and social factors' but are constrained by the 'lack of empirical data for testing concepts' (H.J. Deacon 1975: 565). Good dating controls are also vital to this investigation of variability and change, especially now that arguments about 'gradualism' versus 'punctuated evolution' are being applied to the archaeological record. However, as often occurs in archaeological research, the modest gains in information achieved over the past decade have mainly served to complicate our understanding of the changes involved.

In Clark's (1970) formulation, the Sangoan and late Upper Acheulean provide the first evidence for regional adaptations, however broad, in sub-Saharan Africa. Acheulean adaptations continued in relatively open environments while the Sangoan represented a widespread, new adaptation to more wooded regions. Later, H.J. Deacon (1975) considered the possibility that the latest Acheulean, too, may represent a new grade of regional adaptation. From these perspectives, the latest ESA is viewed as a take-off point for increasing regional differentiation of adaptations throughout the rest of the Stone Age. This topic will be considered further after the MSA is discussed. The remainder of this section will consider recent information on the change from ESA to MSA, which pertains almost exclusively to dating and industrial changes.

For the period under consideration, the only widespread, convenient marker for distinguishing ESA from MSA seems to be the presence or absence of large cutting tools. Although the utility and validity of this distinction may be questioned, at least it provides a framework for discussion. Moreover, it is not assumed that the makers of late ESA artifacts left handaxes and cleavers at every site they occupied, but only that these tools were part of their repertoire. Thus, assigning individual assemblages to the ESA or MSA may pose problems, as was indicated above for Mwanganda.

Given the distinction adopted here, the change from ESA to MSA may simply be one of attrition. Large cutting tools, for whatever reason, could have gradually declined in importance until they were seldom, then never, produced. This may have occurred in Zimbabwe, where the proportion of handaxes and cleavers in late Sangoan (Charama) assemblages is considerably lower than in earlier Sangoan ones, and even heavy-duty tools may be relatively rare (C.K. Cooke 1966a). Although more detailed studies are needed, it also appears that over the course of the Sangoan in Zimbabwe, the flake tool component increasingly resembles that in the following MSA. At Pomongwe Cave, for instance, published descriptions (C.K. Cooke 1963) suggest little difference between the flake tools in the lowest, Charama levels and in the MSA (Bambata) levels which immediately overlie them.

At present, the time span of the Sangoan is poorly established, though it seems increasingly likely that it began well before the end of the Middle Pleistocene. Sampson (1974) suggests that Charama assemblages and even earlier Sangoan ones from Zambia and Zimbabwe are younger than any Acheulean assemblages farther south. Additionally, it is my impression that the earliest MSA assemblages known from Zimbabwe may be comparable to MSA assemblages from south of the Limpopo which are younger than the beginning of the Upper Pleistocene.

If the possibility exists, at least in Zimbabwe and farther north, that the change from ESA to MSA was gradual and not completed until the early Upper Pleistocene, south of the Limpopo it is increasingly clear that MSA assemblages had replaced ESA ones by the beginning of the Upper Pleistocene and perhaps considerably earlier. Although no 'transitional' assemblages have been identified in southern Africa (and the existence of such assemblages may be questioned on theoretical as well as empirical grounds), Sampson (1972) has suggested that an assemblage 'transitional' between Acheulean and MSA might contain very small handaxes and very long flake-blades, especially when the raw material is hornfels (indurated shale, lydianite). This suggestion is based in part on Sampson's belief that the earliest MSA assemblages are characterized by large flake-blades, but flake-blades are generally rare and small in what appear to be the oldest MSA assemblages known in southern Africa. The cores and flakes in these early MSA assemblages do not suggest any advances over Acheulean techniques, and there is so little formal retouch of any sort that, beyond simple recognition of this fact, comparisons with other assemblages are not very informative. Many younger MSA assemblages do contain a more standardized and diversified range of flake tools than late Acheulean collections, and, as will be discussed later, suggest new manufacturing and technological skills. If the change from Acheulean to MSA seems more abrupt than that from Sangoan to MSA, it is not clear whether this is only the trivial reflection of an uneven data base or perhaps involves more profound differences in environmental, demographic and other factors, which certainly are implicated in later distributions of archaeological materials.

Although the present evidence tends to suggest that the change from ESA to MSA was time transgressive in southern Africa, with the Sangoan persisting north of the Limpopo after the MSA had replaced the Acheulean farther south, it is not at all clear whether ESA and MSA assemblages were made concurrently in restricted regions. The possibility of such a situation is suggested by recent research on the change from Acheulean to Mousterian much farther north, where typological distinctions are, if anything, less clear-cut than those between ESA and MSA. Overall, the replacement of ESA by MSA in southern Africa may be compared with this change from Lower Paleolithic to Middle Paleolithic, which took place in Europe and the circum-Mediterranean lands at about the same time and is much better documented, though it is just as poorly understood.

Some of the most interesting observations come from northern France, where assemblages that are typologically and technically 'Mousterian' appear in the next-to-last glacial and are broadly contemporary with other assemblages containing variable proportions of classic Acheulean handaxes (Tuffreau 1979). As yet, the behavioral implications of this variability have hardly been addressed. Bordes (1977) has attempted to deal with the problem of nomenclature by suggesting that 'the' Mousterian cannot be defined, but it may be possible to define a Mousterian stage or level of evolution, with emphasis on the flake tool component of assemblages. As more information becomes available on the later ranges of the Middle Pleistocene in southern Africa, such an approach may help provide better understanding of what is implied by the terms ESA and MSA.

More and more, the later Middle Pleistocene stands out as a time of major developments in human cultural and biological evolution (see also Rightmire, this volume). If the preceding discussion has focused almost exclusively on culture stratigraphy, this is not because artifacts are all important. Rather, it highlights our virtual lack of other relevant information about the latest ESA and earliest MSA. New research projects, such as the reinvestigation of Florisbad by the National Museum, Bloemfontein, are needed throughout southern Africa to establish a broad, high quality data base and a reliable dating framework.

THE MIDDLE STONE AGE

Introduction

The MSA assemblages which replace ESA ones in southern Africa lack handaxes and cleavers and are characterized by flake and blade tools, such as points, scrapers and denticulates, although any of these types may be rare or absent in a particular assemblage. Backed and/or truncated pieces are a prominent component in the Howieson's Poort group of assemblages. If these pieces were hafted as parts of composite tools, which seems likely but still needs to be demonstrated, they may be the earliest evidence for composite tools yet found in the Old World. In any case, retouch on stone artifacts in earlier MSA assemblages strongly suggests that MSA peoples were hafting both carefully retouched and otherwise untrimmed pieces by the early Upper Pleistocene.

As orginally defined by Goodwin (1928; 1929a), the MSA was distinguished by triangular flakes with convergent dorsal scars and faceted butts. Such flakes do occur in most MSA assemblages, but usually are infrequent in unselected samples. Two broad categories of cores are represented in most MSA assemblages: cores for the production of flakes with intersecting dorsal scars (mainly radial/discoid and change of orientation cores) and cores for the production of flakes with parallel, sub-parallel or, rarely, convergent dorsal scars (mainly single and double platform cores, some of which conform to the European concept of Levallois blade cores.[1] An advance over ESA primary stoneworking techniques may be indicated by some of the flake-blades produced from this second kind of core. The narrow, thin butts and diffuse bulbs of percussion on some of these pieces suggest the use of an intermediate punch (indirect percussion) in their manufacture (see Figures 8-10). Levallois flake cores are very rare in the southern Cape, but occur in higher frequencies in the southern African interior and farther north.

MSA sites are found in the same variety of locations as ESA ones, including river banks, floodplains, lake or pan margins, and spring deposits. Especially in recent decades, archaeological attention has focused on cave sites which contain long occupation sequences and preserve organic materials, primarily faunal remains. However, in contrast to the limestone caves of Europe, in which occupation units often are well-separated and easily distinguished due to rapid disintegration of the parent cave rock, most southern African caves are formed in rocks which weather slowly. In many cases a major portion of the cave sediment is externally derived. Eolian sands may make up the bulk of the deposits in coastal caves, and at both coastal and inland sites the activities of animals and especially people probably made a substantial contribution to sediment accumulation. At some sites, occupation horizons are readily identified because they are darker than surrounding deposits, a result of their higher organic content derived from animal and plant sources, such as food waste, bedding and fuel. At many cave sites, slow deposition and acidic groundwater runoff have led to the removal of fine sediments, leaching of organic materials, and in extreme cases have reduced the MSA horizons to lag deposits characterized by very high artifact densities. In the compressed sequences that commonly result from this compaction and leaching, even the thinnest recognizable MSA horizons probably represent several, formerly distinct, occupation units.

Culture stratigraphy and dating, with particular reference to demographic implications and the end of the MSA

A similar pattern of assemblage change through time occurs in MSA sequences south of the Limpopo (Volman 1981). At some sites chronological controls are provided by contextual evidence which allows correlations with the deep-sea oxygen isotope record and with changes in regional climates for time spans beyond the range of conventional radiocarbon dating (Butzer, this volume; Klein 1980). Contextual data and typological comparisons make it possible to tentatively place some isolated assemblages into the pattern. A different pattern may occur north of the Limpopo and will be discussed later.

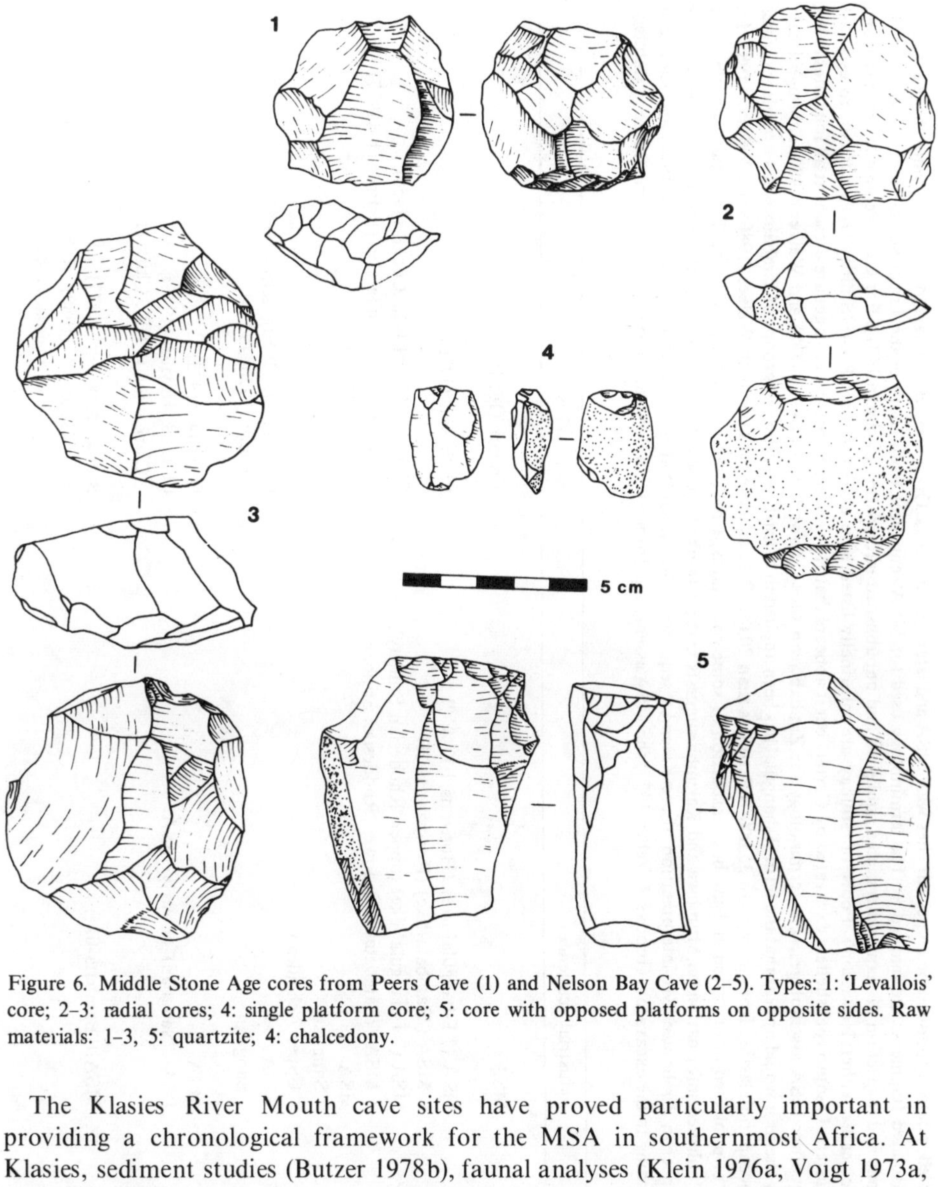

Figure 6. Middle Stone Age cores from Peers Cave (1) and Nelson Bay Cave (2–5). Types: 1: 'Levallois' core; 2–3: radial cores; 4: single platform core; 5: core with opposed platforms on opposite sides. Raw materials: 1–3, 5: quartzite; 4: chalcedony.

The Klasies River Mouth cave sites have proved particularly important in providing a chronological framework for the MSA in southernmost Africa. At Klasies, sediment studies (Butzer 1978b), faunal analyses (Klein 1976a; Voigt 1973a, b, 1975; D.M. Avery 1982a; G. Avery, in prep.) and oxygen isotope analyses of marine shells (Shackleton 1982) suggest correlation of the long MSA sequence, with most of oxygen isotope stage 5 and the beginning of oxygen isotope stage 4. This correlation is supported by amino-acid determinations (Bada & Deems 1975), preliminary ionium dates (Kronfeld & Vogel 1980; Vogel, pers. comm.) and infinite radiocarbon dates (Vogel 1979). Contextual data and radiocarbon dates at other sites with long MSA sequences, especially Boomplaas, Die Kelders 1, Border Cave and Apollo 11, support the Klasies framework and extend it through isotope stages 4 and 3.

It must be emphasized that the four stage MSA sequence outlined below (see also

Table 2. Cave, rockshelter and sealed, relatively undisturbed, open-air sites with MSA and later ESA artifacts. Open-air sites are marked * and sites which have provided faunal samples other than a few isolated fragments are marked ×. The terminology used for MSA sequences from south of the Limpopo (see text) is based on my studies of southern Cape collections and inspection of many others, plus published and unpublished information (Volman 1981). For Cape Peninsula sites excavated before the 1960s, selected samples called either Howieson's Poort or Stillbay are informally listed here as Howieson's Poort/Stillbay. It is difficult to establish individual site sequences for the eastern Lesotho rockshelters excavated by Carter, but it appears that Howieson's Poort assemblages are present at all four sites listed, preceded or followed, or both, by other MSA assemblages. The terminology for Zimbabwean sites follows C.K. Cooke (1979b) and Walker (1980) although I have applied it to sites not discussed by them. It would appear that the term 'Tshangula' has been applied to at least two different groups of Zimbabwean assemblages which may be separated by a hiatus in occupation lasting from more than 35,000 B.P. to less than 20,000 B.P. at some sites (see text, Walker (1980) and Volman (1981) for further discussion). On the basis of published information and my brief inspection of some collections, I have tentatively assigned some 'Tshangula' samples from Zimbabwean sites to the MSA. Published information on Mumbwa and Kalemba suggests that portions of the sequences in these Zambian caves are comparable to some Bambata-'Tshangula' sequences in Zimbabwe. My interpretation of the Kalemba sequence differs from that of the excavator (Phillipson 1976) and should be considered highly tentative. According to the excavators (Brooks & Yellen 1977) the MSA assemblage from ≠Gi in Botswana is typologically comparable to Bambata materials from Redcliff.

Site	Industrial sequence	References
ANGOLA		
Humpata ×	MSA	França (1964)
ZAMBIA		
Kalemba ×	MSA (?'Tshangula'-like) (Horizons I-K & ?L)	Phillipson (1976)
	MSA (?Bambata-like) (Horizons G-H)	
Mumbwa ×	MSA (?'Tshangula'-like) (upper 'Rhodesian Stillbay')	Clark (1942); H.B.S. Cooke (1950);
	MSA (?Bambata-like) (lower 'Rhodesian Stillbay')	Dart & Del Grande (1931); Protsch (1977)
Kalambo Falls*	MSA	Clark (1969, 1974)
	Sangoan	
	Upper Acheulean	
Kabwe ×	?MSA	Clark *et al.* (1950); Klein (1973);
(Broken Hill)	Sangoan	Clark (1970, 1977b)
MALAWI		
Malowa	MSA	Denbow (1973)
Mwanganda Area 1*×	ESA (?Sangoan) or early MSA	Clark & Haynes (1970);
		Clark *et al.* (1970); Clark (1981)
BOTSWANA		
≠Gi*×	MSA (?Bambata-like)	Brooks & Yellen (1977);
		Brooks *et al.* (1980); Rubin (in prep.);
		Helgren & Brooks (1983)
NAMIBIA		
Apollo 11×	MSA 3 (layer E)	Wendt (1972 1974, 1976);
	Howieson's Poort (layer F)	J.F. Thackeray (1979)
	MSA 2b (layer G)	

Zebrarivier×, Pockenbank 1×, Schlangengrotte, Zais, Fackelträger, Bremen 1C×, Haalenberg	MSA 3 ?Howieson's Poort (Pockenbank 1) MSA 2 ?Upper Acheulean (Bremen 1C)	Wendt (1972); J.F.Thackeray (1979); Vogel & Visser (1981); Freundlich *et al*. (1980)
Namib IV *×	Upper Acheulean	Shackley (1980)
ZIMBABWE		
Pomongwe ×	?'Tshangula' Bambata Charama	C.K. Cooke (1963, 1971); Brain (1981)
Tshangula	Bambata	C.K. Cooke (1963); Sampson (1972); Walker (1978)
Zombepata	?'Tshangula' Bambata	C.K. Cooke (1971); Sheppard & Swart (1973)
Redcliff ×	?'Tshangula' Bambata	Brain & Cooke (1967); Brain (1969b); C.K. Cooke (1978); Klein (1978a, 1980)
Duncombe Farm	Bambata	Hitzeroth (1973)
Nswatugi ×	'Tshangula' Bambata	Walker (1978, 1980); N. Jones (1933)
Bambata	?'Tshangula' Bambata Charama	Armstrong (1931); N. Jones (1940, 1949)
Gumani	Bambata	Armstrong (1931)
Nyazongo	Bambata	Martin (1938)
Pfupi	?'Tshangula' Bambata	K.R. Robinson (1952)
Mtemwa Rocks	?'Tshangula' Bambata	K.R. Robinson (1952)
Amadzimba	Bambata	K.R. Robinson & Cooke (1950); C.K. Cooke & Robinson (1954)
Inyanga Site XIXB	'Tshangula'	K.R. Robinson (1958)
Inyanga Site XXVII	Bambata	K.R. Robinson (1958)
SWAZILAND		
Lion Cavern	MSA (?2b)	Boshier & Beaumont (1972); Dart & Beaumont (1968, 1971); Beaumont (1973b, 1979a)
Sibebe	MSA (?3)	Price Williams (1981)
LESOTHO		
Melikane ×, Ha Soloja ×,	MSA 3	Carter (1969, 1976, 1978);

Table 2 (*Contd.*)

Site	Industrial sequence	References
Moshebi's, Sehonghong ×	Howieson's Poort MSA 2	Carter & Vogel (1974); Aldworth (1979)
SOUTH AFRICA (outside southern Cape)		
Cave of Hearths ×	?Howieson's Poort (Bed 9) MSA 2b (Beds 6-8) MSA 2a (Beds 4-5) Upper Acheulean (Beds 1-3)	Mason (1957, 1962b, 1967a, 1971); Sampson (1968b, 1972); H.B.S. Cooke (1962, 1963)
Olieboompoort	MSA 2b (Bed 2) Upper Acheulean (Bed 1)	Mason (1957, 1962b); Sampson (1968b)
Rufus and Pietkloof	MSA	Mason (1951, 1962b)
Kalkbank *×	MSA	Mason (1962b, 1967b, 1979a); Mason *et al.* (1958); Netterberg (1974); Welbourne (1971); H.B.S. Cooke (1962)
Mwulu's Cave	MSA 2b (Mwulu 3) MSA 2a (Mwulu 1 & 2)	Tobias (1949); Mason (1957); Sampson (1972)
Bushman Rockshelter ×	MSA (perhaps MSA 3 in levels 15-18, MSA 2 in levels 19-30 and upper portion of levels 31-107, and MSA 1 in lower portion of levels 31-107)	Plug (1978, 1981); Louw (1969); Brain (1969a, 1981) Vogel (1969); Eloff (1969); Butzer & Vogel (1979)
Witkrans ×	MSA (?3)	Clark (1971a); Peabody (1954); Butzer, Stuckenrath *et al.* (1978); Klein (1980)
Wonderwerk Cave ×	MSA Upper Acheulean	Beaumont (1979b, pers. comm.); Butzer, Stuckenrath *et al.* (1978); Malan & Cooke (1941); Malan & Wells (1943)
Doornlaagte *	Upper Acheulean	Mason (1966, 1967b); Butzer (1974b)
Rooidam *	Upper Acheulean	Fock (1968); Butzer (1974b)
Kathu Pan *×	MSA (?2) Upper Acheulean	Beaumont (pers. comm.); Klein (pers. comm.)
Orangia 1 *	MSA (?2)	Sampson (1968a, 1972)
Zeekoegat 27*	?MSA 3 (or ?LSA)	Sampson (1968a, 1972)
Florisbad *×	MSA (?2) ?ESA	Dreyer (1938); Hoffman (1955); Meiring (1956); Butzer (1971); H.B.S. Cooke (1963); Clark (1955); Gentry & Gentry (1978); Beaumont *et al.* (1978)
Highlands	MSA 3 (Lower Member)	H.J. Deacon (1976)

Site	Industry	References
Grass Ridge ×	MSA (?3)	Opperman (pers. comm.)
Oakleigh Farm QOB	Howieson's Poort (levels 11/9)	Derricourt (1977)
Rose Cottage Cave	?MSA 3 Howieson's Poort	Malan (1946, 1952); Mason (1969a); Beaumont & Vogel (1972b); Butzer & Vogel (1979)
Border Cave ×	MSA 3 (2BS.UP-WA) Howieson's Poort (3BS-1RGBS.B) MSA 2b (1GBS.UP-BACO.B) MSA 2a (BACO.C-BACO.D)	H.B.S. Cooke *et al.* (1945); Beaumont (1973a, 1978, 1979a, 1980); Butzer, Beaumont & Vogel (1978); Beaumont *et al.* (1978); Klein (1977b); Avery (1982b)
Holley Shelter	MSA	Cramb (1961)
SOUTH AFRICA (southern Cape)		
Klipfonteinrand	Howieson's Poort (spits 5-6) MSA 2b (spits 7-9)	Parkington & Poggenpoel (1971b); Parkington (1977a); Volman (1981); A.I. Thackeray (1977)
Diepkloof ×	Howieson's Poort (orange-black, organic loam complex)	Parkington (1977a); Butzer (1979a)
Elands Bay Cave	MSA 1	Parkington (1977a); Butzer (1979a); Volman (1981)
Sea Harvest *× & Hoedjies Punt*×	MSA 3	Volman (1978); Klein (1979a); Butzer (in prep.)
Elandsfontein 'Cutting 10' *×	Upper Acheulean	Singer & Wymer (1968); Klein (1978c)
Duinefontein 2*×	MSA 1	Klein (1976b); J. Deacon (1976a); Butzer (in prep.)
Montagu Cave	MSA 3 (layer 2: surface 6 and above) Howieson's Poort (layer 2: below surface 6) Upper Acheulean (layers 3 & 5)	Goodwin (1929b); Keller (1970a, 1970b, 1973); Butzer (1973e); Volman (1981)
Peers Cave (Skildergat; Fish Hoek)	?MSA 3 Howieson's Poort/Stillbay ?MSA 2 MSA 1 ?Upper Acheulean	Goodwin (1929a, 1946a); Peers (1927, 1928, 1929); Anthony (1963, 1967, 1972); Keith (1931); Jolly (1947, 1948); Volman (1981)
Tunnel Cave	Howieson's Poort/Stillbay	Malan (1955)
Skildergatkop *	Howieson's Poort/Stillbay	Malan (1955)
Trappieskop (Peers Shelter B/102)	Howieson's Poort/Stillbay	Goodwin (1953b)
Die Kelders 1 ×	MSA 3 (levels 4-15)	Schweitzer (1970, 1979); Tankard & Schweitzer (1974, 1976); Tankard (1976); Klein (1975a); Avery (1982a); Butzer (1979b); Volman (1981)
Gatboskloof	MSA	Schweitzer (pers. comm.)
Welgevonden	MSA	Binneman (pers. comm.)
Cape St. Blaize Cave	MSA (?2) (layer C)	Leith (1898); Keller (1969); Goodwin & Malan (1935)

Table 2 (*Contd.*)

Site	Industrial sequence	References
Buffelskloof	MSA	Opperman (1978)
Boomplaas ×	MSA 3 (Members BP, OLP & BOL) Howieson's Poort (Members OCH & LOH)	H.J. Deacon (1978, 1979); H.J. Deacon & Broker (1976); H.J. Deacon *et al.* (1976); J. Deacon (1979a); Klein (1978b); Avery (1982a)
Paardeberg	MSA (?2) (units BSL, KAD, MAD & LAD)	H.J. Deacon & Brooker (1976); H.J. Deacon & Klein (n.d.) Butzer (n.d.); Volman (1981)
Kangkara	MSA (?3) (units BHL2, BHL2C & BHL3)	H.J. Deacon & Brooker (1976); H.J. Deacon & Klein (n.d.) Butzer (n.d.); Volman (1981)
Howieson's Poort Rockshelter	Howieson's Poort	Stapleton & Hewitt (1927, 1928); J. Deacon (1979b)
Amanzi *	Upper Acheulean	H.J. Deacon (1970b)
Herolds Bay Cave ×	MSA 2a	Binneman *et al.* (1979); J. Deacon (1979a); Brink & Deacon (1981)
Nelson Bay Cave	Howieson's Poort (levels crust-7) MSA 2b (levels 7-8) MSA 2a (levels 9-10)	Inskeep (1965, 1972); Klein (1972a); J. Deacon (1978, 1979a); Butzer (1973a); Volman (1981)
Klasies River Mouth ×	MSA 3 ('MSA IV': Cave 1, layer 13) ('MSA III': Shelter 1A, layers 1-9) Howieson's Poort (Shelter 1A, layers 10-21; Cave 2, layers 1-5) MSA 2b ('MSA II': Cave 1, layers 14-17b; Shelter 1A, layers 22-36) MSA 2a ('MSA I': Cave 1, layers 37-40; Shelter 1B, layers 1-15; Cave 1C, layers 36-37; ?Cave 5, layers 5-7)	Singer & Wymer (1982); Wymer & Singer (1972); Klein (1975a, 1976a); Butzer (1978b); Voigt (1973a, 1973b, 1975); Avery (1982a)

Table 2) is based on relatively few observations, given the large area and long time span involved, and that there is considerable variability among assemblages assigned to a particular stage in such features as artifact morphology, core types, raw materials, and so forth. At present, detailed comparisons of morphological variability have not been undertaken, and such studies are hampered by the very low proportion of formally retouched pieces, usually much less than 1%, in most MSA assemblages. Moreover, assemblage comparisons will have to control for the great variability in primary forms and flaking properties of the different raw materials used to make MSA artifacts.

The terminology used for the culture-stratigraphic sequence is informal, except for the Howieson's Poort group of assemblages. These are characterized by certain retouched forms first recovered in quantity from Howieson's Poort Rockshelter and the term Howieson's Poort is commonly used for similar assemblages in the southern Cape Province, South Africa. Here, the term Howieson's Poort is extended to include comparable assemblages throughout southern Africa south of the Limpopo.

MSA 1. This is the most problematic of the MSA stages proposed here, but it definitely seems to exist in the southern Cape at Duinefontein 2, Peers Cave and Elands Bay Cave, and perhaps occurs at Bushman Rockshelter in the eastern Transvaal. Some or all of the assemblages assigned to the MSA 1 may be considerably older than any assemblages placed in the following MSA 2. The southern Cape assemblages are characterized by very little formal retouch, small, broad flakes with few faceted butts, and a high proportion of cores for the production of flakes with intersecting dorsal scars.[5] Retouched points are absent in the assemblages and scraper retouch is very rare. Most carefully trimmed flakes are denticulates, which appear to represent the well-retouched end of a continuum dominated by pieces with minimal, denticulate-like trimming. The Peers Cave sample is distinguished by a high proportion of cobbles, slabs and even some large flakes with rather crude, alternating and/or bifacial retouch. There is one similar specimen in the small Elands Bay sample.

My inclusion of the lower Bushman Rockshelter samples in the MSA 1 is highly tentative and subject to revision once I have completed detailed analyses of the site sequence. My preliminary inspection of the Bushman Rockshelter materials and brief reports on the artifact sequence (Eloff 1969; Aldworth 1979) suggest that higher levels in the site may contain artifacts comparable to those assigned here to the MSA 2 and MSA 3. The lower Bushman Rockshelter samples are similar to the southern Cape MSA 1 collections in their very low frequencies of formally retouched pieces, but seem to contain higher proportions of relatively narrow flakes with parallel dorsal scars and faceted butts.

Suggested dating: Oxygen isotope stage 6. Archaic forms in the Duinefontein 2 fauna and the local sediment sequence suggest occupation during a period of falling sea level prior to the end of the Middle Pleistocene. The Elands Bay Cave basal rubble and the lower portion of the Bushman Rockshelter sequence are assigned to stage 6 on the basis of geomorphic/sedimentologic correlations, supported in the case of Bushman Rockshelter by infinite radiocarbon dates for levels much higher in the sequence. The MSA 1 from Peers Cave comes from the lower part of what was formerly a long artifact sequence. Selected artifact samples from higher in the

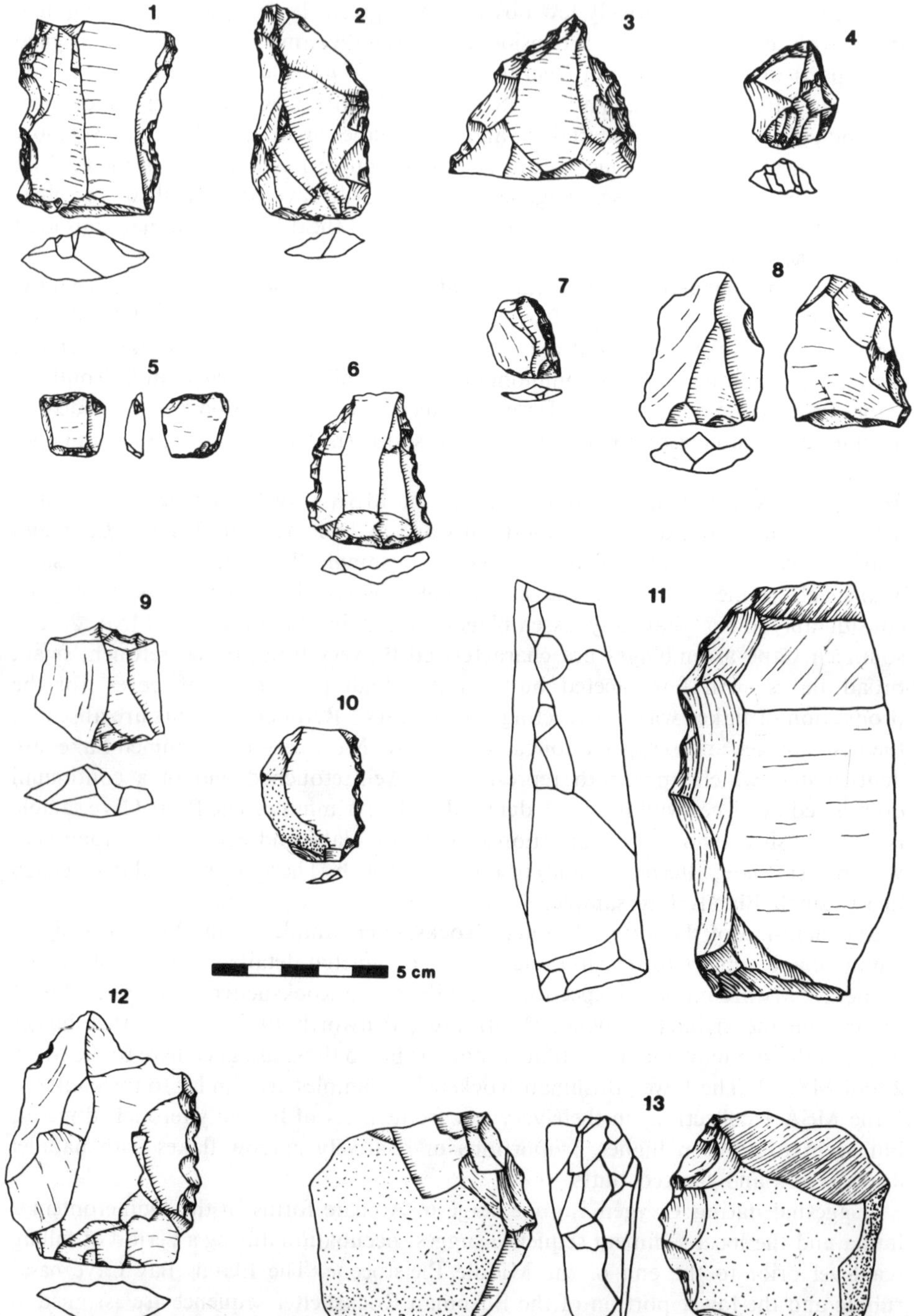

Figure 7. Middle Stone Age (MSA 1) artifacts from Duinefontein 2 (1–4) and Peers Cave (5–13). Types: 1, 3, 6, 10, 12: denticulates; 2: endscraper with lateral retouch; 4, 7: scrapers; 5: scaled piece; 8: flake with ventral retouch; 9: denticulate endscraper; 11: slab with heavy-duty unifacial retouch; 13: weathered slab with heavy-duty bifacial retouch. Raw materials: 1–6, 8: silcrete; 7, 10: quartz; 9, 11–13: quartzite.

sequence appear to be comparable to those assigned here to the MSA 2 and Howieson's Poort.

MSA 2. Although there is continuity within sequences assigned to this stage, differences between the older and younger samples suggest a division into an earlier MSA 2a and a later MSA 2b. Overall, the MSA 2 is characterized by relatively large, narrow flakes (including flake-blades and true blades) which decrease in average length through time. In general, both the relative abundance and variety of formally retouched pieces increases from the MSA 2a to the MSA 2b. Denticulates are common in MSA 2a assemblages from the southern Cape, at Apollo 11, and at some sites in the interior. Denticulates are much less common or absent in MSA 2b assemblages and, when present, are not so carefully made as the better MSA 2a examples. Unifacial and bifacial points appear for the first time in the MSA 2, but are rare or absent in many MSA 2a assemblages. At many sites, sidescrapers increase relative to endscrapers within MSA 2 sequences. The presence of some tanged artifacts, and of many more with extensive proximal retouch, suggests that hafted tools made their appearance by this time (Figures 8 and 9). Large, partially backed flake-blades occur in some MSA 2 assemblages (e.g. Cave of Hearths and Border Cave), but not in the southern Cape. Some MSA 2b collections also contain occasional small backed and/or truncated pieces which would not appear out of place in the following Howieson's Poort.

Suggested dating: oxygen isotope stages 5e-5c, with some MSA 2a assemblages perhaps dating to the end of stage 6. Typological comparisons between the Cave of Hearths and Klasies sequences suggest that the Bed 4 assemblages from the Cave of Hearths are older than the earliest assemblages at Klasies (correlated with the end of isotope stage 5e or the beginning of stage 5d) and may therefore be Middle Pleistocene in age. Artifacts from part of the Bushman Rockshelter sequence and from the basal level at Apollo 11 also may be comparable to those from Cave of Hearths Bed 4, and sediment studies suggest that the relevant portion of the Bushman Rockshelter sequence precedes the Upper Pleistocene. Sediment studies may indicate that MSA 2 assemblages from Border Cave are Middle Pleistocene in age, but typological comparisons suggest correlations with early Upper Pleistocene sequences elsewhere.

Howieson's Poort. This term is used here for MSA assemblages with relatively numerous segments, trapezoids and other backed and/or truncated forms. It includes assemblages called Howieson's Poort in the southern Cape and similar assemblages from other sites, some of which have been called Epi-Pietersburg and Modderpoort. Such assemblages once were considered local expressions of the 'Magosian' or 'Second Intermediate' and thought to be transitional between the MSA and LSA. However, recent excavations have demonstrated conclusively that the Howieson's Poort is an integral part of the MSA.[6] At Klasies, Border Cave, Apollo 11 and perhaps Peers Cave, the Howieson's Poort assemblages are both preceded and followed by MSA assemblages in which segments and allied forms are rare or absent. Howieson's Poort assemblages at Nelson Bay Cave, Klipfonteinrand and perhaps at the Cave of Hearths are preceded by MSA 2b ones. At Boomplaas, probably at Montagu Cave, and possibly at Rose Cottage Cave, Howieson's Poort assemblages

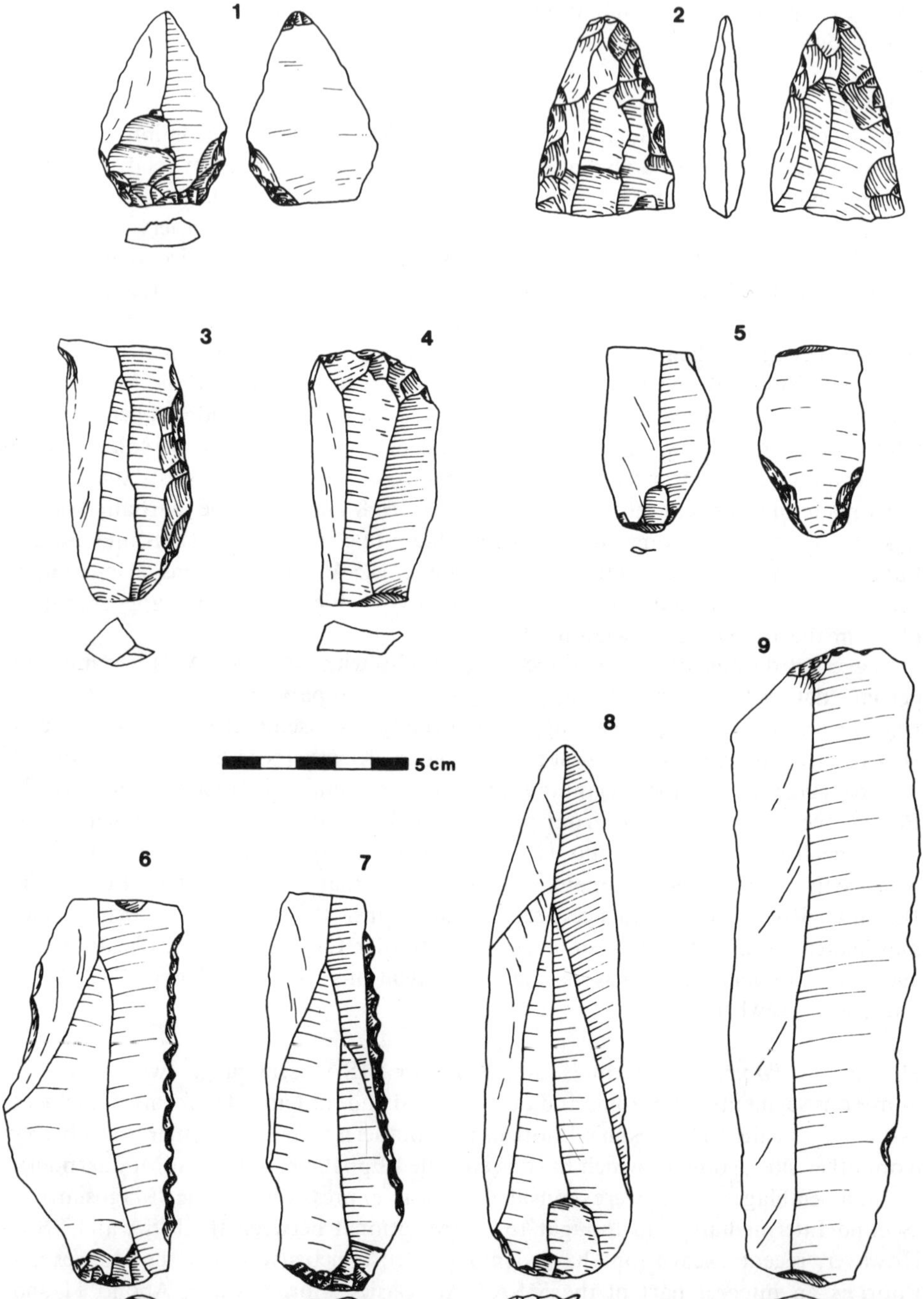

Figure 8. Middle Stone Age (MSA 2a) artifacts from Klasies River Mouth (1, 6–9) and Nelson Bay Cave (2–5). Types: 1: triangular flake with extensive proximal retouch; 2: distal tip of bifacial point; 3: sidescraper; 4: endscraper; 5: broken blade with proximal retouch on ventral; 6–7: denticulates; 8: triangular blade; 9: blade from opposed platform core. Raw materials: 1: silcrete; 2–9: quartzite.

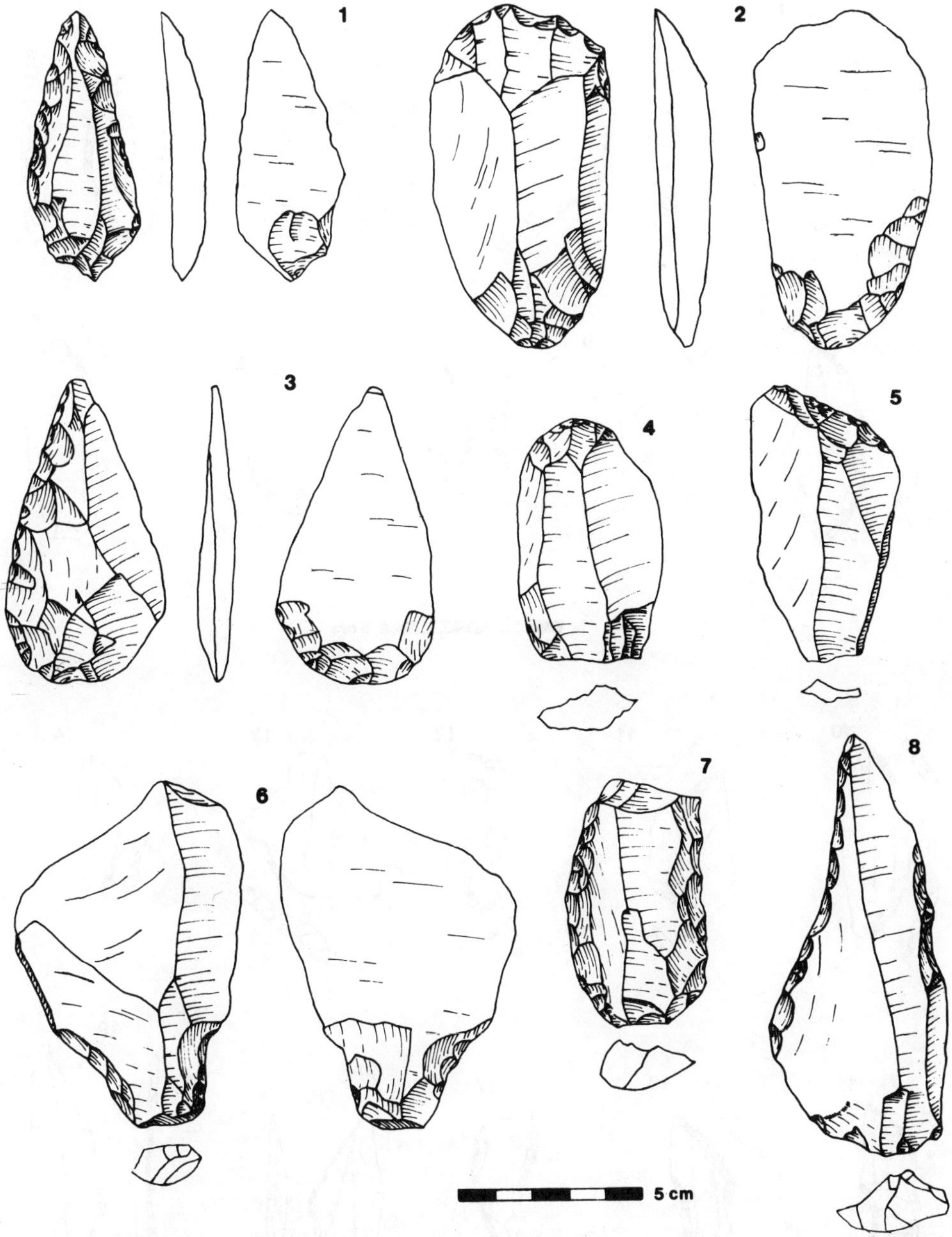

Figure 9. Middle Stone Age (MSA 2b) artifacts from Nelson Bay Cave. Types: 1, 3: unifacial points with proximal retouch; 2: endscraper with proximal retouch; 4–5: endscrapers; 6: 'tanged' flake; 7–8: sidescrapers. Raw material: quartzite.

are overlain by MSA 3 ones. The eastern Lesotho sequence is not entirely clear, but appears to include sites with Howieson's Poort assemblages above, below and between other MSA assemblages.

At sites for which numerical data are available, the flakes in Howieson's Poort

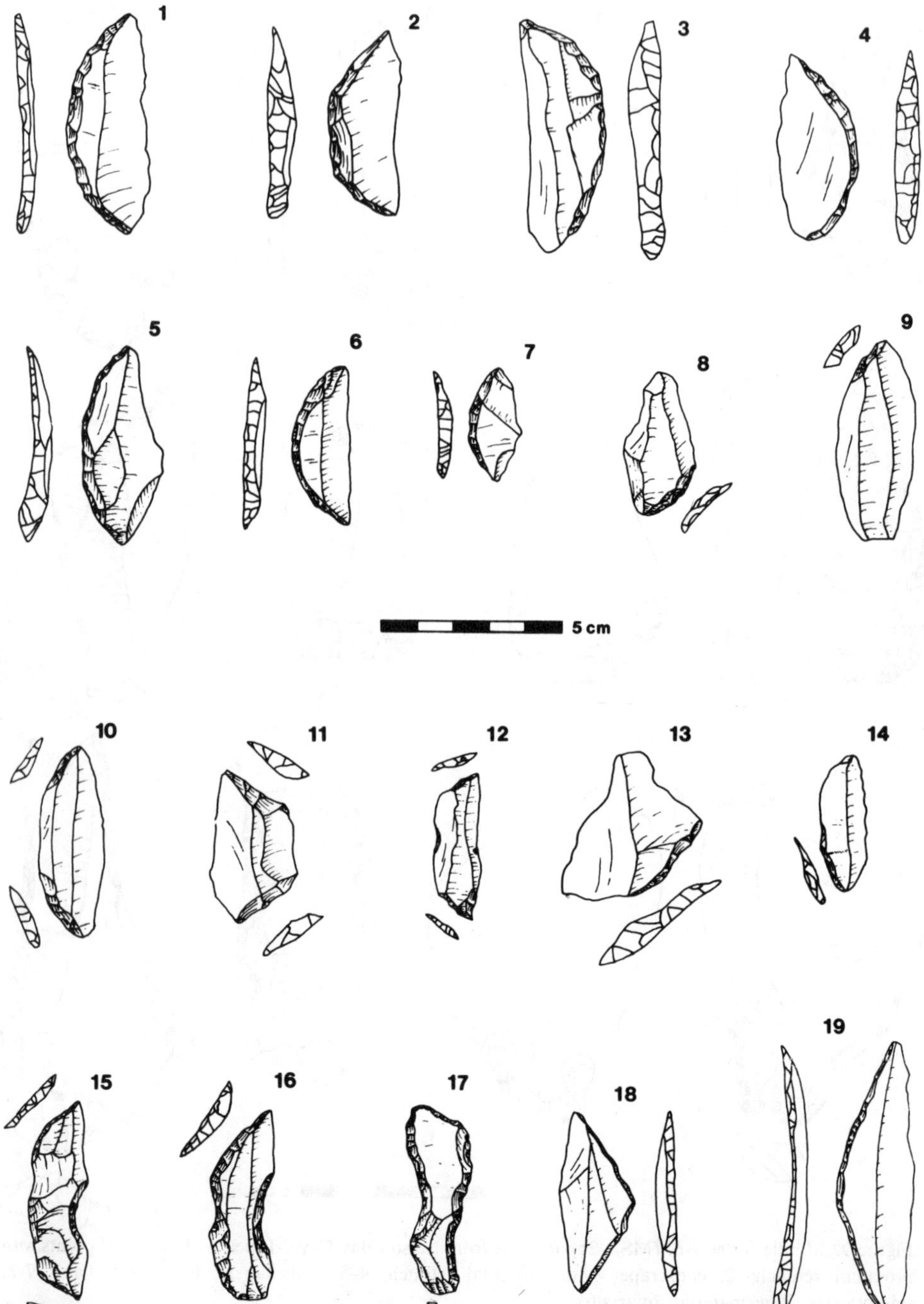

Figure 10. Middle Stone Age (Howieson's Poort) segments and allied forms from Nelson Bay Cave. Types: 1–7: segments; 8–9, 13–14: truncated pieces; 10–12: trapezoids; 15: trapezoid with notch; 16: strangulated and backed piece; 17: multiple notched piece; 18–19: triangles. Raw materials: 1–6, 9–11, 13–14, 18–19: quartzite; 7: quartz; 8: chalcedony; 12: silicified shale; 15–17: silcrete.

assemblages are, on average, smaller, relatively broader and have lower proportions of faceted butts than flakes in overlying or underlying assemblages. Fine-grained rocks, especially cryptocrystalline ones, generally were preferred for segments and allied forms. At sites where these raw materials are uncommon in earlier MSA assemblages, their relative abundance increases in the Howieson's Poort. Since LSA peoples are known to have mounted one or more small backed and/or truncated blades of fine-grained rock in mastic as parts of composite tools, it seems likely that the comparable Howieson's Poort pieces were also hafted, and perhaps were parts of composite tools. Unfortunately, no traces of mastic are preserved on any Howieson's Poort artifacts. Detailed studies, including microwear analyses where this is feasible, are needed for the Howieson's Poort geometrics, especially because composite tools are generally thought to appear much later in time.

There is considerable variability in the representation and relative frequencies of most other formal tool types found with the characteristic Howieson's Poort pieces. Denticulates, though, appear to be uniformly rare. Some assemblages contain well-retouched unifacial and bifacial points; in others, bifacial points are absent and even minimally trimmed unifacial points are absent or very rare. So far as I know, scrapers occur in all Howieson's Poort assemblages, but relative proportions of sidescrapers and endscrapers are highly variable. At southern Cape sites, rare, small, strangulated or multiple notched blades are virtually restricted to the Howieson's Poort. Some typological and technological trends exist in the southern Cape Howieson's Poort, such as the tendency for triangles, if present, to be restricted to older Howieson's Poort levels. However, few sequences have been studied in detail, and sampling and correlation problems prohibit any firm conclusions about widespread Howieson's Poort trends.

Suggested dating: oxygen isotope stage 5b, with some sequences perhaps lasting until the beginning of isotope stage 4. At Klasies and Border Cave, the contextual data indicate that the Howieson's Poort is replaced by MSA 3 assemblages at the beginning of isotope stage 5a. The Nelson Bay Cave sediments indicate cool external climate (stage 5b) during most of the Howieson's Poort sequence, with an incipient warming trend, correlated with the beginning of stage 5a, at the top. Sediments associated with some other Howieson's Poort assemblages also indicate cool climate, which on the basis of the Klasies, Border Cave and Nelson Bay Cave sequences, seems reasonable to correlate with stage 5b. At Boomplaas and Rose Cottage Cave, the Howieson's Poort sequences may begin in isotope stage 5b and persist through the beginning of stage 4.

MSA 3. The MSA assemblages which either overlie Howieson's Poort ones or can confidently be dated to a younger time range are quite varied, both typologically and in primary stoneworking techniques. On the basis of their artifacts alone, they are not readily distinguished as a group from MSA 2 assemblages, especially MSA 2b ones. They are placed here in the MSA 3 for convenience, even though temporal variability within some of the few long sequences suggests that subdivision will be necessary when more information and better dating controls are available.

In the southern Cape, the youngest MSA 3 assemblages from Boomplaas and Die Kelders 1 probably date to the end of isotope stage 3 and contain larger flake-blades than earlier MSA 3 assemblages from these sites. The fairly large flake-blades in

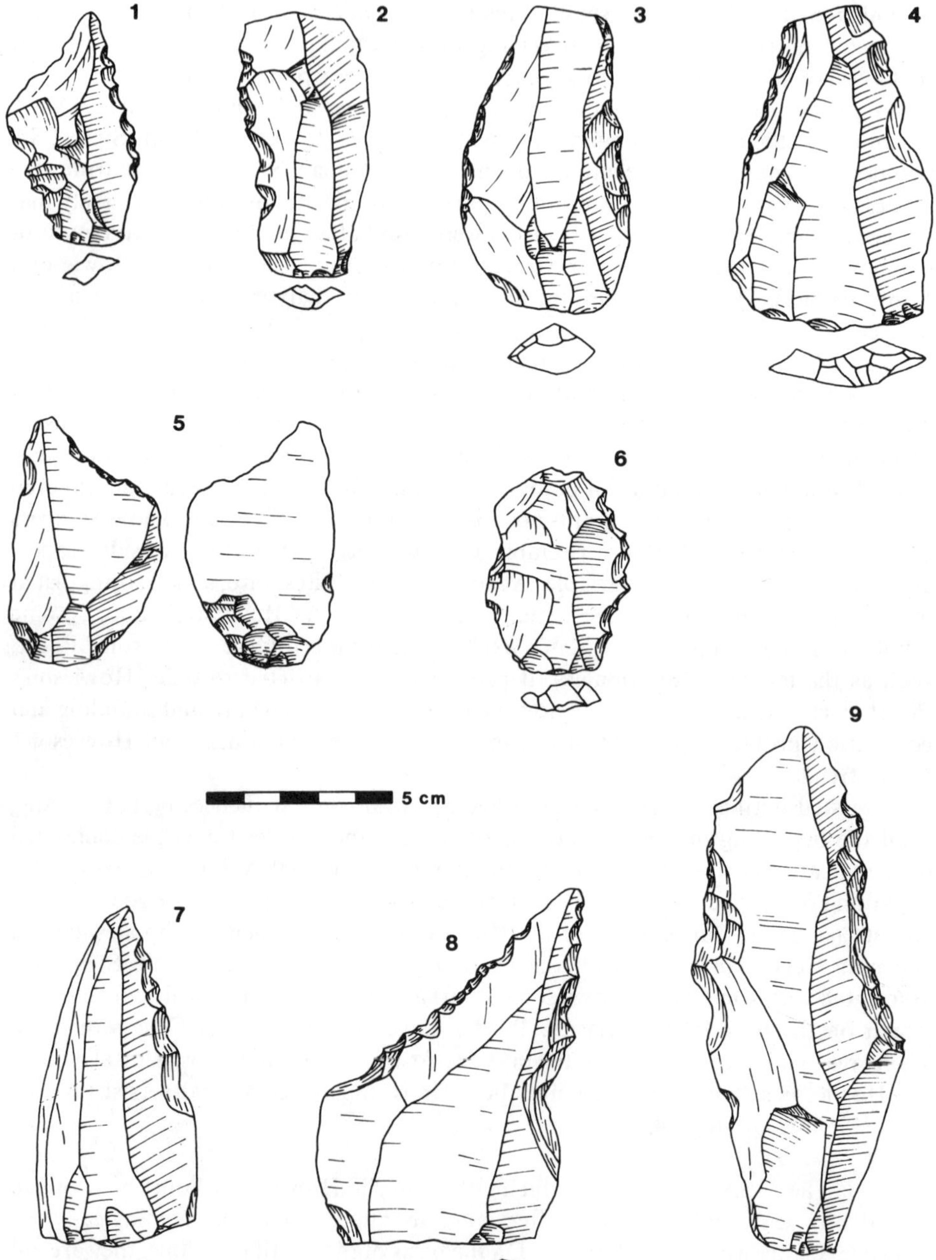

Figure 11. Middle Stone Age (MSA 3) artifacts from Die Kelders 1. Types: 1–2, 6–7, 9: denticulates; 3–4: sidescrapers; 5: utilized/retouched flake with proximal retouch on ventral; 8: convergent denticulate. Raw materials: 1–2: silcrete; 3–5, 7–9: quartzite; 6: quartz.

roughly like-aged assemblages from Highlands are consistent with this apparent trend towards larger flake-blades in the final MSA. Otherwise, no widespread technological or typological trends are discernable in the MSA 3 as it is presently known.

Suggested dating: oxygen isotope stages 5a-3. Late MSA 3 assemblages are within the range of conventional radiocarbon dating, but many inconsistent or obviously incorrect dates have been published for these and earlier MSA assemblages and further determinations are needed. Indeed, it is only in the last decade or so that the MSA has been shown to be the equivalent in sub-Saharan Africa of the European Middle Paleolithic and not the Upper Paleolithic. In the southern Cape, MSA 3 assemblages appear to last until about 30,000 years ago, with the best temporal controls provided by Boomplaas. The latest MSA assemblages at Apollo 11 may be slightly younger and the same is possibly true at some eastern Lesotho rockshelters.

Overall, it appears that MSA 3 assemblages are less abundant than MSA 2 and Howieson's Poort ones, at least in the southern Cape, and that few MSA 3 assemblages can confidently be assigned an age younger than the early Last Glacial. In conjunction with the scarcity of LSA occupations before the latest Pleistocene, these observations suggest a widespread reduction in population and restriction of settlement during most of the Last Glacial, presumably in response to widespread aridity during this time (Butzer, this volume).

The long Border Cave MSA sequence ends before 49,000 B.P. and here and at Heuningsneskrans in the eastern Transvaal, early LSA assemblages appear between 40,000 and 30,000 B.P. Thus it is possible that the MSA in the southern Cape lasted several thousand years longer than it did in the eastern interior. On the other hand, Price Williams and his associates (Price Williams & Watson 1982; Price Williams *et al.* 1982) argue that the MSA persisted in Swaziland and much farther afield until about 20,000 B.P. Price Williams notes the widespread occurrence of MSA artifacts, comparable to those assigned here to the MSA 2b and MSA 3, in colluvial sediments exposed in many dongas (dry eroded watercourses). Finds of chipping stations, with hammerstones, cores and flaking debris, show that the MSA artifacts were manufactured while the colluvia accumulated. Radiocarbon dates on underlying calcretes and on calcrete nodules in the colluvia bracket MSA occupation to between 30,000 and 12,000 B.P. Price Williams suggests that the colluvia and other geomorphological features reflect a climate less moist than today, but one that may have encouraged increased settlement through environmental changes, such as an expansion of open savanna and an increase in larger mammals. He further suggests that the open sites may complement the abandonment of cave sites elsewhere in southern Africa.

Price Williams's discovery of a widespread and perhaps lengthy period of colluviation during the MSA is an important contribution to archaeological and paleoenvironmental research in southern Africa. I find it difficult, though, to reconcile his dating with the evidence from sites in the same general region, such as Border Cave and Heuningsneskrans, for an earlier replacement of MSA by LSA. Moreover, many other cave sites in the eastern interior show the same hiatus in occupation during much of the Last Glacial that is recorded at cave sites elsewhere in southern Africa. If the MSA persisted throughout a large portion of the eastern interior, alongside a sparse LSA, this would have important implications for studies of late Upper Pleistocene demography and industrial variability in southern Africa. However, until the colluvial sequences provide radiocarbon dates on materials other than calcretes, which often give unreliable results, I would prefer to reserve judgment on the dating of Price Williams's assemblages.

The Upper Pleistocene assemblages which follow Charama ones in Zimbabwe are presently called Bambata and Tshangula (C.K. Cooke 1979b; Walker 1980). Bambata assemblages seem similar in many respects to MSA 2b and some MSA 3 assemblages from south of the Limpopo, and commonly include a variety of scrapers, unifacial and bifacial points, and denticulates, as well as occasional backed pieces. Except for Redcliff and Bambata Cave, there is no indication of typological or technological change through time within Bambata sequences. Unfortunately, the Bambata Cave collections were selected and are now dispersed. However, the published information does suggest an increase through time in the amount and refinement of retouch on some points. Changes undoubtedly occur in the Bambata sequence at Redcliff in such aspects as raw materials, artifact sizes, and perhaps in relative proportions of tool and core types, but no clear trends are apparent and some of the differences between various lenses may be idiosyncratic.

Cooke's descriptions indicate that Tshangula assemblages are distinguished from Bambata ones primarily on the basis of their smaller artifact sizes, more prominent small blade/backed artifact component, and higher proportions of fine-grained materials. However, recent work by Walker (1980:23) indicates that 'the term Tshangula is being applied to at least two different industries, and that the more recent of the two in the Matopos is a less formal scraper dominated industry. Bone tools and shell ornaments, bladelets and rare ground and bored stone artifacts are noteworthy elements, while backed tools are not common.' Radiocarbon determinations suggest that these younger Tshangula assemblages in the Matopos are less than 20,000 years old, although similar assembalges from Duncombe Farm and perhaps some other Zimbabwean sites may be older. The small 'Tshangula' assemblage from Nswatugi in the Matopos is older than 42,000 B.P. and, in addition to a variety of characteristic MSA retouched pieces, contains a few relatively large segments and allied forms which seem comparable to those in Howieson's Poort assemblages from farther south. Tshangula collections from Zombepata and some Matopos sites may represent mixtures of older and younger assemblages, although it is not clear whether the older assemblages involved are similar to the 'Tshangula' ones from Nswatugi or to Bambata ones. Although they have been called Tshangula by Cooke (1978), artifacts from the top of the Redcliff sequence seem more like Bambata ones, or perhaps the 'Tshangula' from Nswatugi, than they do to the younger Tshangula group. My brief inspection of the artifacts from lower levels at Inyanga Site XIXB suggests to me that they may also be referable to the older 'Tshangula' group.

Cooke suggests that Tshangula assemblages (except at Nswatugi) appear after a rockfall event which he believes he can identify in cave sequences throughout Zimbabwe. An alternative explanation (Walker 1980; Volman 1981) is that these 'rockfalls' represent prolonged weathering accumulations. Walker suggests that the 'rockfalls' reflect a hiatus in occupation during a time of unfavourable climate (?cold and arid), which lasted at least 15,000 years. However, roof blocks and spalls also accumulated during Bambata occupations and after 20,000 B.P. in some caves, and radiocarbon dates suggest that the breaks in occupation were not synchronous throughout Zimbabwe. Cooke (1971, 1979b) has also suggested that microlithic blade assemblages, similar to the ones he calls Tshangula, occur within Bambata sequences at Bambata Cave, Zombepata and Redcliff. A detailed examination of the published information (Volman 1981) indicates that the proposed 'intrusions' do not represent the same phenomenon, and that the evidence for intrusions of any kind at each site is not conclusive.

Overall, poor chronological controls and uncertainty about stratigraphic sequences limit present understanding of the Zimbabwean archaeological record before 20,000 B.P. Walker's recent work in the Matopos, inconsistencies in published information, and my own examination of several collections have convinced me that Cooke's characterization of the Upper Pleistocene sequence in Zimbabwe needs revision. It would appear that MSA assemblages were not manufactured in Zimbabwe after 20,000 B.P., if not considerably earlier. Evidence for typological and technological change within the Zimbabwean MSA is difficult to evaluate, but the sequences at Nswatugi and some other caves may indicate that artifacts in MSA assemblages become smaller through time, and that later MSA assemblages contain higher proportion of backed pieces, which are more carefully trimmed to standardized shapes.

Outside Zimbabwe, there are few sites in southern Africa north of the Limpopo which have provided sealed, unselected samples of MSA artifacts. Two cave sites in Zambia, Mumbwa and Kalemba, contain what appear to be Bambata-like assemblages overlain by 'Tshangula'-like ones, perhaps comparable to those in the upper portion of the Nswatugi sequence. Radiocarbon dates on bone apatite from Kalemba are inconsistent, but if the older dates are correct, they indicate that the MSA sequence at the site is older than 31,000 B.P. Early LSA artifacts appear between 24,000 and 21,000 B.P. in the lowest levels at Leopard's Hill Cave in northern Zambia (Miller 1969) and possibly comparable early LSA artifacts may be even older at Kalemba.

Subsistence

Information on MSA subsistence comes almost entirely from faunal remains preserved at some cave sites and a few open-air ones. A few carbonized seeds and nuts have been recovered from MSA deposits, most notably at Bushman Rockshelter and Border Cave, but virtually nothing is know about the exploitation of plant resources. The distribution of grindstones, a rare item in MSA assemblages, may reflect regional differences in plant foods and their processing requirements. In particular, grindstones are more common in the Transvaal and neighboring areas, where seeds, nuts and other above-ground plant foods are relatively common, than they are in the southern Cape, which provided recent hunter-gatherers with a wealth of underground bulbs and corms.

Remains of micromammals are often plentiful in caves and rockshelters where other bones are preserved. Although MSA peoples may have eaten the occasional mouse or shrew, there is no evidence that micromammals were systematically exploited. Instead, most micromammals in MSA deposits probably were brought into the sites by birds of prey, especially owls (Avery 1982a). In some instances, large raptors or small carnivores may have contributed substantial numbers of small mammals. At Die Kelders 1, for example, Cape hares, dune mole rats and very young grysbok, steenbok and vaalribbok probably were prey of eagle owls (Klein 1975a, 1981c).

MSA peoples generally relied little on reptiles or birds, though they occasionally collected tortoises and ostrich eggs in considerable quantities. Seabirds and marine mammals, especially penguins and seals, were eaten by the occupants of coastal caves. These may have been dead or disabled animals that washed ashore, a not uncommon occurrence along portions of the southern African coast today. As G. Avery has pointed out (Avery & Sigfried 1980), the seabirds are mainly island-breeding species, such as the jackass penguin and Cape ganet, and it is unlikely that MSA peoples could reach offshore rookeries. Cape fur seal rookeries were located on many near-shore islands and occasionally on the mainland. These animals were an important resource at Klasies River Mouth and Die Kelders 1, though it is not clear whether they were scavenged or actively hunted, or both.

These two sites, along with Herolds Bay Cave and two open-air middens at Saldanha Bay, contain the oldest evidence known in the world for the systematic exploitation of marine resources. At Klasies River Mouth, remains of molluscs, seals and penguins are relatively abundant throughout most of the Last Interglacial sequence and continue into the early Last Glacial levels. Last Interglacial deposits at Herolds Bay Cave contain mollusc shells in association with MSA artifacts similar to those in the lower levels at Klasies. The Die Kelders 1 MSA sequence begins during the early Last Glacial, but unlike Klasies and Herolds Bay, the sea remained fairly close to the cave throughout much of the Last Glacial. Penguin and seal bones are common at Die Kelders, but if mollusc shells were present they have since weathered away. Finally, Sea Harvest and Hoedjies Punt are two open-air shell middens of terminal Last Interglacial age which preserve abundant mollusc shells in association with MSA artifacts. Other open-air MSA shell middens may exist near Maputo in Mozambique (Barradas 1965, 1967), but the shells may be natural accumulations rather than human food waste.

Very few remains of fish or flying birds occur at the MSA sites with marine food debris. In LSA faunas from the same or comparable contexts, remains of fish and of flying birds are abundant to superabundant. The most economical explanation for this contrast is that MSA peoples were technologically incapable of active fishing and fowling. Implements such as gorges and line (or net) sinkers, which are reasonably interpreted as fishing and fowling gear, are so far known only in LSA contexts. Mollusc shells provide additional evidence that MSA peoples exploited marine resources less intensively than their LSA successors. At Klasies River Mouth the MSA shells are much less densely packed than the shells in overlying LSA levels. Similarly, shellfish densities at Herolds Bay Cave, Sea Harvest and Hoedjies Punt are much lower than in nearby LSA middens. Moreover, diameter measurements for the most common mollusc at Sea Harvest, *Patella granatina*, show that the shells are significantly larger on average than specimens from nearby LSA sites, and are almost as big as the largest individuals sampled from live populations in the area (Buchanan *et al.* 1978). Unfortunately, due to the Last Glacial drop in sea levels, most of the evidence needed to trace developments in marine resource exploitation during the Upper Pleistocene is now located well offshore.

The most extensive, detailed studies of larger mammals from MSA sites in southern Africa are by R.G. Klein (1974, 1975a, c, 1977a, 1978d, 1979a, 1980, 1981c, 1982a, in press-a, in press-b; see also Table 2). According to Klein (1977a: 120), 'the relatively substantial faunal information drawn from MSA sites suggests a common hunting pattern – MSA hunters preyed mainly on medium-to-large ungulates and generally avoided both the larger carnivores and the largest, most dangerous herbivores (rhinoceroses and elephants).'

Throughout the MSA, differences in the species composition of archaeological faunas seems mainly to reflect regional and temporal differences in environmental conditions. As yet, there is no evidence for regional differences in hunting techniques, nor for any changes over time in hunting capabilities. Most of the animals hunted by MSA peoples occur in southern Africa today. The few large mammals that became extinct after the beginning of the Upper Pleistocene all disappeared after the appearance of LSA industries (Klein 1980). Climatic/environmental changes probably played a major role in the extinction of these animals, but because they survived comparable changes during the MSA, the increased competency of LSA hunters may well have been a significant factor in their demise (Klein 1980, in press-b). New technologies, such as the bow and arrow and various traps and snares, would have enabled LSA hunters to deal more effectively with more dangerous animals, such as pigs, and to take a greater proportion of the locally available game. Klein's analyses of faunal remains from Stone Age sites in the southern Cape show that the eland, a relatively docile animal, is proportionately much more common in MSA assemblages than it is in LSA assemblages that accumulated during comparable climatic/environmental conditions.

Furthermore, Klein (1978d, 1981c, 1982b) has used measurements of dental crown heights to establish mortality profiles for several bovid species that are well-represented in MSA faunal assemblages. Klein found that most species are poorly represented by prime-age adults. Instead, MSA hunters focused on those individuals most susceptible to endemic diseases, accidents or carnivore predation – the old and especially the very young. Active hunting of these most vulnerable and reproductively irrelevant individuals posed no threat to the long-term viability of the prey species. Eland mortality profiles, however, are comparable to the age structure of living herds. These profiles are often called 'catastrophic' because they can result from events, such as flash floods and epidemics, which suddenly wipe out entire groups of animals. Because it is improbable that MSA peoples relied heavily on eland killed by natural disasters, the implication is that MSA hunters were taking entire herds by driving them over cliffs, surrounds, and the like (Klein 1979a, in press-a, in press-b). These techniques could have seriously affected eland population densities and reproduction rates and, quite literally, driven the eland to extinction. Nonetheless, eland continued to prosper in southern Africa, even though LSA faunal assemblages and rock art show that drives continued until recently (Klein 1979a). However, because MSA peoples relied much more heavily on eland without seriously affecting them, it would appear that MSA hunters were seldom in a position to drive eland, and were even less successful at killing other large mammals. Thus, Klein's analyses of larger terrestrial mammals complement the other evidence concerning MSA and LSA resource exploitation in consistently pointing to the lower overall effectiveness of MSA hunter-collectors.

Unfortunately, comparisons of MSA and ESA resource utilization are hindered by our poor knowledge of ESA subsistence practices. Still, MSA faunal remains suggest a more systematic utilization of a wider variety of animals, with marine resources a particularly notable example. MSA people may also have placed greater emphasis on active hunting than earlier populations. MSA faunal assemblages provide strong evidence that entire herds of eland and perhaps bastard hartebeest were occasionally taken, as well as the more vulnerable members of many other species. For the latter, the body part representation of both smaller mammals and very young individuals of larger species indicates that entire animals were brought back to cave sites. Active hunting is implied because bones of these animals are usually destroyed by carnivore feeding, and it is highly unlikely that people often found carcasses before scavengers did (Klein 1982a). A greater reliance on scavenging by ESA peoples may be indicated by higher proportions of very large or dangerous animals in some faunal assemblages associated with ESA artifacts.

Finally, MSA faunal assemblages provide the first clear evidence in the southern African archaeological record for size related differences in game processing. Klein (1977a: 121) found that 'the larger the species is, the smaller the variety of body parts is by which the species is likely to be well represented.. [which] indicates that smaller animals were often returned to base camps intact, while larger ones were generally butchered where they killed and only certain parts were regularly brought home.' Klein (1977a; see also 1976a, b) suggests that the selective removal of 'meat- and marrow-bearing limb bones' of larger bovids is reflected in the relative abundance of these bones compared to vertebrae at Klasies Cave 1, an intensively occupied base camp, as opposed to the higher ratio of vertebrae to limb bones at Duinefontein 2, an open-air, 'kill/butchery' site.

Other aspects of material culture and behavior

The MSA archaeological record consists overwhelmingly of stone artifacts, sometimes accompanied by animal food waste and remains of fires. Scattered bits of charcoal are common in MSA deposits and definite hearths, containing charcoal and ash, are preserved at many cave sites. The hearths are often discrete, isolated features, but sometimes occur as intercalated lenses of charcoal and ash. The Howieson's Poort levels at Klasies River Mouth provide particularly striking examples of thick accumulations of multiple, intercalated hearths.

Worked or utilized organic materials are very rare (Table 3), providing a major contrast between MSA and LSA assemblages. Because charcoal, occasional twigs and traces of humified plants are practically the only macrobotanical remains preserved in southern African sites before the Holocene, faunal remains provide the best comparisons. Whereas items made from faunal remains are relatively common at LSA sites well before the end of the Pleistocene (see J. Deacon, this volume), shaped implements from MSA contexts are limited to a bone point from Klasies and several 'daggers' of pig tusk from Border Cave.[7] The presence of bone fragments with separate, small notches at Klasies, Border Cave and Apollo 11, and of a similarly notched piece of ochre at Howieson's Poort Rockshelter, indicates a rare but widespread practice. The purpose of these pieces is unknown, but the possibility that some kind of notation system may be involved is intriguing.

Table 3. Unusual worked or utilized organic materials from southern African MSA sites.

Site	Item
	MACROBOTANICAL REMAINS
Wonderwerk Cave	Bedding
Border Cave (MSA 3)	Hematite stained bedding
Border Cave (MSA 3)	Utilized acacia thorn points
Apollo 11 (MSA 3)	Mastic traces along basal portion of a utilized blade
Florisbad (?MSA)	Broken, curved wooden implement (?throwing stick) with roughly parallel cut markings at tapered end
	IMPLEMENTS MADE FROM FAUNAL REMAINS
Border Cave (MSA 3)	Seven 'daggers' of split and ground tusks of warthog or bushpig
Klasies River Mouth (Howieson's Poort)	Thin bone point, ground and then polished, either intentionally or through use
	OTHER WORKED FAUNAL REMAINS
Klasies River Mouth (MSA 2b)	Two rib fragments with notches along an edge and striations on one or both surfaces
Klasies River Mouth (MSA 2b)	Small bone fragment on which four parallel lines are scratched
Border Cave (MSA 2b)	Perforated *Conus* shell associated with possible MSA infant burial (BC3)
Border Cave (MSA 3)	Notched rib fragment
Apollo 11 (MSA 2b)	Two notched bone fragments
Apollo 11 (Howieson's Poort)	Incised fragments of ostrich eggshell

Perhaps the most common non-utilitarian finds at MSA sites are pieces of hematite that are ground or show other signs of abrasion. At Howieson's Poort Rockshelter, a trihedral piece of soft ochre has one surface ground flat and small, V-shaped notches along the three edges. Otherwise, all of the hematite found in MSA sites appears to have served as a source of pigment, or, in the case of a few, very hard pieces, as a raw material for tools. At Lion Cavern in Swaziland, where MSA peoples were mining for specular hematite by perhaps 110,000 years ago, a large adit was cut into the side of a specularite-rich portion of the hematite cliff face. The artificial bedrock floor inside and outside the present cave is overlain by deposits with MSA artifacts, including heavy-duty mining tools.

Beyond the fact that hematite was ground, little is known about its uses or preparation. One of the five grindstones from Bed 2 at Olieboompoort preserves traces of ochre, but it is unlikely that the rare MSA grindstones served primarily for processing pigments. In its powdered form, ochre may have been used for body decoration or on skins. There are no indications that MSA peoples painted on cave walls, though preservation of any parietal art would be highly unlikely at most sites. At Apollo 11 in Namibia, several fragments of painted slabs were recovered from the very top of a compressed layer containing MSA artifacts and perhaps an indeterminate macrolithic industry. The slabs are at least 19,000 years old and may well date from a late MSA occupation at roughly 27,000 B.P.

Except for the occasional rock or mineral with an unusual shape or color, the only other non-utilitarian stone artifacts claimed from MSA context are portions of three small bored stones found at Zombepata (C.K. Cooke 1971). The largest piece is made from chlorite schist while the others are a soft micaceous schist. One of the latter has marks in the perforated area that may be decoration, or perhaps formed during use. None of the stones was very heavy when whole; the largest probably weighed about 200 grams and the others less than 100 grams. At least one of these bored stones may come from a Bambata or 'Tshangula' level that is more than 40,000 years old. However, there are inconsistencies in the dates and descriptions of the industrial sequence at the site (Volman 1981), and I am not convinced that any of the stones is definitely an MSA artifact. It should be noted that small bored stones similar to the Zombepata pieces were found in the early LSA at Border Cave, including one with incisions around the perforated area (Beaumont *et al.* 1978).

The most extensive evidence for MSA constructions comes from Orangia 1 in the Orange Free State where Sampson (1968a) excavated seven curvilinear arrangements of dolerite cobbles and boulders. The structures, which may have anchored windbreaks made of branches, occur on an intact occupation horizon about 45 cm below the present surface. Since about equal numbers of stone artifacts lie among the structures and on surrounding surfaces, Sampson estimates that roughly half of the site has been destroyed by erosion. The structures appear to be of two sorts. One complete and perhaps one partially collapsed structure are about 4 m long, and consist of a slightly arched alignment of stones with one end strongly curved. The other five structures are much smaller (about 1.2-1.5 m long) and semi-circular in shape. The larger arrangements and one of the smaller structures are made of loosely packed rocks which lie on, or slightly embedded in, the ancient land surface. The rocks making up the other small structures are more densely packed, and are stacked against what seem to be the sides of excavated hollows, about 30-60 cm deep. These structures were relatively free of artifacts, and may have been used for sleeping (Sampson 1968a).

Artifacts were not uniformly or randomly distributed on the site and Sampson's plans show several dense scatters of pieces. Three core clusters, two of which are associated with the larger structures, are particularly striking. These seem to be intentional accumulations with no relationship to the distribution of manufacturing waste. On the basis of San ethnography, Sampson (oral report to the Southern African Association of Archaeologists, Cape Town, June 1979) suggests that the site may have been occupied by an all-male hunting party.

Zeekoegat 27, another partially eroded site excavated by Sampson in the Orange River Basin, may also be an MSA site with a structure. Sampson (1968a:63) describes the structure as 'a large circular scatter of dolerite cobbles and sandstone blocks together with smaller scree fragments.' The arrangement has a diameter of about 11m, and 'there is a sparse scatter of large blocks and smaller scree contained within the circle.'

Except for the occasional shallow hearth, evidence for modification of cave sites is also very limited. One square excavated in the MSA 2b deposits at Apollo 11 contains what Wendt (1976:7) describes as 'an artifical accumulation of blocks, slabs and other debris.' A concentration of unmodified rocks, perhaps artificially arranged, also occurred in 'Surface VI' at Montagu Cave. Additionally, both Wonderwerk and Border Cave may contain MSA sleeping hollows.

The only definite burial attributed to the MSA is the Border Cave infant (BC3), which appears to have been interred in a shallow grave originating from the MSA 2b levels at the site (Beaumont 1980; Butzer, Beaumont & Vogel 1978). A perforated

Conus shell was associated with the skeleton and ochre may have been scattered over the body. Burials containing Neanderthals as well as other constructions attributed to ritual activities occur in Europe during the Middle Paleolithic. Thus, the virtual lack of similar occurrences in southern Africa, where anatomically modern human fossils may date from the Last Interglacial (see Rightmire, this volume), is somewhat surprising. Beaumont (1980) has suggested that all of the Stone Age human remains at Border Cave may have been buried, because they are less fragmented than other faunal remains at the site. The anatomically modern mandible (BC5), excavated by Beaumont in a Howieson's Poort level, came from a possible grave at the edge of his trench. The BC1 cranium and the BC2 mandible came from a farmer's fertilizer pit. Assignment to the MSA, much less a specific level, depends on circumstantial evidence.

A number of modern-looking fossils, such as the open-air finds at Boskop Tuinplaas (Springbok Flats) and Cape Flats, were uncovered during commercial activities in the early decades of this century (Rightmire 1975). They are often assigned to the MSA and may also be burials, but, if so, it is not at all clear that they are MSA burials. Another burial commonly assigned to the MSA is Fish Hoek Man from Peers Cave, but this is probably a terminal Pleistocene, LSA burial (Rightmire 1978b). These fossils, and the ones excavated at Border Cave in the 1940's, are prime candidates for accelerator radiocarbon dating once this technique is developed beyond the experimental stage (Bennett 1979). Aside from the BC5 mandible, human remains from recent, well-controlled excavations consist mainly of isolated teeth or, in the case of Klasies, of scraps that are in no better shape than the rest of the faunal remains from the site.

The perforated shell found with the Border Cave infant is that of a marine gastropod and must have come from the Natal coast, more than 80 km to the east. Several other sites also provide evidence for long-distance transport of materials during the MSA. However, most items found at MSA sites have local sources and few detailed searches for distant raw material sources have been attempted. Seashells found at inland sites are obvious examples of long-distance transport. For instance, a few seashell fragments in the MSA 2b at Apollo 11 indicate connections with the Atlantic coast, over 100 km away. Identification of distinctive raw material sources should provide some indication of the movements of MSA populations. For a few sites, such as Border Cave and Sibebe, searches already indicate that MSA peoples were bringing back stones from distances of 20 km or more.

SUMMARY AND DISCUSSION

Hominids have lived in southern Africa for at least 3 million years. From about 2 million years ago, the archaeological record provides glimpses of their toolmaking and subsistence activities, but knowledge about human behaviors and capabilities is very limited before the Upper Pleistocene. The earliest artifact assemblages from Swartkrans and Sterkfontein Extension may be comparable to Oldowan ones from East Africa, and include utilized bones that may have been used to dig up underground plant foods. Throughout the Lower and Middle Pleistocene, a persistent difficulty at sites preserving artifacts and faunal remains is to sort out the effects of human behaviors, animal activities and geological processes. This problem also exists for materials from MSA occupations, but the large samples and contextual information available from a number of MSA sites makes it considerably easier to test hypotheses.

The few Lower Acheulean assemblages in southern Africa may date from about 1.5-1.0 Myr ago. They are followed by much more numerous and geographically widespread Upper Acheulean ones. An overall increase in population almost certainly occurred. The expansion of peoples into previously unoccupied areas suggests an ability to adapt to a greater range of habitats, though the stability of populations in various areas needs to be established.

Some Upper Acheulean assemblages show considerable refinement in toolmaking capabilities, including the use of a variety of prepared cores and the production of elegant handaxes, perhaps finished by soft hammer techniques. However, these refinements do not suggest any major improvements in the effectiveness of Acheulean technology. The earliest MSA assemblages, probably dating from at least 150,000 years ago, do not show technological improvements over ESA ones, but by the early Upper Pleistocene MSA peoples were hafting a variety of stone artifacts and may have begun making composite tools. The use of indirect percussion techniques for producing tool blanks may have appeared at this time.

MSA sites provide the earliest evidence yet found for the systematic exploitation of marine resources. Comparisons of the limited faunal evidence from ESA contexts with that from MSA sites suggests that MSA peoples systematically exploited a greater variety of food resources and perhaps relied more on active hunting. There is much better evidence to indicate that MSA peoples utilized both marine and terrestrial faunas far less effectively than did their LSA successors.

There is widespread evidence that MSA peoples ground hematite, and even mined it extensively at Lion Cavern. However, in contrast to the LSA, art objects and ornaments are very rare, as are shaped bone tools, structures and graves. Indeed, claims for MSA provenience of some unique finds would be considerably strengthened by similar discoveries in future excavations.

The geographical distribution of late ESA assemblages (Upper Acheulean and Sangoan) may indicate the beginnings of major regional differences in adaptations. Increasing differentiation during the MSA is not evident, however, perhaps in part because it is difficult to control for the effects of raw materials and to distinguish geographical from temporal variability. Nonetheless, MSA sites from south of the Limpopo show very similar industrial sequences during the Last Interglacial and the same pattern of industrial changes may occur in Zimbabwe, southern Zambia, Botswana, and perhaps elsewhere.

Possibly the best contrast is between these MSA assemblages and the poorly known 'Lupemban' and other MSA assemblages of the Zaire (Congo) River Basin and surrounding areas (Clark 1971b). This large-scale patterning seems comparable to that characterizing LSA assemblages from north and south of the Zambesi in the Holocene (Deacon & Deacon 1980), and probably reflects similar broad ecological divisions. If more subtle markers exist for localized populations, such as H.J. Deacon (1976) has identified for the LSA in the southern Cape, they have yet to be identified.

Southern Africa contains one of the longest, best known archaeological records in the world. In the near future, it can be anticipated that our understanding of cultural developments during the later Pleistocene will greatly increase with the continuing, large-scale excavation of cave and open-air sites, better paleoenvironmental reconstructions, and the application of accelerator dating techniques. It is likely, though, that our knowledge of various areas will be very uneven for many years to come. Progress is likely to be much slower for earlier time ranges, and will depend much more on the discovery of informative sites and the development and application of new analytical techniques.

FOOTNOTES

1. According to Bordes (1980), the term 'Levallois' applies to any flake whose form is predetermined by careful core preparation before its detachment. Levallois cores include those prepared for the removal of a single flake (e.g. Victoria West and Levallois flake and point cores), or for the single or multiple production of blades (e.g. Levallois blade cores). Unfortunately, the term 'Levallois' has not been used with any consistency in the African archaeological literature. 'Levallois' and 'prepared' core technique often are equated, although the latter concept implies a greater variety of approaches to flake production. Among prepared cores, I would include radial (or discoid) ones, which may be prepared initially like Levallois flake cores, but from which many flakes are detached. These cores are common in many African ESA and MSA assemblages. Bordes (1961) notes that Levallois flake cores are often further worked as radial ones. In some southern African ESA and MSA assemblages, the last removal from a radial core may give it the appearance of a Levallois flake core. Other core types that are 'prepared' but not 'Levallois' include the prismatic blade cores of the European Upper Paleolithic, which grade into Levallois blade cores (Bordes 1980). Prepared blade and flake-blade cores with a single straight or curved platform occur in the South African interior during the ESA. Similar cores, plus prepared cores with opposed platforms in the same or parallel planes, are widespread in southern Africa during the MSA.
2. I thank H.J. Deacon for calling my attention to many of the problems discussed in this paragraph.
3. Whether the Mwanganda Area I assemblage should be considered ESA or MSA is not clear. Clark and Haynes (1970:402) suggest that the artifact assemblage from Mwanganda "antedates the full development of the regional 'Middle Stone Age'" and that it may be an activity facies of either the MSA or the Sangoan. More recently, Clark (1981) has assigned the Mwanganda occurrence to the Middle Pleistocene and noted that the artifacts are similar to Developed Oldowan ones. For purposes of the present discussion, the Mwanganda assemblage is tentatively assigned to the ESA (?Sangoan) primarily because of its potential high antiquity and its apparently crude flake production techniques.
4. Extracting meaningful information on early human behavior from such complicated archaeological occurrences is a high priority for ESA research. Comprehensive, interdisciplinary projects, like those being conducted by Isaac and his associates at East Turkana in Kenya (Bunn *et al.* 1980), are needed throughout the African continent.
5. For MSA assemblages following the MSA 1, relative core types and faceting proportions are poor 'cultural markers.' Statistical studies of southern Cape collections (Volman 1981) suggest that while multiple faceted butt proportions, mean flake lengths and breadth/length ratios are fairly highly correlated, they are not unique to particular portions of the MSA sequence. Moreover, these aspects of flake variation are more clear-cut within site sequences than among sequences or individual samples. My measurements of complete samples from various levels at Nelson Bay Cave contradict Beaumont's (1979a) suggestion that a decrease in flake relative thickness characterizes the MSA 2-Howieson's Poort sequence.
6. Metrical data for southern Cape assemblages indicate a clear size difference between Howieson's Poort and Wilton (LSA) geometric backed pieces, although there is a slight overlap in ranges for various measurements (Volman 1981). Segments in Howieson's Poort assemblages usually are between 25 and 50 mm in length, while Wilton segments are rarely more than 20 mm long. These metrical data also indicate that the trapezoids in Howieson's Poort assemblages are not unfinished segments. Average breadth measurements are less for trapezoids (backed at both ends on one lateral, but not in the middle) than they are for segments (completely backed along one lateral).

7. Several bone implements and even a small decorated piece (tube-shaped with deep, encircling incisions) were recovered from the uppermost levels with MSA artifacts at Bushman Rockshelter (Plug 1978, 1983). However, the worked bones from these MSA levels, including the decorated piece, have counterparts in the lower LSA levels at the site and several inconsistent radiocarbon dates suggest some mixing of MSA and LSA materials. It would be premature, I believe, to assign any of the worked bone from the uppermost MSA levels at Bushman Rockshelter to the MSA.

ACKNOWLEDGEMENTS

I thank G. Avery, P.B. Beaumont, C.K. Brain, H.J. Deacon, J. Deacon, A.R. Hughes, R.G. Klein, R.J. Mason, D. Price Williams, J.C. Vogel, E.S. Vrba, L. Wadley and N.J. Walker for discussions and helpful comments on various topics covered in this paper. The National Science Foundation, the Hinds Fund of the Committee on Evolutionary Biology, University of Chicago and the Department of Archaeology, University of the Witwatersrand provided financial support for my own research reported here. The research was conducted in facilities kindly provided by the South African Museum and by the Department of Archaeology, University of the Witwatersrand, where I was a Postdoctoral Research Fellow in 1980 and 1981.

JANETTE DEACON
Department of Archaeology, University of Stellenbosch, South Africa

7

Later Stone Age people and their descendants in southern Africa

INTRODUCTION

Artefacts characteristic of the Later Stone Age (LSA) appear in the archaeological record during the last 40,000 years of a 2 million year-old Stone Age tradition of artefact manufacture and hunter-gatherer subsistence in southern Africa, a tradition that survives today amongst a few thousand people who are still mostly hunter-gatherers, but who no longer make stone tools. The character of the LSA was defined in the late 1920s by A.J.H. Goodwin and by Goodwin and Van Riet Lowe (1929:7) as including those assemblages similar to the European Mesolithic (i.e. with a microlithic toolmaking tradition), and the Neo-Anthropic Cultures, meaning those relating to modern man. The term 'Later Stone Age' therefore incorporates assemblages with microlithic stone artefacts, as well as those with items of material culture that are, or were until very recently, made and used by modern southern African hunter-gatherers. In contrast to the Middle Stone Age (MSA), therefore, we have good ethnographic and historical records that can be used to generate direct ethnographic analogues.

The elements of material culture characteristic of the LSA are listed below. Not all these items are found at all LSA sites, nor are they all of equal antiquity. A few (rock paintings, formal burial, shell amulet and a polished bone point) are known in context with MSA stone artefacts, but in each case the find is unique to one site (Table 11). While these few elements, then, are older than the LSA stone toolmaking tradition, they are found more often during the LSA and the gradual accretion of these and other technological traits can be used to trace the development of such aspects of modern behavior as symbolic and representational art, personal adornment and the design of a wider range of specialized tools. The main factor which distinguishes LSA from MSA assemblages, however, remains the technique of stone tool manufacture as evident in the range of core types and the size and shape of the flakes. Items characteristic of the Later Stone Age are:

- Art and items of personal adornment, for example paintings, engravings (petroglyphs), decorated ostrich eggshell, decorated bone tools, and beads, pendants and amulets of ostrich eggshell and marine and freshwater shells.
- Specialized hunting and fishing equipment in the form of bows and arrows (recognized archaeologically in the occurrence of polished bone points and fore- or

linkshafts, backed microliths and, more rarely, fragments of bows and arrows), fish hooks and sinkers.

– A greater variety of numerous formal scrapers apparently used mostly for skin-working, associated with bone skin-working tools, bone needles and awls and, more rarely, fragments of leather with sewn seams.

– Specialized food gathering tools and containers such as bored stone digging stick weights, carrying bags of leather and netting, storage pits, ostrich eggshell water containers, tortoiseshell bowls and scoops and, later, pottery and stone bowls.

– Formal burial of the dead in graves sometimes covered with painted stones, grindstones or unmodified cobbles and boulders, and sometimes accompanied by grave goods.

– The miniaturization of some stone tools linked to the practice of hafting stone artefacts in mastic to handles. In the absence of hafted microliths this practice is assumed from the occurrence of microbladelets and the cores from which they were struck.

– A characteristic range of formal stone tools designed for making some of the items listed above, for example borers for making beads, reamers for making bored stones, adzes for woodworking, and grooved stones for smoothing beads, straightening arrows and applying poison to arrowheads.

The nature of the archaeological record and of sampling make it impossible to reconstruct in any detail the process by which these traits were adopted and adapted in southern Africa, but the bulk of the evidence summarized in this paper shows that it is only within the last 40,000 years that ethnographically known elements of material culture appear in LSA assemblages and it is only within the last 10,000 years that they are at all common at most sites.

The similarity between LSA material culture and that of the historically known San (Bushmen) and Khoi (Hottentot) inhabitants of southern Africa has been frequently described (Gooch 1881; Péringuey 1911; Hewitt 1921; Jones 1926; Burkitt 1928; Goodwin & Van Riet Lowe 1929; Clark 1959, 1970; Inskeep 1967, 1978a, b; Deacon, H.J. 1972, 1976; Parkington 1972, 1977a, b; Phillipson 1977; Runder 1979). San-like *Homo sapiens sapiens* people were present in southern Africa as much as 115,000 years ago (Beaumont *et al.* 1978; Beaumont 1980), however, indicating that there is no close correlation between the development of physically modern people and the artefacts used by modern hunter-gatherers that appear only after 40,000 years ago. Thus the constellation of attributes that make up the anthropology and material culture of present-day hunter-gatherers in southern Africa is the result of the gradual introduction and adoption of a range of items and habits that has been constantly changing throughout the LSA. Neither the LSA nor the ethnographic present can be seen as a static entity with a fixed range of traits, but should be recognized as different slices of the Stone Age time span, the LSA representing part of the prehistory of the historically documented Khoisan people. Ethnographic analogies should therefore be used with caution in interpreting the archaeological record because assumptions of uniformity throughout the LSA and even beyond the historic time span are almost certain to be in need of qualification, particularly when environmental conditions were different from those of the present. The safest analogies are those that assist in the identification of the uses of particular artefacts; those that draw on the habits and subsistence strategies of present-day San in the Kalahari are less so.

The southern African LSA has traditionally been described in terms of the stone artefact content of assemblages that have been classified into industrial complexes (cultures), industries and phases. While this methodology can be criticised because it tends to emphasise 'breaks' and overlooks the basic continuity of the LSA, it is nevertheless useful to find some basis for subdividing the time span into meaningful units. In this review I have chosen to describe the data within four broadly defined units on the basis of time, the characteristic method of tool manufacture and the range of formal tools. These units are: Late Pleistocene microlithic assemblages (ca 40–ca 12,000 B.P.), terminal Pleistocene/early Holocene non-microlithic assemblages (ca 12,000–ca 8000 B.P.), Holocene microlithic assemblages (ca 8000–ca historic times, and late Holocene assemblages with pottery (ca 2000 B.P. to the present). These subdivisions are also broadly coincident with shifts in subsistence strategy brought about by habitat changes at the end of the Pleistocene and by the introduction of domestic stock within the last 2000 years. To put these data into perspective I have made use of ethnographic analogy where possible to give an indication of the amount of detail that is lost in the archaeological record, and to indicate where ethnographic data can profitably provide testable hypotheses for future archaeological work. The geographical area covered in this paper extends northward to the Zambesi River and the Namibian/Angolan border and is bounded on its other sides by the Indian and Atlantic oceans. Natural subdivisions of southern Africa are assigned to ecozones as defined by De Vos (1975) and used by H.J. Deacon (1976, in press) and Klein (1980 and this volume).

This review of the southern African Later Stone Age is designed to be a summary of current data and viewpoints and is not directed at criticism, although contrary opinions are expressed at times. The earlier literature (pre-1960) has been adequately summarized by a number of writers (Clark 1959; Inskeep 1967; Deacon, H.J. 1972; Sampson 1972, 1974) and will not be dealt with in any detail here. The paper has been divided into three main parts: the first deals with the culture-stratigraphic record, the second cross-cuts culture-stratigraphy and examines the ethnographic and archaeological data to interpret them in terms of artefacts and activities, and the third traces temporal and spatial variability observed in several major formal tool classes. Each section gives a different perspective on the interpretation of artefacts and lifestyles of LSA people who lived in southern Africa during the last 40,000 years by analysing the data from an alternative standpoint. Each method has both advantages and disadvantages, but the combination of the three will hopefully provide a broader panorama for Later Stone Age research than has been possible to date.

THE CULTURE-STRATIGRAPHIC SEQUENCE

Terminology

In the terminological scheme proposed by Goodwin and Van Riet Lowe in 1929 the LSA was subdivided into two main 'industries' (now preferably referred to as

Industrial Complexes), the Wilton and the Smithfield. The latter was further subdivided into the Smithfield A (thought to be the oldest), Smithfield B (the youngest) and Smithfield C (found in cave sites) and later the Smithfield N was recognized in Natal, the Smithfield P in Pondoland and the Coastal Smithfield was recognized in the southern and western Cape (for a fuller description see Inskeep 1967, 1978a; Deacon, H.J. 1972; Deacon, J. 1972, 1978; Sampson 1972, 1974). The Wilton was the more widespread and was recognized in Zambia, Zimbabwe, Botswana, Namibia and the western and southern regions of South Africa. The Smithfield, on the other hand, was largely confined to the interior of South Africa where hornfels (=lydianite= indurated shale) was the most abundant raw material and was used almost exclusively in the Smithfield A, B, and P, and to a lesser extent in Smithfield N.

As part of the worldwide microlithic toolmaking tradition in which small (less than 25 mm long) bladelets with a sharp cutting edge were blunted or backed to facilitate hafting onto a shaft or handle with mastic, the Wilton formal tool assemblages (Figure 5) were dominated by small convex and other scrapers, backed bladelets and segment-shaped backed microliths, adzes, borers for making ostrich eggshell beads, and a range of bone, shell and wooden tools and ornaments. Smithfield assemblages differed in the rarity or absence of segments and the subdivisions were based on the size and shape of the scrapers: Smithfield A (Figure 4) was characterized by large circular concavo-convex and end scrapers, Smithfield B (Figure 7) by long end scrapers, Smithfield C by small convex scrapers and Smithfield P and N by end and hollow (strangulated) scrapers (Goodwin & Van Riet Lowe 1929; Goodwin 1930; Van Riet Lowe 1936; Clark 1959).

It was not until excavations on the southern Cape coast at Oakhurst and Matjes River were conducted in the 1930s that the stratigraphic relationship between the Smithfield and Wilton was demonstrated. The picture appeared confused, however, as at Oakhurst the Smithfield B and C were thought to precede the Wilton (Goodwin 1938), while at Matjes River nearby the Smithfield C was absent and the Smithfield B overlaid the Wilton (Dreyer 1933; Louw 1960). This latter pattern was repeated at Ncinginwinde in Transkei (Laidler 1933) and although Goodwin (1946a) rejected it as an unlikely interpretation, the reason for the anomaly did not become clear until the re-excavation of the Wilton name site in the 1960s (Deacon, J. 1972). Analysis of the metric attributes of scrapers from Wilton showed that through the Holocene, in the southern Cape at least, large scrapers both preceded and succeeded assemblages with smaller scrapers and that the occurrence of significant numbers of segments was restricted in time to the mid-Holocene. More tantalizing was the observation (Deacon, J. 1974) that in the drier interior of South Africa, the type area of the Smithfield, there were no radiocarbon dated assemblages from the mid-Holocene. Thus the lack of segment-dominated assemblages in that region may be the result of a lower incidence of occupation sites there during the mid-Holocene or it may be due to the fact that almost all the radiocarbon dated assemblages come from rock shelters and the mid-Holocene segment-dominated assemblages may be more prominent at open sites. In either case the low density of assemblages in which segments are prominent led to the classification of all material in the interior of South Africa as 'Smithfield C' rather than as Wilton. Recent excavations at Wonderwerk in the northern Cape to the west of the region dominated by Smithfield assemblages show that the mid-Holocene assemblages there are indeed dominated by segments (P.B. Beaumont and Anne I Thackeray, pers. comm.).

Thus, instead of the Smithfield and Wilton being broadly contemporary, they seem to be chronologically differentiated. A similar dichotomy between the Zambian Wilton and the Nachikufan in Zambia is also considered to be a chronological one (Sampson & Southard 1973; Phillipson 1977:41). Holocene LSA industries in southern Africa are therefore characterized both by changing parameters of scrapers through time and by changes in the relative frequency of backed microlith sub-types, particularly segments. In some instances, these changes are accompanied by shifts in the relative frequencies of raw materials.

In 1974 Sampson took note of these observations and proposed a new classification scheme for southern Africa in which the assemblages formerly grouped within the LSA were described in terms of three time-successive Complexes, namely the Oakhurst Complex (formerly the Smithfield A), the Wilton Complex (formerly the Smithfield C and Wilton) and the Smithfield Complex (formerly the Smithfield B). There is, in addition, an informally defined Strandloper or Coastal Complex that incorporates late coastal assemblages. While details of this scheme need closer scrutiny (Deacon, J. 1976b), it provides a broad-based scheme to consolidate the southern African data.

With a great increase in the number of excavated sites in southern Africa over the last 15-20 years the LSA succession has become somewhat more complicated than the relatively simple Wilton/Smithfield sequence described by Goodwin and Van Riet Lowe and modified by Sampson. Indications of the greater antiquity of the microlithic toolmaking tradition were apparent from the dating of the Nachikufan in Zambia and the Tshangula in Zimbabwe to between 20,000 and 12,000 B.P., but dating of the Matupi sequence in Zaire (Van Noten 1977) and of Border Cave and other sites in South Africa (Beaumont *et al.* 1978) suggest that this tradition was already present in southern Africa by about 40,000 years ago. Beaumont and Vogel (1972a,b) grouped these late Pleistocene microlithic assemblages into an 'Early LSA' to distinguish them from the Holocene microlithic ones, but they and Phillipson (1977) have included the Early LSA with the Oakhurst Complex of Sampson. The differences between the Oakhurst Complex industries and the early microlithic ones were made explicit in the southern Cape, however, where the local expression of the Oakhurst was termed the Albany Industry and the Robberg Industry was named to describe the preceding microlithic assemblages dating between 18,000 and 12,000 B.P. (Klein 1974; Deacon, H.J. 1976; Deacon, J. 1978).

Sampson's scheme of Complexes was partly designed to put into effect recommendations made in 1966 (Clark *et al.* 1966; Bishop & Clark 1967:867) that the terms Earlier, Middle and Later Stone Age be dropped. Although there is still some support for retaining the term Later Stone Age at a higher hierarchical level than the Complexes (see, for example, Masao 1979; Shaw 1979), this is not given in recent syntheses by Phillipson (1976, 1977) who has proposed that the industries formerly grouped within the LSA would be better described within J.G.D. Clark's mode terminology as 'mode 5'. While this may be appropriate for the sequence north of the Zambesi, mode 5 should include only assemblages 'characterised by microlithic components of composite artefacts' (Phillipson 1977:23) and thereby excludes terminal Pleistocene/early Holocene and late Holocene non-microlithic assemblages which form an integral part of the LSA sequence to the south.

In assessing the relevance of terms that have been proposed it is important to remember that the LSA is first and foremost a technological unit. Although this is not always explicitly stated, the classification schemes that have been devised group at the top of the hierarchy those assemblages in which a similar technique of flaking stone has been used and a similar range (but not necessarily a similar frequency) of formal tools has been made. This is reflected in the range of cores and formal tools found and in the size and shape of the flakes produced. Thus the LSA is differentiated from the MSA because of the smaller size of flakes in the former and the incidence of blades and prepared cores in the latter and because the range of formal tools is different in each. At this level the range of formal tools is so distinctive that different typological schemes are necessary to describe MSA and LSA assemblages. In the differentiation of Complexes such as the Wilton and Oakhurst, however, while there are differences in the range of cores and the size and shape of flakes, the same typological scheme can be used for both although the relative frequencies and size and shape of the formal tools may differ markedly. Regional industries within the complexes represent variations on the same theme with differences in the relative frequencies of formal tools and raw materials that also occur through time within each industry and complex.

It is important to keep this hierarchy in mind when explanations for change are offered. Changes from one complex to another, for example, imply an adjustment in the size and shape requirements for tools that is reflected in the core sub-types, the size and shape of flakes and formal tools and in the range of formal tools made. Where such a change is widespread geographically and cross-cuts environmental boundaries, it is likely that new ideas were spread through social contact rather than developed *in situ* as a response to environmental stimuli. Within a complex, where contemporary samples in different regions have distinctive formal tool frequencies, these are likely to reflect functional adaptations to particular environmental stimuli or activities. If we wish to define the boundaries of socially related communities using archaeological data, however, we may be better advised to look for 'markers' that have no obvious adaptive advantage (Sampson 1980: 10). It is assumed that changes in the metric attributes of formal tool classes (where function is assumed to be constant) may be one such marker while others may be found in the layout and location of camp sites or distinctive non-lithic artefacts. These are questions of style that define groups by the way in which they do things rather than by the things they do (*pace* Yellen & Harpending 1972).

Terminology used in this review

With the plethora of names at large in southern African Stone Age terminology, it has been difficult to decide which are the most appropriate. Additional data have meant that the schemes of Sampson and Phillipson need modification, but there has not been sufficient inter-regional comparative analysis of the samples to allow new formal terms to be used with confidence. The culture stratigraphic record will therefore be described here under generalized descriptive headings that are similar to Sampson's Complexes and are hierarchically intermediate between LSA, or mode 5, and industries. They group informally together assemblages that are broadly contemporary and that have been made using a similar stone flaking technique that has yielded a similar range of core types, flake shapes and sizes and formal tools. This should not be seen as a new classification scheme, but rather as a useful device that will enable similar assemblages to be described together under one heading. I believe that we need many more observations and a standardized description method before a widely acceptable classification scheme can be proposed.

The headingsto be used in this paper are summarized in Table 1 together with their synonyms. The sample of assemblages on which this scheme is based consists of several hundred radiocarbon-dated units from over 160 sites in southern Africa (Figure 1, Tables 2 and 12). The dating of the units is summarized in regional groups in 1000-year periods in Table 2a demonstrating that Holocene assemblages are much more common than those of the late Pleistocene. It is therefore to be expected that we know more about Holocene assemblages than we do about those of the late Pleistocene. There is also a regional bias as the data in Table 2a show that some regions have more excavated sites than others and here the southern Cape receives some emphasis.

Late Pleistocene microlithic assemblages

Late Pleistocene microlithic assemblages are known from deposits dating between about 40,000 and 12,000 B.P. (Table 2a, b). The oldest samples are from Border Cave in KwaZulu (Beaumont *et al.* 1978) where they date to ca 38,000 B.P. and thereafter, in order of antiquity, are known from Rose Cottage in the eastern Orange Free State, Heuningneskrans in the eastern Transvaal (Beaumont 1981), Boomplaas, Nelson Bay, Melkhoutboom and Kangkara in the southern and eastern Cape (Deacon, H.J. 1976; Deacon, J. 1978), Elands Bay in the western Cape (J.E. Parkington, pers. comm.) and Byneskranskop in the south-western Cape (Schweitzer & Wilson 1979 and 1982). In Lesotho a similar assemblage has been described from Sehonghong (Carter 1978) and Beaumont (1978) has suggested that the Lower Occupation at Shongweni in Natal (Davies 1975) may relate to these assemblages as well. In Zimbabwe assemblages assigned to the Tshangula Industry (Cooke, C.K. 1963, 1971) date to between 25,000 and 12,000 B.P. but retain a number of formal tools that are typologically MSA as well as backed microliths and ostrich eggshell beads typical of the LSA; their precise relationship to contemporary assemblages to the south and north is not yet fully understood. Also contemporary with the late Pleistocene microlithic assemblages are non-microlithic ones from Namibia (Wendt 1976 and pers. comm.) that date between about 40,000 and 10,000 B.P.

Late Pleistocene microlithic assemblages, with the exception of those from Border Cave and Heuningneskrans where the bladelet element is not well represented, are characterized by the systematic production of small bladelets from standardized single-platform bladelet cores and by the occurrence in some quantity of bi-polar cores that have sometimes been so reduced by flaking that they are classed as *pièces esquillées, outils écaillés* or scaled pieces. Quartz and quartz crystal are preferred raw materials, but where these are not available other fine-grained rocks have been used.

Very few of the bladelets were retouched into formal tools, but where these are present they take the form of segments and casually backed bladelets. Scrapers tend to be

Table 1. The dating, characteristics and equivalent terms of Later Stone Age culture-stratigraphic units used in this review.

Culture-stratigraphic unit	Dating	Characteristics	Terms included
Late Pleistocene microlithic assemblages	40,000-12,000 B.P.	Clear evidence for the use of a microlithic flaking technique with single platform bladelet cores and bi-polar cores yielding numerous bladelets less than 25 mm long. Few formal tools, but including unstandardized scrapers, and a few backed microliths. Includes bored stones, polished bone points, ostrich eggshell beads, tortoiseshell bowls. Middle Stone Age formal tools absent.	Robberg Industry of the southern Cape (Klein 1974; Deacon, H.J. 1976; Deacon, J. 1978) and the late Pleistocene industry from Sehonghong (Carter 1978). Broadly equivalent to the Nachikufan I in Zambia (Miller 1969), the early microlithic from Matupi in Zaire (Van Noten 1977) and the Second Intermediate from Kisese in Tanzania (Inskeep 1962). Also includes the Early LSA from Border Cave, Heuningneskrans and Rose Cottage (Beaumont 1978). Excludes proto-LSA of Leopard's Hill in Zambia and the Tshangula in Zimbabwe because both have MSA-type formal tools.
Terminal Pleistocene/early Holocene non-microlithic assemblages	12,000-8000 B.P. Two sites in Namibia date between 20,000 and 12,000 B.P.	Microlithic flaking techniques absent or very rare. Cores generally large and irregular, flakes quadrilateral and larger than preceding and succeeding assemblages. Few formal tools, but scrapers predominate with large circular and end scraper forms. Bone tools numerous.	Smithfield A (Goodwin & Van Riet Lowe 1929), Early and Middle Smithfield of the Transvaal (Mason 1962), Albany Industry of the southern Cape (Klein 1974; Deacon, H.J. 1976; Deacon, J. 1978), the Oakhurst Complex encompassing the Oakhurst, Lockshoek and Pomongwe Industries (Sampson 1974), the Pomongwe Industry of Zimbabwe (Cooke, C.K. *et al.* 1966), the Early LSA at Apollo 11 and Pockenbank (Wendt 1972, 1976), the Bushman Rock phase of the Transvaal LSA (Plug 1978) and the early Holocene assemblage at Good Hope shelter in Natal (Cable *et al.* 1980).
Holocene microlithic assemblages	8000–a few hundred years B.P. One site over 10,000 B.P. in Zimbabwe and one in Namibia in 10th millennium B.P.	Fully developed microlithic tradition with numerous formal tools; backed microliths and small convex scrapers highly standardized; bone, shell and wooden artefacts; rock art clearly associated.	Wilton & Smithfield C (Goodwin & Van Riet Lowe 1929) Wilton Industry of the southern and eastern Cape (Deacon, J. 1972, 1974; Deacon H.J. 1976), Wilton Complex of Sampson (1974) incorporating the Coastal and Interior Wilton, the Khami and Matopo Industries (Cooke, C.K. *et al.* 1966; Cooke, C.K. 1971, 1979); the Inland Wilton of Humphreys (1979) and the LSA II of Lesotho (Carter 1978).

Late Holocene assemblages associated with pottery	Within the last 2500 years, with most sites more recent than 2000 B.P.	In some areas the pottery is associated with microlithic assemblages, but in the northern Cape the Transvaal and Natal pottery is associated with assemblages dominated by long end scrapers and few backed microliths. Metal (copper), glass beads and glass artefacts also occur. Another facies is expressed in coastal midden assemblages in which formal tools are either absent or very rare, but pottery is present.	*With microlithic tools:* Post-Climax Wilton (Deacon, J. 1972; Inskeep 1978a), Ceramic Wilton (Sampson 1974), LSA I in Lesotho (Carter 1978), late microlithic in Namibia (Wendt 1972; Wadley 1979), Sandy Bay industry of the south-western Cape (Rudner & Rudner 1954; Sampson 1974; Mazel & Parkington 1978) and the final expression of the Khami/Wilton/Matopo sequence in Zimbabwe (Cooke, C.K. 1971, 1979; Walker, N.J. 1980). *With long end scrapers*: Smithfield B of Goodwin & Van Riet Lowe (1929), Smithfield Complex (Sampson 1974), Smithfield P and N (Goodwin 1930; Van Riet Lowe 1936), the Later Smithfield of the Transvaal (Mason 1962). *Coastal*: Coastal Complex of Sampson (1974), Strandloper of Rudner (1968). *Macrolithic with pottery:* Brandberg Culture (Rudner 1957) and late macrolithic in Namibia (Wendt 1972; Wadley 1979).

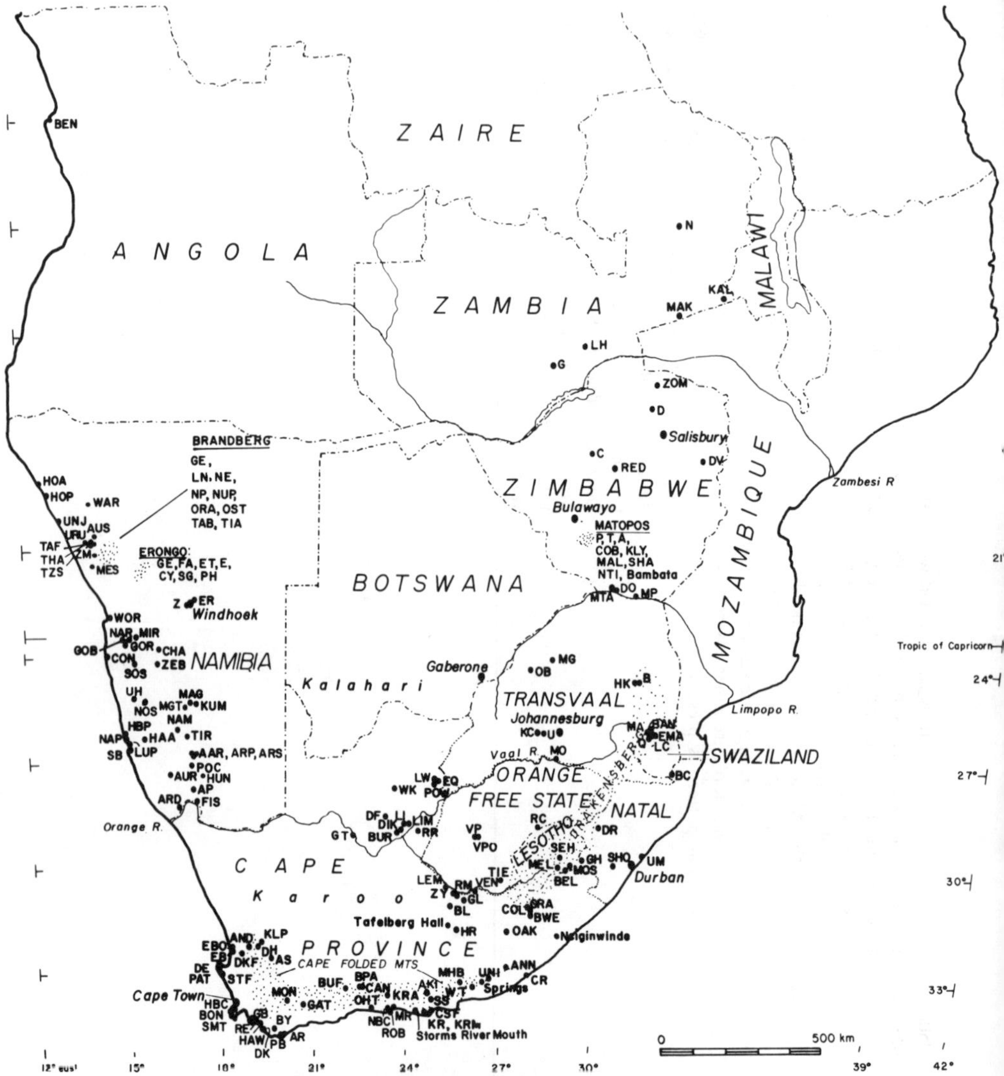

Figure 1. Location of southern African sites mentioned in the text. Key to abbreviations of site names can be found in Table 12.

small (less than 25 mm long) and relatively high (thick), but there are also rare and well made large forms reminiscent of the concavo-convex types described by Goodwin and Van Riet Lowe (1929) for the Smithfield A (Figure 2). In some assemblages (Nelson Bay and Boomplaas) a few of the single platform bladelet cores have polish on the striking platform consistent with their having been used as scrapers as well as cores (J.N.F. Binneman, pers. comm.). In all cases the percentage frequency of formal tools is less than 1% of the total assemblage and in general they are less standardized than those in Holocene assemblages. Most of the formal tool types made during the Holocene were made during the late Pleistocene but occur very rarely in the excavated samples. Where measurements on flake lengths have been

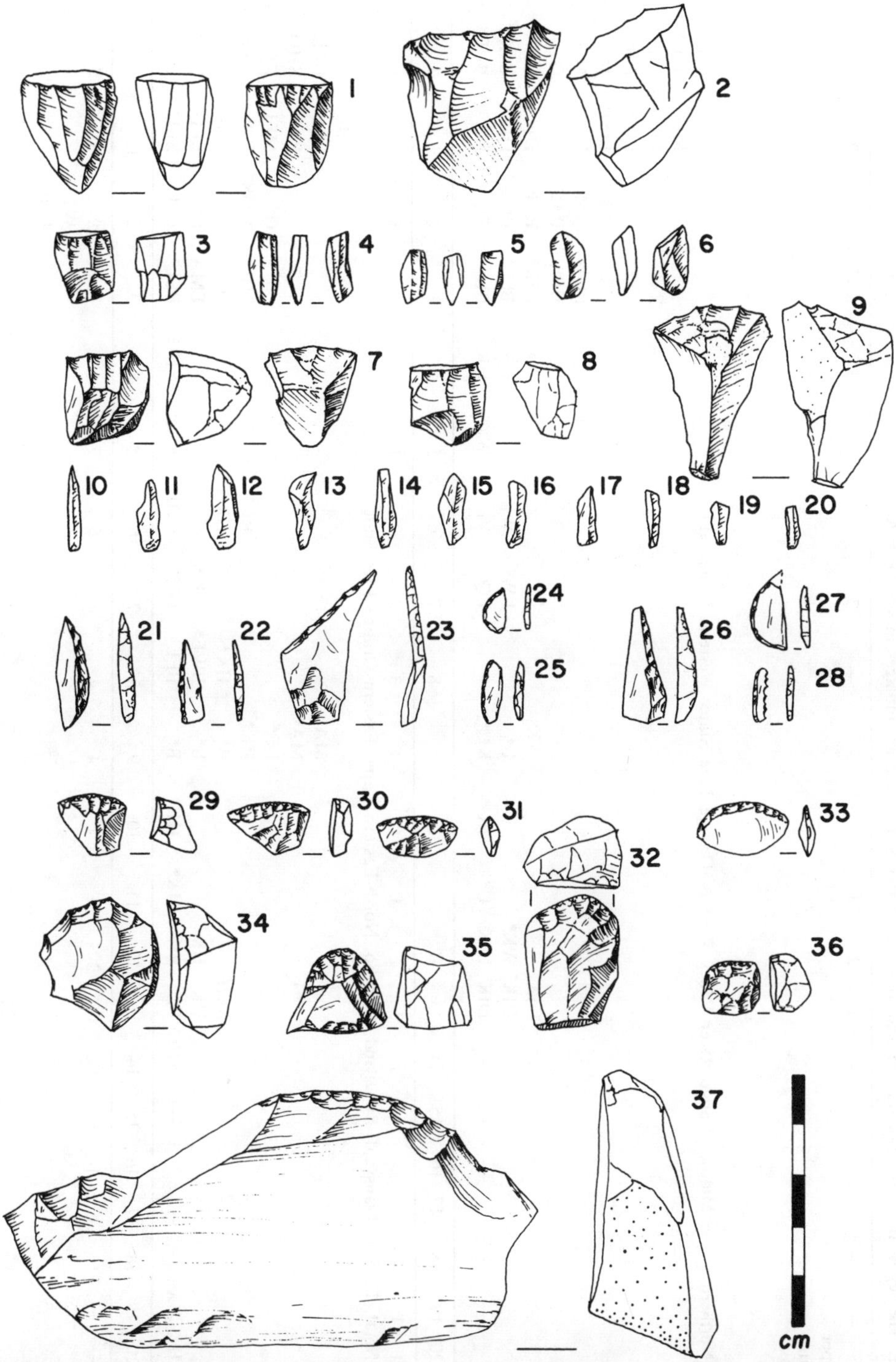

Figure 2. Stone artefacts from Late Pleistocene microlithic assemblages in the southern Cape. 1-5, 21-23, 29-30: Nelson Bay Cave YSL and YGL units; 6-20, 24-28, 32-37: Boomplaas Cave CL1 and CL3 units. Bladelet cores: 1-9; bladelets (unretouched): 10-20; backed microliths: 21-28; scrapers: 29-37.

Table 2a. The regional distribution of radiocarbon-dated Later Stone Age assemblages in southern Africa, and their industrial affinities.

	20	19	18	17	16	15	14	13	12	11	10	9	8	7	6	5	4	3	2	1	0
Number of sites per millennium	—	—	—	—	—	—	—	1	2	1	1	3	—	2	2	1	3	6	4	9	10
Northern Cape, Middle Orange River Valley Western Orange Free State, Western Transvaal																					
× 1000 years	—	—	—	—	—	—	—	DIK DIK DIK	WK* DIK	U*	KC*	WK° U* KC*	—	LW° EQ°	WK° VP	WK°	WK° LW° HR°	ZY° ZY° RR POW° HR° DIK° BL° BL°	WK° RM° RM° RM° POW° EQ°	WK° VP^{z} MO° MG^{z} LW° LIM° LIM° LI° DF^{z} DF^{z} BL°	ZY° ZY° TIE° TIE° TIE° TIE° OB^{z} MO MO GT GL° DIK2^{z} BUR°
B.P.	20	19	18	17	16	15	14	13	12	11	10	9	8	7	6	5	4	3	2	1	0
	1	1	—	—	—	—	1	3	2	1	1	4	3	2	3	1	1	4	5	7	6
Natal, Eastern Transvaal, Swaziland, Lesotho, North-Eastern Cape, Eastern Orange Free State																					
× 1000 years	HK	HK	—	—	—	—	*SHO*	*SEH* HK BC	HK B*	HK*	HK*	MA* MA* LC* HK* B* B* B*	HK* BWE* BEL	HK* GH*	RC° GRA* GRA* COL°	BAN	SHO	SHO Q EMA BEL°	Q MOS° GH° BWE BEL°	SHO SHO SEH° RC° MEL° DR° COL° BEL° BEL°	UM° SHO RC° MOS° COL° COL° BEL°
B.P.	20	19	18	17	16	15	14	13	12	11	10	9	8	7	6	5	4	3	2	1	0
	—	—	1	1	1	1	1	1	2	3	4	4	4	6	5	7	5	7	9	5	5
Southern and South-Eastern Cape																			WT° UNI° UNI° OAK° NBC° NBC°		

× 1000 years B.P.	20	19	18	17	16	15	14	13	12	11	10	9	8	7	6	5	4	3	2	1	0
	—	—	*NBC* *NBC*	*BPA*	*NRC*	*MHB*	*BPA*	*BPA*	*KRA* *BPA* *BPA*	NBC* NBC* NBC* MR* BUF*	NBC* NBC* NBC* NBC* NBC* MR* MR* MHB* BPA*	NBC* MR* MR* MR* KRA* BPA*	WT OHT* OHT* NBC* NBC* NBC* BUF*	OHT° MON° MHB° MHB* MR° MR° KRA* BUF* BUF°	NBC° NBC° NBC° MHB° KRA° CAN BPA°	NBC° NBC° NBC° MR° MHB° KRA° CSF° BUF° ANN°	WT° ROB° NBC° NBC° KRM° KRM° KR°	ROB° ROB° OHT° NBC° NBC° NBC^{Z} MR° CAN BUF°	NBC^{Z} NBC^{Z} NBC^{Z} MHB° KRM° KRM° KR° CR^{Z} ANN^{Z}	ROB° OAK° OAK° NBC^{Z} NBC^{Z} BPA° BPA° BPA° BPA°	SS^{Z} OAK^{Z} NBC^{Z} GAT^{Z} CR^{Z} CR^{Z} ANN^{Z} ANN^{Z}
	—	—	—	—	—	—	—	—	2	1	1	3	1	1	3	2	1	4	6	16	8
Western and South-Western Cape	—	—	—	—	—	—	—	—	*EB* *BY*	EB*	EB* EB* EB* EB*	EB* EB* BY* AS	EB* EB*	EB*	RE° BY* BY° BY° AS	KLP° AS	BON°	PAT° GB^{Z} EB° EB° EB° BY° BY° BY°	PB^{Z} GB^{Z} GB^{Z} EB^{z} DK° DK° DE BON°	STF^{Z} SMT^{Z} SMT^{Z} PB^{z} PB^{z} PAT^{Z} HBC^{Z} HBC^{Z} HAW^{Z} HAW^{Z} EBO^{Z} EB° EB° DKF° DKF° DK^{Z} DK^{Z} DK^{Z} DK^{Z} DK° DH° DE^{Z} DE° DE BY° AR AND°	PAT^{Z} PAT^{Z} EBO^{Z} EBO^{Z} EB° DKF° DKF° DH° DH° DH° DH° DH° DH° BY° BY° AS^{Z} AND°

Table 2a (*Contd.*)

× 1000 years B.P.	20	19	18	17	16	15	14	13	12	11	10	9	8	7	6	5	4	3	2	1	0
	—	—	—	—	—	—	—	—	2	1	3	3	3	3	1	1	3	—	4	4	2
Zimbabwe																					
	—	—	—	—	—	—	—	—	T* COB* P		NTI* DV° COB* COB*	T* P* P* NTI*	T° SHA° P°	SHA° P° P° NTI°	NTI°	KLY°	MAL° COB° A°	—	ZOM T° DV° D°	T° DV° DO° C°	MTA° MTA° MTA° MP° MP°

× 1000 years B.P.	20	19	18	17	16	15	14	13	12	11	10	9	8	7	6	5	4	3	2	1	0
	—	2	1	—	—	—	1	1	2	1	1	1	2	2	6	7	2	6	12	11	37
South West Africa / Ņamibia																					ZM^{Z} ZM^{Z} ZM^{Z} ZM WOR^{Z} WOR^{Z} UH° TIR° TIR° THA° THA° TAB^{Z} TAB^{Z} TAB^{Z} TAB^{Z} SOS^{Z} SG° POC° POC° PH° ORA° ORA° NUP NP° NOS^{Z} NOS^{Z}

× 1000 years B.P.	20	19	18	17	16	15	14	13	12	11	10	9	8	7	6	5	4	3	2	1	0
																					NAP
																					MES°
																					MAG°
																					MAG°
																					LUP
																					KUM°
																			WAR		KUM°
																			UNJ		HUN
																			TIA°		HOP
																			SB°		HOA^{z}
																			SB°		HBP
																			SB°		GOR
																			OST°	URU^{z}	GI°
																			OST°	NAR	GI°
																			MES°	MIR°	FIS^{z}
																			MES°	MES°	FIS^{z}
																			MES°	GOR^{z}	FA°
																			LNU°	GE°	ET°
																			LN°	GE°	ES°
																			HAA°	ER°	CON
															POC°				HAA°	E°	CON
															MIR°	Z°			GI°	AUR	CHA°
															MIR°	TZS°			GE°	ARD^{z}	AUS°
															MIR°	NAM°		TAF°	GE°	ARD^{z}	ARA°
															GI°	MIR°		SG°	FA°	ARD^{z}	ARP
								AP*						NAM°	CHA°	MIR°		PH°	FA°	ARD^{z}	AP^{z}
		POC*						AP*					NAM°	NAM°	AP°	MAG°		MGT°	FA°	ARD^{z}	AP^{z}
		POC*	AP*					AP*	GOB				MIR°	AP°	AP°	CY°	SG°	LN°	FA°	AP°	AAR°
	—	AP*	AP*	—	—	—	AP*	AP*	AP*	ZEB	AP	AP°	MIR°	AP°	AAR°	CHA°	LN°	GE°	FA°	AP°	AAR°
Total sites per millennium, all regions:	1	3	2	—	1	1	3	6	12	8	11	18	13	16	20	19	15	27	40	52	68

Dates without a symbol have uncertain associations
Italics: Late Pleistocene microlithic assemblages
° Holocene microlithic assemblages
*Terminal Pleistocene/early Holocene non-microlithic assemblages
^{z}Late Holocene non-microlithic assemblages
Dates associated with pottery are summarized in Table 9
For abbreviations and references see Table 12.

Table 2b. Dated Late Pleistocene assemblages from southern Africa with Later Stone Age affinities, including the 'Early LSA', the Tshangula and non-microlithic/non-MSA assemblages. For abbreviations and references see Table 12.

× 1000 years	BC	—	—	—	—	>RC HAA°	—	BC	BC	BC	BC	BC	BC	BC	POC° (HK) BC	—	RC POC°	—	BC	—	RED* RC	(HK) HK	SHO HK	NOS° (HK)	ZOM* P* HK BPA° BPA°	ZOM* HK EB°	—	D*	—	—	P*	P*	ZOM* ZOM*
B.P	45	44	43	42	41	40	39	38	37	36	35	34	33	32	31	30	29	28	27	26	25	24	23	22	21	20	19	18	17	16	15	14	13

> Non-finite date
° Non-microlithic
* Tshangula
( ) Amino acid calibrations

Note: Assemblages other than Tshangula ones dating to between 19,000 and 12,000 B.P. are included in Table 2a.
Border Cave (BC) dates from 45,000 to 32,000 B.P. are almost all from the 1WA layer. The best estimate of its age is ca 38,000 B.P. (Beaumont *et al*. 1978).

Table 3. Dimensions of untrimmed flakes from the Boomplaas Cave sequence of the last 22,000 years (after Deacon, J. in prep.).

Member	f	Mean	STD deviation	Range	Approximate date B.P.
a) Length					
DGL	316	18,8	8,65	6–80	1500–1700
BLD	210	20,5	8,33	10–57	2000
BLA	131	19,0	7,43	8–55	6500
BRL	149	26,1	11,21	9–64	9000–10,500
BRL 7	62	24,7	10,43	9–64	10,600
CL 1–2	134	25,5	13,20	9–78	12,000–12,500
CL 3–4	133	20,8	8,49	10–62	14,200
GWA	52	20,5	6,92	11–38	?17,000
LP	144	17,6	8,05	7–50	?20,000
LPC	413	23,3	12,44	6–80	22,000
b) Height					
DGL	316	4,5	2,88	1–30	1500–1700
BLD	210	4,4	2,44	1–18	2000
BLA	131	4,0	1,92	1–12	6500
BRL	149	5,8	3,10	1–19	9000–10,500
BRL 7	62	5,1	2,69	1–14	10,600
CL 1–2	134	5,3	3,92	1–26	12,000–12,500
CL 3–4	133	3,5	2,57	1–15	14,200
GWA	52	3,7	2,04	1–9	?17,000
LP	144	4,4	2,84	1–18	22,000
LPC	413	6,3	4,12	1–27	22,000
c) Width/length ratio (× 100)					
DGL	316	94,4	33,72	33–283	1500–1700
BLD	210	84,4	27,98	29–188	2000
BLA	131	85,8	26,00	33–238	6500
BRL	149	97,9	33,15	38–200	9000–10,500
BRL 7	62	91,6	40,46	32–206	10,600
CL 1–2	134	81,6	39,01	22–193	12,000–12,500
CL 3–4	133	67,4	35–42	25–156	14,200
GWA	52	67,7	35,77	18–156	?17,000
LP	144	80,3	36,57	18–225	?20,000
LPC	413	90,31	36,56	25–250	22,000

taken (Table 3) the late Pleistocene samples are shorter than those of the preceding MSA and are also shorter and more blade-like than those from terminal Pleistocene/early Holocene assemblages (Beaumont 1978, 1981).

Although the frequency of formal tools is low, the LSA character of these late Pleistocene microlithic assemblages is reflected in several associated lithic and non-lithic artefacts that are directly comparable with items in Holocene assemblages and the ethnographic present. The dating and location of these finds are summarized in Tables 11 and 12, but they include a small bored stone (too small for use as a digging stick weight), ostrich eggshell beads and a polished bone point from Border Cave at ca 38,000 B.P. and ostrich eggshell water containers, tortoiseshell bowls, beads and polished bone points/linkshafts from ca 20,000 B.P. and 14,000 B.P. at Boomplaas.

Dating

The transition from clearly MSA to clearly LSA assemblages is not easy to pinpoint in time, but in the rare instances where late Pleistocene microlithic assemblages are found stratified between MSA and fully developed LSA horizons the shift takes place between 40,000 and 30,000 B.P. (Beaumont *et al.* 1978; Deacon, H.J. 1979). In Zimbabwe the situation may have been more complex with MSA characteristics persisting in the Tshangula to ca 12,000 B.P. and in Namibia MSA assemblages are dated to ca 26,000 B.P. (Wendt 1976) at Apollo 11 where they are associated with painted slabs. The overall picture is therefore one of considerable variability in timing, but once the shift towards microlithic toolmaking was made at a site or in a region, there is no evidence for the reintroduction of MSA techniques.

At the other end of the time scale the shift to non-microlithic toolmaking seems to have occurred over a wide geographic area over a period of about 1000 years around 12,000 B.P. There are no assemblages of late Pleistocene microlithic character dated more recent than 12,000 B.P. with the exception of one level at Heuningneskrans where it is dated within the 12th millennium (Beaumont 1981).

The late Pleistocene microlithic assemblages are broadly contemporary with the latter half of the last glacial cycle and the warmer intervals within it after 40,000 B.P., and with the last glacial maximum and subsequent warming of the southern oceans after 14,000 B.P. There is thus no convincing correlation between this toolmaking tradition and a particular climatic regime except that the entire period was somewhat cooler than at present. The dates for the late Pleistocene microlithic assemblages are summarized in Table 2.

Evidence for subsistence

Climatic conditions during the time when late Pleistocene microlithic assemblages were made in southern Africa were between 3° and 5° C cooler than they are today. The time period under discussion (40,000-12,000 B.P.) falls within oxygen isotope stages 3 and 2 (Shackleton & Opdyke 1976) and the coolest conditions pertained around 18,000 B.P. Data on past rainfall levels during this time period are limited both geographically and temporally (see Butzer *et al.* 1978; Lancaster 1979b; Vogel & Visser 1981), but those that are available, together with simulation studies generated for the last glacial maximum (Gates 1976; Manabe & Hahn 1977), suggest that while most of southern Africa probably received less precipitation at 18,000 years ago than at present, some areas, notably the northern Cape and Botswana, probably had a higher effective precipitation. This situation may not have pertained throughout oxygen isotope stage 2, however (Heine 1978a), and precipitation levels during time periods leading up to and following the glacial maximum may have been different from those of the maximum itself (Williams 1978).

Another significant effect of the expansion of high latitude ice sheets was the global lowering of sea level. At the maximum extent, the coastline would have been between 100 and 200 km south of its present position off the southern tip of Africa where the continental shelf is at its widest, but only a few km offshore along the Transkei/Natal coast where the shelf is narrower. With a drop in sea level of between −90 and −130 m, however, the coastal cliffs would have been very steep (Dingle & Rogers 1972; Vogel & Visser 1981) and few sites that are on the coast at present would have been near to the seashore between 40,000 and 14,000 years ago. However, as the sea level rose rapidly after 14,000 B.P., they would have come closer to the coast.

The effect of lower temperature and a different rainfall from that of the present differed from one ecozone to another and is most clearly reflected in the faunal remains from occupation sites which are assumed to give a reasonably good indication of the range of animals available in the vicinity of the site. Assemblages in the western, southern and eastern Cape, for example, show more differences between late Pleistocene and Holocene faunas than do assemblages elsewhere because the combination of cooler temperatures and the exposure of the continental shelf led to the grass element in the vegetation becoming more dominant that at present in the modern fynobos. The presence of springbok, ostrich, wildebeest, quagga, bontebok and warthog at Nelson Bay argues for grassy plains in the vicinity of the site which are not present today. A similar increase in grass cover is apparent inland in the Cape Folded Mountains at Melkhoutboom and Boomplaas (Deacon, H.J. 1972, 1976, 1979; Klein 1972a, b; 1974, 1978b, 1980). While this somewhat crude differentiation between grassland and shrubland environments may mask more subtle differences in site environs, it is clear that most of the shifts noted in the faunal remains at occupation sites are explicable as due to changes in habitat potential for the animals rather than due to changes in cultural preferences of the hunters.

A feature of late Pleistocene faunal assemblages in the Cape ecozone is the presence of a number of large mammals that became extinct at the end of the Pleistocene or early in the Holocene, probably partly as a result of climatic and habitat changes, and partly because of the role of man (Klein 1980: 272). These include the giant buffalo, *Pelorovis antiquus*, the giant Cape horse, *Equus capensis*, the giant alcelaphine, *Megalotragus priscus*, a large warthog, *Metridiochoerus* sp. and two springbok, *Antidorcas bondi* and *A. australis.* Klein (loc. cit.) suggests that Stone Age people in the Cape intensified their hunting of the large plains game at the end of the Pleistocene when habitat changes made these species less numerous and, together with the technological innovation of the bow and arrow, some species were driven below a critical threshold and could not maintain their reproductive capacity.

Outside the Cape ecozone late Pleistocene faunal assemblages are less common. Two older samples, from Border Cave (ca 38,000 B.P.) and from the Tshangula levels at Redcliff in Zimbabwe (30-20,000 B.P.), however, indicate that cooler conditions allowed some species to increase their geographical range (Klein 1977b, 1978a; Beaumont *et al.* 1978).

No significant differences have been noted in the range of animals hunted by MSA and late Pleistocene LSA people, but when late Pleistocene faunal assemblages, regardless of cultural association, are compared with Holocene ones there are differences both in the range of animals (due to differences in availability) and in the relative frequency of gregarious and non-gregarious species (apparently influenced by changes in preference and hunting methods (Klein 1980, 1981c)). There are no striking differences apparent between MSA and LSA hunting methods and capabilities judging from mortality profiles of small bovids (Klein 1981c), but LSA people showed greater competence and were able to kill more dangerous animals (for example, bushpig) than were their MSA counterparts, enabling them to increase the range of animals hunted. Klein (1976a, 1979a) has also argued that Holocene people were more efficient in fishing and fowling than their MSA counterparts but the lack of coastal sites with evidence for fishing and fowling as well as shellfish collecting between ca 90,000 and 12,000 years ago makes it impossible to test whether this efficiency was coincident with the advent of the LSA or whether it developed gradually throughout the late Pleistocene.

Poor preservation is probably responsible for the lack of plant remains at most late Pleistocene sites with microlithic assemblages. An exception is Border Cave where grass bedding material, leaves, twigs and seeds are present in the units dating to ca 38,000 B.P. (Beaumont 1978, 1980). There are some 95 leaves from 15 species, of which 14 leaves from 5 species are known to be edible (*Strychnos spinosa, Vitellariopsis marginata, Encephalartos lebomboensis, Bridelia micrantha* and *Encephalartos vollosus*), and two other species are known to be of medicinal value. Of the 55 seeds from 13 species, 26 from 5 species are edible and 7 from 2 species are medicinal (*Diospyros dichrophylla, Pappea capensis, Ricinus communis, Ekebergia capensis, Harpephyllum caffrum, Ziziphus mucronata, Encephalartos* spp., *Vangueria infausta* and an unidentified cucurbit). Carbonized plant materials have also been found beneath stones in late Pleistocene levels at Boomplaas and Elands Bay but they do not include edible species.

The limited data suggest that the hunting of large migratory plains game was common in the Cape and Transvaalian ecozones during the late Pleistocene and that plant food collecting was contributing to the diet. Ethnographic analogy suggests that hunting of large migratory game is generally associated with relatively large groups of people to facilitate game drives and the sharing of meat (Deacon, H.J. 1976). This could have meant larger territories per group and therefore lower archaeological visibility of sites as the population was more dispersed, and may account in part for the small number of known sites associated with microlithic assemblages in the late Pleistocene.

Terminal Pleistocene/Early Holocene non-microlithic assemblages

South of the Zambesi a group of assemblages occurs post-dating the Tshangula in Zimbabwe and the late Pleistocene microlithic assemblages south of the Limpopo and pre-dating the Holocene microlithic assemblages in Namibia. They are characterized by the absence or extreme rarity of bladelet cores and the associated backed microliths, and the presence of a variety of medium to large-sized scrapers with a mean length greater than 20 mm (Figures 4 and 11). Other formal tools are rare. Cores tend to be casual multi-platform types and untrimmed flakes are quadrilateral or irregular in shape and somewhat longer than those in preceding and succeeding assemblages (Table 3). A feature of some assemblages is the sophisticated range of associated bone tools. The coastal sites of Elands Bay and Nelson Bay in particular have finely polished bone 'fish gorges', points, spatulas and grooved bone beads (Deacon, J. 1978; J.E. Parkington, pers. comm.) and the bone tools from Bushman Rock Shelter in the eastern Transvaal include flaked bone artefacts as well (Plug 1978). Artefacts known in the ethnographic present and associated with stone artefact assemblages of the terminal Pleistocene/early Holocene include bored stone digging stick weights at Tshangula (N.J. Walker, pers. comm.) and Matjes River (Louw 1960), an eyed matting needle from Pomongwe (N.J. Walker, pers. comm.) and engraved stones from the nearby site of Bambata (N.J. Walker, pers. comm.) and from Wonderwerk in the northern Cape (Thackeray, A.I., *et al.* 1981) (Figure 3) where one is associated with charcoal dated to 10,000 B.P. (Pta-2786). Deliberate burials

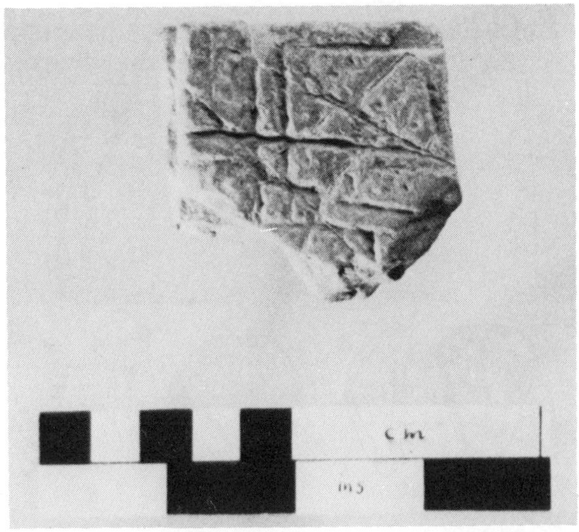

Figure 3a. Engraved stone from Bambata Cave, Zimbabwe. Photo N.J. Walker.

Figure 3b. Engraved stone from Wonderwerk Cave, north-western Cape. Photo Anne I. Thackeray.

are dated to this period at Matjes River (Louw 1960) and Nswatugi in Zimbabwe (Walker 1980).

The scrapers (often the only formal tools in these assemblages) are consistently larger than those of both the Holocene and the late Pleistocene and show some changes in size and shape through time within the 12,000-8000 B.P. period. In earlier assemblages (where these have been dated, for example at Boomplaas, Nelson Bay

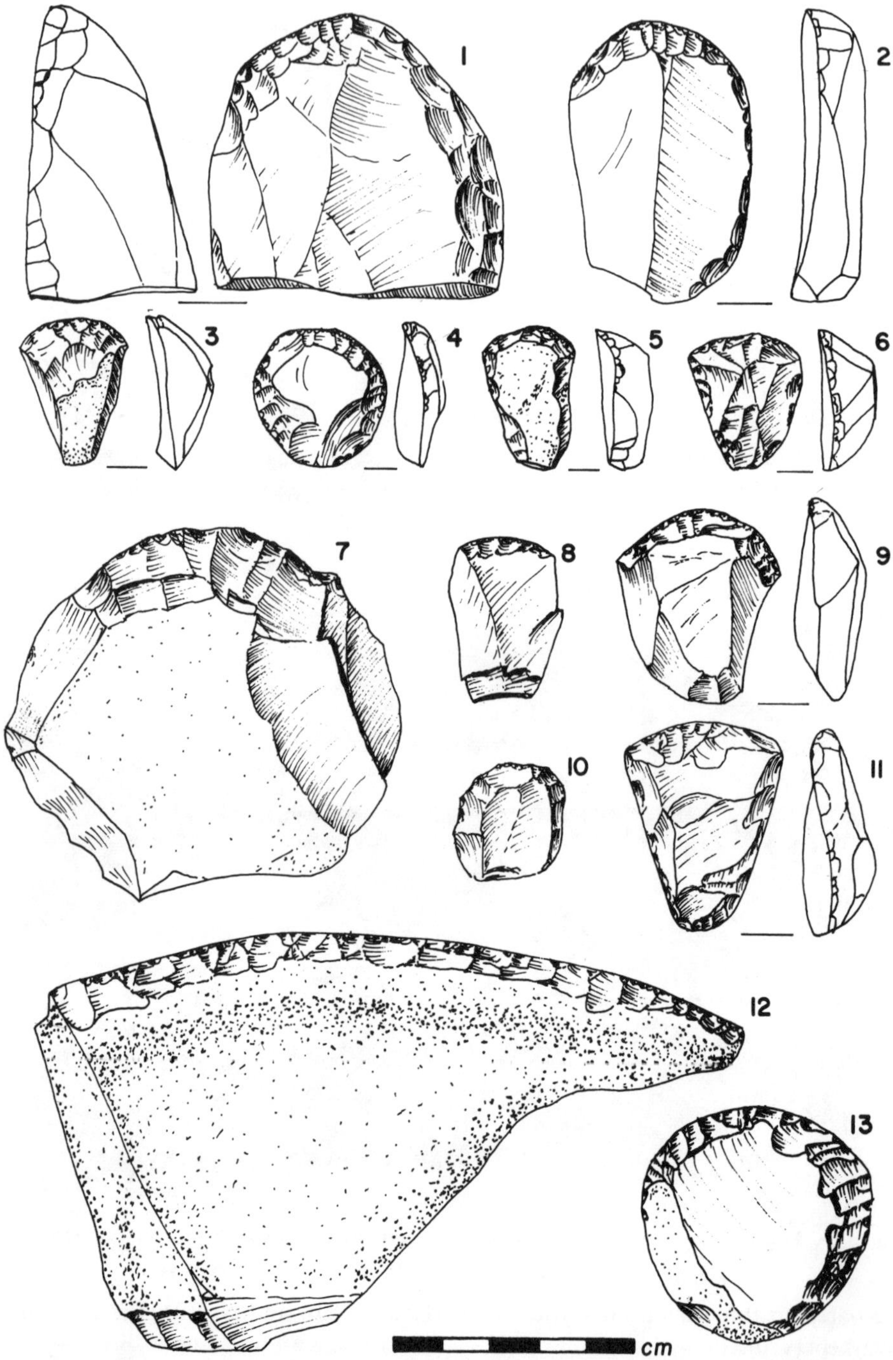

Figure 4. Artefacts from non-microlithic assemblages of the terminal Pleistocene/early Holocene in southern Africa. 1-3: Nelson Bay Cave; 4-6, 9: Boomplaas Cave; 7, 8, 10: Pomongwe (after Cooke, C.K. 1963: Fig. 16); 11: Lemoenfontein 10-11 (after Sampson 1970: Fig. 3); 12: Kopje 18, middle Orange River valley (after Sampson 1970: Fig. 23); 13: Zeekoegat 13, lower level (after Sampson 1967b: Fig. 12). All these artefacts have been classified as scrapers.

and Elands Bay) the classic Smithfield A concavo-convex forms and large side scrapers (Figure 4) that are also occasionally found associated with late Pleistocene microlithic assemblages, are relatively common (Deacon, J. 1978 and in prep.). In the more recent horizons, however, elongated end scraper forms reminiscent of Goodwin and Van Riet Lowe's (1929) Smithfield B are numerically prominent and are particularly characteristic of assemblages dating between 9000 and 7000 B.P. in the southern Cape. They are made on relatively thick flakes with scraper retouch on the curved working end and steep, almost adze-like, retouch down one or both sides (Figure 4:6, 11). Some specimens from undated Lockshoek surface sites in the northern Cape (Sampson 1972:184, 1974) are steep-sided but have natural facets rather than retouch down the sides (Figure 4:3, 5). Sampson has called this class 'frontal scrapers' (1972:186) to distinguish them from the flatter more regular 'end scrapers' that date to the end of the Holocene in the northern Cape.

The reason for the absence or rarity of backed microliths and bladelets in these terminal Pleistocene/early Holocene assemblages is not clear, but it may mean that, if the bow and arrow was in use at that time, bone points were used in place of stone armatures. Backed microliths continue to occur in assemblages of this time period north of the Zambesi.

Dating and distribution

The distribution of dated terminal Pleistocene/early Holocene non-microlithic assemblages is summarized in Table 2 where it can be seen that the majority date between about 12,000 and 8000 B.P. The general lack of dated assemblages in the northern Cape is the result of the absence of stratified and datable deposits in this region where deep cave deposits are rare. However, the occurrence of numerous open-air assemblages assigned to the Lockshoek Industry (Sampson 1974, 1980) demonstrates that the same range of non-microlithic assemblages occurs there and it is likely that they date to the same time range.

Non-microlithic assemblages of this time range have not been reported north of the Zambesi. The fact that they are also not known north of the Matopos in Zimbabwe may be an artefact of sampling as few sites of this time range are known. The one sequence that covers this time range is from Diana's Vow (Cooke, C.K. 1979a) and here no larger scrapers are in evidence in the early Holocene. This limited distribution endorses the conclusions of Goodwin and Van Riet Lowe (1929) who saw the Smithfield as an essentially local development in South Africa, but recognition of the Pomongwe Industry in the Matopos and the Albany or Oakhurst Industry in the southern Cape has extended its distribution beyond the hornfels region to which they thought it was restricted. The earlier dating of non-microlithic assemblages from Pockenbank and Apollo 11 in Namibia (Wendt 1974; Freundlich *et al.* 1980) is not easy to assess at the moment because detailed descriptions of the artefacts are still in preparation.

The appearance of non-microlithic assemblages at the end of the Pleistocene is fairly abrupt throughout southern Africa within the 13th millennium B.P. (with the exception of Namibia and possibly Heuningneskrans (Beaumont 1981)), but their replacement by Holocene microlithic assemblages is spread over a wider time period between ca 10,000 and ca 7000 B.P. Of interest here is the fact that the earlier Holocene microlithic assemblages are located to the north in northern Zimbabwe,

Namibia and in the northern Cape at Wonderwerk (Cooke, C.K. 1979; Wendt, pers. comm.; Thackeray, pers. comm.) while the more recent ones are in the south as, for example, at Byneskranskop (Schweitzer & Wilson 1979, 1982). Should this pattern be confirmed by more detailed data and more dates it may be possible to test the competing hypotheses that either the Holocene microlithic toolmaking tradition was introduced from north of the Zambesi and spread southwards (Goodwin & Van Riet Lowe 1929; Beaumont & Vogel 1972a,b) or that it developed *in situ* as a response to changing environmental conditions (Phillipson 1976, 1977). In the southern Cape the observation that the beginning and ending of the non-microlithic tradition coincides with habitat changes as reflected in the fauna associated with the artefacts, has led to the suggestion that the artefact changes were stimulated by adjustments to environmental shifts (Deacon, H.J. 1972, 1976; Klein 1972a, b, 1974), but no source for the artefact changes has been suggested. Phillipson's view is that the appearance of microlithic toolmaking (mode 5) was an indigenous innovation as a response to climatic amelioration which changed faunal distributions (Phillipson 1976:204-5, 1977:55). More specific data on the dating of environmental changes in different regions of southern Africa are sorely needed to resolve this issue, but it is likely that the picture will be a complicated one and one solution may not be applicable everywhere.

Evidence for subsistence

The majority of the occurrences grouped together here are associated with food remains different from those of earlier assemblages at the same sites. On the coast this is particularly noticeable in the appearance of quantities of marine shell, fish bones and the bones of marine birds and mammals. Away from the coast it is evident in a smaller range of large gregarious game than is present in older samples, but a larger percentage than in Holocene samples. The coastal pattern is a direct result of the rising sea level at the end of the Pleistocene bringing present-day coastal sites within a few km of the coastline by 12,000 B.P. Even inland, for example at Boomplaas which is now some 80 km from the coast, marine shells are more common after ca 13,000 B.P. than they were before.

Marine shells at early Holocene sites show evidence for cooler water and changes in coastal morphology as well as providing information on exploitation patterns. At Nelson Bay the black mussel, *Choromytilus meridionalis*, is present between 12,000 and 8000 B.P., but is absent thereafter and is not found in the area today (Klein 1972a). At Elands Bay on the west coast, however, the black mussel is common on the coast today, but is absent before 10,000 B.P. (Parkington 1976, 1977a, b). The early presence and later absence of the black mussel at Nelson Bay in the south coast is likely to be a reflection of the waters of the offshore Agulhas current becoming warmer in the early Holocene for this species prefers cooler waters, but its presence at Elands Bay only after 10,000 B.P. may rather reflect the changing configuration of the coast or that the site was occupied in the summer months prior to 10,000 B.P. because black mussels have an increased risk of toxicity at that time of year and would have been avoided (Parkington 1976, 1977a, b).

The exploitation of marine resources in the terminal Pleistocene and early Holocene was noticeably more efficient than it was during the Middle Stone Age (Voigt 1973; Klein 1976a) and presumably this expertise was built up during the last glacial cycle when the sea level was lower. These occupation sites are now under water, however, and cannot be studied. Mussels and limpets are best represented at most sites, but whelks and rock lobster also occur. Flying birds such as gannets and cormorants are common, as are seals, dolphins and fish, including estuarine species at Elands Bay. Analysis of the age at death of seals from early Holocene levels at Elands Bay suggests that the site was occupied during the winter and spring (Parkington 1976), but a wider seasonal pattern is evident at Nelson Bay Cave (Parkington, pers. comm.). Two *Patella tabularis* shells from Nelson Bay middens dating to ca 9000 B.P. were submitted for oxygen isotope analysis to establish water temperature at the time of death of the shellfish and results indicate that both were collected in the winter or early spring (Shackleton 1973: 138-9).

Mammalian faunas dating to the terminal Pleistocene/early Holocene at Nelson Bay and Elands Bay include the most recent occurrences of extinct forms such as the Cape horse, giant buffalo and an extinct springbok (Klein 1980). Inland these species do not occur later than 12,000 B.P. at Boomplaas and are absent from terminal Pleistocene deposits at Bushman Rock Shelter and Heuningneskrans in the Transvaal, Apollo 11 in Namibia and Pomongwe in Zimbabwe. While these latter samples include only species present in their respective ecozones in historic times, there is an emphasis towards larger rather than smaller bovids when compared with mid- and late Holocene samples (Brain, cited in Sampson 1974: table 45; Plug 1978, 1981; Thackeray, J.F. 1979; Beaumont 1981). A higher incidence of larger bovids when compared with mid- and late Holocene samples is also present in terminal Pleistocene/early Holocene assemblages in the southern and eastern Cape (Hewitt 1931a; Deacon, H.J. 1972, 1976; Klein 1974, 1980) as can be seen in the data from Boomplaas summarized in Table 4. The terminal Pleistocene/early Holocene faunal assemblages can therefore be seen as intermediate between those of the late Pleistocene and those of the Holocene.

As in the late Pleistocene occurrences, plant food remains are rarely preserved in deposits dating between 12,000 and 8000 B.P. Notable exceptions are sites where hardier

Table 4. Changes in the size of bovids hunted at Boomplaas Cave over the last 20,000 years (after Klein 1978b). The figures show a shift from large gregarious equids and alcelaphines to smaller non-gregarious antelope from the Late Pleistocene to the late Holocene.

	Wilton	Albany	Robberg	Late Pleistocene
Mountain reedbuck, vaalribbok, klipspringer, grysbok/steenbok	38 (81%)	62 (65%)	17 (22%)	6 (14%)
Equids and alcelaphines	9	33	59	38

Selected chi-square values	chi-square	p (two-tailed)	
Wilton vs Albany	3,65	,10–,05	(significantly different at one-tailed probability of 0,5–0,25)
Wilton vs Robberg	40,14	,001	
Wilton vs Late Pleistocene	41,04	,001	
Albany vs Robberg	31,22	,001	
Albany vs Late Pleistocene	32,01	,001	
Robberg vs Late Pleistocene	1,33	,3 – ,2	(not significant)

seeds rather than the softer residues of plant foods have accumulated, as at sites in the Matopos in Zimbabwe (Cooke, C.K. 1963; Walker, N.J. 1980), at Zebrarivier in Namibia (Wendt, pers. comm.), and Bushman Rock shelter (Plug 1978, 1981), Heuningneskrans (Beaumont 1981: 142) and Kruger Cave (Friede & Pienaar 1974) in the Transvaal. Marula seeds, *Sclerocarya caffra*, are the most common in the Matopos and Transvaal samples and are still widely eaten today both for the fruits and the kernels. At Bushman Rock Shelter many of the seeds were broken to extract the protein-rich kernels and a minimum of 1 422 seeds is represented by over 7000 fragments (Plug 1978: table 56). Other fruits represented by seeds at the site include stamvrugte (*Bequaertiodendron magaliesmontanum*), the sour plum *Zimenia caffra* that has tasty fruits and a kernel that provides an oil used for softening hides and leather (Smith 1966:447), and the cycad, *Encephalartos* sp., some species of which are known as 'broodboom' or 'bread tree'. Indigenous people are known to remove the pith of the stems, bury them for about two months to aid fermentation and softening of the fibres, and then to knead them into a bread (Smith, C.A. 1966:179). The Kruger Cave samples include marula and sour plum as well as *Strychnos pungens* or monkey orange and *Bauhinia* sp. of which one species is the widely eaten gemsbok bean. At least 25 species of edible fruits have been recorded at sites in the Matopos (N.J. Walker, pers. comm.). In Namibia the most common seed is that of the nara melon at Zebrarivier in a horizon dated to between 10 and 11,000 B.P. Modern hunter-gatherers use the melon and the seeds which can be dried and stored for long periods.

The available evidence for the subsistence pattern associated with terminal Pleistocene/early Holocene non-microlithic assemblages in southern Africa shows a gradual trend away from the hunting of large grazers towards a higher percentage of smaller non-gregarious antelope (Table 4). Marine resources at coastal sites are well represented and show that fish, marine mammals and shellfish were used in quantity. There is evidence for the use of a variety of edible fruits and nuts in the Transvaalian ecozone and for use of the Nara melon in Namibia, and circumstantial evidence for the gathering of underground plant foods in the finding of bored stones suitable for use as digging stick weights, together with the reamers for their manufacture. Seasonal occupation of coastal sites is suggested in the evidence for winter and spring occupation at Elands Bay and regular transhumance or contact with coastal people is indicated by finds of marine and estuarine shells inland. An increase in the density of sites is suggested by the increase in the number of radiocarbon dates after 12,000 B.P. (Table 2). Whether this is a 'real' increase in terms of a higher density of population or simply the result of the reorganization of people into smaller groups with a higher archaeological visibility is not yet clear, but its coincidence with a change in the artefact record is tantalizing and needs investigation.

Holocene microlithic assemblages

Holocene microlithic assemblages are characterized by a wider range of formal stone tools than appear at any other time during the LSA and by a return to a microlithic flaking technique (see Figure 5). It should be remembered, however, that during the Holocene and during the late Pleistocene when microlithic tools were made that not all the artefacts in an assemblage were made with this flaking technique. Indeed, the

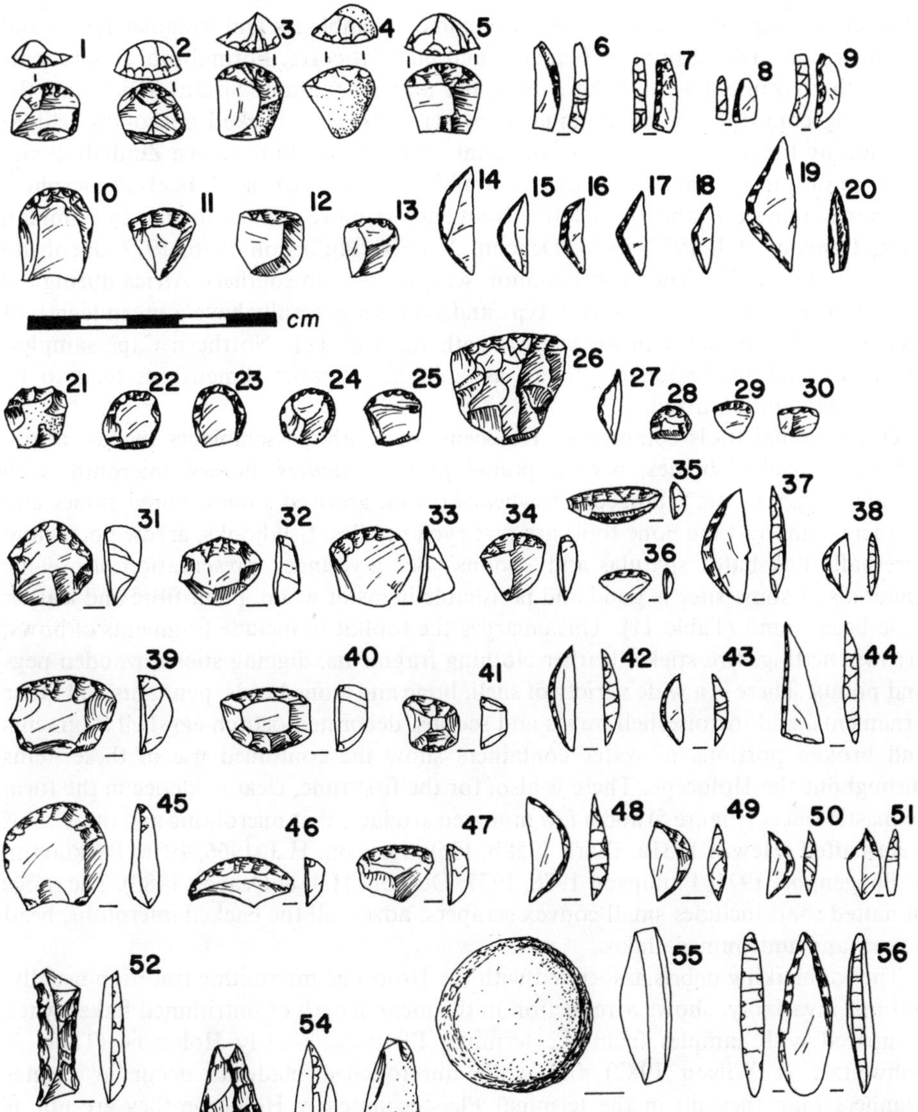

Figure 5. Stone artefacts from Holocene microlithic assemblages in southern Africa. 1-9: Glen Elliott level VIII (after Sampson 1967a: 18-22); 10-20: Diana's Vow (after Cooke, C.K. 1979a: Fig. 17); 21-30: Pomongwe (after Cooke, C.K. 1963; Fig. 17); 31-38, 52, 56: Boomplaas Cave (after Deacon, H.J. *et al.* 1978); 39-44: Wilton name site, layer 3F (after Deacon, J. 1969, 1972); 45-51: Nelson Bay Cave, BSC unit (after Deacon, J. 1978: Fig. 7); 53, 54, 57, 58: Melkhoutboom (after Deacon, H.J. 1976).

Scrapers: 1-5, 10-13, 21-26, 28-36, 39-42, 45-47, with mastic: 57. Backed microliths: 6-9, 14-20, 27, 37, 38, 42-44, 48-50, with mastic: 58. Borers: 51, 56. Adzes: 52-54. Milled edge pebble: 55.

number of microlithic pieces is small in some assemblages and irregular flakes and irregular cores far outnumber bladelets and bladelet cores. Formal tool components in southern Zimbabwe, southern Namibia and throughout South Africa and the adjoining countries are all dominated by scrapers, with backed microliths seldom accounting for more than 30% of the total formal tools. In northern Zimbabwe and at some sites in central and northern Namibia, on the other hand, backed microliths are more numerous and the relative frequency is more similar to that in Zambian sites (Deacon, H.J. 1972, 1976; Deacon, J. 1974; Phillipson 1976, 1977; Jacobson 1978; Wadley 1979). The most common scraper form in southern Africa during the mid-Holocene is the small convex type and samples generally have a mean length of less then 20 mm and a mean width/length ratio of 1:1. Northern Cape samples, however, tend to be longer. In all assemblages, scraper dimensions tend to be remarkably standardized.

Other formal tools common in Holocene microlithic assemblages include adzes, segments, backed blades, backed points, *petits tranchets*, borers, microliths with serrated edges ('saws'), tanged arrowheads (rare), grooved stones, bored stones and reamers. Amongst the bone tools are rare eyed needles, fish hooks, arrow points and foreshafts/linkshafts, spatulas and 'spoons'. Being younger, preservation of organic materials at some sites is good and perishable items of wood, plant fibre and leather have been found (Table 11). This enlarges the toolkit to include fragments of bows, arrows, netting, fire sticks, leather clothing fragments, digging sticks, wooden pegs and points. There is a wide variety of shell, bone and stone beads, pendants and other ornaments, and tortoiseshell bowls and scoops, decorated ostrich eggshell fragments and broken portions of water containers show the continued use of these items throughout the Holocene. There is also, for the first time, clear evidence in the form of mastic traces (Figure 5) and a few mounted artefacts that microlithic and other tools were hafted (Hewitt 1931a, Clark 1958b, 1959; Deacon, H.J. 1966, 1976; Parkington & Poggenpoel 1971; Phillipson 1976, 1977; Deacon, H.J. & Deacon 1980). The range of hafted tools includes small convex scrapers, adzes, all the backed microliths, bead borers and untrimmed flakes.

The toolmaking debris associated with the Holocene microlithic tradition usually, but not invariably, shows a reduction in the mean length of untrimmed flakes when compared with samples from the terminal Pleistocene/early Holocene (Table 3; Schweitzer & Wilson 1982) and while unretouched bladelets occur in greater numbers than they do in the terminal Pleistocene/early Holocene they are not as numerous as they were in samples dating between 18,000 and 12,000 B.P. at some sites (for example, Nelson Bay, Boomplaas, Sehonghong). The cores include single and double platform bladelet forms, small disc cores and less standardized types, but again the small bladelet cores so typical of the late Pleistocene microlithic assemblages are not as numerous in the Holocene. Where data are available there is some change through time in the relative frequencies of scraper and backed microlith sub-classes (Deacon, J. 1972; Sampson 1972; N.J. Walker, pers. comm.).

There is a marked increase in the incidence of burials and human skeletal remains in living sites during the Holocene and in some cases burials are covered with ochre and, more occasionally, painted gravestones. This pattern is more common in the western, southern and eastern Cape than elsewhere, however. Rock paintings and engravings are clearly associated with mid- and late Holocene microlithic assemblages in the form of painted stones dated from as early as 6400 B.P. in the southern Cape (Deacon *et al.* 1976) and engraved stones throughout the Holocene sequence at Wonderwerk in the northern Cape (Thackeray, A.I. *et al.* 1981).

Dating

The earliest dated occurrences of the Holocene microlithic tradition have been reported from Diana's Vow in Zimbabwe at 10,650 ±80 B.P. (Pta-1857) (Cooke, C.K. 1979) and from Apollo 11 in Namibia at 9430 ±80 B.P. (Kn-I-610) (Wendt, pers. comm.). By 6000 B.P. microlithic tools were widespread in southern Africa, but there does seem to be a tendency for the earliest dates to come from sites to the north suggesting perhaps that the stimulus to re-adopt a microlithic technology came from that direction.

The dating of the 'end' of the Holocene microlithic tradition is not easy to pinpoint either. Assemblages of this type persist at some sites through into the Christian era in Lesotho (Carter 1979) and are associated with pottery at Boomplaas (Deacon, H.J. *et al.* 1978) and De Hangen (Parkington & Poggenpoel 1971), while at other sites the microlithic tools are replaced by longer scrapers and fewer backed microliths during the last 2000 years. The diversity of late Holocene assemblages has led to the naming of a number of contemporaneous industries (Sampson 1974) and to the suggestion that they are all of the same complex, the Wilton (Humphreys 1979). For the moment it would be prudent to acknowledge that southern African herders and hunter-gatherers were still making microlithic tools on occasion during the last 1000 years, but that other toolmaking traditions were also in vogue after about 2000 B.P.

The dated assemblages in various regions of southern Africa are summarized in Table 2. In contrast to the distribution of dates in the terminal Pleistocene/early Holocene and the late Holocene time ranges, there is a general dearth of dated mid-Holocene assemblages from the South African interior (with the exception of the Drakensberg region and foothills), and it is not until after 4500 B.P. that dated sites become anything like as numerous as they are in the adjacent ecozones in the southern and eastern Cape, Zimbabwe and parts of Namibia. It has been suggested that this distribution may reflect a lower population density in the interior during the mid-Holocene climatic optimum (Deacon, J. 1974). Data on microfauna (Avery, D.M. 1981) and other sources (Butzer *et al.* 1978) support the contention that conditions were drier after ca 6500 B.P. in the northern Cape, but data are not available as yet from other regions, in particular from Natal where little is known about the mid-Holocene and the dearth of dated sites may be an artefact of modern archaeological research rather than a reasonably good reflection of site density. In Namibia, where dated and excavated sites are relatively numerous, the picture is more complicated with mid-Holocene assemblages present in some areas but not in others (Vogel & Visser 1980; Wendt, in prep.). The distribution of radiocarbon dates available at present would support an hypothesis that population fluxes occurred during the last 10,000 years. In some areas where rainfall or other environmental factors played a critical role in maintaining population levels, some regions were at times more densely populated than at other times. Unknown factors are the extent to which the preference for open sites (which are currently undated) may have fluctuated in relation to rock shelter sites, and the bias that is introduced by the sampling programmes of present-day archaeologists.

Environmental changes during the mid-Holocene are also seen at coastal and inland sites in the western, southern and eastern Cape. The mid-Holocene marine transgression (Martin 1968) was probably responsible for the absence of some early and mid-Holocene coastal shell middens (Robertshaw 1978a), and the absence of the estuarine shell, *Nassa kraussiana,* used for ornaments in the late Pleistocene, early Holocene and late Holocene at numerous southern and eastern Cape sites, can be attributed to the flooding of some river estuaries and lagoons which possibly made the shells less easily obtainable during the mid-Holocene.

Ethnographic evidence for subsistence

As with the cultural record, there are two sources of information on the subsistence ecology of the Holocene peoples of southern Africa – the historic and ethnographic accounts of traditional lifeways on the one hand, and the analysis of archaeological materials – in this case, food waste – on the other. The ethnographic studies of traditional hunter-gatherer, herder and agricultural economies supply a useful corpus of information on adaptations to particular plant and animal taxa and provide the sort of detailed basis for understanding how indigenous food resources were used that would be impossible to reconstruct from the archaeological record alone. The archaeological record, on the other hand, can give us some idea of the time depth of these practices, and of their geographic spread. The evidence for subsistence from both the Holocene microlithic assemblages and the late Holocene assemblages with pottery will be discussed together in this section in order to explore the time depth of the ethnographic tradition. It is realized, however, that subtle shifts in emphasis have occurred within the last 8000 years that would be difficult if not impossible to detect archaeologically and that it is only for the last 2000 years or so that the historic and ethnographic record is really pertinent (see Ebert 1978). It is also realized that many of the ethnographic records relate to the Kalaharian ecozone and may not be applicable to the rest of southern Africa; the focus is therefore on general patterns of subsistence rather than on particularistic detail.

Schapera (1930) has provided a useful synthesis from the early literature of the ethnography of the Khoi and San in parts of the Cape Province, Namibia and Botswana, but it is only recently that in-depth studies have been undertaken in which subsistence has been considered in detail. An impetus to a more ecological approach was provided by Story (1958, 1964) who accompanied the Smithsonian-Peabody Expedition to study the Kalahari San in 1955 and 1959 and gave an outline of the main plant foods used by the !Kung and G/wi. Story was able to show that not only were there regional differences in the availability of particular food plants, and thus differences in the importance of specific plants to specific groups of people, but that there was a high dependence on plant foods relative to animal foods in the Kalahari hunter-gatherer diet (Story 1964:99). This latter point had been appreciated by Schapera (1930), but he ascribed it to the decimation of game in the Kalahari region in recent times. It is not possible to tell from the archaeological record whether there has indeed been a recent increase in plant foods relative to meat, but Parkington (1980:83) has suggested that there was shift towards plant food exploitation in the western Cape during the last 2000 years.

Ethnographers have built on the initial observations of Story and there are several studies of individual communities such as, for example, the !Kung (Lee 1965, 1968,

1969, 1972, 1979), the G/wi (Silberbauer 1965, 1972; Tanaka 1969) and the Nharo (Bleek 1928; Steyn 1971, 1981). What emerges from these studies in an admittedly restricted range of habitats that are more arid than the regions for which we have Holocene archaeological observations, is that the modern hunter-gatherers have an extensive knowledge of economic botany and animal behaviour and that this is a prerequisite for survival. Their daily, seasonal and annual movements are geared to the availability of water and plant and animal foods that are not uniformly distributed through the landscape nor through the year. Successive years of drought or good rainfall can have a marked effect on availability of foods. Residential groups tend to have rights to resources in restricted territories of the order of 3000-12,000 km^2 in extent. Within these territories social mechanisms to cope with variations in availability in the seasonal round and the dispersed but clumped distributions of resources involve the fragmentation of residential groups into smaller 'family' foraging groups when resources are scarce (Yellen & Harpending 1972). The schedule of movements from one resource nexus to the next is not fixed in that there are year-to-year changes in productivity, but they are constrained by the general character of the habitat (Silberbauer 1972:300). For example, in the drier areas where surface water may be available at only a few localities in a given season, the choice of possible camp sites is limited. Consequences of the necessity for groups to move to resources are reflected in the need to limit possessions to what is portable, and in the reciprocal sharing of food and gift exchange within and between groups. While sharing has other complex functions, it also aids in the spreading of foods to where they are needed.

As is traditional in most other hunting and gathering societies, there is a sexual division of labour with men concerned mostly with hunting and women responsible for the bulk of the plant food collecting. The high reliance of all the present-day San in southern Africa on plant foods is explicable in that plants are a more easily guaranteed source of food than is meat, but the fact that most of the underground plant foods (which provide the bulk of the plant foods used) are generally high in carbohydrates and low in protein, makes it necessary that they be supplemented periodically by foods with higher protein values, of which meat is the obvious choice. Historical observations of hunter-gatherers consuming seemingly incredible quantities of meat at a single sitting are related to a pattern of periodic rather than regular meat eating which is reflected also in ethnographic records of hunting successes (Table 5). Lee (1979:450) recorded the relative importance of vegetable foods and meat at a Dobe !Kung camp in July 1964 and found that vegetable foods comprised 69% by weight and 71% of the calories consumed by the group, although in the late summer/autumn when the camp moved to a mongongo nut grove this proportion rose to 80%. In terms of energetics or the effort expended for the return in calories, gathering was more lucrative because men spent 21,6 work hours per week on subsistence activities, whereas women spent only 12,6 (Lee 1979:278). These figures would be different in other situations, however, where mongongo nuts (which are a high protein foodstuff) are not available.

In a different area in Botswana, for example, there is a marked seasonal fall-off in the availability of plant foods collected by the G/wi. Figures presented by Silberbauer (1972:298-300) show that in the summer rainy season from about December to March the plant food collected by one out of 16≠ Xade households in a

Table 5. Annual and seasonal meat yield estimated from data given for animals killed by one ≠Xade band in one year in Botswana (after Silberbauer 1973).

Species	Meat yield per individual animal (kg)	Total meat yield per annum/summer/winter			Number of individuals killed Summer						Winter					
					Dec	Jan	Feb	Mar	Apr	May	June	July	Aug	Sept	Oct	Nov
Giraffe	136	136	136	—	—	1	—	—	—	—	—	—	—	—	—	—
Eland	86	774	688	86	1	2	1	—	1	3	1	—	—	—	—	—
Kudu	45	135	135	—	—	—	1	—	1	1	—	—	—	—	—	—
Gemsbok	68	1360	816	544	3	2	4	2	1	—	3	3	2	—	—	—
Hartebeest	36	396	216	180	1	—	—	2	1	2	2	2	1	—	—	—
Wildebeest	36	432	252	180	1	1	—	3	2	—	2	2	1	—	—	—
Springbok	11	352	253	99	1	4	2	6	6	4	6	2	1	—	—	—
Duiker	4,5	293	112	181	4	5	4	3	7	2	3	1	2	12	10	12
Steenbok	3,6	245	98	147	3	3	4	3	6	8	2	4	2	8	10	15
Springhare	0,7	155	119	36	32	30	24	28	30	26	22	15	4	3	—	8
Porcupine	4,5	9	9	—	—	—	—	1	—	1	—	—	—	—	—	—
Warthog	16	16	16	—	—	—	—	—	1	—	—	—	—	—	—	—
Fox	1,8	33	8	25	—	—	—	—	4	—	3	1	3	2	1	4
Jackal	2,3	32	6	26	—	—	—	2	—	1	2	2	3	4	—	—
Rodents	0,06	18	5	13	30	20	30	30	30	40	20	20	15	15	15	30
Birds	0,23	37	18	19	8	18	22	13	14	6	12	4	12	16	16	20
Tortoises	0,12	53	47	6	90	90	90	80	40	—	—	—	—	—	—	50
Snakes	0,23	11	8	3	4	8	4	—	8	10	8	2	—	—	—	6
Total		4.487	2,942	1,545												
% Total		100	65,57	34,43												
Ants (litres)		2,5	2	0,5	0,5	0,5	0,5	—	—		—	—	—	—	—	0,5
Termites (litres)		8	8	—	2	4	2	—	—		—	—	—	—	—	—
Caterpillars		no figures given			—	—	—	×	×	×	—	—	—	—	—	—

band of about 80 people was in the region of 32 kg per 2,6 km² per day, but in May in a year when the tsama melons were in good supply this figure rose to 334 kg, but fell to 58 kg in a poor tsama season. By contrast, at the end of the summer at the hottest and driest time of year, only 0,38 kg of plant food was collected over 2,6 km² per day.

All ethnographic studies show that while the present-day hunter-gatherers may know of a wide variety of edible and medicinal plants (several hundred species in many cases), only a few of these are used as staples. Amongst the !Kung, for example, 14 out of 105 species used regularly constitute almost 75% of the calories in the plant food diet (Lee 1979: 159), while amongst the G/wi only 13 species are of major dietary significance (Silberbauer 1972: 183). It is the natural storage organs of plants, such as bulbs, corms, roots, rhizomes, seeds and nuts, that are mostly used for staple foods (Table 6) because they keep well and are available over long periods during the year. Fruits and leaves are more seasonal and do not generally last as long unless they are dried. Underground plant foods are more widely available to the south of southern Africa while nuts and prostrate runners such as the tsama and nara melons are used as staples further north.

Table 5 summarizes the estimated meat yield from animals caught and hunted by a

Table 6. Relative adundance of various types of plant foods from Holocene archaeological sites and ethnographic sources, grouped in ecozones as used by Klein (1980 and this volume).

Sites and regional surveys with references	Melons and cucurbits	Tubers, roots, corms and bulbs	Seeds and fruits	Leaves
CAPE ECOZONE				
Buffelskloof (Opperman 1978)		XX	X	
Boomplaas (Moffett & Deacon 1977)		X	XX	
Paardeberg (H.J. Deacon, unpubl.)		XX	X	
Scotts Cave (Wells, M. 1965)		XXX	X	X
De Hangen (Parkington & Poggenpoel 1971)		XXX	X	
Andriesgrond (Parkington, unpubl.)		XXX		
Melkhoutboom (Deacon, H.J. 1976)		XXX	X	
Springs (Deacon, H.J. 1976)		XX	XX	
Diepkloof (Parkington 1977a)		XX	XX	
BASUTOLIAN ECOZONE				
Highlands (Deacon, H.J. 1976)		XX		
Tafelberg Hall (Hewitt 1931b)		XX		
Tienfontein 2 & 7 (Brooker 1980)		XX	X	
TRANSVAALIAN ECOZONE				
Bushman Rock Shelter (Plug 1981)			XXX	
Nswatugi (Walker 1980)			XXX	
Kruger Cave (Friede & Pienaar 1974)			XXX	
KAROO-NAMAQUALIAN ECOZONE				
Mirabib (Sandelowsky 1977)	XXX		XX	
Big Elephant Shelter (Wadley 1979)		XXX	X	
Apollo 11 (Wendt, pers. comm.)	X			
MODERN SAN IN KALAHARIAN ECOZONE				
!Kung (Dobe) (Yellen & Lee 1976:38-41)	XX	XX	XXX	X
Nharo (Steyn 1971, 1981)	XX	XX	X	X
G/wi (Siberbauer 1965:44-6)	XX	XX	XX	X
≠Kade (Tanaka 1976:117-8)	XX	XX	XX	X
Korana (Engelbrecht 1936)	X	XXX	X	X

≠ Xade band during a year (Silberbauer 1972: table 7-2) and reflects several factors. Firstly, although fewer large animals were successfully killed, they provided the main source of meat by weight; the smaller game provided less meat, but kills were spread throughout the year. The smaller animals must therefore be seen as providing a more regular, if less spectacular, source of protein. Secondly, there is a clear difference between the amount of meat consumed during the months of summer and the amount consumed in the leaner winter months. In this summer rainfall area 65,5% of the annual meat yield is taken in the six months from December through May, compared with only 34,5% from June through November.

The seasonal rainfall and hunting pattern affect other parts of the system too. During the summer, for example, the G/wi mostly hunt with poisoned arrows because this is restricted to times of the year when poison grubs are available and for 4-6 months thereafter. Beyond this time the poison is either weakened by age, or supplies have run out (Silberbauer 1972:292). In the winter months, on the other hand, trapping or snaring are more suitable than they are in the wet summer months when rain interferes with the strength of the twine (Lee 1979:208). Lee notes also that snaring is not as efficient as bow and arrow hunting in the region in which his observations were made. Only one small animal was caught per hunter per week from July through August in 1964, although each person regularly set 20-25 snares. Furthermore, animals caught in snares accounted for 61% of the kills, but only 20% of the meat during the four weeks that observations were made (Lee 1979:209). While snaring proficiency may have been better in different regions and at other times in the past, this observation nevertheless suggests that the introduction of the bow and arrow probably increased the meat yield of LSA peoples considerably.

Another aspect of the Kalahari hunter-gatherer system affected by the difference in the availability of food and water in summer and winter is the settlement pattern. Amongst the G/wi, for example, the people tend to reside in larger groups during the summer when there is standing water available, melons are in season and larger animals are hunted, but in the leaner winter months when surface water is scarce, the groups fragment and disperse. The !Kung, on the other hand, fragment into smaller groups during the summer when mongongo nuts and fruits (*Ricinodendron rautanenii*) are plentiful and aggregate during the dry winter season. There seems to be a direct correlation between the availability of water, which can vary quite considerably from one part of the Kalahari to another, and the fluidity of groups in socio-territorial organization (Barnard 1979:140-1). Barnard points to a ranking of San groups from !Ko through G/wi, !Kung and Nharo in terms of most nucleated to least nucleated that corresponds with the ranking of least water to most water available in each territory. Thus, although the Kalahari may seem to be a 'uniform environment' to the outsider, it is the local variations in topography and substrate that control the amount and seasonal availability of surface water and, ultimately the socio-territorial structure and seasonal movements of the people living in that vicinity. For this reason it is unwise to assume that a particular seasonal pattern of aggregation and dispersal correlates with a particular rainfall regime. The reality is a complex interplay of a number of factors, not all of which are predictable from the standpoint of the 20th century.

While some of the correlates of seasonal patterning are variable amongst the present-day San, the general picture that emerges is that there is a clear contrast between the quality and quantity of food hunted and collected in the wet summer and the dry winter seasons. In some instances the nutritional stress does not have a marked effect on the people, as for example amongst the !Kung. In studies undertaken by Howell and Lee (Lee 1979:294ff) no significant weight loss could be detected from summer to winter. Amongst other groups, however, the San are malnourished in comparison with modern standards (Silberbauer 1972).

A strong pattern that emerges from the ethnographic studies is the regional adaptation to a preferred set of staples, despite the fact that the plant lore includes knowledge of a much wider range of plants. The essential element here is security. Hunter-gatherers need to know which plants can provide the most assured yield throughout the year and it is these which become the staples. The growth regimes of individual taxa determine when they are usable, which in turn imposes a seasonal rhythm on exploitation making it necessary to know what is available, in what quantity and at what location. Together with the knowledge of where secure water supplies are to be found through the year, the collecting of plant foods requires continual evaluation of information on availability that could affect and indeed structure the residential pattern and the size of groups.

Archaeological evidence for subsistence

Studies of the archaeological evidence for Holocene subsistence were begun in the 1930s (Hewitt 1931a; Goodwin 1938), but have become common only since the early 1960s (Deacon, H.J. & Deacon 1963; Hendey & Singer 1965; Wells, M. 1965). There is now considerable emphasis given in archaeological research in southern Africa to the recovery of food waste, its quantitative and qualitative analysis and the interpretation of aspects of subsistence ecology. While faunal and floral studies are limited by preservation factors at open sites, the majority of caves and rock shelters as well as some open middens have bone preserved, while dry cave deposits generally preserve some plant food remains. The most important results of Holocene food waste analyses will be discussed below under the headings of faunal evidence, plant remains and marine resources, with a final section on the evidence from various sources for seasonal transhumance.

Faunal evidence. The fauna from all sites that post-date ca 8000 B.P. includes only those animals that were present in southern Africa in historic times and the 'mix' of larger mammals is usually what might be expected from analogies with present-day hunter-gatherers and our knowledge of animals present in the vicinity of the sites in historic times (Table 7). From the range of taxa represented it seems likely that both trapping and projectile hunting methods were used and that there was some degree of seasonal change in hunting patterns in some areas, particularly the Drakensberg. In line with the functioning of these Holocene subsistence economies within territorial ranges of limited size, migratory herd animals such as springbok, zebra, wildebeest and buffalo are not prominently represented, but there is good evidence for the regular hunting of small and small-medium bovids such as grysbok/steenbok, bushbuck, duiker, vaalribbok, mountain reedbuck and southern reedbuck on a regular basis. This gives the Holocene faunas a smaller mean body weight per individual animal than those of the late Pleistocene (Deacon, H.J. 1972, 1976; Klein 1974, 1977b, 1980; Thackeray, J.F. 1980), particularly in the western, southern and eastern Cape where habitat changes were apparently more drastic than in the grassland regions to the north. The scale of change observed between late and terminal Pleistocene faunal assemblages and those of the last 8000 years can be seen in Table 4. Table 7 summarizes the frequencies of particular species at a number of Holocene sites to illustrate the regional differences and general reliance of smaller bovids. It should be noted, however, that the absence or low frequencies of small animals (rodents, lizards, frogs, birds) is due to the fact that some faunal analysts have not included them in their counts partly because they may occur naturally in caves and rock shelters and may therefore not relate to hunting and collecting and partly because of difficulties involved in their identification to species level.

The generally smaller size of the animals hunted during the Holocene (accompanied by a corresponding increase in the minimum number of individuals) has received considerable comment because of the implications of this shift in terms of hunting weapons and methods, seasonal movement and a stable meat supply. The habits of the bovids are of particular interest in assessing the consequences of this shift. The grassland herd animals such as springbok, impala, wildebeest, zebra, buffalo, and hartebeest tend to move seasonally with the grass growing regime, congregating in large herds at the end of the dry season and dispersing to better grazing during the summer, but still in fairly large herds. Their constant mobility would necessitate an equally mobile hunting strategy. By contrast, the smaller non-gregarious browsers such as grysbok, steenbok, bushbuck, duiker and klipspringer and the solitary grazers such as the oribi, mountain reedbuck, southern reedbuck and vaalribbok, live in fixed territories that do not change seasonally. They live in small groups, in pairs or singly and tend to hide during the day and feed in the early morning, evening or at night (Dorst & Dandelot 1970). Their predictable daily movements and lifestyle make it easier for them to be caught through trapping and snaring than through projectile hunting, but they are also sometimes run down with the aid of dogs. The result is that non-gregarious animals can be hunted at any time of the year and are more successfully hunted by people working on their own or in small groups. Larger browsers and mixed feeders such as the eland, Cape buffalo, kudu and blue antelope are intermediate in habit between the gregarious grazers and non-gregarious smaller antelope as they move in small herds with some seasonal movement to take advantage of the best grazing.

The expectation of Story (1958) was that hunting would have played a more prominent role in the economy of grassland systems which supported large herds of gregarious grazers in historic times. He mentions in particular the plains of the Orange Free State, but unfortunately we have very few observations from archaeological samples in this area. Story's hypothesis has not been tested and may be incapable of being tested because of poor bone preservation at the known sites, the majority of which are in the open. Where samples are available in the south near the Orange River and in the northern Cape, however, plains game is present at some localities. Analysis of the faunal remains from 13 late Holocene sites in this region showed that there is a contrast in the range of taxa represented at sites in or near bushveld vegetation and those near grassveld, with the obvious corollary that animals common in the grassveld occur more frequently in the grassveld sites (Klein 1979b). This can be seen in Table 7 that shows an increase from 22,9% of small-medium bovids in the bushveld sites to 51,2% in the grassland sites due to a higher proportion of springbok in the latter. Within the same group of sites, those situated in areas of greater topographical relief include more mountain reedbuck, vaalribbok and klipspringer which live in broken country, in contrast to the springbok, zebra and alcelaphines which are more prominent at sites close to open plains.

Despite the observations by Klein (*op. cit.*) and J.F. Thackeray (1979:30) that the species represented in the deposits are a good reflection of those available in the immediate vicinity of the site (but not necessarily in the same proportions), the range of animals listed and the minimum numbers of individuals are also the result of preservation and sampling factors. The result is that the data represented in Table 7 offer only a very general idea of the relative frequencies and range of taxa hunted by Holocene people in southern Africa. They do not include, for example, an accurate reflection of the numbers of small animals such as rodents, birds, and reptiles and although we know that baboons, jackals and small carnivores are hunted by modern San, these animals also live in caves and rock shelters and some of their remains may be the result of natural deaths rather than LSA hunting. In spite of these problems, there are few surprises in the archaeological data and most of the regional variation is explicable in terms of local availability. Equids, for example, are relatively more common in the Cape Folded Mountain sites, in the northern Cape, Orange Free State and Namibia than they are in the southern Cape coastal sites, while the Cape buffalo accounts for the high percentage of large bovids in the southern Cape coastal sample. Seen in broad overview, the emphasis on smaller bovids that is observed ethnographically is also clear in all the archaeological samples.

Plant remains. As in the case of faunal remains, archaeological sites cannot provide complete information on the range and quantity of plant foods eaten by Holocene people because only the inedible residues of plants collected and processed at the home base are preserved there. Many plants have little or no inedible residue, others are consumed away from the home base, and those that do have inedible residues have different quantities per unit consumed and different preservation properties that make it difficult, if not impossible, to estimate accurately the minimum numbers eaten. What plant remains can show, however, are inter-regional differences in the range of plant foods and medicinal plants and at least a partial view of the taxa of economic importance. The information that can be gained from ethnographic and historical sources on the uses and methods of preparation of plants is invaluable (Quin 1959; Smith, C.A. 1966; Heinz & Maguire 1974; Steyn 1981).

One of the largest and best studied assemblages of plant remains has come from horizons dating within the last 7500 years at Melkhoutboom Cave in the eastern zone of the Cape Folded Mountain belt (Deacon, H.J. 1976). It shows that the most common plant residues preserved are from monocotyledenous geophytes of the iris family (Iridaceae) represented by corm bases and tunics of the underground storage organs. Melkhoutboom is some 40 km inland and 800 m above sea-level, yet the plant remains include all the taxa found at a comparably preserved but much younger occurrence at Scott's Cave situated less than 100 m above sea-level in the Gamtoos Valley close to the southern Cape coast (Deacon, H.J. & Deacon 1963; Wells, M. 1965). This evidences a stable adaptation to a particular set of plants with wide geographical and seasonal availability in this habitat, as is the case amongst present-day San in the Kalahari. The southern Cape pattern shows that most of the collecting activity was directed at the open heathland and grassland areas in the vegetation mosaic, in contrast to the more closed shrubland-woodland habitats exploited by the hunters living at the same sites. A spring and summer collecting pattern is apparent from the maturity of inflorescences, fruits and availability in the two areas around Melkhoutboom and Scott's Cave and is repeated in the western Cape at De Hangen and Diepkloof (Parkington 1976, 1977a, b) and inland in the Cape Midlands at Highlands and Tafelberg Hall (Hewitt 1931b; Deacon, H.J. 1976). Thus in both the selection of a few staple plant foods and in the way in which the availability of these plants is responsible for the seasonal round, the plant food collecting pattern of the mid and late Holocene in the Cape ecozone is essentially similar to the strategy followed by present-day hunter-gatherers.

Correlation of the data on plant food remains from a number of western, southern and eastern Cape sites (Table 6) shows that geophytes are the most commonly represented, and of these *Watsonia* sp. corms are ubiquitous. The dominance of this genus is related to its high visibility through much of the year as the leaves above the ground are generally longer and heavier than those of other genera and do not blow away at the end of summer. In addition the individual corms are somewhat larger than those of other species such as *Cyperus usitatus, Moraea* sp., *Freezia/Tritonia* sp. and *Hypoxis* sp., all of which have a shorter seasonal visibility. Ethnographic and historic accounts record that the corms were roasted in the coals or ground and then formed into 'cakes' for roasting. As the palatability of the corm changes through the growth cycle (H.J. Deacon, pers. comm.), preparation may have been different at different times of the year. At their best in early summer, they look and taste somewhat similar to chestnuts.

The ecology of the Iridaceae and *Watsonia* sp. in particular, is of interest in assessing the role it played in the subsistence regime of LSA people and a study of a natural field of *Watsonia pyramidata* at Jonkershoek near Stellenbosch is illuminating. The species has the common name of 'bitter corm' and may not have been particularly favoured for eating, but the growth cycle is typical for the genus. The patch studied is on the flanks of a hill covering an area of less than 0,5 km^2 and includes an estimated one million individual plants, each with a corm weighing an average 50 g. The plants have a low rate of reproduction, flowering at irregular

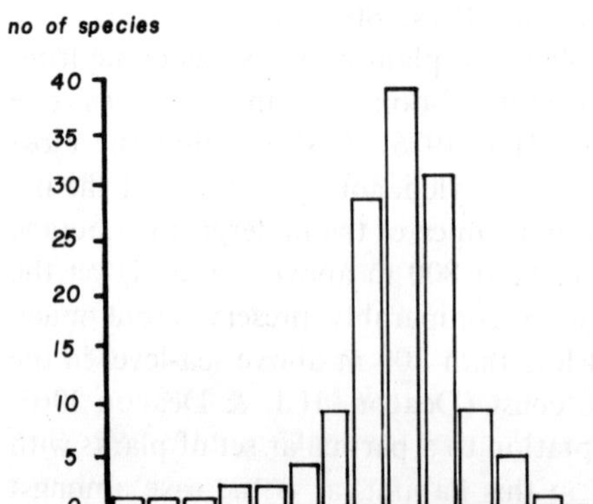

Figure 6. The number of species of Iridaceae that flower per month in the south-western Cape (after Parkington 1976b:132) showing the abundance during spring. The corms are best gathered a month or two after flowering.

intervals from 5 to 30 years (Kruger 1977, 1979), but this can be 'controlled' by burning. In this and other genera of the same family (Iridaceae) a flush of flowering, and therefore reproduction by seeding and corm division, follows a burn and the implication is that fire management was a necessary technique for exploitation and was practised by the indigenous people who made use of the plants (Deacon, H.J. 1972, 1976). Observation of the *W. pyramidata* patch showed an increase in flowering from 5% of the plants in a normal year to some 40% after an autumn burn. Another point of interest to *Watsonia* consumers is that during the growing season when the new corm is developing, waste products accumulate in the old corm as nutrients are gradually used up through the year. The old corm becomes unpalatable or even poisonous and cannot be used until the new corm has developed. There is therefore a period in the annual growth cycle from April/May to about July/August in the Cape winter rainfall area when exploitation is uneconomic (although it is still possible), but the timing of this period is variable in other rainfall areas and in some other species. The annual cycle is illustrated in Figure 6 for corms collected in the southwestern Cape.

Regional differences can be seen in the western Cape where the analysis of plant remains from the sandveld site of Diepkloof shows a slightly different emphasis from that at Melkhoutboom and De Hangen in the mountains (Parkington 1976). *Gladiolus* corms replace those of *Watsonia* in numerical importance and the seeds of fruits are better represented, in particular those of the tortoise berry, *Nylandtia spinosa*. This difference reflects a greater abundance of plants with berries in the vegetation mosaic and confirms summer occupation of the site. In the summer rainfall area of central Namibia, however, Wadley (1979) notes the occurrence of *Cyperus fulgens* corm tunics at Big (Great) Elephant Shelter which would indicate winter exploitation as these flourish after the summer rains have fallen.

Plant foods are not abundant everywhere. At Boomplaas Cave, for example, plant food residues are virtually absent in the younger Holocene deposits and are restricted to two geophyte taxa, *Watsonia* and *Hypoxis*, both of which are of widespread importance, and to seeds of *Pappea capensis* fruits (Deacon, H.J. *et al.* 1978; Deacon, H.J. 1979). A vegetation survey of the valley in which the site is situated (Moffett & Deacon 1977) showed few harvestable plant food resources in the vegetation established in the limestone substrate and it is clear that gathering could not have supported any residential group in the area over more than a minimal period. Some 2000 years ago, however, there is evidence for the specialized use of the cave as a place for the collection and storage of *Pappea capensis* fruits in pits lined with grass and the bulbar leaves of *Boophone disticha*. The fruits are both edible and rich in oil used by indigenous people for cosmetic purposes (Smith, C.A. 1966).

Seasonal data are not always obtainable from plant food resdiues, however. Seeds of the nara melon are found in southern Namibia at Apollo 11 Cave where they are in levels dated to between 7 and 8000 B.P. (Wendt 1976 and pers. comm.), at Mirabib in central Namibia where they date back to ca 6500 B.P. (Sandelowsky 1977) and at several other late Holocene occurrences which may be as much as 100 km from the nearest known present-day occurrence of the melons. Wendt (in prep.) notes that the seeds keep well and may be used as a food supplement for some months after they have been collected making it impossible to use their occurrence for the reconstruction of seasonal patterns, although where they occur in quantity they are likely to relate to the harvesting season between December and May/June.

There are few archaeological observations of plant food remains from the dry interior of South Africa. Corm-dominated samples have been recorded from Highlands and Tafelberg Hall, as mentioned above, with *Cyperus usitatus* and *Freezia corymbosa* predominating. They are available in summer and autumn. Dikbosch I preserved a small quantity of plant material, but this is mostly bedding grass (Humphreys 1974a, 1979). The best evidence for the use of plant foods in this ecozone comes from ethnographic observations of the Korana by Engelbrecht (1936) who notes that although fruits, berries and leaves were collected, bulbs, corms and roots supplied the bulk of the plant foods eaten. They are most plentiful in June and July (Engelbrecht 1936: 113-4) in contrast to the summer availability further south and later winter abundance further north.

The lack of suitable sites makes it impossible to test whether the emphasis on fruits and nuts persisted in the Transvaal during the mid-Holocene, but it seems likely from the ethnographic record (Quin 1959) that this was the case. Their importance in the diet of the !Kung in Botswana has been widely discussed (Lee 1965, 1979) and the data summarized in Table 6 show that they are more important than underground plant foods or leaves in the Transvaalian and Kalaharian ecozones.

Other gathered foods. Other foods for which we have limited archaeological evidence, but which are regularly mentioned by early travellers and ethnographers as providing important but seasonally abundant sources of food, include honey, locusts, termites, caterpillars,tortoises, birds' eggs (including the ubiquitous and nutritious ostrich eggs), land snails, lizards and snakes. Honey leaves no archaeological trace but rock paintings illustrate the raiding of hives (Vinnicombe 1976). Portions of termitaria have been found in archaeological context at Melkhoutboom and Springs Rock Shelter (Deacon, H.J. 1976) and at De Hangen (Parkington 1977a: 62) where two complete and several fragmentary ovoid 'cakes' packed with the bodies of termites were also found. These may be the result of the collection of clay from termitaria for making pottery rather than for food (M.L. Wilson pers. comm.). Caterpillars also leave no archaeological trace, but the thriving mopane caterpillar trade in the Transvaal and Zimbabwe today leaves no doubt as to their suitability as a foodstuff and widespread use. They are gutted, roasted and dried for use in the winter months (Quin 1959).

Marine resources and freshwater fish and shellfish. The third important element of the Holocene subsistence strategy was fish and shellfish. Substantial numbers of shell middens dating to the terminal Pleistocene and Holocene are known along the southern African coast from Angola in the north-west to Mozambique in the north-east. Open midden sites postdating ca 3000 B.P. are more numerous than those in caves and rock shelters, but mid-Holocene marine transgressions along stretches of the coast have eroded away some of the early Holocene open sites making these less numerous. Inland, particularly in Natal, Lesotho and the north-eastern Cape, middens of freshwater mussels are known and a few rock shelters sites include the remains of freshwater fish. Seashells have been transported as much as 100 km or more inland for use in making ornaments.

It was the expectation of Lee (1965) that hunter-gatherers with access to coastal resources would have relied on these in the same way as the !Kung San rely on mongongo nuts for their staple foodstuff. All archaeological evidence, however, suggests more seasonal use of coastal resources (Deacon, H.J. 1969, 1970a, 1972, 1976; Parkington 1972, 1976, 1977a, b), complementing hunting and gathering activities in the annual round. Their importance seems to lie in offering alternative food resources in the stress period for gathering which in most of the Cape coastal areas is in the winter months.

Following on Goodwin's pioneer studies at Oakhurst (1938) and elsewhere (1946b) interest was revived in the study of midden occurrences in the 1960s (Maggs & Speed 1967) and the numbers of observations on midden deposits have grown significantly (Deacon, H.J. 1970a; Parkington 1972, 1976; Klein 1972a; Voigt 1973; Avery, G. 1974, 1975, 1976; Schweitzer 1974, 1979; Buchanan *et al.* 1978; Robertshaw 1977, 1978a, 1979; Poggenpoel & Robertshaw 1981). Research has centred on the range of species represented, their relative quantities, the relationship of these to local availability and the food values of individual species and communities.

In general it is the intertidal zone along the sea shore that is most heavily exploited with larger shellfish caught below the low water mark being rarer overall and possibly needing spring low tides to allow exploitation. As with hunting and gathering, the 'mix' of species present in an archaeological deposit is a reasonably good reflection of what was available in the immediate vicinity of the site. This has been confirmed by several studies (Deacon, H.J. 1970a; Avery, G. 1974; Parkington 1976; Robertshaw 1977, 1979; Buchanan *et al.* 1978). Thus middens along sandy shores will contain sand-burrowing mussels such as *Donax serra*, while those near rocky promontories will contain limpets and mussels which prefer a rock substrate. Rock shelter sites and middens further from the sea, however, tend to have a wider range of species suggesting that they were inhabited because of their preferred micro-environment rather than their location close to the source of the shellfish. Middens with larger shells, such as *Haliotis*, tend to be located closer to the sea as the shells were probably shucked there (Avery, G. 1974: 111). The frequency of middens along a shore may be influenced by the nature of the shoreline. Where rocky promontories are more isolated, as around Elands Bay, the middens are larger, deeper and more localized adjacent to rocky intertidal shores, and they have remarkably similar contents. By contrast, middens around Paternoster along a coast where short stretches of rocky shores are separated by short sandy beaches, have more variable contents (Buchanan *et al.* 1978).

Along the south-western Cape coast Parkington (1976) and others have attempted to estimate how long prehistoric shellfish collectors may have remained at the coast and whether the resources could have sustained long-term predation. They have assumed that the mean length of limpet shells is a reasonably good reflection of the upper size range of limpets available on the rocks at the time of collection. Data collected along the coast between the mouths of the Berg and Olifants Rivers suggest that predation as reflected in shell middens dating withing the last 3000 years was of the order of 40% of the live population at any point in time. Depending on the size of the LSA groups, this implies that intense exploitation could not have been sustained for much longer than 3-4 months a year (Parkington 1976: 135-7). A similar pattern is reported from the coast south of the Berg River mouth at Paternoster (Robertshaw

Table 7. Composition of faunal assemblages from Holocene Later Stone Age sites in southern Africa compared with estimated annual kills of modern G/wi. For abbreviations and references see Table 12. Data for modern G/wi are from Silberbauer (1972).

	Western Cape coast		Southern Cape coast		Cape Folded mountains		Northern Cape–Orange Free State a) Bushveld		Northern Cape–Orange Free State b) Grassveld		Namibia		Natal		Modern G̀/wi	
	f	% Total bovids	f	% Total bovids	f	% Total bovids	f	% Total bovids	f	% Total bovids	f	% Total bovids	f	% Total bovids	f	% Total bovids
Bovids																
small	333	86,0	162	47,5	278	52,6	35	42,2	63	30,1	33	56,9	18	36,7	133	60,5
small-medium	16	4,1	85	24,9	158	29,9	19	22,9	107	51,2	19	32,8	18	36,7	32	14,5
large-medium	19	4,9	27	7,9	68	12,9	25	30,1	35	16,7	6	10,3	9	18,4	46	20,9
large	19	4,9	67	19,7	25	4,7	4	4,8	4	1,9	—	—	4	8,2	9	4,1
Total	387	99,9	341	100,00	529	100,0	83	100,0	209	99,9	58	100,0	49	100,0	220	100,0
	f	ratio to bovids	f	ratio to bovids	f	ratio to bovids	f	ratio to bovids	f	ratio to bovids	f	ratio to bovids	f	ratio to bovids	f	ratio to bovids
Equids	—	—	—	—	20	1:26	23	1:25	20	1:10	9	1:6	1	1:49	—	—
Hippopotamus	2	1:194	10	1:34	1	1:529	1	1:83	—	—	—	—	1	1:49	—	—
Rhinoceros	2	1:194	—	—	2	1:265	6	1:14	1	1:209	1	1:58	—	—	—	—
Elephant	1	1:387	—	—	3	1:176	—	—	—	—	—	—	—	—	—	—
Giraffe	—	—	—	—	—	—	1	1:83	—	—	—	—	—	—	—	—
Bushpig/warthog	1	1:387	29	1:12	49	1:11	8	1:10	23	1:9	1	1:58	5	1:10	1	1:220
Aardvark	—	—	—	—	1	1:529	—	—	10	1:20	—	—	2	1:25	—	—
Clawless otter	—	—	12	1:28	—	—	—	—	3	1:70	—	—	—	—	—	—
Pangolin	—	—	—	—	1	1:529	—	—	1	1:209	—	—	—	—	—	—
Porcupine	7	1:55	23	1:15	14	1:38	6	1:14	16	1:13	—	—	2	1:25	2	1:110
Honey badger	4	1:97	—	—	2	1:265	2	1:42	—	—	—	—	—	—	—	—
Hyrax (dassie)	45	1:9	162	1:2	193	1:3	18	1:5	80	1:3	79	1:0,7	16	1:3	—	—
Hare	10	1:39	—	—	46	1:12	32	1:2,5	70	1:3	35	1:1,7	11	1:4	222	1:1
Tortoise	+58	------	------	------	data incomplete or absent				------	------	52	1:1	+3	—	390	1:0,6
Lion	—	—	—	—	—	—	1	1:83	—	—	—	—	—	—	—	—
Leopard	2	1:194	8	1:43	5	1:106	—	—	3	1:70	—	—	—	—	—	—
Caracal	3	1:129	8	1:43	14	1:38	4	1:21	5	1:42	1	1:58	1	1:49	—	—

Wildcat/genet	15	1:26	5	1:68	21	1:25	3	1:28	7	1:30	5	1:12	1	1:49	—	—
Jackal/fox/ aardwolf	11	1:35	9	1:38	7	1:75	10	1:8	18	1:12	2	1:29	3	1:16	32	1:7
Mongooses	10	1:39	23	1:15	43	1:12	7	1:8	19	1:12	3	1:19	3	1:16	—	—
Rodents	1449	1:0,3	data incomplete or absent												295	1:0,7
Sheep	35	—	8	—	20	—	3	—	—	—	?1	—	—	—	—	—
Cattle & ?cattle	?2	—	—	—	6	—	5	—	9	—	—	—	2	—	—	—
? Domestic dog	?2	—	?1	—	—	—	—	—	—	—	—	—	—	—	—	—
Seal	146	1:3	235	1:1,5	—	—	—	—	—	—	—	—	—	—	—	—
Cetaceans	+9	—	22	1:16	—	—	—	—	—	—	—	—	—	—	—	—
Marine fish	numerous		numerous		—	—	—	—	—	—	—	—	—	—	—	—
Freshwater fish	—	—	—	—	present		—	—	present		—	—	133	—	—	—
Birds	+106	data incomplete or absent									15	1:1,4	no data		161	1:1,4
Sites represented (for references see Table 12)	EBC, PAT, DK, GB, BY		NBC, KR, KRM		WT, MHB, BUF, BPA, AK1, SS		DF, POW, LW, LI, LIM, DIK, DIK2, BUR		BL, ZY, RM, GL, TIE, LEM, VEN		AP, FA, TIR, NAM, NOS, AAR, HA, POC		DR, GH			

Bovid size categories: small (0-23 kg live weight): duiker, grysbok, klipspringer, oribi, steenbok
small-medium (23-84 kg): bushbuck, mountain reedbuck, southern reedbuck, vaalribbok, blesbok, bontebok, impala, springbok
large-medium (84-296 kg): kudu, blue antelope, hartebeest, wildebeest, gemsbok
large (more than 296 kg): eland, buffalo.

Note: Primates have not been included.

1977) and Duiker Eiland (Robertshaw 1979:10). At the latter site age estimates of rock lobster remains indicate a random collecting pattern because the age curve of the archaeological sample is similar to that of a modern living population.

Fish were caught by lines, by spearing and in tidal fish traps and cold water upwelling may also have provided people with an upredictable quantity of fish from time to time. There is no physical evidence for the use of nets, but the large number of small fishes caught suggests they may have been used (Poggenpoel, in prep.). For the deposits dating within the last 3000 years at Nelson Bay, for example, 95.8% of a sample of over 10,000 individual fishes are estimated to have weighed less than 117 g each. As with shellfish, it is clear that the range of fish species in an occupation deposit is a fairly accurate reflection of the habitats for fishes in the immediate vicinity of the site and that people brought back to the site the fishes that were most abundant. Changes in the size range and species of fish through time at a site probably reflect the changing morphology of natural pools in the vicinity rather than changes in cultural preferences.

In addition to fish, shellfish and crustaceans, marine mammals such as seals and occasional cetaceans are also represented in the coastal occupation deposits. From the analysis of bird remains G. Avery (1977, 1981) has been able to estimate the season of occupation of some sites and from modern surveys of beached birds has suggested that many of the birds and marine mammals were probably scavenged from the beaches rather than actively hunted by LSA people.

Remains of terrestrial mammals at coastal sites attest to the fact that the inhabitants still derived an important part of their diet from hunting and trapping (Table 7). Plant food remains, on the other hand, are rare at coastal sites even where other plant remains are preserved and it has been suggested (Deacon, H.J. 1972; Parkington 1976) that shellfish gathering probably replaced plant food gathering to a large extent at times when coastal sites were occupied. Because plant food gathering was largely the preserve of women and hunting of men, it is therefore highly likely that the shellfish were gathered by women.

Inland, spawning runs are known on some rivers and were exploited in historic times by the indigenous people living along the Orange River and its tributaries. Rock paintings of inland fishing scenes are also known in the Drakensberg region (Vinnicombe 1965). Fish hooks and bones of freshwater fish have been found at LSA sites in Lesotho (Carter 1978) and Natal (Maggs & Ward 1980) and fish and freshwater mussels, crabs and frogs have been identified from sites in the middle Orange River area (Sampson 1967a: 147) and at sites in the Cape Folded Mountains (Deacon, J. 1972; Deacon, H.J. & Deacon 1963; Deacon, H.J. 1976; Parkington 1977b:154). Stone fish traps are also known inland (Willcox 1965). There is, however, no archaeological evidence that the inland riverine resources of southern Africa were exploited on a large scale during the LSA. This is perhaps understandable in that many southern African rivers are highly seasonal in their flow, are subject to long periods of drought in many areas, and do not have a rich fauna. The freshwater fish and molluscs have therefore not apparently been more than a supplement to hunting and gathering inland, in contrast to the far more substantial coastal exploitation pattern. What is of interest is that the dating of inland middens and assemblages which include fish bones and fish hooks tends to be late in the Holocene, within the last 2000 years. It has been suggested that this exploitation of an additional resource may have been stimulated by local readjustments in group territories and annual movements as a result of the influx of pastoralists and Iron Age people and the consquent demographic changes (Deacon, J. 1972). The shift has been seen by Parkington (1980:83) as part of a general trend in the later Holocene, towards 'broad spectrum' gathering necessitated by higher population density in southern Africa at that time.

Evidence for seasonal transhumance during the Holocene. Assessment of the season of occupation of sites from the floral and faunal remains has been a subject of interest since Bowker (1884), Goodwin (1953:152) and others pointed out the implications of seasonal movement. But while it has long been realized that LSA people moved to take advantage of seasonal abundances (Clark 1959:217), it is only recently that good data have been produced to demonstrate it. The plant remains from Scott's Cave (Wells, M. 1965), Melkhoutboom, Springs and Highlands (Deacon, H.J. 1976) in the eastern Cape and from De Hangen and Diepkloof in the western Cape (Parkington 1972, 1976, 1977a, b) provided the first clear evidence for spring and summer occupation inland in the Cape Folded Mountains and near the coast. A local record from the Longkloof in the southern Cape describes seasonal movements of the last bands of San in the 19th century and reports that they moved to the coast in summer and to the moutains in winter (Deacon, H.J. 1969, 1970a, 1976), but a careful analysis of the historic records and archaeological data in the western Cape shows that the mountain zone was favoured there in the summer and the coast in the winter (Parkington 1972, 1977a, b; Avery 1975). This apparent contradiction between the southern and western Cape data is explicable, however, in terms of the different rainfall regimes of the two areas, the western Cape having predominantly winter rain and the southern Cape having rain at all seasons of the year, but predominantly in spring and autumn (Avery, G. 1976). The rainfall regime has a strong influence on the seasonal abundance of underground plant foods, particularly the Iridaceae, and as these appear to have been the staple plant foods of LSA people it is logical that their movements would have been governed by the timing of the season of maximum abundance, as is also the case amongst present-day San in the Kalahari.

The pattern would have been different again inland. In the northern Cape, for example, where rainfall is generally low and falls more often in summer than in winter, resources are more evenly, if sparsely, distributed, and coastal resources were not available as a release from stress in seasons of low plant food abundance. Humphreys (1979:383-4, 1980:94) has suggested that the more varied adjacent habitats are and the greater the seasonal contrast, the more patterned would be the response of hunter-gatherers to this structured environment. Thus the unstructured environment of the northern Cape would show less patterned seasonal movements. In the more varied Drakensberg mountain region, however, Carter (1970, 1978) hypothesises that hunter-gatherers would have followed game into the higher pastures in spring and summer and down to the foothills in the winter. The archaeological evidence for these hypotheses is discussed below.

The southern and eastern Cape with a relatively high rainfall with no marked seasonal peak shows no clear contrast in the seasonality of archaeological remains between mountain and coastal plain sites. There are instances of spring and summer occupation both inland at Melkhoutboom (Deacon, H.J. 1976) and near the coast at

Scott's Cave (Wells, M. 1965). Few data are so far available on the season of occupation at coastal sites in this region, but what is available suggests that occupation of the sites could have taken place at several times during the year. Analysis of the $^{16}O/^{18}O$ ratio in a small sample of two shells from ca 6000 year-old levels at Nelson Bay indicates that both were collected in the winter or early spring (Shackleton 1973). The age at death of seals from the Holocene levels on the other hand, shows a less seasonal pattern and individuals could have been caught or collected at several different times of the year (Parkington 1976:138-9 and pers. comm.) while historical records indicate that herders were met in that area during both spring (Sparrman 1785) and summer (Storrar 1978:16).

Assessment of a regular pattern of transhumance in the southern and eastern Cape may be further hampered by the possibility that changes occurred through time and that radical adjustments were necessitated after the introduction of domestic stock. The data on seals from Nelson Bay have not yet been subdivided according to level, but the samples are small and even if shifts are noted through time these may not prove to be significant. That a change in the seasonal pattern may have occurred was first suggested after a shift was observed after 2500 B.P. in the relative frequency of marine and freshwater shells at the Wilton name site in the Cape Folded Mountains, the latter becoming more common. This intimated that either the coast was not being frequented as often as before, or that there had been a change in the season of occupation of the site (Deacon, J. 1972). The latter possibility was considered more likely with the identification of the late summer/autumn flowering *Cyperus usitatus* in the surface levels at the nearby site of Melkhoutboom where it was not found in the pre-pottery levels despite excellent preservation of plant remains (Deacon, H.J. 1976).

Independent evidence on dietary changes was sought in the analysis of the $^{13}C/^{12}C$ ratio in human bones from 67 burials in the eastern Cape (Silberbauer 1979), but the results have proved to be complicated by the fact that the isotopic composition of milk from sheep and cattle enriches the ^{13}C content of bones in the same way as marine foods and the consumption of meat from mixed feeders (as opposed to purely browsers) in that region. Nevertheless, from the very small number of dated individuals (2) it seems that those from ca 8000 to ca 5000 B.P. had a lower intake of marine foods/mixed feeders/milk than those of the later Holocene, and that, although there was some overlap, individuals judged from burial style and other factors to have been hunter-gatherers differed from those judged to have been herders in the mean ^{13}C value of their bones. Thus, while we have some intimation that changes occurred in the seasonal pattern and the relative importance of marine foods through the Holocene, more intensive work is needed on this problem.

In the western Cape the seasonal contrasts between mountain, coastal plain and coast are apparently greater than in the southern and eastern Cape and this is reflected in the archaeological data. The relative paucity of staple plant foods on the coast and adjacent sandveld of the coastal plain, and the high risk of toxicity amongst marine mussels and other filter feeders during the summer combine to suggest that people would have been better off inland amongst the mountains and intermontane valleys during the summer (Parkington 1972, 1976, 1977a, b), while the relative lack of plant food staples everywhere during the winter encouraged people to move to the coast to exploit marine resources at that time. That people were present in the mountains during summer is confirmed by analysis of the age at death of dassie (rock hyrax) remains at De Hangen, by the occurrence of corm remains indicating summer collection and by the relatively high incidence of tortoise bones (tortoises tend to hibernate in winter and are therefore more easily collected in spring and summer) (Parkington 1977b: 154). Along the coast, on the other hand, Parkington (1972, 1976) has shown that seals from the Elands Bay terminal Pleistocene and early and late Holocene sequence were killed or collected over a very short time span between July and October. This highly restricted time period seems to be confirmed by the assessment of growth rings on seal teeth (Fletemeyer 1977). In addition, tortoise remains are relatively rare and the large quantities of mussels here and at other sites in the vicinity also suggest winter occupation. Combined with observations on limpet shell sizes which indicate a short-term annual exploitation pattern, there is good evidence for the hypothesis of seasonal mobility with regular transhumance involving summer occupation in the mountains and winter occupation along the coast. A functional interpretation of formal tool frequencies by Mazel and Parkington (1978, 1981) and Parkington (1980) gives some support for mobility in that the relative frequencies of formal tools differ at mountain, sandveld and coastal sites. A similar pattern of artefact frequency changes occurs in Natal and Lesotho (Cable *et al.* 1980) where it has also been linked to seasonal occupation.

The pattern of seasonal mobility documented from the archaeological evidence in the western Cape may not have applied everywhere in the western and south-western Cape, however. Analysis of the shellfish and marine bird remains at Paternoster to the south of Elands Bay indicates summer occupation at the coast from about 3500 B.P. to about the time that pottery was introduced, while the uppermost level, dated just after 900 B.P., has evidence for autumn occupation (Avery, G. 1977; Robertshaw 1977:71-2). Robertshaw (*op. cit.*) suggests that the alternative strategy adopted by the people who occupied the Paternoster site is the result of a narrower coastal plain in the vicinity bringing a more varied set of habitats closer to the coast and reducing the necessity for a clear-cut division between summer and winter habitation sites. Data from analyses of seal, tortoise and shellfish remains at another coastal midden at Duiker Eiland a few kilometres from Paternoster combine to suggest occupation from June to November with bird remains possibly extending through to summer, but nearby Duiker Eiland shows a variety of seasonal occupations while another site in the vicinity at Stofbergsfontein has evidence for summer occupation (Robertshaw 1978a). The variability both spatially and through time needs to be documented further before any firm patterns can be established.

Inland and at some distance from the coast a highly seasonal pattern of transhumance has been hypothesised for the Lesotho and Natal Drakensberg region and the adjacent Natal Midlands that is based more on game movements than on plant food cycles. Carter (1970, 1978; Cable *et al.* 1980) notes that the historically documented movement of game into the higher mountain pastures in summer would have encouraged the movement of hunter-gatherers into the mountains to make use of this resource, while the severe winters in the higher mountains would have discouraged hunter-gatherers at that time of year. Support for the hypothesis is given in the occurrence of bones of the freshwater fish *Barbus natalensis* in the late Holocene levels at Sehonghong which could only have been collected during summer (Carter 1978:225). Winter occupation of Melikane and Belleview situated lower down the mountain slopes, on the other hand, is assumed on climatic data that show no extremes of winter cold thus allowing the growth of winter grazing and ensuring the availability of wood for fuel. Circumstantial evidence for summer occupation of Good Hope shelter in the Highland Sourveld of Natal is proposed on the occurrence of a higher frequency of convex scrapers there. The assumption is that these indicate a lesser emphasis on plant food collecting than is apparent at sites in the Thornveld/Mixed *Podocarpus* forest regions of the Natal Midlands where notched scrapers are more numerous. The notched scrapers are thought to reflect woodworking and in particular the manufacture and maintenance of digging sticks used for collecting plant foods (Cable *et al.* 1980).

In another, drier, inland region, Humphreys (1979:378) has stressed the uniformity of resources and the general similarity of faunal remains and formal tool frequencies at sites in the northern Cape and suggests that this reflects a diversified hunting strategy that was not geared to a seasonal pattern because the habitats of the animals did not warrant it. This assumption, however, does not take plant foods into account because no plant food remains have been recovered from the sites. That seasonal abundance and scarcity did affect the plant food collecting pattern in this region is clearly evident from the ethonographic study of the Korana by Engelbrecht (1936). The uniformity in the sites studied by Humphreys (which include those from the Gaap Plateau as well as the middle Orange River) may thus be present in the fauna only. Modelling of seasonal movements in the middle Orange River catchment area is being currently investigated by Sampson (1980) and the role of plant foods and their seasonal pattern is to be give due weight.

Data on seasonal occupation at Namibian sites are not yet published in full (Wendt 1976), but analysis of the plant materials, birds' eggs and fauna from the Big (Great) Elephant Shelter in central Namibia has established that this site at least was utilized for short periods at all times of the year (Wadley 1979). Data are also not yet available for analysis of seasonal movements in Zimbabwe and Botswana but would be particularly interesting for the latter because modern collecting patterns are well documented and can be used for comparison (Yellen 1977). Some summer occupation at sites in the Matopos in Zimbabwe is indicated by the finding of marula endocarps (Walker, N.J. 1980).

Evaluation of the evidence for seasonal transhumance during the Holocene needs to take a number of variables into account. Barnard (1979) has pointed out that the availability of water (both its abundance seasonally and the location) has an important influence on the movements of modern hunter-gatherers in the Kalahari and it undoubtedly affected prehistoric people in better-watered areas too. The most important point that he makes, however, is that local differences in availability of water cross-cut other seasonal regularities such as the abundance of plant foods and movements of game making it difficult to predict a prehistoric seasonal pattern with knowledge only of the rainfall regime and staple plant foods. What the archaeological data show is that changes also occur through time with shifts in the subsistence economy probably coincident with the introduction of domestic stock. While gross variability may be introduced with changes from one rainfall regime to another and with inter-regional differences in staple plant foods, the viability of hypotheses and archaeological data also depends on estimates of the size of the territory over which it is suggested that people may have moved during a year. Some of the models offered are unrealistic in this regard. A final point is that several independent lines of evidence need to be sought in order to test a seasonal hypothesis fully. Evidence from plant food remains that a site was occupied at a particular season is not necessarily evidence that the site was unoccupied at other seasons. The archaeological data that have been amassed so far bear ample witness to the complexity of both the prehistoric patterns of seasonal movement and the evaluation of the remains.

Late Holocene assemblages with pottery

Southern African stone artefact assemblages which include pottery and, in many cases, the bones of domesticated sheep/goat and/or cattle, have been grouped together here because the advent of these new elements within the last 2000 years or more evidences the spread of a new technological development in the manufacture and use of baked clay pots, and a modification of the traditional hunter-gatherer-fisher economy with the introduction of domesticated stock exotic to the sub-continent. It must be stressed, however, that here as in eastern Africa (Phillipson 1977:81), the new additions were not generally accompanied by a change in the stone artefact manufacturing technique although there are gradual changes within the sequences involving the relative frequency of some formal tools and the size and shape of scrapers. The overall impression is that pottery and domestic stock were added to the pre-existing Stone Age tradition and that there is continuity in the range of stone tools made in each region. Within southern Africa as a whole there is regional differentiation in the styles of pottery made and in the size of scrapers (Deacon, H.J. 1972; Sampson 1972, 1974).

This regional differentiation is expressed in the recognition (Sampson 1974) of a Smithfield Complex, a Ceramic Wilton and a Strandloper Complex that are broadly contemporary. The Smithfield is found in the northern Cape, the Transvaal and Natal, with some sites in the Brandberg in Namibia; the Wilton is found in Zimbabwe, parts of Namibia and the western, southern and eastern Cape as well as the Drakensberg region, and the Strandloper Complex is found along the coast. Humphreys (1979), however, considers the Smithfield to be an integral part of the Wilton Complex and the characteristic long end scrapers of the Smithfield to be merely the result of the use of hornfels in some areas, while Beaumont (pers. comm.) recognizes three industries in the northern Cape: the Ceramic Wilton, the Smithfield and a Ceramic LSA with no microliths and uncommon informal scrapers. The latter would be an inland equivalent of Sampson's Strandloper Complex.

Sampson's Smithfield Complex (formerly Smithfield B of Goodwin and Van Riet Lowe (1929)) is characterized by late assemblages in which pottery is associated with stone artefact assemblages that include long end scrapers made in hornfels in the northern Cape and Natal (Figure 7) and somewhat smaller scrapers in other raw materials in the Transvaal and Namibia (Figure 8). Backed microliths are rare or absent. Added to this toolkit in Natal are numbers of spokeshaves (notched, hollow or strangulated scrapers), the type tool of the original Smithfield N and thought to be functionally equivalent to the adzes of the western Cape. At the top of the northern Cape and other sequences, glass, glass beads, metal (usually copper or brass) and other European trade items may be found. In 1974 Sampson reported that all the Smithfield assemblages post-dated A.D. 1000, but more recent research suggests that

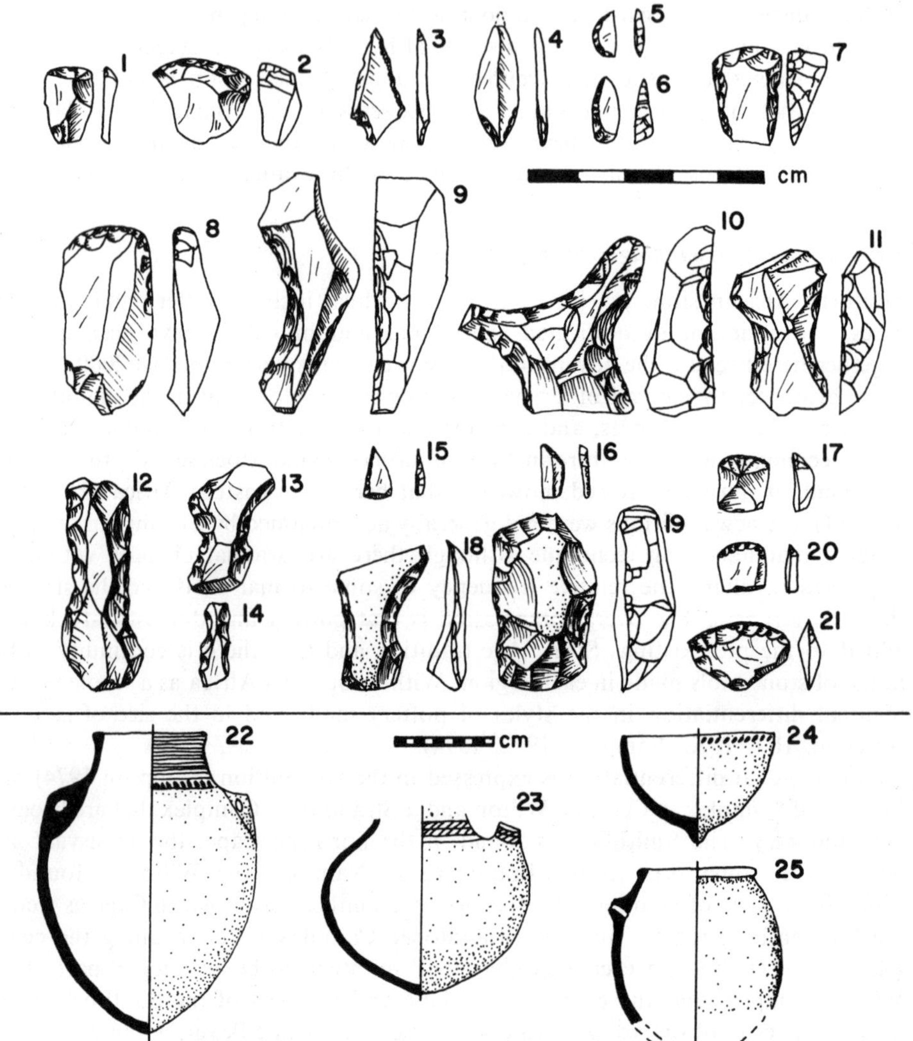

Figure 8. Artefacts from late Holocene assemblages with small scrapers, adzes/spokeshaves and pottery. 1-11: Driel, Natal (after Maggs & Ward 1980); 12-14: Sandy Bay, south-western Cape (after Rudner & Rudner 1954); 15-21: Boomplaas cave (after Deacon *et al.* 1978).

Scrapers: 1, 2, 7, 8, 17, 20, 21. Tanged points: 3, 4. Backed microliths: 5, 6, 15, 16. Spokeshaves: 9, 10, 11, 13, 18. Adzes: 12, 14, 19.

22-25: Cape coastal pottery from Robberg (22), Jeffreys Bay (23) and Kleinsee (24, 25) (after Rudner 1968:634, 651, 654). 26, 27: Inland LSA pottery from Avalon, Orange Free State and Joubert's Gif near the middle Orange River sites (Sampson 1970:166-7).

this dating may be extended to post-date A.D. 100 (Maggs & Ward 1980), at least in Natal.

The Ceramic Wilton is seen by Sampson (1974) at a different hierarchical level from the Smithfield Complex in order to stress the continuity in stone artefacts in pre- and post-pottery Wilton Complex assemblages. The reason is that assemblages grouped into the Ceramic Wilton do not show an increase in scraper length within the last 2000 years and retain more backed microliths than contemporary assemblages in the Smithfield Complex. A good example of this continuity can be seen in the assemblages from Boomplaas in the southern Cape (Deacon, H.J. *et al.* 1978) that are summarized in Table 8. Ceramic Wilton assemblages may show a reduction in the frequency of formal tools, a reduction in the relative frequency and variety of backed microliths and in some sites an increase in the range and number of bone tools (Schweitzer 1979). Copper and brass are associated with stone artefacts in the upper levels at some sites (e.g. Boomplaas (Deacon, H.J. *et al.* 1978) and Andriesgrond (Parkington, pers. comm.)) as are glass beads in some areas (Wendt 1972; Wadley 1979). In the western Cape there is some evidence for an increase in the frequency of adzes (Mazel & Parkington 1978; Parkington 1980).

The third set of assemblages associated with pottery are linked together by Sampson (1974) as 'Strandloper' sites. At most coastal sites there is an element of

Table 8. Metric and raw material data for scrapers from pre- and post-pottery levels at Boomplaas Cave, southern Cape (after Deacon, H.J. *et al.* 1978 and Deacon, J. in prep.).

Member	f	Mean mm	STD DEV mm	Range mm	Approximate dating B.P.	Pottery
DGL	157	12,4	5,42	5–39	1500–1700	Present
BLD	377	11,8	4,78	4–35	2000	Absent
BLA	213	11,5	3,97	5–41	6500	Absent
b) Width						
DGL	157	15,2	5,07	9–47	1500–1700	Present
BLD	377	14,6	4,66	8–43	2000	Absent
BLA	213	14,9	3,66	9–30	6500	Absent
c) Height						
DGL	157	4,5	2,20	1–16	1500–1700	Present
BLD	377	4,7	2,09	1–17	2000	Absent
BLA	213	4,7	1,82	2–15	6500	Absent
d) Width/length ratio ($\frac{W}{L} \times 100$)						
DGL	157	130,7	36,8	61–250	1500–1700	Present
BLD	377	133,3	42,2	46–325	2000	Absent
BLA	213	137,9	43,3	55–262	6500	Absent

e) Raw material

Member	f	Quartz	Chalcedony	Shale	Silcrete	Quartzite	Dating	Pottery
		%	%	%	%	%		
DGL	157	79	14	3	4	—	1500–1700	Present
BLD	377	75	16	8	1	—	2000	Absent
BLA	213	80	16	4	—	—	6500	Absent

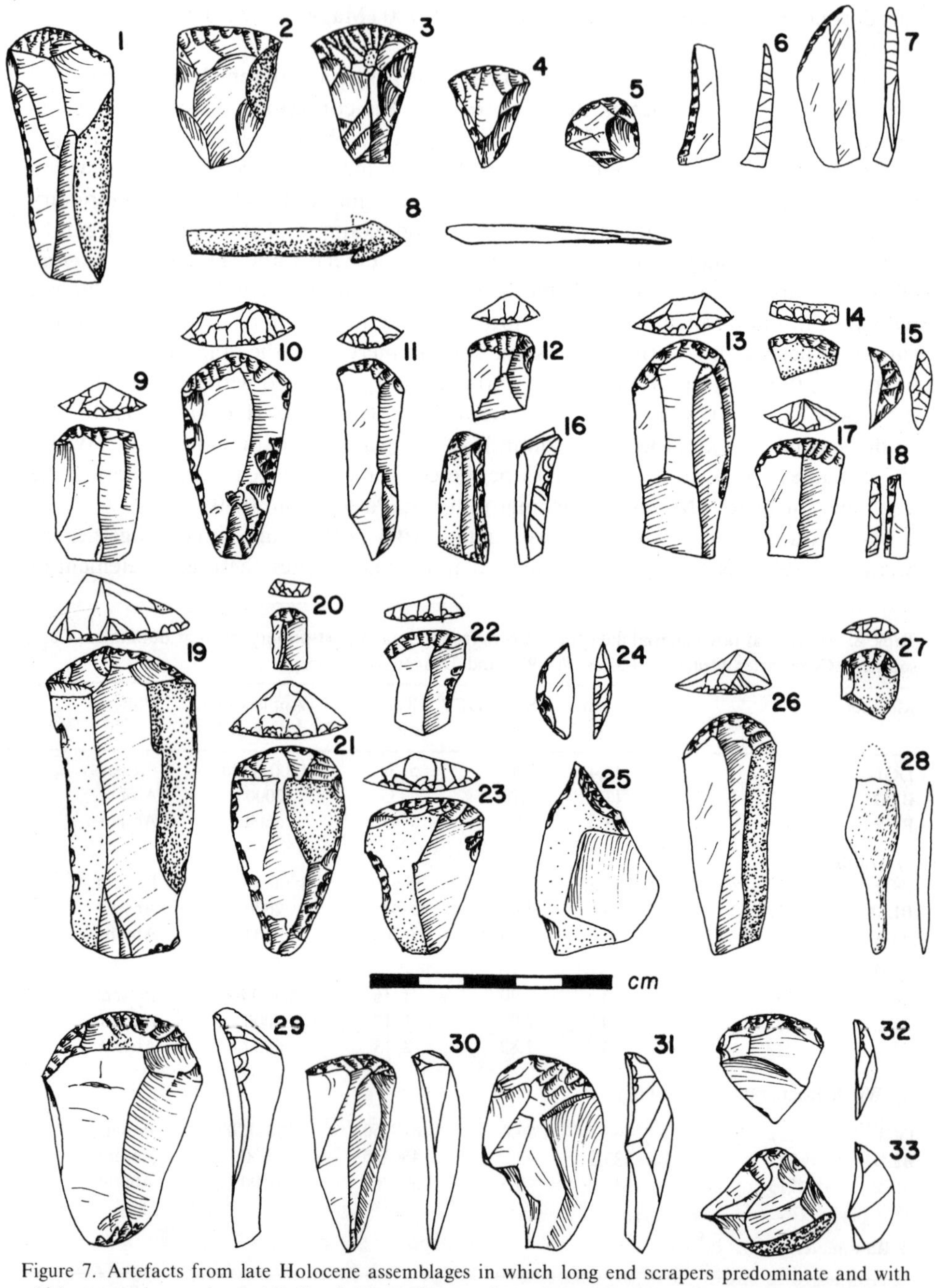

Figure 7. Artefacts from late Holocene assemblages in which long end scrapers predominate and with which pottery is associated. 1-7: Burchell's Shelter, northern Cape (after Humphreys 1979: Fig. 34); 8-12, 16: Glen Elliot level II; 13-15, 17, 18: Glen Elliott level III (both after Sampson 1967a); 19-28: Zaayfontein level V (after Sampson 1967c: 87-90, 1970: 165); 29: Tafelberg Hall; 30: Highlands (both after Deacon, H.J. 1976: Figs. 60, 63); 31-33: Wilton large rock shelter layer 1 (after Deacon, J. 1969: Plate 2). Scrapers: 1-5, 9-14, 17, 19-23, 26, 27, 29-33. Backed microliths: 6, 7, 15, 18, 24. Adze: 16. Borer: 25. Tanged bone arrowhead: 8. Iron arrowhead: 28.

large unretouched flakes and flaked cobbles associated with shell middens and a generally low frequency of formal tools. This pattern is sometimes exaggerated and assemblages may then contain no formal tools at all as at Scott's Cave (Deacon, H.J. & Deacon 1963) and a number of shell middens (Avery, G. 1974; Robertshaw 1977, 1978a 1979). While many researchers consider that such assemblages represent activity variants of those containing formal tools as well, Sampson (1974:436 ff) considers that they may represent the occupation sites of the so-called Strandlopers or Fishermen described by early European settlers such as Jan van Riebeeck (Thom 1952:80) who regarded them as ethnically and economically distinct from other local herders and hunter-gatherers. While this may be the case, there seems little way in which this hypothesis could be tested, but it is of interest that assemblages without microliths and with only a few informal scrapers are associated with pottery inland as at Doornfontein in the northern Cape (Beaumont & Boshier 1974) and in Namibia (Jacobson 1980). These would presumably not be relatable to the Strandlopers.

In comparing pre- and post-pottery assemblages in general the change most commonly observed is a reduction in the frequency of backed microliths and segments in particular. It is possible that this may relate to the introduction of metal at least after A.D. 300 when Iron Age people were already settled south of the Limpopo, and the replacement of stone inserts by metal ones. Alternative explanations have been offered by Humphreys (1979:366) and Sampson (1974:376) and are discussed later in this paper. Other new items which appear in this time range are smoking pipes of stone and bone as at Highlands (Deacon, H.J. 1976) and Gordonia (Rudner 1971a:139) in the northern Cape and in the Brandberg of Namibia (Rudner 1957), bone hooks assumed to have been used for fishing in the Drakensberg region (Carter 1978; Maggs & Ward 1980; H. Opperman, pers. comm.), and stone-walled fish traps along the southern Cape coast (Goodwin 1946b; Avery, G. 1975) and inland (Willcox 1965). Matting, presumably linked to the manufacture of mat-covered huts by herders, is preserved at sites where preservation is good and other items such as wooden tools, cordage, netting and leather are more common because of a higher chance of preservation in more recent deposits (Grobbelaar & Goodwin 1952; Deacon. H.J. 1969, 1976; Parkington & Poggenpoel 1971; Humphreys 1974a; Wadley 1979). Despite the presence of bones of domesticated stock, however, there do not seem to be many artefacts specifically linked to a herding economy, with the exception of pottery, used for cooking and for storing milk and milk products, and matting for huts.

As mentioned above there is some regional differentiation of pottery styles. Rudner (1968) has recognized a Cape coastal ware that is similar to pottery from Namibia (Sydow 1967), along the coast from southern Namibia to the eastern Cape and inland along the lower reaches of the Orange River and to the inland margin of the Cape Folded Mountains. It is characterized by a range of pointed based pots with internally reinforced lugs and incised decoration on the neck and rim (Figure 8). Rarer forms include spouts and bowls are also known. The temper is usually quartz grains. The larger pots can be as high as 400-500 mm and the smaller pots and bowls are about 50-150 mm high.

Inland, in the northern Cape, smaller bowls with thicker walls and heavily impressed decoration (Sampson 1972, 1974; Figure 8) are commonly made with grass rather than quartz temper and are regarded by Rudner (1979) as having been made by San hunter-gatherers rather than herders or Iron Age farmers. In Zimbabwe (Cooke, C.K. 1979), the Transvaal (Mason 1974) and Natal (Maggs & Ward 1980) the pottery recovered from LSA contexts has been limited in sample size, but the affinities in decoration and pot shapes are considered to lie with the Iron Age tradition rather than with the Cape coastal or the northern Cape wares.

There has been much speculation about the ethnic affinities of the people who made the pottery and artefacts dated within the last 2000 years. While some researchers maintain that there is a close correspondence between the historically known Khoi herders and the Ceramic Wilton on the one hand, and the historically known San hunter-gatherers and the Smithfield Complex on the other (Rudner 1968, 1979; Rudner & Rudner 1970; Sampson 1974), others are less convinced (Deacon, H.J. 1976; Deacon *et al.* 1978; Inskeep 1978a; Schrire 1980), if only because we have no factual basis for assuming a one-to-one correlation between stone artefact traditions, pottery styles, subsistence, and ethnic and language groups. In a recent review of Khoi settlement at the Cape Elphick (1977) has used linguistic data collated by Westphal (1963) to suggest that herders met by early Europeans in the southern Cape from the early 16th century onwards were linguistically more similar to Tshu-Khwe speakers in present-day Botswana (a group that includes the G/wi and Nharo hunter-gatherers) than they were to the hunter-gatherers who lived alongside them at the Cape. The implication drawn from this observation is that the southern Cape herders were descendants of people who had moved with their stock from Botswana at some time in the past, introducing pottery, sheep and cattle to South Africa and Namibia. However, as pointed out by Inskeep (1978b:35), we have no basis for assuming that *all* pastoralists belonged to the same language group, particularly since it has been established that hunter-gatherers are known in both the 'Bush' and 'Hottentot' language families. Furthermore, all four 'Bush' language families and all but one of the 'Hottentot' languages that Westphal has studied were recorded north of the Orange River. The one exception is Griqua which was spoken in the northern Cape and Transkei in the last century, but for which there is no adequate documentation (Westphal 1963:248). Westphal, considers that the small number of differences between Hottentot languages recorded between Angola and the northern Cape argues for a remarkably homgeneous language family with no important variants between the Natal-Transkei border and the borders of Angola (1963:249). While this still does not give us confidence to assume that the herders met along the Cape coast after 1496 A.D. spoke the same language as the herders who spoke Griqua in the northern Cape in the 19th century, it does give some support for a relatively rapid and recent population movement by Tshu-Khwe-speakers southwards from Botswana. Whether this was the same population movement which introduced domestic stock and pottery further south to the southern and south-western Cape, however, remains a matter for speculation.

Several archaeologists (Deacon, H.J. *et al.* 1978; Inskeep 1978a, Robertshaw 1978b) have recently attempted to use archaeological data, particularly radiocarbon dated assemblages with pottery and sheep, to test Elphick's (1977) hypothesis based on historical data. The latter suggests the route followed by immigrant Tshu-Khwe people with domestic stock southwards from Botswana was initially directly southwards to the northern Cape in the vicinity of the confluence of the Orange and Vaal Rivers. From this centre groups moved south and east to the eastern Cape and thence westwards along the coast to the south-western Cape. A second major movement from the northern Cape centre was westwards along the Orange River to the west coast and thence northwards into Namibia and southwards along the coast where they met the other branch between the Olifants and Berg Rivers. The long period of separation implied by the independent movements of the two branches from the northern Cape centre would account for the differences noted in historical times between the Nama of the west coast and the Khoi groups in the southern and south-western Cape (Elphick 1977:21). In testing this hypothesis archaeologically we would expect to find the earliest pottery and domestic stock in the northern Cape and Botswana, but as the data discussed in the next section will show, the pottery in Namibia is as early as that from the southern Cape coast, no dates are as yet available for Botswana, and the data from the northern Cape are ambiguous.

A second difficulty exists in distinguishing early herder sites from those of hunter-gatherers. There is little to differentiate the stone toolkits in pre-and post-pottery assemblages, skeletal remains are both rare and ambiguous, where the bones of domesticated animals are not present there is nothing to differentiate the hunting and gathering pattern, and it is very difficult to date and indeed to locate open sites where herders preferred to camp. All these factors account for the low visibility of herder sites and it is clear that the chances of tracing the routes taken by the hypothetical first herders are extremely slim.

The high mobility of individuals in herder and hunter-gatherer communities in historic times (Marks 1972; Schrire 1980) may be a recent result of disruption by Europeans, but the spread of domestic stock and pottery making was undoubtedly assisted as much by the acculturation of indigenous hunter-gatherers as by the migration of people from one part of the sub-continent to another. This would account for the lack of distinction between the stone toolkits of hunter-gatherers and herders and makes one cautious to accept a correlation between Ceramic Wilton and herders and Smithfield and hunter-gatherers.

On present evidence, a hypothesis involving the fairly rapid southward migration of people with pottery and sheep from Botswana to the northern Cape is tenable, but needs to be tested against new archaeological data. Acceptance of this hypothesis would not rule out the possibility, however, that acculturation of local hunter-gatherers also took place and may have been the mechanism by which pottery and domestic stock was introduced to the eastern, southern and south-western Cape. In this instance the source of the domestic stock and pottery-making techniques need not have been the same as that of the Botswana/northern Cape herders. A third route southwards along the west coast from northern Namibia to the south-western Cape may have been followed by ancestors of the Nama.

Dating

The speed with which pottery was introduced into southern Africa can be seen in Figure 9 which summarizes the earliest occurrences on non-Iron Age pottery in Africa plotted against their latitudinal positions. Pottery vessels were made at least as early as the 10th millennium B.P. in north Africa (Alessio *et al.* 1978; Close 1980), but the technique filtered south of 5° S into southern Africa only some 7000 years later. On present evidence it appeared virtually simultaneously about 2000 years ago over a latitudinal distance of some 3000 km from Angola to the southern Cape coast. The

Table 9. Radiocarbon dates associated with pottery, sheep and cattle in Later Stone Age contexts in southern Africa. For abbreviations and references see Table 12.

			ZM
			ZM
			ZM
			ZY*
			WOR
			WOR
			TIR
			TIR
			TIE*
			TIE*
			TIE
			THA
			SS*°
			SOS
			SG°
			POC
			POC
			ORA
			ORA
			OAK*°
			NUP
			NP
			NOS
			NOS
			NBC°
			NAP
	VP		MO
	URU		MTA
WK	T		MTA
STF	SS*°		MOS
SHO	SMT		MAG
PB	SMT		LUP
NBC°	SHO		HUN
MIR°	SEH	ZY	HOA
MEL°	RC	SHO	HOP
LIM	MO	RC	HBP
LI*°	MES	PAT	GT
HAW°	LW	PAT	GL
ER	LIM	OB	GL
EB	HBC	NE	FA
DR	GOR	MTA	ET
DKF	GE*°	MO	ES
DK°	EBO	MES	EB°
DK°	EB°	GI	DKF°
DK°	DV	EBO	DH*°
DK	DKF°	EBO	DH*°
DE	DK*°	DKF°	DH*°
COL	DF*	DIK2*°	DH*°
C	DF*	CR	DH*°
BY	BEN	CR	CON
BPA°	BEL	CON	COL
BPA°	ARD	COL	BY°
BPA°	ARD	BY°	BUR

Table 9 (*Contd.*)

			ZOM	BEN	ARD	BEL	ARS
		RM	*WAR*	BEN	ARD	AUS	ARP
		GE	*UNI*	*AP*	ARD	AS	AP
		GE	*UNI*	AP	AR	ANN	AP
	DIK	*FA*	*MOS*	AND	AP	AND°	ANN*°
Radiocarbon years B.P.	3000-3499	2500-2999	2000-2499	1500-1999	1000-1499	500-999	0-499

NB: includes Angola but excludes Zambia and Malawi
° Date associated with sheep remains
*Date associated with cattle remains or cf cattle
Italics indicate dates where association with pottery is uncertain, according to the excavator.

rapidity of this dissemination, its close association with a new herding economy and its dating only a few hundred years prior to the Early Iron Age suggest that these new features appear as a bow-wave ahead of the Iron Age mixed farmers. The impetus for the southward movement can be seen as a result of the expansion of Early Iron Age communities into southern Africa prior to 2000 B.P. and the pottery, sheep and cattle as deriving from the same source, namely early Iron Age communities in central and eastern Africa.

Table 9 summarizes the geographic spread of radiocarbon dates associated with pottery, sheep and cattle in southern Africa. The earliest date considered to be reliably related to pottery is from Zombepata in Zimbabwe at 2110±50 B.P. (SR-185; Cooke, C.K. 1971, 1979a). There are earlier dates in the 3499-2500 B.P. time range from Big Elephant Shelter (GE) in Namibia (Wadley 1979) and from Dikbosch and Riversmead in the northern Cape (Humphreys 1974a; Sampson & Sampson 1967:18), but the excavators consider the dates may be anomalously early. A similar degree of caution has been expressed about dates between 2499 and 2000 B.P. from Uniondale in the eastern Cape (Mary Leslie Brooker, pers. comm.), Warmquelle in Namibia (Vogel & Visser 1981) and Moshebi's Shelter in Lesotho (Carter 1978), but future research may yet prove them to be acceptable. Within the first 500 years after 2000 B.P., on the other hand, there are 23 sites with 30 dates, all of which are clearly associated with pottery and at least 9 of which are associated with sheep remains as well. As it is unlikely that all these represent the 'first' appearance of sheep and pottery in southern Africa, there is a good chance that earlier well dated occurrences will be found.

Stone Age pottery-making in southern Africa shows regional differences in form and decoration between Zimbabwean Bambata ware, Cape coastal ware and the pottery at inland sites in the northern Cape and Natal/Drakensberg region. Changes through time are apparent at sites as far afield as the northern Cape (Sampson 1972), East London and Matjes River (Rudner 1968:598) and the Wilton name site (Deacon, J. 1972) in the eastern and southern Cape, and Die Kelders in the south-western Cape (Schweitzer 1979) where relatively thick-walled pottery was preceded by thinner and more finely made ware suggesting that pottery was initially made by people who were experienced in the art and was then adapted locally with regional styles developing. Apart from the thickness of the sherds, the earliest pottery at Die

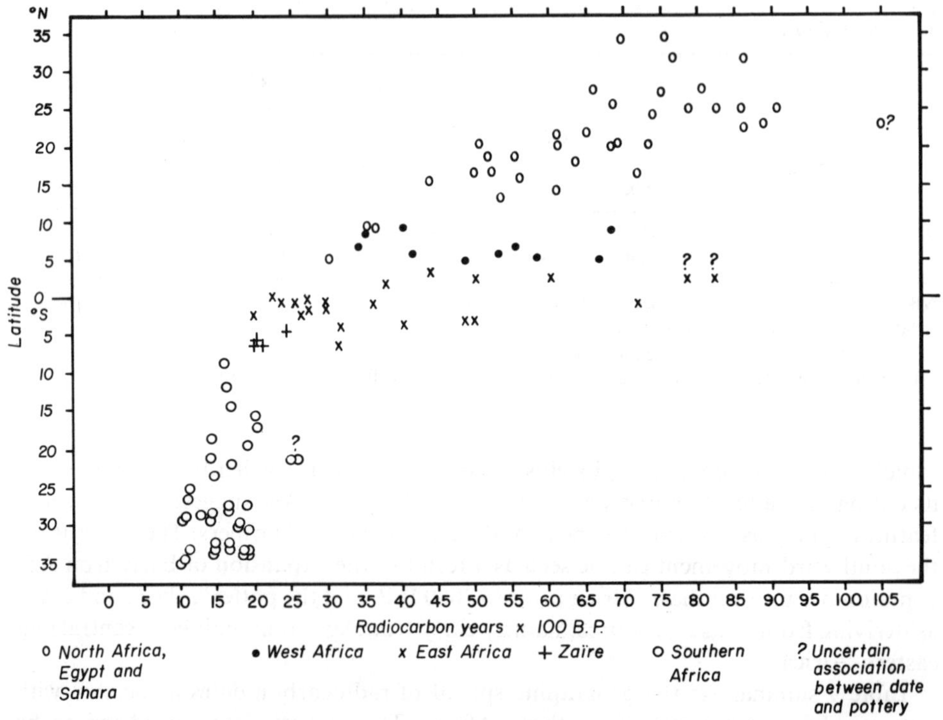

Figure 9. Early radiocarbon dates for Stone Age assemblages with pottery vessels in Africa. Time (horizontal axis) is plotted against latitudinal position of the sites (vertical axis) to demonstrate the spread of pottery southwards. Data are from date lists published in *Radiocarbon*. Southern African sources are listed in Table 9. Queries indicate that the association between the dated sample and pottery is uncertain.

Kelders and Boomplaas (Deacon, H.J. *et al.* 1978:58) does not include the typical reinforced lugs and pointed bases of Cape coastal ware which appears at both sites after about 1600 B.P. The wide geographic spread of the earliest dates for pottery and the apparently later development of the distinctive Cape coastal ware makes the suggestion (Inskeep 1978a:116) that pottery could have been independently invented in the southern Cape no longer tenable.

Subsistence

As the hunting-gathering-fishing part of the economy of late Holocene people has been discussed in the previous section, only information relating to the herding aspect of their economy will be dealt with here, but it must be remembered that this formed only a part of the total subsistence. Exotic livestock was introduced to the sub-continent within the last 2000 years or more and was adopted by many of the later Stone Age people before the time of European contact. Cattle and sheep (goats were probably introduced somewhat later, possibly as late as the 17th century (Elphick 1977:57-8)) were accumulated as a form of wealth, however, and apart from providing their owners with milk and its by-products, did not usually provide the bulk of their diet. Khoi herders practised no agriculture although some contemporary reports claim that dagga (a narcotic related to hemp and marijuana), which was used as an important trade item, was cultivated both in the western Cape (Moodie 1841:215) and along the Orange River (Elphick 1977:63). The result of a reluctance to slaughter livestock made it vital for the herders to maintain their hunting, gathering and fishing skills and these were relied on even during times of plenty.

Our knowledge of Khoi animal husbandry as it was practised at the time of European contact in the 15th, 16th and 17th centuries at the Cape is limited by the lack of detailed contemporary descriptions. Those we have may be found in official Dutch documents such as the Journal of Jan van Riebeeck (Moodie 1841; Thom 1952, 1954, 1958) or in travelogues such as those of Sparrman (1785) and Paterson (1790). Other shorter reports have been translated and collated by Moodie (1841) and Raven-Hart (1967, 1971). More recent accounts of Nama herders in the north-western Cape and Namibia have been published by Schultze (1907) and Hoernlé (1925). Other sources frequently quoted, such as Theal (1896), Stow (1905), Schapera (1930), Maingard (1931), Engelbrecht (1936), Epstein (1937, 1971), Wilson (1969) and Elphick (1977), have drawn on the older literature, while Kolben (1731) compiled much of his account from hearsay. Details of interest to archaeologists include the daily and annual cycle which would have controlled movements, the size and composition of the herds, the products associated with herding, items of material culture that were exclusive to the herders, and the stock breeds and their origins.

As is the case amongst most herding peoples, the Khoi were well versed in management techniques and had a great affection for and intimate knowledge of the individual histories of all their animals. Cattle, in particular breeding stock, were highly prized. They were trained to answer to a whistle (Raven-Hart 1967:82) and oxen in particular were used for riding, as pack animals when moving camp or transporting veldkos, and would assist in warding off enemies, both animal and human, by forming a barrier between their owners and the enemy (Engelbrecht 1936:114; Raven-Hart 1967:82, 1971:119; Wilson 1969:56; Elphick 1977:56). Riding oxen were guided by a rawhide 'bridle', the bit being a piece of wood or leather passed through a hole in the nose cartilage (Schapera 1930:297).

Khoi cattle were the ancestors of the modern Afrikander breed, classified today as a member of the cervico-thoracic-humped Sanga group that has evolved through inter-breeding between humped zebu and cattle characterized by long or gigantic horns (Epstein 1971:556). Epstein (*op. cit.*) suggests they began their evolution in Ethiopia or Uganda after the introduction to eastern Africa of humped zebu from south Baluchistan and southern Arabia during the second millennium B.C. A study of the haemoglobins of Afrikander cattle by Singer and Lehmann (1963:125), however, indicates that they are decisively of African rather than Asiatic orgin (see also Epstein 1971:551). The hump was not always well-developed amongst Khoi cattle (Kolben 1731), but it is mentioned by several observers in the 16th and 17th centuries (Raven-Hart 1967:19,20, 1971:213) and Epstein (1937) suggests that it would only have developed prominently if the cattle were well fed. Contemporary observers estimated that Khoi cattle were on average larger than European breeds and oxen weighed between 225 and 275 kg (Kolben 1731). They were mostly reddish in colour with rare yellow-brown individuals and rarer white ones. The Nama are known to have selected their breeding stock for particular colours and to have traded or exchanged animals in order to obtain herds of the same colour (Schultze 1907).

Cattle were owned by families and jointly by husband and wife, with the result that they were passed on to children and other relatives when the owners died. All the cattle belonging to families living together as a residential unit were herded together, owners taking it in turns to tend the herd daily by sending young boys and children out with them (Schapera 1930). Wealthy families with very large herds would dispense their cattle to client herdsmen (often San hunters) who would care for them and earn cattle of their own in return (Schapera 1930: 296). Such a practice could well complicate the task of the archaeologist in establishing who were the herders. Additions to herds could be made by breeding, raiding other groups, trading and exchange. Estimates of the size of Khoi herds are difficult to find and differed from region to region, but one made in November 1658 of the holdings of the Cochoqua in the western Cape suggested about 20,000 cattle and 15-16,000 sheep (Elphick 1977: 118, footnote 3) to have been owned by 16-18,000 people. The ratio of cattle to sheep was apparently higher in the southern Cape where grazing was better than to the north of Cape Town (Table 10a).

Cattle grazed around the camp during the day and were brought in (with the exception of oxen and old cows) to an enclosure of branches or stones surrounded by huts at night (Epstein 1937: 650ff), but it is only when stones were used that these camps would be visible archaeologically. Bulls were not separated from cows and breeding took place when the cows were in good condition; as this was usually at the end of the rainy season the time of year was different in summer and winter rainfall areas. Young bulls were castrated when about a year old (Epstein 1937:653).

Cows were milked morning and evening by the women and the milk was generally eaten when sour by both men and women (men were not allowed to drink ewe's milk, however). The milk was stored in a variety of containers from leathern sacks with the hair inside in the south-western and southern Cape (Raven-Hart 1971: 130, 350), to finely woven baskets in the eastern Cape (Sparrman 1785), gourds and clay pots. Butter was used mostly for smearing on the body together with soot, herbs and even dung. The practice was the subject of much pious comment from European observers who found the smell 'unsavoury' (Raven-Hart 1971:395), but it is interesting to note that the purpose of smearing the body was to demonstrate wealth and prosperity (Elphick 1977:60); those without greasy bodies were considered to be poor (Raven-Hart 1971: 16). It may, therefore, have served as an outward sign or status symbol for herders making it possible to identify them as distinct from hunters who did not have access to butter (Stow 1905:314-5).

Neither sheep nor cattle were slaughtered simply for meat, but were kept for special occasions such as weddings, farewells, sacrifices to combat illness or celebrations of victories, and even then it was only old or 'otherwise damaged' individuals that were eaten (Raven-Hart 1967:101, 162, 1971: 113, 119, 343; Wilson 1969: 55; Elphick 1977: 38). This does not correlate with the archaeological evidence, however, which indicates that numbers of immature individuals were also killed. Men were responsible for slaughtering and cutting up of the carcass and although women were allowed to drink the blood, they were not allowed to eat the meat of cattle (Schapera 1930). Intestines were eaten with relish and were often hung about the body (Raven-Hart 1971).

Two breeds of sheep were seen at the Cape in historic times, both hairy rather than woolly, and both with fat tails. They are the larger, less rangy Ronderib Afrikaner

Table 10a. Sources of livestock obtained by the Dutch colonists from Khoi herders at the Cape from A.D. 1662–1713 (after Elphick 1977:160).

Khoi Group	Location	Number of animals obtained		Ratio of		
		Cattle	Sheep	Sheep	:	Cattle
Peninsulars	Adjacent to Cape Town	613	1 951	3,2	:	1
Cochoqua	Immediately north of Cape Town	2 310	9 328	4,0	:	1
Mainly Chainouqua & Hessequa	East of Cape Town to the Gouritz River	6 634	10 475	1,6	:	1
Guriqua and Namaqua	West coast north of Saldanha	99	372	3,8	:	1
Unknown or combinations of above		4 707	10 682	2,3	:	1
Total		14 363	32 808			

Table 10b. Approximate areas of 'territories' occupied by Khoi observed by the Dutch ca A.D. 1700. Calculated from a map compiled from historical literature by Le Roux (1945).

Khoi Group (from north-west to south-east)	Area in km²
Klein Namaqua	47 124
Guriqua (Grigriqua)	21 432
Cochoqua	9 411
Peninsulars	3 019
Chainouqua	12 242
Hessequa	6 574
Guriqua	4 025
Attaqua	6 708
Outeniqua	4 528
Gamtours	3 689
Inqua	14 590
Damaqua	4 864
Damasonqua	6 540
Hoengeyqua	7 378
Gonaqua	11 404

represented today by small herds in the north-central Cape, and the Namaqua Afrikaner sheep of the north-western Cape that is well adapted to drier conditions (Mason, I.C. & Maule 1960:101-4; Epstein 1971:147-60). The live weight of adult Ronderib rams varies between 60 and 80 kg and that of Namaqua rams between 50 and 64 kg. The fat tail, which some sources claim could be cut from the animal while it was alive and would grow again, is variously estimated as weighing between 4 and 14 kg (Raven-Hart 1867:35, 1971:166, 343; Epstein 1971). The ancestral stock of the Ronderib were owned by all Cape Khoi except the Namaqua (who kept Namaqua sheep), while the sheep kept by the Iron Age Bantu-speaking people of southern Africa had more in common with the Namaqua breed than with those of other Khoi (Epstein 1971:156). The thin-tailed sheep kept by the Bantu speaking people occur in

Zambia and along the east coast of Africa from Kenya to Mozambique, Swaziland and Zululand, but fat-tailed Ronderib have close relatives today in the sheep of East Africa (Epstein 1971:157).

Historic information on management of flocks is seldom differentiated from that on cattle herds. Rams were not kept separate from the flock and ewes had lambs twice a year when and where grazing conditions were favourable. Women were again responsible for milking and children were allowed to drink straight from the ewes. Young rams and older ones no longer used for breeding were castrated (Epstein 1937). Young lambs were kept in a special kraal away from their mothers except at morning and evening milking time and were allowed with the rest of the flock once they were weaned (Epstein 1937:652). Archaeologically, then, we would expect to find smaller enclosures to house them. Slaughtering of sheep was again limited to special occasions and ceremonies. The hides were favoured for cloaks and harboured numerous lice which were shaken from the cloaks from time to time and eaten (Raven-Hart 1971:509).

Dogs were kept to help with rounding up the herds and flocks and to warn off predators. They were observed by the first European visitors (Raven-Hart 1967:3,18, 1971:119), but the Khoi do not seem to have valued their dogs very highly (Epstein 1937:654). Some controls against indiscriminate breeding were observed (Epstein 1937). Khoi dogs found today in Namibia are of African pariah stock, possibly with a strain of greyhound blood carried south from East Africa (Epstein 1971:49-50). Those noted by Kolben (1731) as the Cape were seldom higher than 0,45 m with an ash-grey coat, a small head, sharp muzzle and erect pointed ears. Dogs observed with Khoi in the north-central Cape still had the characteristic ridge of hair along their backs turned forward, as is present in their hybrid modern descendants, the Rhodesian ridgeback (Epstein 1971). All these breeds, however, are considered to belong to the same subspecies, *Canis familiaris palustris,* found throughout southern Africa and closely related to eastern African domestic dogs.

Historic evidence for seasonal transhumance is often vague and may be ambiguous in that some Khoi movements may have changed with the attraction of the trading post at Cape Town. Nevertheless, it is clear that a regular grazing regime was followed, even if the Dutch could not predict precisely where the Khoi would be at a particular time (Avery, G. 1976; Elphick 1977:58). In the north-western Cape the availability of standing water for stock was a critical factor and led to groups congregating around reliable water holes during the dry season (Le Roux 1945). Data on territorial ranges of Khoi 'tribes' at the Cape are again difficult to estimate from the brief historic records, but Maingard (1931) and Le Roux (1945) have attempted to delineate the geographic distributions of these people at about A.D. 1700. Sixteen apparently distinct Khoi 'tribes' (i.e. with allegiance to a leader and presumably speaking the same dialect) were noted between southern Namibia in the north-west and the Kei River in the south east and between the coast and the inland margin of the Cape Folded Belt. Le Roux (1945) has plotted the location of kraals visited by Europeans and from his map a rough calculation of the territorial range of each group is given in Table 10b. It can be seen that the Klein Namaqua, who occupied the driest region, had the largest area of over 47,100 km^2, while in the southern Cape where the rainfall is higher the smallest territory is that of the Gamtours at about 3,600 km^2. While the details of this distribution may well be faulty due to the vagaries of both the historic records and our interpretations of them, it would nevertheless seem that there is a close relationship between rainfall and territorial size amongst herders at least. Archaeologically, therefore, we would expect greater lateral variability along the southern Cape coast than in the north eastern Cape.

Archaeological evidence for herding is limited to the remains of domesticated animals amongst the faunal remains (Tables 7, 9), accumulations of dung and the occurrence of stone walled kraals with no evidence for their having been used by Iron Age farmers (Humphreys 1972a; Carr *et al.* 1978). The largest and best preserved assemblage of sheep remains so far excavated (34 individuals) comes from Die Kelders in the south-western Cape (Schweitzer & Scott 1973; Schweitzer 1974, 1979) where 78% of the sample that could be aged is represented by sexually immature individuals, most of which were males. This suggests that they may have been culled from flocks rather than raided by opportunistic hunters, but this interpretation is open to doubt (Schweitzer 1974; Avery, G. 1976). At Boomplaas in the southern Cape, where burnt sheep dung in the floor of the cave suggests the site itself was used as a kraal, younger sheep are under-represented with only 33% of the sample being sexually immature (Klein 1978b). Klein, however, considers that this may be the result of high temperatures in the burning deposit having destroyed the bones and teeth of the younger individuals thereby biasing the age distribution. At all other sites where sheep have been identified (Nelson Bay, Dikbosch, Big Elephant Shelter and ?Aar 1) the frequencies are too low for age group statistics to be relevant. The preservation of quantities of sheep dung at Mirabib (Sandelwsky 1977) and Boomplaas (Deacon, H.J. *et al.* 1978) shows that rock shelters were used as kraals and that the cultural material associated can be firmly linked to herders rather than hunters or Iron Age farmers. The unburnt dung at Mirabib included hairs identified as those of sheep, although the breed could not be ascertained with certainty (Sandelowsky *et al.* 1979: 50). The fact that it was hair and not wool, however, leads to the assumption that they were Namaqua sheep. Analysis of the dung showed that the sheep had obtained most of their food from browsing rather than from grass. Because the dung at Boomplaas had been burnt, no similar analyses were possible there. Both occurrences are dated within the first 400 years A.D.

The incidence of bones that can be securely assigned to cattle, *Bos taurus,* is very low (Table 9) although a relatively large sample has recently been obtained at Blinkklipkop in the northern Cape (Beaumont, pers, comm.; J.F. Thackeray, in prep.) and this may help to redress the balance. Goats appear late in the sequence at Striped Giraffe shelter in Namibia (Plug 1979a; Vogel & Visser 1981) which may confirm the historic record that they were introduced relatively late into southern Africa. The undoubted presence of a domesticated dog is recorded at Cape St Francis where it is associated with a shell midden with pottery (Chappel 1968; Voigt 1978: 183). The absence of dog from pre-pottery sites suggests that they were introduced together with other domesticated animals and pottery, but it should also be remembered that it is difficult to identify domestic dog without key anatomical parts.

Stone-walled kraals without Iron Age affinities are known in the northern Cape (Sampson, pers. comm.), along the Riet River in the Orange Free State (Humphreys & Maggs 1970; Humphreys 1972a) and, with occasional finds of iron, in Namibia

(Carr *et al.* 1978). To date no firm correlations have been drawn between them and their associated pottery and artefacts, but they do indicate the adoption of stone walling techniques that were common amongst Iron Age people in the vicinity and date within the last 400 years.

All occurrences of domestic stock associated with the LSA and so far described in print have come from Namibia and the northern, western and southern Cape, Data from Zimbabwe, Botswana, the Transvaal and Natal are urgently needed to test hypotheses of the routes taken by the earliest herders into southern Africa. Rock paintings of sheep in Zimbabwe (Cooke, C.K. 1969) are tantalizing evidence that they were there. The fact that the breed of sheep kept by Khoi is different from that found amongst Iron Age people in South Africa, and that cattle and sheep were milked by women and not by men, sets the Khoi herding pattern apart from that of the Iron Age and encourages the view that domestic stock filtered southwards a few hundred years ahead of the Iron Age farmers.

The breakdown of Khoi identity and economy within 100 years after the establishment of the Dutch settlement at the Cape in 1652 (Elphick 1977:236-7) was largely due to the fact that they had livestock which the colonists and passing ships wanted for consumption and not for breeding stock. Despite the fact that they valued their cattle and sheep very highly, the Khoi were willing, initially at least, to trade them for very low prices by European standards–for iron bars, nails, brass, copper, beads and tobacco (Harinck 1969). Over the first seven years of occupation at the Cape, the Dutch estimated their livestock needs (probably an under-estimation of what they actually used) at around 4600 cattle and 18,600 sheep (Elphick 1977:153), the vast majority of which were traded from the Khoi and killed for meat. As local supplies dwindled and Khoi became less and less willing to sell their livestock, Khoi entrepreneurs acted as middlemen and obtained sheep and cattle from further afield (Table 10a) and further large numbers of stock were lost in this way. The combined effect of this practice and the competition for land, plus related social factors such as inter-tribal clashes amongst the Khoi, the death of several powerful leaders, the ravages of European diseases and depredations by hunters and dispossessed Khoi, the loss of status and wealth made it necessary for Khoi to join the Dutch in much the same client relationship as had been observed between Khoi and San. They acted as herdsmen on farms and as labourers in towns and unless they lived in areas inhospitable to Europeans, their social order and subsistence economy broke down altogether. The detailed causes of this process have been described by Elphick (1977) and the rapidity with which it occurred is the main reason why we have so little ethnographic data on the economy and material culture of the Cape Khoi.

LATER STONE AGE MATERIAL CULTURE

Descriptions of the culture stratigraphic record of the LSA have been important in demonstrating the distribution in time and space of the major artefact making traditions employed in southern Africa over the last 20,000 years and in defining the range of formal tools found at various times in various places. In order to relate this information to the way of life of LSA people, however, we need to resort again to the ethnographic record to place the artefacts and other finds in their context. We would like to know the extent to which the range of activities of LSA people can be gauged from the artefacts found at a habitation site: are the relative frequencies of stone tools an adequate index of the relative frequency with which particular activities were undertaken? Does the horizontal patterning of tools bear any relationship to areas of activity at a site? Answers to such questions are not easily come by, partly because modern hunter-gatherers and herders live in a different social environment from that which pertained before European contact, they are today largely confined to relatively inhospitable regions, and (with very few exceptions) they no longer make or use stone tools. Nevertheless there are some aspects of their lifestyle and material culture which are relevant and can be used to interpret the prehistoric record and it is these that we shall focus on.

The ethnographic record is most valuable in providing information on items of material culture that do not survive in the archaeological record and in giving an insight on how often and where particular activities were undertaken. The archaeological record, on the other hand, provides a time depth for these items and can give us some idea of the age of the constellation of ethnographically known artefacts. This section will review the range of artefacts made and used in the day-to-day activities of present-day hunter-gatherers in southern Africa. The focus is on recent literature, but extensive descriptions making use of earlier data can be found in Dornan (1925), Schapera (1930), Clark (1959), Gabel (1963), Fagan & Van Noten (1971), Vinnicombe (1976) and Inskeep (1978b).

The relevance of ethnographic data to stone artefact assemblage interpretation

Problems of preservation severely limit archaeologists in reconstructing the toolkits of prehistoric people so that the vast majority of LSA assemblages consist almost entirely of stone artefacts. The most detailed ethnographic data, on the other hand, have been gathered on hunter-gatherers who do not use stone artefacts. The most pessimistic amongst us would therefore reject both the ethnographic evidence (because of lack of stone tools) and the archaeological evidence (because of the absence of perishable items) in reconstructing the activities and artefacts of LSA people.

A more optimistic view, and one to which most of us are obliged to subscribe, would be a belief in uniformitarian principles. These allow us to assume that because many of the present-day items were made in the recent past, and because the hunting-gathering-fishing lifestyle was basically similar in the present, the recent past and the LSA in general, that there was not a great deal of difference in the range of activities and artefacts of present-day and LSA people. Where these assumptions become important is not in the description of material cultural items, but in the interpretation of stone artefact assemblages which represent only a portion of this material culture, and then often only discards and tools used for making other tools. It is therefore often only possible to see where (and, less certainly, how often) artefacts were *made* and not necessarily where, when or how often they were *used*. The nature of archaeological assemblages means that we deal in most cases with a relatively small sample of artefacts from an excavated site (most often a rock shelter or cave) which consists of items accumulated over a portion of a living site floor over an unknown period of time and an unknown number of occasions. This sampling method makes it unwise for us to attempt to recover specific information except in unusual circumstances and we must therefore concentrate on looking for gross patterns of artefact distribution through time and space. It is thus relevant to ask to what extent activities around a living site are patterned: are there task specific areas? Do men and women work in the same areas? Are some sites visited during the annual round reserved for specific activities? Are activities scheduled for areas where particular raw materials are to be found? How often are various items made? These and other aspects of the work load of modern hunter-gatherers have been tackled by ethnographers in recent years.

Dornan's wry comment (1925:95) that the San 'do not believe in unnecessary labour' reflects several interesting features of their lifestyle. Firstly, their possessions are kept to an absolute minimum and do not include items that are difficult to transport. All the artefacts, clothing, containers and possessions regularly used by one person in the Kalahari today weigh less than 10 kg and are easily made, easily transported and easily replaced (Lee 1972:363). Secondly, a limited period of observation of the Dobe !Kung suggests that both men and women together spend less than 15% of their total working time (which is less than 15% of 45 hours a week) on tool manufacture and maintenance, or the tasks that would have involved the use of stone tools in LSA times (Lee 1979:278). This works out at less than 45 minutes a day for women and 65 minutes a day for men (Lee 1979:277), with approximately one third of the time spent on manufacturing and two thirds on maintenance and repair. Wilmsen [quoted in Lee (1979:277)] has estimated that the work load with the addition of the manufacture of stone artefacts would not have been significantly greater than an additional 10-20% of the time taken at present (Lee 1979:438), and that much of this time would have been taken to flake the core and produce a suitable stone tool. A third factor which affects the time spent on manufacture and maintenance is the use-life of individual items; the variable length of time that each item lasts adds to the irregularity of many manufacturing and maintenance tasks. Figures calculated by Lee (1979:274ff) suggest that the artefacts needing most frequent maintenance are the poison kit, the bow and the arrows and these are followed by the adze/axe, knives and the flint-and-steel fire kit. Amongst the present-day Dobe !Kung these all required daily or weekly attention, while the majority of other artefacts required attention once a month (digging stick, bags, spear, probe, snares, mortar and pestle, baby carrier) or once every 2-6 months (net, quiver, club, sandals and store-bought clothing). The use-life of the vast majority of the homemade items was a year or more, with some items such as the spear and net expected to last as long as 10 years (Lee 1979: table 9.10), and a set of oracle discs with a use-life of 15 years. These observations were made over a short period of time on the present-day Dobe !Kung so we would expect that there would be variations on this pattern both geographically and chronologically, but they nevertheless give us some idea of the amount of equipment owned by hunter-gatherers and how often this needs to be maintained and replaced.

In order to test these observations against the archaeological record it is necessary to have a good control on time and to have a detailed geographical coverage. At Boomplass, for example, in the units dating between about 2000 and 1500 B.P., 780 formal tools were recovered over an area of 100 m² or approximately two thirds of the available floor area of the cave. Assuming an even distribution over this floor space, we could calculate that the total number of formal tools from these levels would be about 1170 which works out at 2,34 formal tools per year. If all worked stone is included, the figure is 86,45 per year, including unusable chips and chunks. This would seem to support present-day observations of a relatively low incidence of tool manufacture and maintenance, but time remains the unknown factor in these calculations. The number of tools per year increases dramatically if we assume that the site was visited only two or three times during the 500 years between 2000 and 1500 B.P.

By the same token we cannot expect that activities will be evenly distributed in space either. The Dobe !Kung data suggest that an average of three wooden items (including digging sticks, bows, arrows, quivers, spears, probes and clubs, but excluding worked wood for snares, traps and huts) are made per year by one household (Lee 1979). Thus a group consisting of three households would manufacture less than one item a month. If we assume that this activity was spread out evenly through the annual cycle then the chances of finding evidence for woodworking in the form of stone adzes and woodshavings are very low. If, on the other hand, some sites have many more adzes than others then we could suggest that this evidences a concentration of woodworking at these sites (Mazel & Parkington 1978, 1981; Parkington 1980).

The patterning of artefacts at an intra-site level has been examined amongst the !Kung by Yellen (1977) who has shown that several traditional assumptions are unfounded. At the sites he has examined (which are all in the open and may thus not be strictly comparable to rock shelters and caves which provide most of the archaeological data), activities are not usually separated into task-specific areas and the remains from several activities will be mixed together. This makes untenable the assumption that tools found together were necessarily used for the same task (Yellen 1977: 134), unless the site is task-specific. The lack of patterning in the horizontal distribution of formal tools at most LSA sites would confirm this observation (Deacon, J. 1972), but exceptions do occur as in the distribution of cores at Pomongwe (Cooke, C.K. 1963). Spatial patterning due to sexual division of labour is also rare, but as men may have done most of the stone tool manufacture, woodworking and skin working (at least in present-day Botswana), most of the stone tools probably relate to their activities. That women may have done some of these duties in the Cape, however, is suggested by Kannemeyer (1890) and Dornan (1925).

At the inter-site level, Yellen (1977) notes that there was no seasonal patterning in manufacturing and maintenance tasks as these are undertaken at random intervals. There is, however, seasonal patterning in subsistence activities so that if tools were closely related to subsistence we might expect some seasonal patterning there. There was also no neat correlation between the area in which raw materials for artefacts were collected and the camp at which they were processed (Yellen 1977: 76) and materials were gathered over a period of weeks and over a wide area before an item was finally made (Lee 1972: 362). This would obviously vary with the nature of the raw material, but unfortunately we have no ethnographic data to allow us to compare these patterns of behaviour at an inter-regional level.

A final point of some importance to artefact distributions observed archaeologically is that the system of sharing and gift exchange is very well developed amongst hunter-gatherer societies (Marshall 1976: 287-312; Lee 1979: 456ff) and the high mobility of individuals means that material items of all kinds can travel well outside the area in which they were made. This may apply particularly to the evaluation of finds of sea shells far inland which may not indicate that the site inhabitants had visited the coast themselves, but merely that they were in contact with others who had.

Material culture of the LSA: Ethnographic, historical and archaeological data

In this section the tools connected with various activities will be grouped together to suggest some task-specific sets, and the information we have from ethnographic and historical sources for the range of artefacts in these sets will be contrasted with that from the archaeological record.

Hunting tools and methods

Ethnographic descriptions of hunting methods amongst the modern San and historic observations of the indigenous people are unanimous in stating that the most widespread weapon used amongst non-Iron Age people was and is the bow and arrow, the main power of which lay in the poison smeared onto the arrowhead. An intimate knowledge of local vegetable, insect and snake poisons was therefore of prime importance to hunters and was also used in obtaining game by the contamination of waterholes (Alexander 1838 II: 223; Stow 1905: 42; Dornan 1925: 108; Dunn 1931: 33; Vinnicombe 1976:271). In studies of the 'cost' of various activities in terms of time spent in manufacture and maintenance, Lee (1979: Table 9.10) records that the poison kit was the most expensive item amongst the subsistence, clothing and housing activities of four !Kung San over a period of some weeks during 1964, requiring 9,2 minutes per day from men.

Simple curved bows are used throughout southern Africa and the average length is about 900 mm (Dunn 1931: 100; Vinnicombe 1976: 263). They are strung with sinew after the wood has been carved and heated into shape and, after stringing, are rubbed with the blood of a freshly killed antelope (Lee 1979: 129). When interviewed by J.N.F. Binneman in the Aminuis Corridor area of Namibia in 1980, a Nharo man said his bows generally last 1-1½ years, but Lee (1979: 129: Table 9.10) reports that !Kung bows last several years longer if properly made. The sinew bow string may be replaced four or five times per bow stave (Binneman, pers. comm.) The wood for a bow is carefully selected from a limited number of available species and the bow may be kept in a leather bag to prevent the string from becoming wet (Schapera 1930; Steyn 1971), but this is not always done (Dornan 1925: 95).

Arrows are made by the men for their own use and for gift exchange. Lee (1979: Table 8.8) observed that the arrows in the quivers of individual hunters were not all made by that hunter. Of four quivers examined, one man had 14 arrows, all of which were made by himself, but in another man's quiver there were 19 arrows, none of which he had made himself. The sharing and giving away of arrows amongst the !Kung is confirmed by Marshall (1976: 297) who notes also that men are able to identify the arrows they have made by only slight differences (*op. cit.*145). Arrows are made from a shaft of reed grass, nocked and sometimes fletched, with a foreshaft (which is sometimes called a linkshaft) of bone or wood linked to the arrowhead with a reed coupling bound with sinew. Of the arrowheads made today there are two basic types: those with reversible bone heads and those made from fencing wire beaten flat at one end into a tanged shape (Figures 16 and 17); in addition, some people still make carved wooden arrowheads mostly for shooting birds. Museum collections dating within the last 150 years, however, have additional types with tanged bone heads, split bone foreshafts with small metal inserts, and a few with stone or glass inserts in a mastic-tipped head (Figure 16) (Stow 1905:69-70 and illustration opp. p. 80; Vinnicombe 1971, 1976:267; Clark 1977a). Present-day San carry their arrows in leather, wooden or bark quivers, but when out hunting the quivers are often left at the camp and only a few arrows are taken for immediate use. Steyn (1971) writes that four or five arrows are ready for use in the average Nharo hunting kit, others being left behind or in various stages of manufacture or disrepair. From the rock art of the Drakensberg Vinnicombe (1976:267) observes that 3-9 arrows are illustrated per person in hunting scenes, but the number is greatly increased in paintings of warfare between groups. She quotes Orpen (1908:477) who found 102 arrow in a quiver when a surprise attack was made on a Bushman camp during a skirmish with Europeans in the Orange Free State at the end of the last century to confirm that many more arrows were made for defence and inter-group clashes than were used when hunting.

Wooden spears with fire hardened tips and without metal heads are reported as having been in common use amongst the Cape Bushmen by Dunn (1931:32) and amongst northern Namibian groups by Schapera (1930:132). In more modern times, Silberbauer (1965:59) notes metal-headed spears being used in finishing off animals that have been slowed down by poisoned arrows and a similar observation is made by Lee for the !Kung (1979:137). Spears are also much in evidence in Drakensberg rock paintings (Vinnicombe 1976:271) and were used by Khoi at the Cape (Raven-Hart 1971). Long barbed sticks are used by some people in hunting burrowing animals such as spring hares (Steyn 1971:301; Yellen 1977:142; Lee 1979:139) and sticks are also used for catching snakes and birds (Schapera 1930:133). Methods such as the running down of small antelope need only a dog as an assistant and a wooden club, as well as a lot of practice and stamina (Dornan 1925:103; Schapera 1930:133).

Traps and snares form an important part of the hunting equipment. They are made from sticks of various lengths joined by vegetable twine or sinew to form a noose and spring mechanism (Steyn 1971; Yellen 1977; Lee 1979:141). Larger game was observed to have been caught in elaborate pitfalls disguised with a covering of twigs, grass and soil (Kolben 1731; Sparrman 1785; Barrow 1801:284; Stow 1905:90; Schapera 1930). Barricades and obstructions of sticks and stones, stone hides and walls were also used to lead game into traps or into narrow places where hunters with bows and arrows or clubs could despatch them (Vinnicombe 1976:292), and more elaborate traps were sometimes spiked with poisoned stakes (Stow 1905:92).

This brief description of hunting methods serves to illustrate the wide range of possibilities open to hunters, but it also shows that there is a relatively small number of items relating to these activities that could be expected to appear in a prehistoric habitation site. That the bow and arrow was widely used during the LSA is amply confirmed by rock paintings depicting them being used throughout southern Africa. Physical evidence of their existence in the past, however, is not as abundant. The most common item relating to the hunting toolkit is the bone foreshaft or linkshaft with one or two blunted ends that is often hardly distinguishable in broken form

Table 11. Earliest occurrences of selected artefacts typical of the southern African Later Stone Age. Sites in italics signify Middle Stone Age associations (or Tshangula in the case of Pomongwe (P)). Abbreviations and references are listed in Table 12.

Dating x 1000 BP	Bored stones	Reamers	Digging sticks	Netting	Eyed needles	Tortoiseshell bowls	Bone points	Arrows	Bows	Grooved stones	Fish hooks	Sinkers	Ostrich eggshell beads	Borers	OES water containers	Decorated OES	Painted stones	Engraved stones	Stone hearths	Formal burial	Pottery	Fire sticks
95							*KRM*															
90																						
85																						
80																						
75																						
65																						
60																						
55																*AP*						
50																						
45																						
40																						
	ZOM																					
	BC						BC						BC									
35													HK									
30																	*AP*					
25																						
	LH																					
							BPA						BPA									
	MAT	MAT																				
20																						
19										AP			AP									
													HK									
18													NBC									

17																						
16	N						*P*						*P*									
15						BPA							MHB			MHB						
14	KAL												BPA	BPA	BPA	BPA						
13																						
12							*B* NBC						*B* BY		BY	BY						
11	MR	MR					MR						MR			+		BAM		MR		
10					B		B BPA						+ +		MHB			WK		NTI		
9	X	X			P	P	EB			P			+		AP	+				WT		
8							+			MR			+ +	+		+						
7	X	X											+	+	+	+						
6				MHB			+	MHB		BUF		NBC	+ +				BPA	WK WK				MHB
5	X	KLP	G	X			+	G	G	MIR			+	+	+	+	KRM			CSF		G
4					A		+			+		X	+ +	+	+	+	ROB	WK WK		X		
3	X	X	GE					GE	P GE	+ +	BEL DRI	X	+	+	+	+			DE	+	ZOM DK	GE
2			DKF	SS		DK	+	P	DIK	+	COL		+ +				BPA		BPA	+	BPA	DKF
1	X	X	AP	DH		DH	+	DH	SS	+		X	+	+	+	+				+	++	DH SS

X Fairly common
\+ Found at most sites

from the bone arrow point of the reversible arrow. They are found in small numbers in LSA contexts, from Zimbabwe (Cooke, C.K. & Robinson 1954) to Namibia (Wendt 1972) and South Africa (Deacon, H.J. 1976: Schweitzer 1979). They are most common in deposits post-dating 12,000 B.P., but are known also from Boomplaas dating back to 21,000 B.P., from Border Cave back to ca 38,000 B.P. and one fine example is known from the Middle Stone Age levels at Klasies River Mouth (Singer & Wymer 1982). In general, however, they are not found in deposits older than 20,000 B.P. (Table 11). Broken portions of tanged bone arrowheads are known from Glen Elliott in the northern Cape (Figure 7:8) (Sampson 1972:199) and from Uniondale in the eastern Cape (Mary Leslie Brooker, pers. comm.), both post-dating 2500 B.P.

It is speculative to suggest that the first appearance of backed microliths indicates the introduction of the bow and arrow and the same can be said for the early dating of the bone points from Klasies River Mouth. Although backed microliths are found in a few deposits pre-dating 12,000 B.P. and are most frequent during the Holocene, the only convincing evidence of their having been mounted as barbs or arrowheads comes from instances where mastic is found adhering to them as at Melkhoutboom (Deacon, H.J. 1976) dating from about 5000 B.P. and from a number of other more recent sites where preservation has been good. On the basis of the positioning of mastic traces, Phillipson has suggested a number of ways in which they could have been mounted either as arrowheads or as barbs (Phillipson 1976:217, 1977:26).

Items found more rarely because of preservation problems include broken portions of bow staves (Deacon, H.J. & Deacon 1963; Humphreys 1979:124; Wadley 1979; Cooke, C.K. 1980), broken portions of arrow shafts (Clark & Walton 1962; Deacon, H.J. 1976:43; Cooke, C.K. 1975; Parkington 1977a:63; Wadley 1979:33) and pieces of bark that may have formed part of a quiver (Parkington 1977a:64). Wooden hooks that may have been used as triggers for traps are reported from Aar in Namibia (Wendt 1972:35) and from Windhoek in the south-western Cape (Grobbelaar & Goodwin 1952:103-3) and other pieces of worked wood from sites such as Scotts Cave (Deacon, H.J. & Deacon 1963), Melkhoutboom (Deacon, H.J. 1976) and De Hangen (Parkington & Poggenpoel 1971) may also have been used as parts of traps. Twisted cord or twine made of vegetable fibre has been found at several of the sites mentioned above and this, too, could have been used in the making of snares and traps.

While the number of stone tools used in hunting weapons was limited to backed (and un-backed) microlothic inserts in arrowheads (Clark 1977a), stone tools were used extensively for the manufacture and maintenance of hunting equipment. Unretouched flakes, adzes, spokeshaves and scrapers were used for shaping wood and bone for arrows, arrow parts, wooden sticks and clubs, portions of traps and snares, sharpened stakes for placement in pitfalls and for making quivers from bark and wood. Grooved stones, still in use today amongst some !Kung groups (Marshall 1976:145), were used to straighten bone arrow points and reed shafts when heated in a fire (Kannemeyer 1890:127; Dunn 1931:103; Clark 1959:225) and to apply poison to arrowheads (Stow 1905:66). Grindstones were used for preparing poison and containers for storing poison were made of wood (Steyn 1971:300) and probably other materials as well. A stick with a lump of mastic or poison hanging from it was found in a level dating within the last 1500 years at Melkhoutboom (Deacon, H.J.

1976: Figure 42) and is similar to modern examples often carried in a quiver with the arrows.

Once an animal had been killed, unretouched flakes or a retouched cutting tool were used to skin it and cut up the meat (Kannemeyer 1890; Stow 1905:66). An unretouched flake bound with fibre from Melkhoutboom may have been made for this or a similar purpose (Deacon, H.J. 1976: Figure 42). The meat was either eaten raw, barely cooked or dried for future use, as is the practice today (Lee 1979). Cooking on hot stones is described by Schreyer (Raven-Hart 1971:130) and archaeological evidence for this method has been found in large circular stone-packed hearths with charcoal between and beneath the stones at Boomplaas (Deacon *et al.* 1978:55) and at open sites on the southern Cape coast (Goodwin 1946b:4; Cairns 1975:37), and in the south western Cape (Avery, G. 1974:108; Robertshaw 1979). All date within the last 2500 years. A series of elongated charcoal-filled depressions found in the mid-Holocene levels at Boomplaas are suggested to have been used as smoking pits, possibly for drying meat (Deacon, H.J. 1979) and a similar feature was excavated at Andriesgrond in the south-western Cape (Parkington, pers. comm.).

Fishing tools and methods

Fishing is clearly evidenced in the Later Stone Age after 12,000 B.P. by finds of fish bones at coastal sites and, during the Holocene, at inland sites. Rock paintings in the Drakensberg suggest that inland fish were caught using wooden spears from rafts or reed boats (Vinnicombe 1965, 1976:293) and Stow (1905:72) recorded the use of barbed stone 'harpoons' along the Orange River as well as the use of reed fish traps (1905:93; Dornan 1925:166-7) and bows and arrows (Stow 1905:92). Spears were used by 'Bushmen' along the Vaal River in the early 19th century (Burchell 1822:294), and along the coast (Davernier *in* Raven-Hart 1967:181) where metal hooks and line were also used (Thom 1952:80; Kolben 1731:255-6). Coastal fishing on a large scale was made possible by the use of tidal fish traps still to be seen, and even used, today along the southern and south-western Cape coast (Goodwin 1946b; Avery, G. 1975). They probably date within the last 2000 years and appear to be linked to the activities of herders rather than pure hunter-gatherers (Avery, G. 1975:112). Stone walled fish traps were also built inland in historic times along the Riet River in the Orange Free State (Stamelman 1948; Willcox 1965:139-40).

The archaeological evidence for fishing tools and methods is again rarer than the fish remains themselves. Small stone 'sinkers' have been found at coastal sites dating from at least 6000 B.P. (Deacon, H.J. 1970; Deacon, J. 1978; Inskeep 1978b:43) which may indicate either line fishing or the use of nets. They have not been found inland. Associated with fish remains at terminal Pleistocene deposits at Nelson Bay and Elands Bay are double-pointed polished bone slivers known as 'fish gorges' (Klein 1972b: Parkington 1977a; Deacon, J. 1978). Goodwin (1946b) reported the finding of a similar tool at the end of a fishing line among the stones of a fish trap wall in the southern Cape. He observed that no such bone artefacts had yet been discovered from midden deposits, but they are known now from both the terminal Pleistocene/early Holocene and from late Holocene assemblages (see also Poggenpoel & Robertshaw 1981). The gorges are assumed to have been tied in the middle and would lodge in the fish's mouth when the line was pulled. Thorns are used in this way by present-day Mozambique fishermen (Clark 1959:231; J.L.B. Smith, pers. comm. 1970). At a few sites in the north-eastern Cape (H. Opperman, pers. comm.) Natal (Maggs & Ward 1980) and Lesotho (Carter 1978), finely made bone hooks, assumed to be fish hooks, are known from assemblages dating within the last 2000 years.

The collection, transportation and prepration of shellfish is evidenced by their occurrence in quantity at many midden sites along the coast and at some inland localities where freshwater mussels were available. No formal stone tools are directly associated with shellfish collecting. The suggestion has been made that polished rib bones may have been used to prise shellfish from the rocks (Avery, G. 1976; Poggenpoel & Robertshaw 1981). 'Scraper' retouch around the edge of some *Donax serra* shells, many of which have had a hole made through the middle, has been noted at Elands Bay, Die Kelders, Pearly Beach and Nelson Bay and at several other coastal sites and it has been suggested that they were designed to scale fish (Schweitzer 1979; Poggenpoel, in prep.). The enigmatic shell crescents from Oakhurst (Goodwin 1938) and other sites may also have been used for some purpose connected with marine resources, but we have no ethnographic or historic analogies for them.

Gathering tools and storage methods

In southern Africa the artefact most consistently related to plant food collecting is the digging stick, sometimes weighted with a bored stone. Digging sticks are often referred to in historic record (e.g. Sparrman 1785 I: 306) and are clearly illustrated in the rock art where they are associated with women (Vinnicombe 1976:277 ff). Men are known to use them, however, and to make them for women (Steyn 1971:311). They are and were used for digging up roots, tubers, corms and termite eggs (Sparrman *op. cit.*; Dunn 1931:33), for excavating pitfalls (Stow 1905:67, 91), for digging burrowing animals from their holes (Lee 1979:124) and for digging holes for traps (Steyn 1971:312). Digging sticks used today are made from hard wood about 1-1,5 m long and the tip is sharpened with a knife or is hardened in the fire. There are several descriptions of buck horns being used to tip digging sticks where the ground was stony (Dornan 1925:119; Raven-Hart 1967:3; Bleek quoted in Mazel 1980) and Dunn (1931:105) records that a gemsbok horn was used at times as a substitute for a digging stick.

Bored stone digging weights were usually drilled with stone reamers (or rimers) (Kannemeyer 1890:225; Goodwin 1947:8) by women (Dunn 1931:78-80) and a small wooden peg was used to secure the stone in place on the stick (ibid). Smaller bored stones were used as club heads (Goodwin 1947:3) and weights in traps, and very large ones are known to have been made and used by modern Iron Age agriculturalists to weight their hoes (Goodwin 1947:3). Bored stones have been found in most parts of South Africa, but are apparently unknown in Namibia north of the banks of the Orange River (W.E. Wendt, pers. comm.).

Broken portions of digging sticks have been found in archaeological context in Zimbabwe at Pomongwe (Cooke, C.K. 1980), in Namibia (Wendt 1972:35; Wadley 1979) and in the south-western Cape at Diepkloof (Parkington 1977a). All these finds date within the last 3000 years, although the Pomongwe one may be older. Bored stones are ubiquitous throughout southern Africa although they are relatively rare in dated archaeological sequences. The range in size, shape and distribution is described in detail by Goodwin (1947). The earliest examples of suitable size for digging stick weights so far known come from Matjes River Layer D where they date to about 11,000 B.P. (Louw 1960), but in Zambia one has been found with the Proto LSA at Leopard's Hill dating to about 22,000 B.P. and others are known from the Nachikufan I dating to around 16,000 B.P. (Miller 1969) (Table 11). In addition, much smaller bored stones with grooves around the perforation are reported from Matupi in Zaire at 21,000 B.P. (Van Noten 1977) and from the Early LSA at Border Cave (Beaumont *et al.* 1978); other small examples without grooves are dated from late Pleistocene levels at Zombepata (Cooke, C.K. 1971) and Khami (Cooke, C.K. 1979) in Zimbawe and from Kalemba in Zambia (Phillipson 1976: 152). Thus digging stick weights probably date to within the last 12,000 years south of the Zambesi but extend back as far as 22,000 B.P. or more in Zambia. Digging sticks could be much older, however.

Other items used in the collection of plant foods include skin bags of various kinds and sizes, knotted mesh nets made from plant fibres, and wooden and bark containers (Steyn 1971; Silberbauer 1973; Yellen 1977; Lee 1979: 124ff). Skin bags have not survived in archaeological context (although small pieces of leather have), but they are illustrated in the rock art (Vinnicombe 1976). Small pieces of netting have been preserved at Scott's Cave (Wells, M. 1965: 82), Diepkloof (Parkington 1977a) and Melkhoutboom (Deacon, H.J. 1976). Nets are frequently used today to transport ostrich eggshell water containers, but the mesh size of that from Melkhoutboom is much smaller than these modern examples and suggests that nets were used to transport plant foods, the size of the mesh indicating the size range of these foods (Deacon, H.J. 1976: 42). The Melkhoutboom finds extend the dating of netting back to ca 5000 B.P. (Table 11).

There was and still is extensive use made of stone for the preparation of vegetable foods. Nut cracking stones are used today by the Dobe !Kung in Botswana (Yellen & Lee 1976; Yellen 1977; Lee 1979: 151) and upper and lower grindstones are used for grinding and pounding. Similar grindstones, rubbers and pitted hammerstones are found at many excavated sites but their generalized function makes it difficult to use their presence as an indication of specific activities. Of particular interest was the finding at Montagu Cave of a grindstone at the top of an oval pit which contained carbonized plant remains, suggesting that the stone had been placed in the pit to grind the plant food (Keller 1973: 24). Most of the plant material was from the corms of *Homeria* sp and/or *Moraea* sp.

The most often used storage container throughout southern Africa was the ostrich eggshell, still used today in the Kalahari for storing water (Steyn 1971; Yellen 1977 Lee 1979: 120) and sometimes berries (Dornan 1925: 121). Fragments of ostrich eggshell are found in almost all LSA deposits and are known also from MSA times. The finding of scratched pieces of ostrich eggshell from MSA levels at Apollo 11 may indicate the use of the shells as water containers because they are often decorated, but the earliest clear use of the shells for this purpose comes from finds of fragments of ground apertures preserved at Boomplaas in levels dated at 14,000 B.P. (Deacon, H.J. 1979). Caches of water containers have been found, as have elaborately decorated shells (Rudner 1971a: 139; Sandelowsky 1971), and some shells haye been used for storing specularite and ground ochre. Tortoiseshells were and still are widely used as containers for herbs, tobacco and other materials and were also modified as bowls for eating and drinking (Stow 1905:68; Lee 1979:155). The oldest dated example is again from 14,000 year old levels at Boomplaas (Deacon, H.J. 1979) but others are known from the early Holocene at sites in the Matopos in Zimbabwe (N.J. Walker, pers. comm.), and within the last 2000 years at south-western Cape sites such as Die Kelders (Schweitzer 1979), De Hangen (Parkington & Poggenpoel 1971) and Smitswinkelbaai (Poggenpoel & Robertshaw 1981). Bone 'spoons' were found at Die Kelders (Schweitzer 1979) and Smitswinkelbaai (Poggenpoel & Robertshaw 1981) and a portion of an ivory 'knife' from De Hangen may have been used in a similar way to the bone tsama melon kinves used by people in the Kalahari today (Schapera 1930; Steyn 1971).

The introduction of pottery containers during the last 2000 years or more must have made a considerable difference to the available storage capacity of LSA people and to their cooking methods. Historical records show that pots were made by women from clay and soil from termite mounds, with quartz or sometimes grass temper. The moist clay was moulded to the desired shape, left to dry in the sun, and then fired (Rudner 1968:564ff). A piece of wood was sometimes used to smooth the clay (Rudner 1968:566) and shells were used to impress some of the decorative motifs (Wilson & Halkett 1981). Dating of the pottery has been discussed in an earlier section of this paper. Another method of storage was in the form of pits in rock shelter floors. Those preserved at Melkhoutboom (Deacon, H.J. 1976) and Boomplaas (Deacon, H.J. 1979) were lined with grass and *Boophone disticha* bulbar leaves and were used for storing oil-rich seeds. Localized areas with large numbers of seed casings at Bushman Rock Shelter in the Transvaal may also be the result of pit storage (Plug 1978).

The gathering of foods other than plants, such as tortoises, locusts, honey, birds' eggs and termites would have required no special tools other than collecting bags and possibly digging sticks although Dunn (1931:29) described the use of a multi-pronged stick for dislodging honeycombs and several early travellers remarked on the use of bark containers to transport the honey. Mats were used for winnowing termite eggs, but while portions of matting have been found at Scotts Cave, Melkhoutboom and De Hangen they may have been used for covering huts as well as winnowing.

With gathering, as with hunting and fishing, the number of artefacts specifically linked to gathering activities and likely to be preserved in archaeological sites is limited. Bored stones are the most obvious exception, but it is clear that they were seldom discarded and, when they were, they were not often left at cave and rock shelter sites. Further, the fact that not all digging sticks were weighted makes their absence of little significance when evidence for plant collecting is sought. What is interesting, however, is that the dating of several artefacts closely linked with the Khoi and San of modern times, namely bored stones, tortoiseshell bowls and ostrich eggshell water containers, are of the same general antiquity as the early microlithic technology of the late Pleistocene.

Clothing and decoration

Southern African hunter-gatherers wear the minimum of leather clothing consisting usually of one or two aprons for women, worn from the waist, with a cloak around the shoulders, and a pouched loin cloth or pull-through for men who sometimes also wear a small apron at the back (Schapera 1930:68). Nharo women convert their cloaks into collecting bags and may also have another smaller cloak to cover the back in cold weather if the larger one is used for collecting; a sling is used for carrying a baby (Steyn 1971:311). Khoi herders wore caps and usually had a sheepskin kaross and their aprons were somewhat longer than those worn by the San (Schapera 1930:68). Sandals were worn by some people and were made from the thickest hide available. Dunn (1931:37) records that in an emergency the sandal could provide a meal, for once it had been placed in hot ashes, it became crisp and was then pounded to a coarse powder, mixed with water and boiled into a thick soup! Leather was and still is used for making bangles and anklets and a fillet is sometimes worn around the head, often decorated with strings of ostrich eggshell and other beads. Ostrich eggshell, marine shell and bone beads and pendants were used for anklets, bangles, hair decorations and necklaces and are and were also sewn onto bags and garments. Seeds and dried cocoons are used as rattles and decoration.

Archaeological examples of clothing and decorative items are known from most living sites where preservation is good. A sandal was found at Big Elephant Shelter in Namibia (Clark & Walton 1962) and a part of another was found at Rooiels Cave in the south-western Cape (Smith, A.B. 1981). Portions of sewn garments were found at De Hangen, Melkhoutboom and Scotts Cave, and part of a kaross was found at Roodekrantz in the eastern Cape (Deacon, H.J. 1976:40). Beads and pendants are found at almost every site, with European trade beads in recent deposits. Perforated cocoons for parts of a rattle were found at Fäckeltrager in Namibia (Wendt 1972:59) and at Mirabib (Sandelowsky 1977). Ostrich eggshell beads are particularly numerous in association with burials (Goodwin 1938) where parts of garments can sometimes be reconstructed.

It is difficult to date the advent of leather clothing because of preservation problems, but the well preserved male pubic covering of soft grysbok leather from the Main Bedding layer at Melkhoutboom (Deacon, H.J. 1976:40) is dated to about 5000 B.P. The small convex scrapers assumed to have been used for preparing such leather are only found in quantity in Holocene deposits, but this is not to gainsay that other scrapers had been used for this purpose before. Needles and awls are also only found with any frequency in Holocene deposits. Small numbers of ostrich eggshell beads have been found in deposits older than 20,000 B.P. at Border Cave (Beaumont *et al.* 1978), Heuningneskrans (Beaumont 1981) and Boomplaas (unpublished), but they are more common after 18,000 B.P., and only become numerous in Holocene horizons. Stone borers used for perforating the shell and grooved stones for smoothing the edges are again very rare before 10,000 B.P. Bone pendants occur in terminal Pleistocene levels at Nelson Bay and in Holocene contexts at inland sites such as Wonderwerk (Beaumont pers. comm.) and on the coast shell pendants are found.

Other tools used for the preparation of skins and leather for clothing, bags and decorative items include bone 'spatulas' for working fat and herbs into the skins to make them soft or for removing the hair (Clark 1959:237), and wooden pegs for stretching the skins while drying (Deacon, H.J. 1976:47). Some fine examples of bone spatulas are known from the late Holocene levels at Die Kelders (Schweitzer 1979) and from Nelson Bay (Deacon, J. 1978) where they first appear in the terminal

Pleistocene and early Holocene levels. Wooden pegs are again limited by preservation, but are known from as early as about 5000 B.P. at Melkhoutboom (Deacon, H.J. 1976).

Eyed needles are known from a few scattered localities and while some may have been used for sewing leather, Walker (pers. comm.) has pointed out that they are morphologically much more similar to matting needles from ethnographic context than they are to sewing needles. The earliest example comes from the early Holocene levels at Pomongwe in Zimbabwe (N.J. Walker, pers. comm.) and others are known from Amadzimba (Cooke, C.K. & Robinson 1954:706-8), Bushman Rock Shelter (Plug 1978) and from Umgazana in Transkei (Clark 1959:Figure 52).

Fire and shelter

While the use and control of fire is evident well before the LSA, it is only during this time that the characteristic fragments of fire making sticks are found to indicate the method used. The archaeological examples are exactly similar to modern ethnographic specimens as illustrated by Lee (1979:145). As with other organic materials, the earliest dated occurrence is from the mid-Holocene horizons at Melkhoutboom (Deacon, H.J. 1976:46) and more recent examples are recorded from Scotts Cave, De Hangen, Diepkloof and Big Elephant Shelter where they date within the last 2500 years.

The making of huts or windshelters is one practice in which the San and Khoi differed. The present-day San build flimsy windbreaks or more substantial huts of sticks and grass, but Khoi herders observed in historic times had more elaborate structures or inter-woven sticks and reed mats that were dismantled and carried on the backs of oxen when the camp was moved. There is no clear archaeological evidence for either sort of hut, but stone 'hut circles' have been reported at Pearly Beach on the south-western Cape coast (Avery, G. 1974, 1976) and more elaborate structures are known inland along the Riet River (Humphreys 1972a), and in the Seekoei R. valley (Sampson 1980) where they are associated with possible kraals. The remains of posts and post-holes have been found at Melkhoutboom and Boomplaas from the mid-Holocene levels but no structures could be reconstructed with any accuracy. Bedding hollows filled with grass show that these were used in rock shelters. Early examples are known from the terminal Pleistocene/early Holocene at Wonderwerk (Beaumont, pers. comm.) and from later Holocene contexts at Big Elephant Shelter (Clark & Walton 1962; Wadley 1979), De Hangen (Parkington & Poggenpoel 1971) and Melkhoutboom (Deacon, H.J. 1976).

Burial practices

The most comprehensive source of information on the burial practices of historically known hunter-gatherers and herders in southern Africa is that compiled by Schapera (1930) from the ethnographic and historical literature. Archaeological data, however, suggest that there was greater variation geographically and through time than Schapera was able to indicate, and that the differences between the San and Khoi burial methods may not be as clear as Schapera thought they were.

Amongst the southern San, the Nharo, !Kung and Drakensberg San Schapera (1930:160ff) reports that it was common for the body to be placed in a shallow grave, an old termite mound, or in a shaft grave in a contracted position with the knees drawn up to the chin. Grave goods were sometimes added and buchu and other herbs and ochre may have been placed on the body. In some areas the body was covered with a few stones, in others a small cairn was erected over the grave, and in other instances no stones were used at all. The presence or absence of these features seems to have been variable and some were common to both hunter-gatherers and herders. The main difference that Schapera noted between San and Khoi burials was in the position of the body, the latter preferring a squatting position in a shaft grave with the body sometimes wrapped or sewn into a kaross or skins (Schapera 1930:363).

Archaeological evidence suggests that the position of the body was not clearly linked to ethnic group or physical type, however, as Humphreys (1970) has shown in a series of burials from the Orange Free State where there is some variation in a restricted geographic area. Similar variation is seen in burials which pre-date the advent of herders as, for example, between the shaft burial of a child at Melkhoutboom (Deacon, H.J. 1976) dating to ca 5000 B.P. and a contracted horizontal burial at Cape St Francis dating to about the same time (Thackeray, J.F. & Feast 1975).

The earliest evidence for deliberate burial so far recovered is from Border Cave where a 4-6 month old child was buried in a shallow grave with a perforated *Conus* shell and was associated with the MSA I stratigraphic unit with an inferred dating of ca 105,000 ± 5000 B.P. (Beaumont 1980:31). No burials have yet been recovered from late Pleistocene LSA horizons, however, with the possible exception of that from Peers Cave in the south-western Cape (see Rightmire, this volume). The earliest dated evidence for conscious burial in a Later Stone Age context thus comes from Layer D at Matjes River dating between 11,500 and 10,500 B.P. Louw (1960:91) states that skulls recovered there by Hoffman in 1957 were 'completely encrusted in a shell of ochre at least one inch thick.' The shaft grave dating to ca 5000 B.P. at Melkhoutboom was capped with a thick layer of powdered ochre. The next oldest dated burials are from Nswatugi in Zimbabwe (Walker 1980) and from the Wilton name site in the eastern Cape (Hewitt 1921; Deacon, J. 1972) where they date to the 10th and 9th millennia B.P. respectively. All three are shallow graves with a few stones or slabs covering the body. The Cape St Francis individual was accompanied by a number of shell pendants, ostrich eggshell and shell beads. Painted stones have occasionally been used as burial stones (Rudner 1971).

No human burials have so far been found in rock shelters in Namibia (Wendt 1972:39), but a stone cairn grave in the Walvis Bay district has been dated to 1720 B.P. (Vogel & Visser 1981). Later Stone Age burials in rock shelters are also unknown inland of the Cape Folded mountains and the only well documented graves are from open sites sometimes marked by stone cairns (Humphreys & Maggs 1970; Mason 1974a: Morris 1981) and sometimes accompanied by copper ornaments. This distribution would suggest that it was only in the eastern and southern Cape that cave and rock shelter burials were at all common although this may be the result of the greater depth of deposit in these rock shelters as compared with those inland. Open site burials with and without stone cairns are known throughout South Africa and Namibia, but have not yet been reported from Botswana or Zimbabwe.

Rock art

Southern Africa has more rock art sites than any other region of the world: over 7500 sites have been recorded in Zimbabwe, Botswana, Namibia, South Africa Swaziland and Lesotho, and many more are known but have not been officially reported. This prolific art comprises both paintings and engravings.

Rock paintings are more numerous and more widely distributed than rock engravings; they occur in areas where rock shelters and caves are most common and where rainfall is relatively high (with the exception of southern Namibia). The engravings, on the other hand, are more common in the interior of South Africa where dolerite is the dominant rock surface used and where rainfall tends to be lower. Areas covered by Kalahari Sands are generally devoid of rock outcrops and rock art is consequently rarely found there. The content of the rock paintings and engravings is similar in that both media include human and animal figures as well as geometric designs and so-called 'mythical' figures, but the relative frequency of these different subjects varies from area to area. The differences in medium, technique, distribution and proportions of various subjects between the rock paintings and engravings have encouraged some authors (Stow 1905; Ellenberger & MacGregor 1912; Butzer *et al.* 1979, for example) to suggest that the painters of the Drakensberg region were culturally distinct from the engravers of the Orange-Vaal basin, but this view is not shared by all (see Willcox 1963:72-3). Concerted efforts have been made to order the wealth of art by recognizing successive and regional styles. Classification schemes have usually distinguished between monochrome, bichrome, polychrome and shaded polychrome painting techniques and between formal and dynamic painting styles (Willcox 1963, 1973; Cooke, C.K. 1969; Lee & Woodhouse 1970; Rudner & Rudner 1970).

Despite the variation in regional styles and relative frequencies of subjects, the rock art tradition of southern Africa is remarkably homogeneous and contrasts clearly with that of Zambia where naturalistic paintings are rare and probably relate to stone tool-using hunter-gatherers (Phillipson 1977:280), and where the more common red schematic paintings are attributed to Early Iron Age people and the later white ones to the Later Iron Age (Phillipson 1976:186-7). Painted engravings also occur, as at Chifubwa where the dating was at one time considered to be early-mid Holocene. This has been recently re-assessed (Phillipson 1977:281) and is now considered to be relatable to the Iron Age. Two of the three naturalistic paintings so far recorded in eastern Zambia are, significantly, of eland suggesting an extension of the special role of this antelope in myths, legends and ceremonies of southern Africa north of the Zambesi (Phillipson 1976:186). In addition, the schematic paintings have direct parallels with designs and motifs used by present-day central African people in boys' puberty ceremonies, girls' initiation and funerals of chiefs and headmen.

In southern Africa it is generally believed that the vast majority of the paintings were done by San hunter-gatherers, although some of the art in the south-western Cape is thought to have been painted by Khoi (Rudner & Rudner 1970:231-2). The rock engravings are similarly linked to hunter-gatherers, but there are some recent examples done by Iron Age people.

It is difficult to date this varied art with any precision, but some researchers have suggested that the painters and engravers were more active during the last few thousand years than they were in previous millennia. Problems of preservation may be responsible for the more numerous paintings assignable to this time period, however. The earliest radiocarbon dates on charcoal associated with small painted slabs found at Apollo 11 in southern Namibia indicate an age of between 27,500 and 25,500 B.P. for the paintings associated with MSA artefacts (Wendt 1972, 1976). These paintings are so far unique in their style and antiquity. The next earliest dated paintings are on large cobbles and are sometimes but not always associated with burials. The oldest of these, not associated with a burial, is radiocarbon dated on associated charcoal to 6400 B.P. and comes from Boomplaas Cave in the southern Cape (Deacon, H.J. *et al.* 1976); others come from Matjes River (5400 B.P., Rudner 1971:55-6) and Klasies River Mouth on the southern Cape coast dated at 4110 B.P. and 2285 B.P. (Singer & Wymer 1969) and from Robberg Peninsula at Plettenberg Bay in the same area dated to between 4180 and 1925 B.P. (Rudner 1971b; Rudner & Rudner 1973). Attempts have been made to date the paint directly by paper chromatography using the amino acids (Denninger 1971) but the results all date within the last 1000 years (Pager 1971; Butzer *et al.* 1979) and reservations have been expressed about the validity of the method (Lewis-Williams 1977). The content of some paintings can give a good idea of their age in that they depict mounted horsemen (Vinnicombe 1976), ox waggons accompanied by women in crinoline skirts and men with guns, and ships (Johnson *et al.* 1959; Rudner & Rudner 1970:106).

The rock engravings are more difficult to date because they occur on open sites where occupation deposits are seldom associated with datable materials. In Namibia, Wendt (1977/78:59-61) presents a case for dating some unique human group scenes at Aar to between 8000 and 6000 B.P. on associated archaeological materials, while Willcox (1963:67), using a rate of dolerite weathering of 1 mm per 500 years suggests that some engravings are no older than 500 years. An uncertain identification of an extinct giant buffalo on a northern Cape engraving may mean that some date to the early Holocene (Butzer *et al.* 1979). A more detailed assessment, incorporating geomorphological data as well as analyses of content and style in northern Cape and Orange Free State engravings, argues that they are generally more recent than 4000 B.P. with geometric designs and domestic animal engravings dating within the last 1300 years, and older hairline engravings possibly dating to the early Holocene (Butzer *et al.* 1979:1210; Fock *et al.* 1980). The antiquity of engraving *sensu lato* is confirmed by the finds of engraved slabs dating from 10,000 B.P. at Wonderwerk in the northern Cape (Thackeray, A.I. *et al.* 1981) and from about the same time at Bambata in Zimbabwe (Walker, pers. comm) (Figure 3). The geometric pattern on the Bambata stone, which measures 60×60mm, is reminiscent of others found in undated context in Zimbabwe (Robinson 1975) and South Africa (Goodwin & Van Riet Lowe 1929:plate XXVIII; Malan 1956; Beater 1967). Hairline geometric designs are also present on mid- and late Holocene stones from Wonderwerk (Beaumont 1979; Thackeray, A.I. *et al.* 1981). The oldest of the Wonderwerk hairline engravings (Figure 3) is of particular interest because it depicts an animal as well as a geometric pattern. Intriguing is the observation that all but one of the Wonderwerk engraved slabs have been broken subsequent to their having been engraved (Thackeray, A.I. *et al.* 1981:66). These various lines of evidence point to the engravings and paintings being of the same general antiquity and to their being associated with the cultural traditions of the Later Stone Age.

If dating the rock art is difficult, establishing the reasons for its execution is even more so. The deceptively naturalistic style of rock paintings in southern Africa and the detailed depiction of humans, animals, designs and activities has been the main reason for the widespread view that the motivation behind it was 'art for art's sake' (Burkitt 1928; Battiss 1948, 1971; Willcox 1963, 1973; Cooke, C.K. 1969; Lee & Woodhouse 1970; Rudner & Rudner 1970) or as signs for sympathetic magic (Balfour 1909; Obermaier & Kuhn 1930; Breuil 1931; Frobenius 1931; Holm 1961; Brentjes 1969). A breakthrough came with the development of quantitative techniques, however, and the realization that some subjects have greater significance than the mere depiction of an animal or human. They are either pure symbols or icons acting as symbols (Lewis-Williams 1977:12, 1981:4).

Groundwork for quantitative studies was initiated in the 1960s and 1970s by Vinnicombe (1967, 1972a, b, 1976), Pager (1971, 1975) and Lewis-Williams (1972, 1974, 1977, 1980, 1981) in the Drakensberg. They noted that the eland in particular is more commonly painted than other animals and has had more attention and care devoted to it. Tracing the role of the eland in San mythology and ritual from ethnographic records and from present-day observations on ceremoines and customs of the !Kung San Biesele (1978), Vinnicombe (1976:163ff) and Lewis-Williams (1974a, b, 1977, 1980, 1981) have stressed the fact that the eland is a special favourite of the Manits trickster-deity, /Kaggen, and that it is used as a symbol in several important ceremonies: boys' first kill observances, girls' puberty rituals, marriage and rain-making (Lewis-Williams 1977, 1981). Observation of the dances associated with the ceremonies and still performed by present-day !Kung has helped in the recognition of the symbolism of attitudes and stances depicted in the rock art – hair standing on end along an eland's back, for example, is a sign that the animal is near death – and to suggest that the paintings were not necessarily used to depict specific occasions, but for instruction and inspiration for ceremonies in general. Much of the Drakensberg art is thus related to religious ritual, trance and hallucinations (Lewis-Williams 1980:467).

The next animal in numerical importance in the Drakensberg rock art is the rhebuck. Its frequent depiction with young and in family groups suggested to Vinnicombe that it might have expressed ideas about San family life even though it does not figure prominently in the extant myths and legends of the southern San or the present-day Kalahari San (Vinnicombe 1976:196-8). Lewis-Williams (1980), however, has pointed out that a nineteenth century Maluti informant told J.M. Orpen that the therianthropes with rhebuck heads represented medicine men.

While some antelope like the rhebuck and the eland were selected for frequent depiction, others were ignored by the artists. The wildebeest, for example, appears very rarely in the Drakensberg paintings possibly because it was believed to interfere with hunting as it was an animal which 'lived under false pretences' (Vinnicombe 1976:210).

Although the striking depictions of antelope tend to catch the modern viewer's attention first, human figures outnumber animals in all studies undertaken in the Drakensberg (Pager 1971; Lewis-Williams 1972, 1977, 1981; Vinnicombe 1976). Where sex can be determined from obvious physical features or from associated artefacts, men outnumber women by more than 2:1 (Vinnicombe 1976:245). Rock paintings of human figures have been useful in demonstrating the use of some artefacts such as bored stones and digging sticks and the relative sizes of bows and arrows, while in the south-western Cape the size of groups has been estimated at between 5 and 35 with a mean of 13 (Maggs 1967). These observations help to test the ethnographic record by demonstrating a link between what has been observed amongst modern people and the pictorial record of the past but this is not to say that all rock art depictions can be taken literally.

The generalizations on relative frequency with which certain subjects are depicted in the Drakensberg do not necessarily apply to other regions as well. In Zimbabwe, for example, C.K. Cooke (1969:71) reports that, contrary to much of the South African art, animals outnumber humans in the Matopos where the giraffe rather than the eland is a particularly important subject. The importance of the eland would also seem to have been lower in the south-western Cape than in the Drakensberg for while four Drakensberg studies (Pager 1971:324, 327; Lewis-Williams 1972:51, 1977:349; Vinnicombe 1976:364) show the eland to account for between 43 and 62% of all antelope, it accounts for only 37% of all antelope in the south-western Cape (Van Rijssen 1980), and it is absent in the Brandberg where springbok predominate (Jacobson 1975).

Similarly detailed analyses of rock engravings (petroglyphs) have also been restricted to relatively small areas such as the northern Cape and part of the Orange Free State (Fock 1979), parts of Namibia (Scherz 1975; Wendt 1977/78) and a small area of the Transvaal (Schoonraad 1980). Numerical analyses show that animals (64%) far outnumber human (9,7%) and geometric (26,3%) figures in the Orange-Vaal basin, but in the Richtersveld to the west geometric designs are most common (Butzer *et al.* 1979: Fig. 3). As a further contrast to the rock paintings, the rock engravings seldom depict group scenes (but see Wendt 1977/78) and the geometric designs are more varied and enigmatic. Their symbolic value is unknown, but by analogy with the rock paintings we would expect that they too had some ritual, educational, religious or mythical significance. In addition to recording the range of designs, Fock (1979) has paid particular attention to the location and distribution of rock engravings in the northern Cape/Orange-Vaal basin. He notes that all the engraving sites with high frequencies of geometric designs are found on river beds or adjacent to fountains or natural wells, and that where painted geometrics are found in the same area as the engraved geometrics, they lack certain of the designs which are exclusive to the engraving sites. The hypothesis offered to account for this pattern is that the geometric engravings were made primarily by a single ethnic group that clustered around water sources at a time of lower rainfall (Butzer *et al.* 1979; Fock *et al.* 1980:312); the geometric paintings are suggested to have been done by a later but related people on the southern fringes of the Kalahari (Butzer *et al.* 1979). It is possible, however, that the designs were site-specific and were related to rituals or ceremonies only performed near to water sources.

Although a certain amount of speculation is inevitable in the interpretation of rock art, close attention to the considerale nineteenth century and twentieth century San ethnography is beginning to unravel the complex associations which the depictions held for their original viewers. There now seems little doubt that many of the paintings were symbols for instruction and inspiration and were closely associated with various rituals, especially the trance dance. Certain key metaphors that run through the San myths and rituals are consistently recognizable in the art. The art was, in effect, 'part of the complex and subtle web of San thought and belief' (Lewis-Williams 1974b:252) and its interpretation is heavily dependent on a thorough knowledge of San myths and beliefs.

VARIABILITY IN SOME MAJOR FORMAL TOOL CLASSES

Having outlined the culture-stratigraphic sequence and the role of ethnographic and historical sources in the interpretation of the material culture of Later Stone Age people, this section will focus on some detailed analyses of particular formal tools to demonstrate patterns in their geographic and chronological distributions. As such it draws attention to the role of variability in interpreting the archaeological record. The artefacts selected are scrapers, adzes and backed microliths because they are the most numerous formal tools in LSA assemblages, and for comparison a sample of arrows made by San and Khoi people in southern Africa over the last 150 years has been measured to demonstrate geographic and chronological variability in an ethnographic situation. The perspective offered by these analyses helps to trace unifying features through the time-span of the LSA and cross-cuts the tendency of the culture-stratigraphic framework to stress differences. rather than similarities in the chronicle of LSA material culture. The vast time perspective offered by archaeological data, however, makes it difficult to apply analogies gleaned from the ethnographic record to account for long-term changes and the interpretation of much of the observation pattering needs more sophisticated techniques than those we have been able to apply so far.

Scrapers

Scrapers are characterized by formal retouch to form a convex working edge which may, through utilization, be worked back to a straight, but rarely a concave edge. In side view the ventral surface is flat and the angle of the working edge tends to be steep. Scrapers are made on flakes, blades, cores or chunks and they may be very flat or relatively thick pieces, but the morphology of the working edge remains basically the same whatever the origin of the piece. In samples where metric data are available (for over 2500 scrapers from the southern, eastern and northern Cape), length of the piece, as taken from the mid-point of the working edge, can range from less than 10 mm to over 60 mm, but the width of the working edge remains relatively constant wihtin the middle of the size range, suggesting that this is the more important feature in the design of the tool. In addition to the width, the convexity of the working edge is important for skin scraping as any irregularities on the edge may tear the leather while concavity would reduce its effectiveness. Scraper sub-classes have been defined either on the basis of length [large, medium, small (Deacon, J. 1978)], or on the position of retouch in relation to the long axis of the piece [end, side, side-and-end, circular, etc. (Sampson 1972, 1974)]. Evaluation of changes in scraper morphology through time and between sites however, is best made on metric attribute analysis rather than on relative frequencies in sub-classes (Deacon, J. 1972, and in prep.; Deacon, H.J. 1976; Humphreys 1979).

Throughout most of southern Africa scrapers are the most numerous formal tool in all LSA assemblages, contrasting with Zambia, part of Namibia and northern Zimbabwe where backed microliths predominate (Deacon, H.J. 1972, 1976; Deacon, J. 1974; Nelson 1976; Phillipson 1977; Wadley 1979). Within the Holocene microlithic artefact tradition in particular, the scrapers south of the Zambesi show a greater degree of standardization in metric attributes and show more emphasis on the shaping of the working edge than do samples to the north [compare drawings in Miller (1969), Phillipson (1976), Deacon, H.J. (1976) and Deacon, J. (1978)]. The greater numerical importance and more careful retouch on these southern scrapers suggests that they were related to a specific task more commonly undertaken in the south than in the north. Scrapers are known from historic records (Kannemeyer 1890) and from replication and microwear studies (J.N.F. Binneman, in prep.) to have been used mostly for working skins, and similar tools are still used today in Ethiopia for the same purpose (Gallagher 1977). In maps prepared from ethnographic data in the first two decades of this century under the direction of Frobenius, it is shown (Figure 10) that clothing was made exclusively from leather in most of South Africa, Namibia, Botswana and Zimbabwe, but in Mozambique, parts of Zimbabwe, Malawi, eastern Zambia and parts of western and eastern Africa, the preferred material for clothing was bark cloth. The distribution of higher frequencies of standardized small convex scrapers in Holocene assemblages corresponds closely to the distribution of leather clothing and is low in areas where bark cloth was made (Von Boeckman 1921; Deacon, H.J. & Deacon 1980). More precise interregional comparisons are difficult, however, because where inventories of artefacts from assemblages within southern Africa are compared with those to the north, it becomes clear that the definition currently used for scrapers south of the Zambesi places more emphasis on the contour of the working edge than does that used to the north where

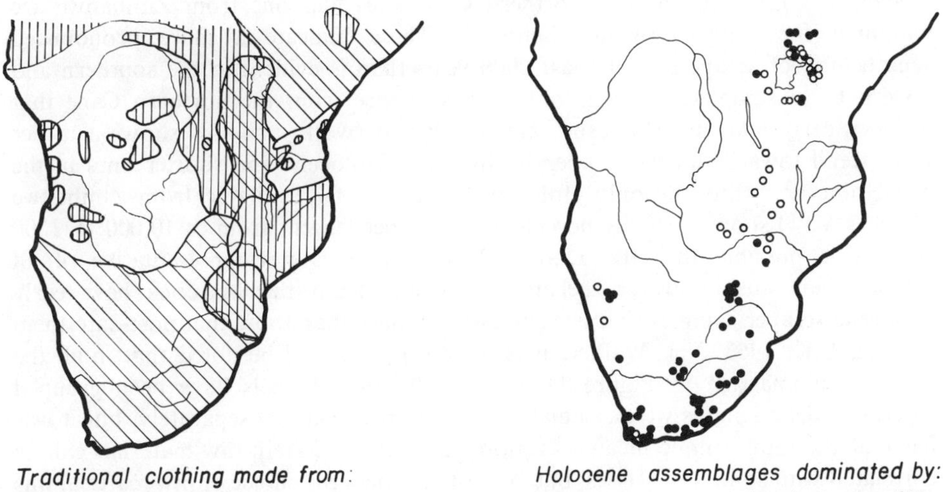

Figure 10. Distribution of traditional manufacture of clothing from leather and bark cloth in sub-Saharan Africa (after Von Boeckmann 1921) compared with the distribution of Holocene assemblages dominated by scrapers or backed microliths (data from sites plotted in Figure 1 and from Masao (1979) for Tanzania, Nelson (1976) and Merrick (1975) for East Africa and Clark (1974) for Kalambo Falls).

all pieces with steep retouch appear to be included in the scraper class. Some standardization of this definition is necessary (see Shaw 1981).

At a more local level the relative frequency of scrapers at contemporary sites can also be integrated in a functional framework. In southern Africa changing mean values in the metric attributes of LSA scrapers show some interesting shifts through time. This was noted in the earliest detailed classification schemes for the LSA with Goodwin and Van Riet Lowe's naming (1929) of the Wilton and the Smithfield A, B and C, and has been confirmed since then, particularly in the eastern and southern Cape (Fagan 1960; Schrire 1962; Deacon, J. 1969, 1972, 1978; Deacon, H.J. 1969, 1976; Opperman 1978; Deacon, H.J. *et al.* 1978), in the northern Cape (Sampson 1972; Humphreys 1979) and in Namibia (Wadley 1979). The longest and most complete sequences are at present known from the former region and from these over 2500 scrapers have been measured in LSA assemblages dating between 20,000 and 1500 B.P. The results are summarized in Figure 11 in which dated assemblages are shown as solid lines and undated ones are interpolated and shown as dashed lines. A clear trend can be seen towards more longer scrapers in the middle of the LSA sequence between about 12,000 and 7000 B.P. than occur before or after.Although the actual lengths may differ from one contemporary sample to another, the rate and timing of the shifts in mean length remain remarkably similar at all eight sites from Buffelskloof in the west to Wilton over 400 km to the east, and from Nelson Bay on the coast to Boomplaas some 80 km inland. A second trend, less well developed, suggests that at some sites (Wilton and possibly Matjes River) there is a tendency for scrapers dating within the last 2,000 years to be longer than mid-Holocene ones, but this is not the case at other sites such as Boomplaas where scrapers associated with pottery are not significantly different from pre-pottery samples (Table 8; Deacon, H.J. *et al.* 1978).

Scraper samples from nine northern Cape sites and one from Zimbabwe are presented for comparison in Figure 12. These indicate a more pronounced lengthening of scrapers in the last 2000 years than is evident in the southern and eastern Cape samples. There is only one sequence from the northern Cape that includes early and mid-Holocene samples (Wonderwerk) and this shows a rather weak trend towards longer scrapers in the early Holocene and shorter ones in the mid-Holocene, while the only Holocene sample so far available from Zimbabwe (Diana's Vow) shows virtually no change in scraper lengths from ca 10,000 to 1500 B.P. It is possible that the Diana's Vow sample is not representative of all Zimbabwean sites, however, because sequences from the Matopos have early Holocene scraper samples that are somewhat longer than those that post-date them (Cooke, C.K. 1963; N.J. Walker, pers. comm.). It should be noted that in all the samples summarized in Figure 11 and 12 all raw materials have been grouped together. Mean lengths were calculated for each raw material separately, but it was felt that the total sample mean was more significant. Certain raw materials can be correlated with some size classes, but in most instances the choice of raw material was open to the toolmakers. The assumption is that raw materials were deliberately selected by the toolmakers to make the size scraper they desired and not that the toolmakers were 'limited' by the raw materials available. This view is not shared by all researchers, however, and Humphreys (1972b, 1979) and Beaumont (1978) prefer to consider different raw materials as 'active' rather than 'passive' ingredients.

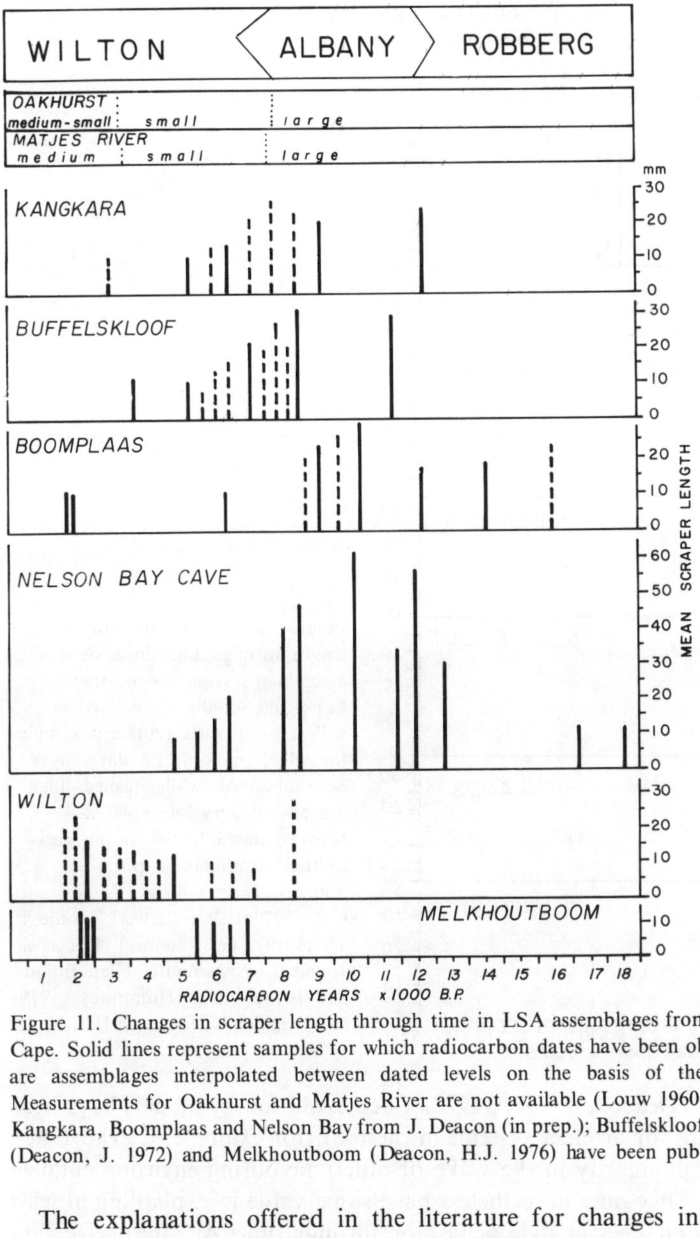

Figure 11. Changes in scraper length through time in LSA assemblages from the southern and eastern Cape. Solid lines represent samples for which radiocarbon dates have been obtained, while dashed lines are assemblages interpolated between dated levels on the basis of their stratigraphic position. Measurements for Oakhurst and Matjes River are not available (Louw 1960; Schrire 1962). Data from Kangkara, Boomplaas and Nelson Bay from J. Deacon (in prep.); Buffelskloof (Opperman 1978), Wilton (Deacon, J. 1972) and Melkhoutboom (Deacon, H.J. 1976) have been published.

The explanations offered in the literature for changes in scraper morphology through time in the LSA are not explanations in the true sense of the word, but are rather frameworks for describing the patterns of change that have been observed. They are what Binford (1981:83) would call 'post-hoc' explanations. They are not derived from the testing of ethnographic analogies because these are not available. When scrapers were first measured from the Wilton name site it was suggested that the shifts in scraper length and shape together with other features of the assemblages conformed to what might be expected in the ontogenetic development of an artefact manufacturing system with phases exhibiting characteristics typical of 'birth',

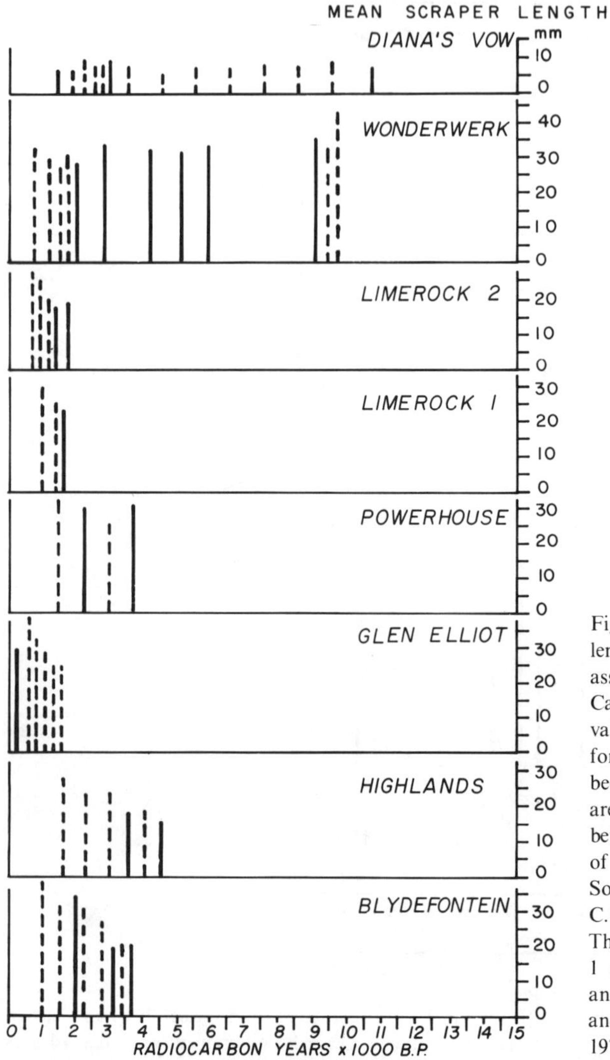

Figure 12. Changes in scraper length through time in LSA assemblages from the northern Cape and middle Orange River valley. Solid lines represent samples for which radiocarbon dates have been obtained, while dashed lines are assemblages interpolated between dated levels on the basis of their stratigraphic position. Sources: Diana's Vow (Cooke, C.K. 1979a), Wonderwerk (Anne I. Thackeray, pers. comm.), Limerock 1 and 2, Powerhouse, Glen Elliott and Blydefontein (Humphreys 1979) and Highlands (Deacon, H.J. 1976).

'maturity' and 'death' (Deacon, J. 1969, 1972). Such evolutionary models have been considered appropriate for artefact systems in the past (for example, Clarke 1968), but have fallen from popularity in the wake of others favouring environmental or functional correlates. They may nevertheless have some value in explaining at least part of the reason for changes in style or fashion through time. An alternative view that took factors besides the artefacts into account was offered by H.J. Deacon (1976) who noted that changes in the artefact assemblages, including the size and shape of scrapers, were broadly coincident with changes in hunting patterns as reflected in the faunal remains at several eastern and southern Cape sites and at others further afield in southern Africa. From this observation it was hypothesised that changes in the artefact system signalled adjustments to changing environmental conditions at about 12,000 and 8,000 B.P. while the periods ca 20,000–12,000 B.P., 12,000–8,000 B.P. and 8,000–2,000 B.P. represented periods of relative stability.

This model of three 'homeostatic plateaux' represented by the Robberg, Albany and Wilton industries in the eastern and southern Cape (Deacon, H.J. 1976:81, 159ff) incorporates the concept that cultural and economic systems are constantly changing and adjusting to the habitat, but that the major environmental changes that took place worldwide at the end of the Pleistocene apparently necessitated more drastic adjustment. In the southern and eastern Cape, and possibly further afield, the Albany industry represents such an adjustment between 12,000 and 8000 years ago.

Implicit in the homeostatic plateau model is the idea that changes in the artefact system were adaptive, i.e. that artefacts were purposefully designed to cope better with the tasks for which they were intended, but the reason why scrapers in the terminal Pleistocene/early Holocene were made larger than those that preceded or succeeded them remains obscure. Implicit in models of evolutionary development of artefact systems, on the other hand, is the assumption that changing parameters of scrapers are the result of changing fashions or norms of tool manufacture that are unplanned yet persist through social reinforcement. The link between style and function remains elusive, however, and we are still faced with the question posed by Sackett (1977:374): is 'style no more than function writ small'? Careful assessment of associated non-artefactual data and use-wear analyses currently being undertaken by J.N.F. Binneman should help to give us some of the answers, as should intensive regional studies (e.g. Jacobson 1978).

Adzes

Adzes are different from scrapers in having a straight or concave working edge rather than a convex one. The angle of retouch is also steep, but the edge is marked by step-flaking and heavy utilization damage rather than the careful shaping flakes evident on scrapers. There are usually three sets of flake scars on an adze: the highest set results from the initial shaping of the tool, the second set has defined the angle of retouch and the third and lowest set has resulted from a chopping action during utilization; the steeper the angle between the tool and the material worked, the heavier the damage (Binneman, in prep.). In areas where hornfels is available, this is the preferred raw material for adzes and smooth hornfels pebbles are often used with cortex left on the central surface of the tool forming a curved base (Figure 5:54). Replication studies have shown that this curved base minimizes fracture during use (Hayden 1979:61). Silcrete and chalcedony are used where hornfels is not available, but quartz and quartzite adzes are rarely found. Mastic traces on adzes from Melkhoutboom (Hewitt 1931a; Deacon, H.J. 1976) show that these tools were hafted and a few specimens still in hafts are known (Clark 1959; Deacon, H.J. 1966; Deacon, H.J. & Deacon 1980). They show that adzes were mounted at the end of a handle with mastic covering most of the tool. Retouch on opposing sides suggests that they were sometimes reversed in the mount.

Artefacts now grouped within the adze class have often in the past been included as sub-classes of the scrapers. The variety of names by which they have been known, such as hollow, notched and strangulated scrapers, spokeshaves, slugs and pebble adzes, is a reflection of the inter-regional variability in the styles of adze manufacture. In the recent literature three terms are used for two basic forms: these are adzes (Deacon, J. 1972; Deacon, H.J. 1976; Humphreys 1979:393; Mazel & Parkington

1978), and notched scrapers (Cable *et al.* 1980) or spokeshaves (Maggs & Ward 1980). Both have similar retouch and step-flaking, but adzes have relatively straight working edges and are more numerous in the western half of southern Africa (Namibia, the northern Cape and the western, southern and eastern Cape), while notched scrapers or spokeshaves have hollow or multiple-hollow working edges and are more numerous in the eastern half of the sub-continent in Zimbabwe, Natal and possibly also in the Transvaal.

That adzes and spokeshaves were used more often for woodworking than for any other activity was recognized by Clark (1958a, b) in comparisons between the southern African artefacts and those used for woodworking by present-day Australian Aborigines. Recent replication experiments and analyses of micro-wear patterns on archaeological specimens essentially confirm this (Binneman, in prep.) but show also that untrimmed flakes, used at a low angle, are as efficient for working wooden tools such as digging sticks as are retouched adzes. Of interest is the fact that the untrimmed flakes show no sign, except under magnification, of having been used for working wood.

Because of the functional connection with woodworking, the geographic distribution of adzes has often been linked to the occurrence of woodland vegetation (Clark 1959:215) and higher frequencies of nŏtched scrapers and spokeshaves are found in the wooded regions of Zambia, Zimbabwe and Natal, but not in adjacent grassland areas. In a recent set of papers (Mazel & Parkington 1978, 1981; Parkington 1980) it has been suggested that the higher incidence of adzes at mountain sites in the south-western Cape evidences seasonal activities involving the manufacture and maintenance of digging sticks during the summer months when underground plant foods were abundant in that area. The relative absence of adzes at coastal sites is taken to indicate that digging sticks were not used in the winter months when the coastal sites were occupied because underground plant foods were either unavailable or unpalatable at that time of year. A functional explanation is also offered for the observation that higher frequencies of notched scrapers occur at Natal thornveld/mixed *Podocarpus* forest sites where wood suitable for artefacts is found (Cable 1980; Cable *et al.* 1980; Carter 1980). In the southern Cape adzes are absent from the coastal site of Nelson Bay but are present in relatively small numbers at mountain sites such as Boomplaas and Kangkara (Deacon, J. in prep.).

Unlike the scraper class, no clear trend through time in the size and shape of adzes has yet been demonstrated. Where measurements have been taken in sequences with high enough frequencies to make conclusions statistically significant the adzes remain remarkably standardized through the mid- and late Holocene (Deacon, J. 1972 and in prep.; Deacon, H.J. 1976), but if adze retouch along the sides of frontal scrapers from terminal Pleistocene/early Holocene assemblages is taken to indicate that these artefacts were used as adzes, then the design of the adze may well have undergone some changes through time. Adzes are present in late Pleistocene samples in the south-western and southern Cape and they may prove to be less standardized than their Holocene counterparts.

Backed microliths

Backed microliths include microlithic bladelets (less than 25 mm long) that have been deliberately and regularly flaked along one or more margins to produce a blunted edge opposite to a sharp one. A variety of shapes occurs (Figure 5), but the main distinction is made between those with backing along straight edges (backed blades or bladelets) and those with a curved back (segments or crescents). A more sophisticated subdivision of the backed blade class has been recognized where sufficiently large numbers permit the various stages of manufacture to be identified as distinct from the formal tools (Deacon, H.J. 1976: figure 61). Backed microliths are assumed to have been designed as inserts in multiple component armatures or cutting tools and mastic traces found on them at some sites confirms the fact that they were mounted.

Backed microliths are more varied in shape and size and are more numerous in assemblages from parts of Namibia, northern Zimbabwe and Zambia than they are to the south (Deacon, H.J. 1972, 1976; Deacon, J. 1974; Phillipson 1976, 1977; Wadley 1979). Thus the number of sub-classes recognized by Miller (1969) and Phillipson (1976) in Zambia is much greater than the segment/backed bladelet/backed flake distinction usually employed in South Africa. Part of the reason may be that backed microliths to the north may have been designed for a wider variety of uses, as is suggested by Phillipson (1976, 1977) who has attempted to reconstruct the way in which various backed tools were mounted as barbs, cutting tools and armatures. There is ethnographic evidence for the more frequent use of barbs north of the Zambesi in present-day and recent historical iron arrows and spears (Frobenius & Von Wilm 1921) and this may imply that they were more widely used there than south of the Zambesi in LSA times as well. Heavy barbing would in part explain the higher frequencies of backed bladelets which may also be explicable in terms of a greater emphasis on hunting relative to gathering north of the Zambesi (Deacon, H.J. 1972).

The assumption that the frequency of backed microliths reflects the relative importance of hunting is also held in the evaluation of the relative frequencies of formal tools at sites in the south-western Cape (Parkington 1980). In contrast to the mountain sites where adzes are numerous, open sites on the sandveld have relatively high proportions of backed microliths and have been interpreted as evidencing arrow manufacture and maintenance there. The use of bows and arrows is thought to have been preferred in the summer because the greater likelihood of rain in winter made the use of snares and traps less reliable when the twine used in them lost elasticity when wet (Mazel 1980).

That at least some of the variation in the relative frequency of backed microliths can be related to changes through time, on the other hand, is demonstrated in Figures 13, 14 and 15. The relative percentage of scrapers vs all backed microliths in dated assemblages in southern Africa is plotted in Figure 13. This shows that the early microlithic assemblages dating between 20,000 and 12,000 B.P. are usually such small samples that the range of relative frequencies is wide and difficult to evaluate. As noted earlier, backed tools are very rare in samples dating between 12,000 and 8000 B.P., but thereafter assemblages from Zambia (large squares) and Namibia (large triangles) regularly include more backed microliths than scrapers while those from South Africa (small circles) include a number of assemblages with over 25% of backed microliths relative to scrapers. After 2000 B.P., however, backed microliths rarely account for more than 15% of the total in South African assemblages. Notable

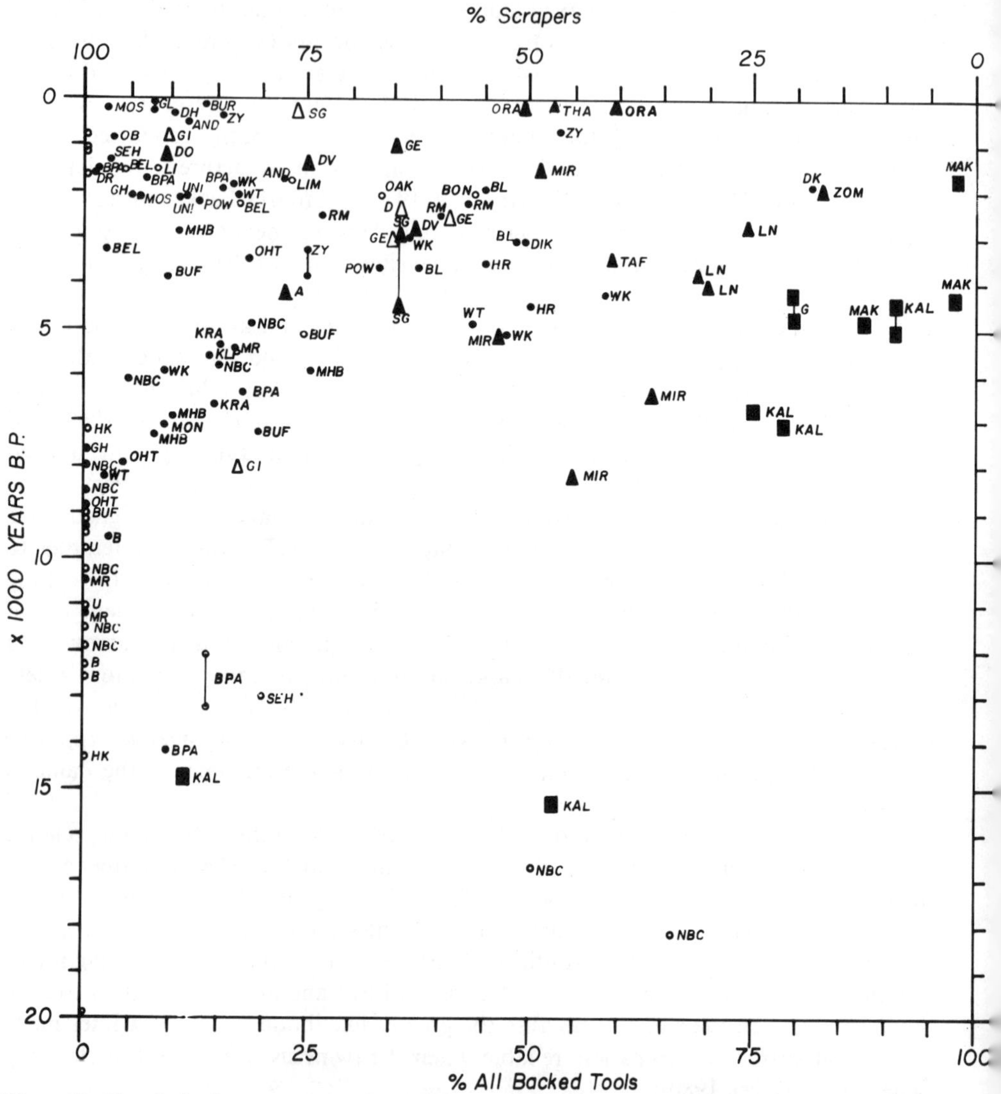

Figure 13. The relative frequency of scrapers vs all backed microliths plotted against time in radiocarbon dated assemblages from southern Africa. A few examples from Zambia (solid squares) are given for comparison. • and O = sites in South Africa, ▲ and Δ = sites in Namibia and Zimbabwe. Solid Symbols=samples with more than 25 scrapers and backed microliths combined, Open Symbols=small samples with less than 25 scrapers and backed microliths combined. Symbols joined by a vertical line indicate that the unit from which the artefact sample derives has two radiocarbon dates. References listed in Table 12. Samples not labelled through lack of space are, from top to bottom, BEL, MG, DF, BPA, WK, BPA, HK, KRA (all against left-hand margin).

exceptions in this time range are Die Kelders (DK) and Zaayfontein (ZY) but they serve to illustrate the fact that the data in Figure 13 only demonstrate a general trend and are not capable of providing a basis for reliable predictions of the age of an assemblage if the relative frequency of scrapers and backed microliths is known.

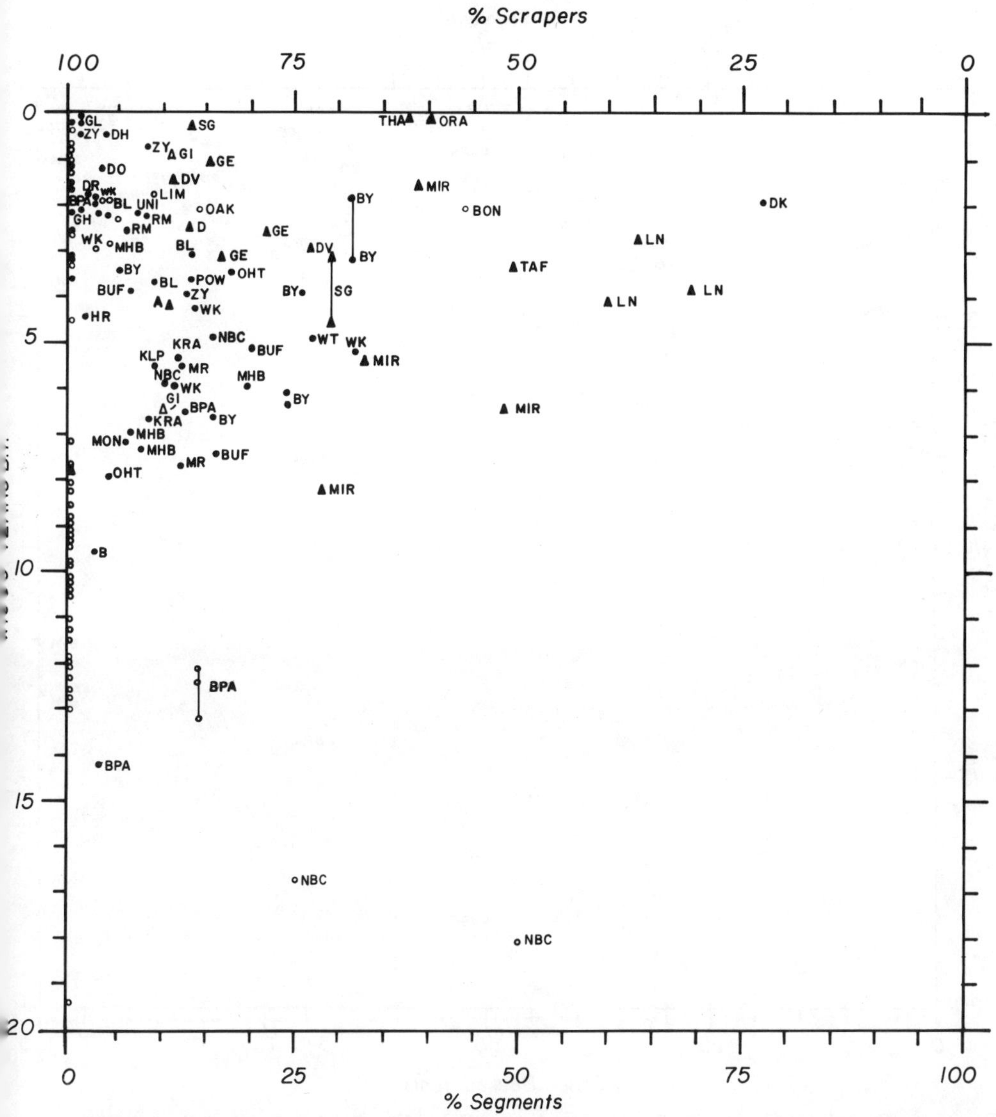

Figure 14. The relative frequency of scrapers vs segments plotted against time in radiocarbon dated assemblages from southern Africa. Symbols as in Figure 13. References listed in Table 12. Samples not labelled through lack of space along left hand margin are (from top to bottom): MOS, OAK, AS, BEL, PAT, MG, BEL, LIM, BPA, MOS, RM, GB, DIK, BEL, GB, HR, BON, HK, GH, P, NBC, WT, NBC, OHT, BUF, HK & WK, BPA, KRA, HK & B, BY, U, NBC, NBC, BPA, MR, NBC, MR, NBC, BUF, NBC, BPA, B, HK, BY, SEH, HK. Circles to right of 2000 B.P. unmarked refer to: BPA, UNI, WT, BEL.

Figure 14 shows that segments, as distinct from backed microliths in general, are more numerous in samples dating between 7000 B.P. and 2000 B.P. than they are in earlier and younger time periods, but, as in Figure 13, there is a wide range of variation. The same phenomenon is contrasted against the incidence of backed microliths other than segments in Figure 15 which shows that after ca 4500 B.P. backed microliths are

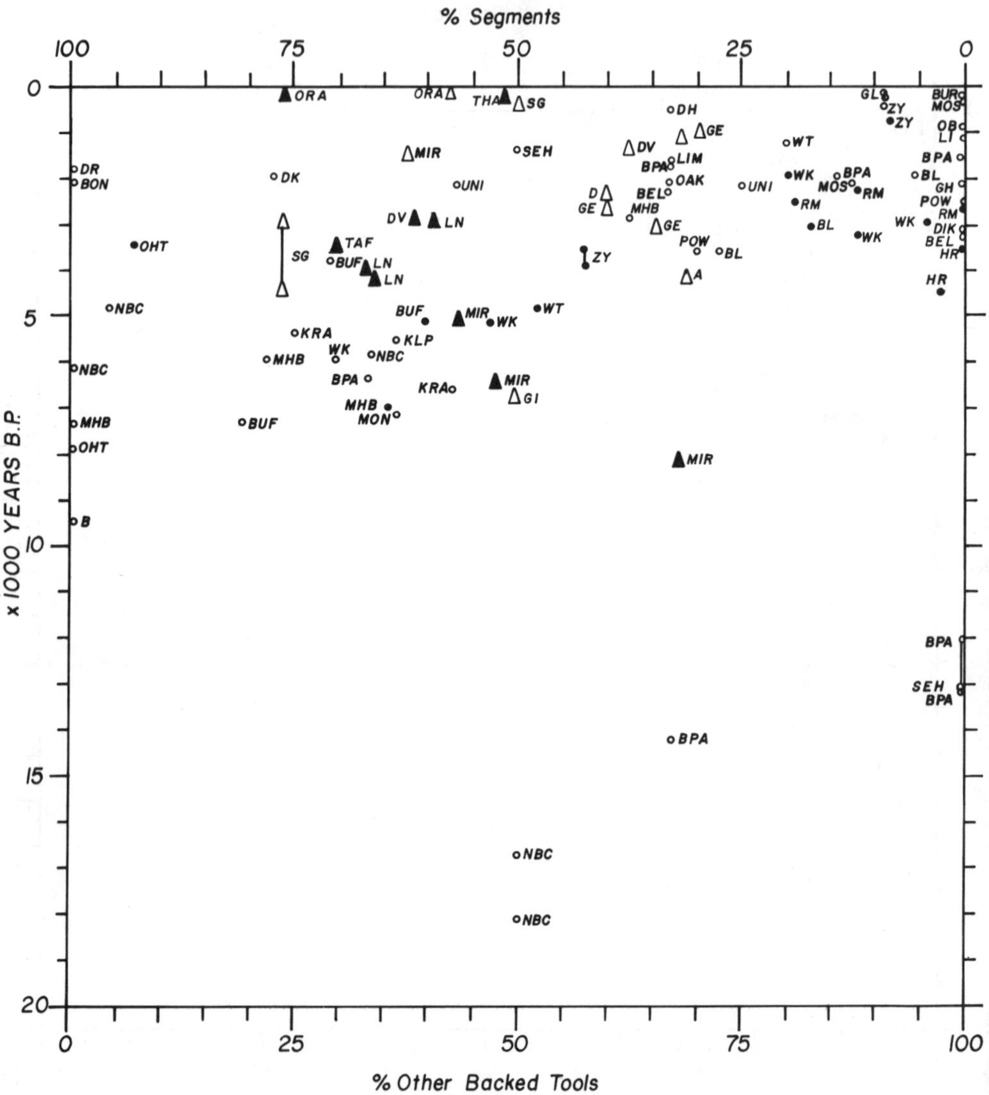

Figure 15. The relative frequency of segments vs other backed microliths plotted against time in radiocarbon dated assemblages from southern Africa. Symbols as in Figure 13. References listed in Table 12.

likely to be more numerous than segments in an assemblage. The pattern that these three figures illustrate is one in which backed microliths are rare and occur in variable proportions prior to 12,000 B.P., are virtually absent between 12,000 and 8000 B.P. and are dominated by segments until about 4500 B.P. Thereafter backed bladelets and other backed pieces tend to be more common than segments and all backed pieces become rarer after 2000 B.P. The reasons for these shifts are probably both functional and stylistic with backed tools indicating the use of stone-tipped arrows and the relative frequencies of segments and backed bladelets indicating changes in arrowhead design. The absence of all backed microliths between 12,000 and 8000 B.P. is possibly the result of the use of bone arrowheads for some reason, while their lower incidence after 2000 B.P. may indicate the use of metal instead of stone inserts. Both these suggestions need to be fully investigated, however. The patterning of these relative frequencies through time makes it necessary to take the dating of assemblages into account when testing hypotheses in which relative frequencies are compared.

Changes in the metric attributes of backed tools have been noted. In Zambia and Zimbabwe 'deep' segments (i.e. those that are wider than they are long) are more common in the mid-Holocene than at other times (Miller 1969; N.J. Walker, pers. comm.) and in the southern Cape the mid-Holocene segments tend to be longer than earlier or later ones (Deacon, J. 1972; Opperman 1978).

The extent to which changes in relative frequencies of tool classes through time can be explained in functional terms is problematic. For example, several explanations have been offered for the lower incidence of backed microliths in assemblages dating within the last 2000 years and the apparently coincidental increase in cylindrical bone 'points'. Sampson (1972:205, 1974:376) has interpreted it as evidence for the arrival of a new ethnic group, the historically documented San, who had apparently been in contact elsewhere with iron and pottery workers and brought with them a knowledge of these crafts. Humphreys (1979:366), on the other hand, suggests that vegetation changes in the late Holocene may have precipitated a change in the design of hunting equipment, the bone points being assumed to be more effective in open terrain. That a reduction in bush cover was coincident with the appearance of bone points is indicated by climatic data inferred from changes in the frequencies of micromammal species in the same general region (Avery, D.M. 1981).

A third possibility is that the stone armatures were replaced by metal ones. Metal-tipped arrows known in historical times from this northern Cape region are illustrated in Figure 16:2, 3. They consist of a cylindrical bone 'point' that fits into the reed shaft. A second bone 'point', linked to the first by a reed coupling bound with sinew, is split at one end to take a small metal insert that is triangular in shape and is made of beaten tin or iron. The bone is then bound with sinew, thickly coated with mastic and poison, and has a quill barb inserted on one side. Many of the cylindrical bone points illustrated by Sampson and Humphreys are incomplete, making it difficult to tell whether they were themselves arrowheads or whether they were linkshafts, but some have a flattened end similar to what one might expect to find had the split end of a metal-headed arrow snapped off. The appearance of cylindrical bone points could therefore indicate the introduction of a more efficient arrowhead, but whether this was precipitated by vegetation changes, population movements or simply trade with Iron Age peoples is not yet clear. It may in fact be impossible to ascertain the prime mover from the archaeological record as all three factors could have played a part.

Modern San arrows

Interest in the measurement of stone artefacts as an aid to distinguishing between assemblages of different ages was started in the 1920s by Goodwin (1929b) who noted changes in the size and shape of handaxes from Montagu Cave. He later used a similar method in tracing what he assumed to be migration routes followed by

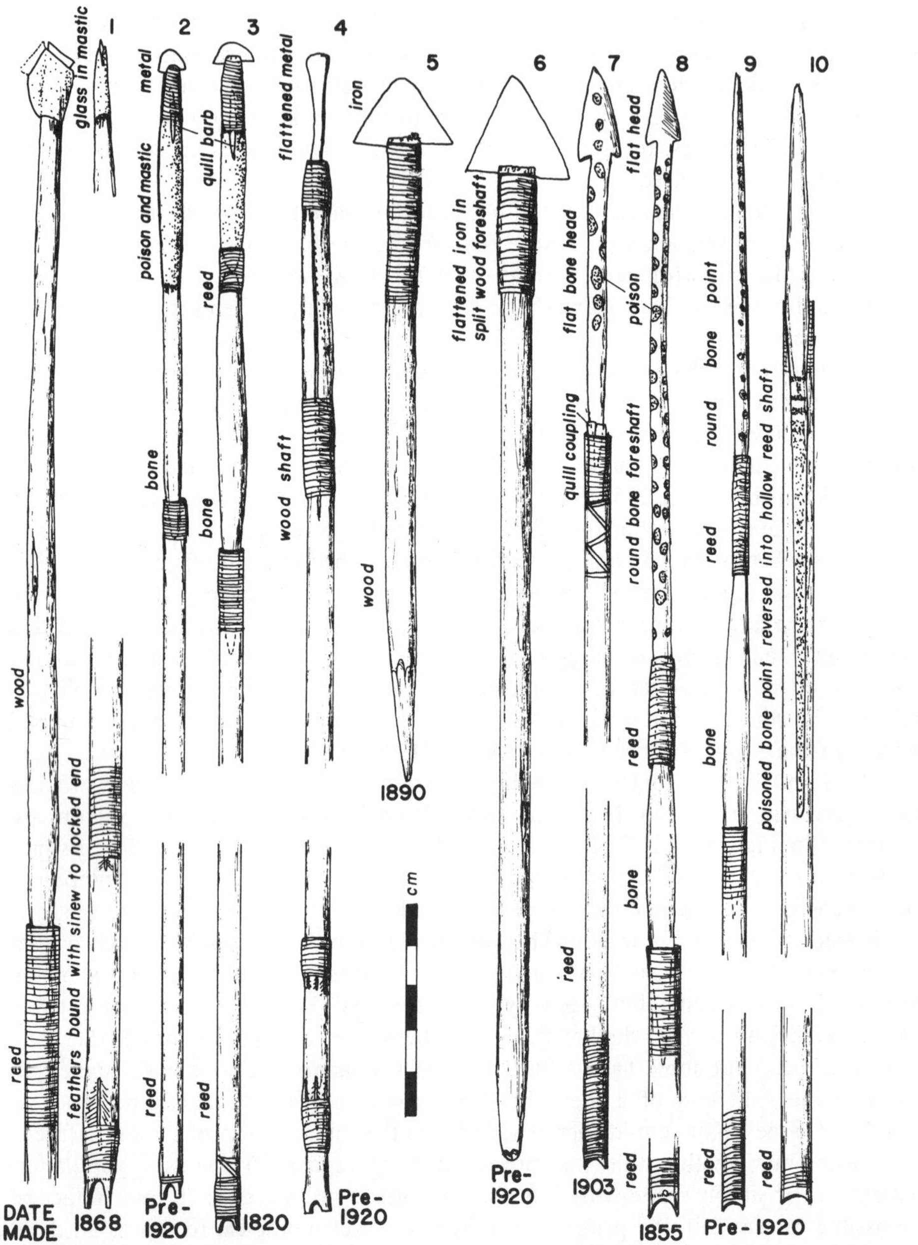

Figure 16. A range of San and ?Khoi arrows made before 1920 in South Africa. 1: glass inserts in mastic in style of backed microliths, from Griqualand West; 2: small metal insert in split bone foreshaft with quill barb, from Natal Drakensberg (Vinnicombe 1971); 3: same style as (2) from Graaff-Reinet; 4: metal head inserted into wooden shaft from Bleek-Lloyd collection of Southern San artefacts; 5: large iron arrowhead in split wood foreshaft from north-west of Kuruman, northern Cape (?Korana/Khoi); 6: same style as (5) from the Prieska district, northern Cape (Bleek-Lloyd collection); 7: tanged bone arrowhead dotted with poison, accessioned as 'Cape Bushman'; 8: same as (7) from Boshoff district, Orange Free State; 9: reversible bone point dotted with poison from the Hei//um, northern Namibia; 10: reversible bone arrow made from one piece of bone, from Barkly West, northern Cape.
1, 3, 4, 6, 7: South African Museum; 2: Natal Museum, Pietermaritzburg; 5, 10: Duggan Cronin Gallery, Kimberley; 8: National Museum, Bloemfontein; 9: National Cultural History Museum, Pretoria.

people making bored stones of different design, by measuring the size and shape of bored stones from South African collections and plotting their geographic distribution (Goodwin 1947). Metric analyses are now standard practice for most stone artefact sequences, but relatively little work has been published comparing contemporary assemblages at an inter-regional level. While such comparisons may be complicated by variations in the availability of raw materials, it has nevertheless been mooted that language (dialectal) differences between neighbouring LSA peoples may be reflected in stylistic differences in their artefacts (Deacon, H.J. 1976:170ff). The assumption that sociocultural boundaries may be detectable in variations in the metric attributes of stone artefacts has been tested on a limited scale in New Guinea (White & Thomas 1972; Thomas 1974:55) and in Kenya (Hodder 1977) where some correlation was noted in a small range of material cultural items. On a broader geographic scale, however, Crosby (1977) has measured museum collections of percussive cutting implements and concludes that although there is some broad correlation between ethnic boundaries, language groups and economic practices, the measurements have little predictive value. We cannot assume that because artefacts are stylistically similar that they must have derived from a common source, nor that communities with similar artefacts were in regular contact with each other. This lack of predictive value was also noted by Hodder (1977, 1980), not only in the measurements of individual items, but also in terms of which items showed significant inter-group variations.

There are no detailed data available on whether or how southern African hunter-gatherers sought to express their identity in material culture. Following the observation that present-day San in Namibia and Botswana fashion beaten wire arrowheads into shapes unique to their group so that !Kung arro̱ws are recognizably distinct from those of the Hei//um, Nharo and G/wi (Figure 17), a sample of over 700 arrows from South African museum collections that date within the last 150 years has been measured (Deacon, J. in prep.). Preliminary results confirm that the group-specific styles amongst present-day San in Namibia and Botswana extend back at least 50 years but older collections that date to the 19th century and first two decades of this century are unfortunately too small and lack sufficient data to enable a similar analysis to be made. Because all the arrows attributed to the southern San in South Africa date to the older time period, there is too little information to enable us to see whether differences in the arrow styles present over a wider geographic area are statistically significant, and the differences noted between arrows from north and south of the Orange River may be the result of changes through time as much as geographic distribution. The main distinction is that arrowheads post-dating ca 1920 from Namibia and Botswana are made of bone or beaten fencing wire, while those from South Africa all pre-date ca 1920 and are made of bone only, bone with small metal inserts, beaten tin or nails, or stone inserts in mastic (Figures 16 and 17). Within South Africa larger metal arrowheads (Figure 16:5, 6) tend to occur in the western half of the country around Kimberley and Prieska, while those with a small metal insert set with sinew, mastic and poison into a split bone point (Figure 16:2, 3) tend to occur in the Drakensberg and surrounding regions. The design of metal arrowheads would thus seem to be largely constrained by the type of metal and the form in which it was available, and it is only once fencing wire was introduced that clear variations in style can be documented.

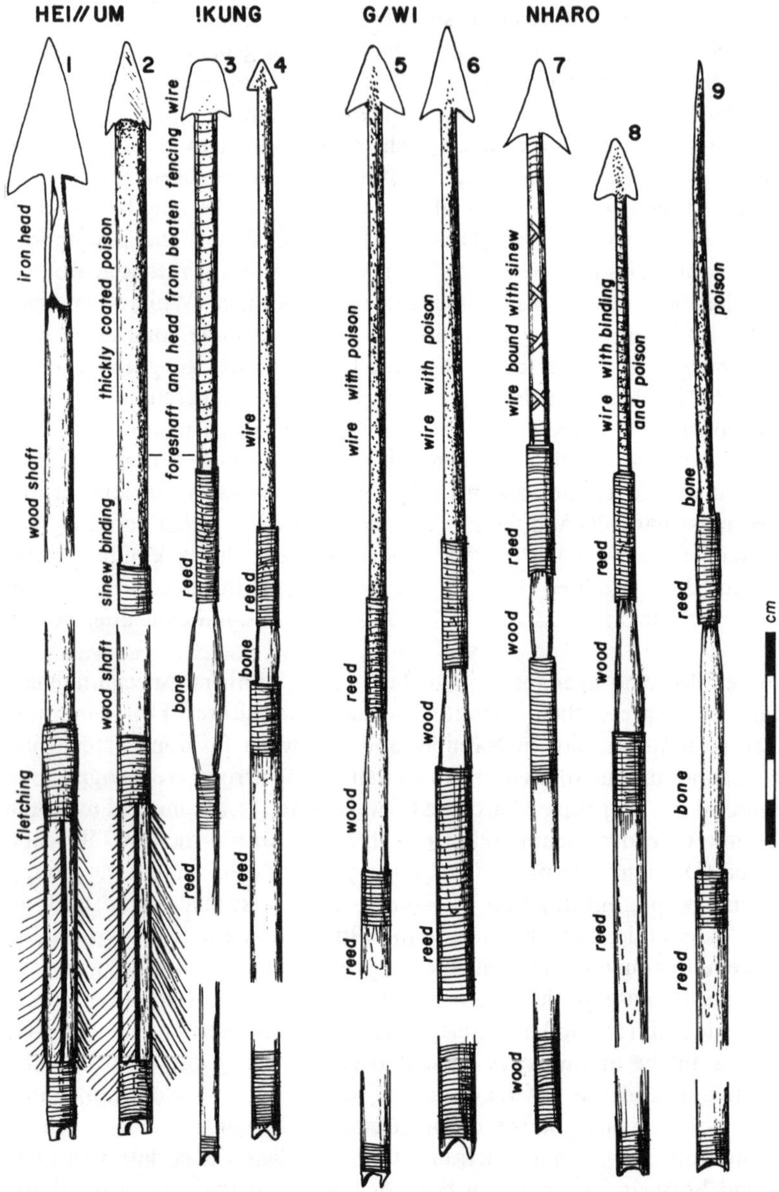

Figure 17. A range of San arrows from known sources, collected after 1920. All arrowheads are made from beaten fencing wire, except 1 (beaten iron wrapped around a wooden shaft) and 9 (reversible bone point), 1, 2, 3, 5-7: National Cultural History Museum, Pretoria; 4, 8, 9: South African Museum, Cape Town.

The bone components of reversible bone arrows that are found in both ethnographic and archaeological samples from northern Namibia to the southern Cape have been measured too, but no significant differences in size or shape could be detected between samples from different regions. This observation seems to confirm the conclusion drawn above, namely that the stylistic variations in arrows made from fencing wire are confined to this medium.

The implications of these observations are that stylistic differences in San arrowheads are present between modern San groups, but that they may be a recent phenomenon resulting from the use of easily manipulated fencing wire. Where they occur, however, they coincide broadly with language boundaries. From this we may conclude that where clear stylistic differences in stone artefact occur between contemporary assemblages in adjacent geographic areas there is a reasonable chance that they may have been made by people that were differentiated linguistically or dialectally, but different artefacts are likely to have different distribution patterns and we cannot predict which will be most useful in plotting dialectal boundries of prehistoric populations. Furthermore an artefact that is useful in differentiating between groups at one boundary zone, may differ in this respect from time to time and from place to place. We have little information at present on the geographic variability of LSA artefacts but this aspect is well worth investigating in the future, if only to demonstrate broad regional similarities and differences.

DISCUSSION AND CONCLUSIONS

In introducing the term 'Later Stone Age' in the late 1920s, Goodwin and Van Riet Lowe saw it as a natural division of the Stone Age sequence. For them it marked the initiation and development of the material cultural tradition of the ancestors of the present-day hunting and gathering peoples of southern Africa because it included ornamental and decorative items as well as art in the form of paintings and engravings. From the perspective of the 1920s the association of mostly microlithic tools with lithic and non-lithic items known to have been used by modern people was evidence for essentially modern behaviour. Research over the intervening 50 to 60 years has provided a much larger data base than was available to Goodwin and Van Riet Lowe. The advent of radiocarbon dating, the accumulation of a much larger sample of stratified sites, more carefully designed ethnographic studies and analyses of associated faunal and floral remains have all shown that the development of the LSA cultural tradition was more complex than could have been envisioned in the 1920s. From the data reviewed in this paper we can see that not all LSA assemblages are microlithic, that individual items of LSA material culture appear at different times in the archaeological record, that some items considered characteristic of the LSA in fact pre-date the introduction of microlithic tools and are found with the MSA, and that the LSA stone artefact tradition sensu lato is probably as much as 40,000 years old. It is the nature and timing of both the first appearance of the LSA and its component parts and subsequent modifications that have been the main concern of this paper.

Despite the increase in the size of the data base, it has proved difficult to date the elusive rubicon between the MSA and LSA, partly because the concept of a 'rubicon' is itself unrealistic, and partly because the small number of observations in the relevant time range are inconsistent. The reason for the small number of observations probably relates to the likelihood that environmental circumstances during the last glacial cycle between ca 90,000 and 12,000 B.P. dictated a different pattern of population distribution and density than that pertaining during interglacial cycles. Our present sampling methods have favoured the location and investigation of the interglacial rather than the glacial sites, but whether it would be possible to locate the glacial occurrences by using a different search strategy remains to be seen. Inconsistency in the dating of the MSA/LSA interface can be seen between interpretations that postulate a late survival of the MSA through to the Holocene in localized areas of southern Africa long after the appearance of LSA assemblages in adjacent regions (Phillipson 1976, 1977), and others that suggest that the transition took place before 40,000 B.P. and is beyond the limits of radiocarbon dating (Beaumont & Vogel 1972a, b; Beaumont *et al.* 1978), and that it dates to between 32,000 and 20,000 B.P. (Deacon, H.J. 1979). I have preferred to keep an open mind on this question; further data will hopefully be forthcoming in the next few decades to enable us to review these dates more realistically.

The timing of the MSA/LSA transition may be poorly understood at present but the nature of the change is more consistent. Where long sequence sites have been excavated the shift from clearly MSA to clearly LSA assemblages is abrupt in South Africa (Border Cave, Boomplaas, Bushman Rock Shelter), although in Zimbabwe there may have been some overlap as apparent in the Tshangula Industry while in Namibia the character of the earliest LSA is not microlithic (Apollo 11). There is no evidence that the initiation of the LSA was the result of a simple response to environmental change nor that it is associated with a new sub-species of *Homo sapiens.* The small number of associated faunal assemblages makes it impossible to assess adequately whether early microlithic toolmakers were more successful hunters than their late MSA counterparts (although there are clear differences between MSA last interglacial and LSA present interglacial hunting patterns), but it would seem that they were not significantly so. The main feature of the transition, then, is in a significant change in the stone flaking technique and the range of stone artefacts thus produced.

The recognition of the LSA as a technological entity is based partly on the appearance of a microlithic flaking technique that is essentially similar to that adopted by people in eastern and northern Africa at about the same time (i.e. between 40,000 and 20,000 B.P.) and partly on the gradual appearance of non-lithic items of material culture such as beads, pendants, polished bone arrow points, linkshafts and needles, ostrich eggshell water containers, tortoiseshell bowls and, later, wooden articles such as portions of bows, arrows and digging sticks. Substages within the timespan of the LSA, on the other hand, are recognized primarily on the basis of noticeable but minor shifts in the method of stone flaking as evident from the range of core types, the size and shape of untrimmed flakes, the range of formal tool designs and the raw materials selected for tool manufacture.

LSA substages are recognizable throughout southern Africa south of the Zambesian ecozone and are for the most part broadly contemporaneous. Early microlithic assemblages date from ca 40,000 to about 12,000 B.P. and were succeeded by non-microlithic assemblages of the terminal Pleistocene/early Holocene. From 10,000 to about 7500 B.P. Holocene microlithic assemblages that represent the classic concept of the LSA, appeared and persisted through to historic times in some areas. A final substage is present in the late Holocene and marks the introduction of pottery and domestic stock. The timing rather than the kind of technological change that took place between 12,000 and ca 8000 B.P. may have been stimulated by shifts in subsistence economy. The similarity of the toolmaking techniques and the range of formal tool designs in a variety of ecozones and local habitats during this time period argues against their having being designed in a conscious effort to cope with specific local requirements. We assume that the same sorts of tasks were undertaken with stone tools throughout the LSA (i.e. skin working, woodworking, bone tool making etc.), but from time to time the tool designs changed. The choice of new designs at times when modification of subsistence strategies was apparently necessitated by climatic change seems to have been influenced by the range of options available amongst LSA communities. The distribution of similar assemblages may thus give us some idea of the limits of information networks that may have operated during the past and the pathways along which such information may have travelled. The data available at present suggest that the LSA network may not have included Namibia during the late Pleistocene because microlithic assemblages did not succeed the MSA there, and that the Zambesian ecozone during the terminal Pleistocene/early Holocene was part of a different network from regions to the south because non-microlithic assemblages are not found there at that time.

The focus of research over the last 20 years has tended to be towards particular sites or relatively small areas and the study of inter-regional correlations in a wider geographical perspective together with related differences in human adaptations and settlement patterns is only slowly emerging. Contrasts that have been drawn between southern Africa and the Zambesian ecozone have highlighted significant differences between adaptations in the deciduous woodland-savanna environments of central Africa and the more open shrubland and grassland habitats to the south. These are reflected in Holocene toolkits that differ in the relative frequencies of backed microliths and scrapers, as well as in the range of subtypes in each of these major formal tool classes. These differences are thought to reflect a preference for bark cloth rather than leather clothing in the Zambesian ecozone leading to a low frequency of scrapers there, and a different range of arrow designs with multiple-barbs leading to a higher frequency of backed microliths. These differences between the Zambesian ecozone and those to the south seem to be of greater magnitude than those observed between regions within southern Africa. Differences in scraper size between the northern and southern Cape, for example, are considered to reflect some degree of social distance and gradients and boundaries between these and other regions are an aspect that should be investigated in future research. It is interesting to note that observations on factors affecting the spatial distribution of modern technological items stress that stationary factors such as soil types or mountain ranges may act as catalysts or retardants in the process of change, but they do not 'explain' a distribution (Hagerstrand 1967:293). If we accept White's (1949) view that such distributions are explicable in cultural terms, then archaeology can offer a unique source of data for testing the role of physical features in the distribution of material cultural items (Sampson 1980).

The relatively large sample of radiocarbon dates available for LSA assemblages has allowed broad distribution patterns to be discerned. This is an interesting development and, if we accept that the distributions are not entirely an artefact of modern research interests, they belie assumptions of an even distribution and continuous long-term settlement in all areas of southern Africa. The low density of late Pleistocene microlithic assemblages and the fact that associated faunal assemblages include more gregarious species than do later assemblages, suggest that people probably organized themselves into relatively large groups that occupied relatively large territories thereby reducing their archaeological visibility. The distribution of sites of this time period in the Cape ecozone and the eastern half of southern Africa suggests that regions of moderate topographic diversity and relatively high rainfall under the present régime were preferred. By contrast, the terminal Pleistocene/early Holocene is marked by a noticeable increase in cave and rock shelter occupations and a shift in the relative frequencies of gregarious and non-gregarious antelope that indicate some re-organization of populations that allowed people to move into areas such as the northern Cape where late Pleistocene occupations are as yet unknown. By contrast, the Holocene hunting pattern was geared to the regular exploitation of small solitary antelope and intensive plant food collecting and the wide site distribution and large number of occurrences suggest that smaller groups of people occupied smaller territories with seasonal movements dictated by the growth cycles of the plant food staples rather than by game movements. Such a re-organization of people would have increased archaeological visibility both because of the large number of sites occupied and because caves and rock shelters, prime targets for archaeological research, became favoured habitation sites in almost all parts of southern Africa. An overall increase in population during the Holocene is not a necessary correlate, but is a likely possibility (Parkington 1980).

Gaps in the Holocene cultural record in the Karoo-Namaqualian and Transvaalian ecozones in particular can be interpreted as evidence for fluctuations in the centres of population density. The reasons for these fluctuations are not easily found, however. It was initially suggested that the lower density of mid-Holocene sites may have resulted from a lower rainfall during the climatic optimum which increased the risk for hunter-gatherers in the drier interior of South Africa. The details of this risk factor are obscure, however, and the rainfall changes are now seen to be more complex than was originally envisaged (Butzer *et al.* 1978; Avery, D.M. 1981) with probably moister conditions in the first half of the Holocene and drier conditions after about 6000 B.P. Other anomalies which have arisen with a larger sample of dated assemblages are the absence so far of dated assemblages from the mid-Holocene in relatively well-watered regions such as Natal and the Transvaal, but their presence in very dry regions such as southern Namibia and the north-western Cape at Wonderwerk. These differences result from the fact that not all areas of southern Africa received a lower rainfall at the same time in the second half of the Holocene (Lancaster 1979b; Avery, D.M. 1981), and from differences in the nature of the food staples used by Holocene people. Variations in reactions of staple plant and animal foods to changes in rainfall and temperature in different regions and the availability or otherwise of suitable substitutes may have been as important to hunter-gatherers as changes in climatic conditions. Regions which seem to have been preferred during the mid-Holocene are the southern and eastern Cape and part of the south-western Cape, the western foothills of the Drakensberg, the Matopos area of Zimbabwe and parts of Namibia. Future research could focus on the factors common to these areas but absent in the Orange Free State, parts of the northern Cape, the Transvaal, Botswana and parts of Namibia.

The final substage within the LSA is that marked by the introduction of pottery and a herding economy to Stone Age people in southern Africa. The relatively sudden appearance of potsherds and bones of exotic domesticated animals over a wide geographic area in excavated deposits dating to about 2000 years ago has stimulated several explanatory hypotheses. The availability of linguistic data for some of the still extant herders and hunter-gatherers in southern Africa has provided the basis for the suggestion that domestic stock in particular was spread from Botswana/Zimbabwe by people who migrated southwards, and who were thus economically and linguistically distinct from the indigenous hunter-gatherers they met in South Africa. Of interest here is the fact that sites where both pre- and post-pottery/domestic stock assemblages occur show no significant difference in the stone artefacts through this sequence. This suggests that acculturation of local hunter-gatherers was also part of the process of change, and/or that the immigrant herders came from an area where the stone artefact tradition was essentially similar to that of the hunter-gatherers in the south. In either event this stresses that prehistoric migrations are difficult to trace archaeologically and any project which seeks to test the migration hypothesis will need to be carefully constructed. Open sites, for example, are likely to be more informative than caves or rock shelters and the possibility that pottery and stock in the southern and eastern Cape may have been introduced from the eastern half of South Africa rather than from the north needs to be kept in mind. The almost total absence of relevant observations in Botswana, Zimbabwe, the Transvaal, Natal and Mozambique makes it difficult to hazard guesses at this stage, but the situation was undoubtedly more complex than simple southward migration of Tshu-Khwe-speakers from Botswana.

Unlike their contemporaries in the northern hemipshere, few of the Holocene peoples of the southern hemisphere domesticated their own indigenous plants and animals. Southern African LSA people were no exception, although there is some evidence for a level of habitat management in the practice of veld burning to promote the growth of geophytes and, later, to stimulate suitable grazing for livestock. The profitability of herding was, nevertheless, appreciated by at least some of the indigenous people when it was introduced and it is probably true that, had the European colonists not arrived in the southern Cape when they did, the hunter-gatherers would have been assimilated by herders in all but the most inhospitable environments. The reason for the absence of domesticated indigenous plants lies in the unsuitability of the staples such as geophytes, melons, fruits and nuts which either have low reproductive rates when planted or disturbed or vary markedly from year to year in their yield depending on the uncontrollable vagaries of the weather. As far as the indigenous animals were concerned, the solitary browsers such as the grysbok, steenbok and bushbuck are not suitable as herd animals while amongst those animals that did herd together in natural circumstances, the supply apparently exceeded the demand and the stimulus to domesticate was not present.

This review has demonstrated that the hunter-gatherer-fisher way of life practiced for so many thousands of years by LSA people and their ancestors in southern Africa was remarkably adaptable to environmental changes, whether through adjustments in population distribution, shifts in staple foods or the adoption of different tool designs. They were not able to maintain their life-style, however, with the influx of settled agriculturalists, both Iron Age and European. Except in a few isolated regions such as the high Drakensberg mountains and the Kalahari desert where farmers spread relatively late, hunter-gatherer-fishers had been mostly acculturated by the middle of the 19th century. It is only small remnant populations who have provided archaeologists with the necessary analogues for Later Stone Age research.

ACKNOWLEDGEMENTS

I would like to thank the following people for assistance in the compilation of data for this paper: B. Adams for laboratory help, P.B. Beaumont for information on northern Cape and eastern Transvaal sites, J.N.F. Binneman for permission to cite unpublished results of his replication experiments and microwear analyses, R.R. Inskeep for unpublished data on Nelson Bay Cave, Mary Leslie Brooker for unpublished information on Uniondale, H. Opperman for unpublished data from the north-eastern Cape, J.E. Parkington for unpublished data on the south-western Cape, H.P. Steyn for information on San arrows, Anne I. Thackeray and J.F. Thackeray for unpublished data on Wonderwerk and for the photograph in Figure 3b, J.C. Vogel for information on radiocarbon dates, T.P. Volman for discussions on the scope of this paper, N.J. Walker for data on Matopos sites and the photograph in Figure 3a, and W.E. Wendt for unpublished data on Namibian sites. I have benefitted from many discussions with H.J. Deacon who allowed me to use unpublished data included in the section on Holocene subsistence.

Draft copies of this paper were circulated to a number of colleagues. I am sincerely grateful to the following people who made pertinent comments, many of which helped in the writing of the final draft: G. Avery, P.B. Beaumont, J.N.F. Binneman, H.J. Deacon A.J.B. Humphreys, R.G. Klein, J.D. Lewis-Williams, A.D. Mazel, J.E. Parkington, C.G. Sampson, A. Scholtz, T.P. Volman, N.J. Walker, W.E. Wendt and M.L. Wilson.

I should also like to thank the Directors and staff members of the South African Museum in Cape Town, The Duggan Cronin Gallery in Kimberley, the National Museum in Bloemfontein, the Natal Museum in Pietermaritzburg, the Killie Campbell Library in Durban, the East London Museum and the National Cultural History Museum in Pretoria for permission to study their collections of San arrows, and the University of Stellenbosch for laboratory space.

Table 12. Key to abbreviations, with references, used in Tables 2a, 2b, 7, 9 and 11 and Figures 1, 13, 14 and 15.

Abbreviation	Name of site	References
A	Amadzimba, Zimbabwe	Cooke, C.K. & Robinson 1954; Deacon, J. 1966
AAR	Aar 2, Namibia	Vogel & Visser 1981
AKI	Andrieskraal 1, s Cape	Hendey & Singer 1965
AND	Andriesgrond, s-w Cape	Parkington 1980:108
ANN	Ann Shaw, e Cape	Derricourt 1973, 1977
AP	Apollo 11, Namibia	Wendt 1972, 1976; Freundlich *et al.* 1980; Vogel & Visser 1981
AR	Arniston, s-w Cape	Maggs 1977:172
ARD	Arrisdrift, Namibia	Vogel & Visser 1981
ARP	Aar pot, Namibia	Vogel & Visser 1981
ARS	Aar 1, Namibia	Vogel & Visser 1981
AS	Aspoort, s-w Cape	Smith & Ripp 1978
AUR	Aurus 6, Namibia	Vogel & Visser 1981
AUS	Austerlitz, Namibia	Freundlich *et al.* 1980
B	Bushman Rock Shelter, e Transvaal	Vogel 1969; Plug 1978; 1981
BAM	Bambata, Zimbabwe	N.J. Walker, pers. comm.
BAN	Banda Cave, Swaziland	Stuiver 1969; Beaumont & Vogel 1972a, b
BC	Border Cave, Natal/KwaZulu	Beaumont & Vogel 1972a, b; Linick 1977; Beaumont 1978; Beaumont *et al.* 1978
BEL	Belleview, Lesotho	Carter & Vogel 1974; Carter 1978
BEN	Benfica, Angola	Dos Santos & Ervedosa 1970; Vogel & Visser 1981
BL	Blydefontein, n Cape	Sampson 1972; Humphreys 1979:305
BON	Bonteberg, s-w Cape	Maggs & Speed 1967; Grindley *et al.* 1970
BPA	Boomplaas Cave, s Cape	Deacon, H.J. *et al.* 1976, 1978; Fairhall *et al.* 1976; Deacon, H.J. 1979; Vogel, pers. comm.
BUF	Buffelskloof, s Cape	Opperman 1978
BUR	Burchell's Shelter, n Cape	Humphreys 1975, 1979:215
BWE	Bonawe, n-e Cape	Opperman, pers. comm.
BY	Byneskranskop, s-w Cape	Schweitzer & Wilson 1978, 1982; Klein 1981a
C	Calder's Cave, Zimbabwe	Cooke, C.K. & Simons 1969; Cooke, C.K. 1979
CAN	Cango Caves, s Cape	Vogel, pers. comm. Deacon, J. 1979
CHA	Cha-Ré shelter, Namibia	Vogel & Visser 1981
COB	Cave of Bees, Zimbabwe	N.J. Walker, pers. comm.
COL	Colwinton, n-e Cape	Opperman, pers. comm.
CON	Conception Bay, Namibia	Vogel & Visser 1981
CR	Chalumna River Mouth, e Cape	Derricourt 1973, 1977
CSF	Cape St Francis, s Cape	De Villiers 1974; Thackeray, J.F. & Feast 1974
CY	Cymot Shelter, Namibia	Sandelowsky & Viereck 1969
D	Duncombe Farm, Zimbabwe	Wadley 1973
DE	Duiker Eiland, s-w Cape	Robertshaw 1979
DF	Doornfontein, n Cape	Beaumont & Boshier 1974
DH	De Hangen, w Cape	Parkington & Poggenpoel 1971
DIK	Dikbosch 1, n Cape	Humphreys 1974a, 1979
DIK2	Dikbosch 2, n Cape	Humphreys 1974a, 1979
DK	Die Kelders, s-w Cape	Schweitzer 1970, 1979
DKF	Diepkloof, w Cape	Parkington 1977a
DO	Dombozanga, Zimbabwe	Robinson 1964
DR	Driel Shelter, Natal	Maggs 1977; Maggs & Ward 1980
DV	Diana's Vow, Zimbabwe	Cooke, C.K. 1979a

Table 12 (*Contd.*)

Abbreviation	Name of site	References
E	Etemba 2, Namibia	Wendt 1972; Freundlich *et al.* 1980
EB	Elands Bay Cave, w Cape	Parkington 1977a and pers. comm.
EBO	Elands Bay Open w Cape	Horwitz 1979
EMA	Emambeni, Swaziland	Stuiver 1969; Beaumont & Vogel 1972a, b
EQ	Equus Cave, n Cape	Beaumont, pers. comm.
ER	Eros Shelter 1, Namibia	Beaumont & Vogel 1972a, b; Wendt 1972
ES	Eros Shelter 2, Namibia	Beaumont & Vogel 1972a, b; Wendt 1972
ET	Etemba 14, Namibia	Wendt 1972; Freundlich *et al.* 1980
FA	Fackeltrager, Namibia	Wendt 1972; Vogel & Visser 1981; Freundlich *et al.* 1980
FIS	Fish River, Namibia	Vogel & Visser 1981
G	Gwisho, Zambia	Fagan & Van Noten 1971
GAT	Gatboskloof, s Cape	Wilson, pers. comm.
GB	Gordons Bay, s-w Cape	Van Noten 1974
GE	Great (Big) Elephant Shelter, Namibia	Clark & Walton 1962; Wadley 1976, 1979
GH	Good Hope Shelter, Natal	Carter 1978:167; Cable *et al.* 1980
GI	Girls School Shelter, Namibia	Jacobson 1978; Vogel & Visser 1981
GL	Glen Elliot, n Cape	Sampson 1972
GOB	Gobabeb, Namibia	Vogel & Visser 1981
GOR	Gorob skeleton, Namibia	Vogel & Visser 1981
GRA	Grassridge, n-e Cape	Opperman, pers. comm.
GT	Grootdrink pot, n Cape	Rudner 1971, 1979:24
HAA	Haalenberg, Namibia	Vogel & Visser 1981
HAW	Hawston, s-w Cape	Avery, G. 1975:112
HBC	Hout Bay Cave, s-w Cape	Buchanan 1977
HBP	Hottentot Bay pot, Namibia	Vogel & Visser 1981
HK	Heuningneskrans, e Transvaal	Vogel & Marais 1971; Beaumont 1978, 1981
HOA	Hoarusibmund, Namibia	Vogel & Visser 1981
HOP	Hoanibmund pot, Namibia	Vogel & Visser 1981
HR	Highlands Rock Shelter, n Cape	Deacon, H.J. 1976:222, 224
HUN	Huns, Namibia	Freundlich *et al.* 1980
KAL	Kalemba, Zambia	Phillipson 1976:27
KC	Kruger Cave, Transvaal	Mason & Vogel 1974
KLP	Klipfonteinrand, w Cape	Parkington 1980:108
KLY	Kalanyoni, Zimbabwe	N.J. Walker, pers. comm.
KR	Klasies River Mouth 5, s Cape	Singer & Wymer 1969
KRA	Kangkara, s Cape	Vogel, pers, comm.
KRM	Klasies River Mouth main site, s Cape	Sampson 1972; Singer & Wymer in prep.
KUM	Kumakams, Namibia	Vogel & Visser 1981
LEM	Lemoenfontein, n Cape	Klein 1979b
LC	Lion Cavern, Swaziland	Dart 1967; Dart & Beaumont 1968
LH	Leopard's Hill, Zambia	Miller 1969
LI	Limerock 1, n Cape	Humphreys 1979:240
LIM	Limerock 2, n Cape	Humphreys 1979:240
LN	Lower Numas 1, Namibia	Rudner 1957, 1972; Vogel & Marais 1971; Jacobson 1978; Vogel & Visser 1981
LUP	Luderitz potsherd, Namibia	Vogel & Visser 1981
LW	Little Witkrans, n. Cape	Beaumont, pers. comm.
MA	Mlaula, Swaziland	Beaumont & Vogel 1972a, b
MAG	Maguams, Namibia	Freundlich *et al.* 1980; Vogel & Visser 1981
MAK	Makwe, Zambia	Phillipson 1976:72

Table 12 (*Contd.*)

Abbreviation	Name of site	References
MAL	Maleme Dam, Zimbabwe	N.J. Walker, pers. comm.
MAT	Matupi, Zaïre	Van Noten 1977
MEL	Melikane, Lesotho	Carter 1978:167
MES	Messum 1, Namibia	Freundlich *et al.* 1980; Vogel & Visser 1981
MG	Magabengberg, w Transvaal	Mason 1962b
MGT	Maguams Terrace, Namibia	Freundlich *et al.* 1980
MHB	Melkhoutboom, s-e Cape	Deacon, H.J. 1976
MIR	Mirabib, Namibia	Sandelowsky 1977; Vogel & Visser 1981
MO	Munro's site, s Transvaal	Vogel & Marais 1971; Mason & Vogel 1974
MON	Montagu Cave, s Cape	Keller 1970a, 1973
MOS	Moshebi's Shelter, Lesotho	Carter & Vogel 1974; Carter 1978
MP	Mpato, Zimbabwe	Cooke, C.K. Simons 1969
MR	Matjes River Cave, s Cape	Louw, J.T. 1960; Vogel 1970
MTA	Mtanye, Zimbabwe	Cooke, C.K. 1979a:146
N	Nachikufu, Zambia	Miller 1969
NAM	Namtib, Namibia	Freundlich *et al.* 1980; Vogel & Visser 1981
NAP	Nautilus pot, Namibia	Vogel & Visser 1981
NAR	Narob grave, Namibia	Vogel & Visser 1981
NBC	Nelson Bay Cave, s Cape	Inskeep 1965 and pers. comm; Vogel 1970; Klein 1972a, b; Switsur & West 1973; Fairhall *et al.* 1976; Deacon, J. 1978 and in prep.
NE	Numas Entrance Shelter, Namibia	MacCalman 1965
NOS	Nos, Namibia	Vogel & Visser 1981
NP	Numas Plateau, Namibia	Vogel & Marais 1971
NTI	Nswatugi, Zimbabwe	Walker, N.J. 1980:19, 21
NUP	Numas pot, Namibia	Vogel & Visser 1981
OAK	Oakleigh Farm, e Cape	Derricourt 1973, 1977
OB	Olieboompoort, w Transvaal	Mason 1962b
OHT	Oakhurst, s Cape	Deacon, J. 1979:36; Vogel, pers. comm.
ORA	Orabes Upper & Lower, Namibia	Jacobson 1978; Vogel & Visser 1981
OST	Ostrich Shelter, Namibia	Jacobson 1978; Vogel & Visser 1981
P	Pomongwe, Zimbabwe	Cooke, C.K. 1963; Beaumont & Vogel 1972a, b; Walker, N. J. 1980:22
PAT	Paternoster, w Cape	Robertshaw 1977
PB	Pearly Beach, s-w Cape	Avery 1974 and pers. comm.
PH	Philip Cave, Namibia	Martin & Mason 1954
POC	Pockenbank, Namibia	Freundlich *et al.* 1980; Vogel & Visser 1981
POW	Powerhouse Cave, n Cape	Humphreys 1979:173
Q	Castle Quarry, Swaziland	Dart & Beaumont 1968; Vogel 1970
RC	Rose Cottage, e Orange Free State	Vogel 1970; Vogel & Marais 1971; Beaumont & Vogel 1972; Beaumont 1978:637
RE	Rooiels Cave, s-w Cape	Smith, A.B. 1981
RED	Redcliff, Zimbabwe	Beaumont & Vogel 1972
RM	Riversmead, n Cape	Sampson & Sampson 1967; Sampson 1972
ROB	Robberg caves D, E & F, s Cape	Vogel & Marais 1971; Rudner & Rudner 1973; Hausman 1980: table 2
RR	Riet River burial, Orange Free State	Humphreys 1974b
SB	Steenbras Bay, Namibia	Vogel & Visser 1981
SEH	Sehonghong, Lesotho	Carter & Vogel 1974; Carter 1978

Table 12 (*Contd.*)

Abbreviation	Name of site	References
SG	Striped Giraffe Shelter, Namibia	Sandelowsky & Viereck 1969; Jacobson 1978; Wadley 1979; Vogel & Visser 1981
SHA	Shashabugwa, Zimbabwe	Walker, N. J. 1980 and pers. comm.
SHO	Shongweni, Natal	Davies 1975:657-8
SMT	Smitswinkelbaai, s-w Cape	Poggenpoel & Robertshaw 1981
SOS	Sossusvlei, Namibia	Vogel & Visser 1981
SS	Scott's Cave, s Cape	Deacon & Deacon 1963; Deacon, H.J. 1967; Klein & Scott 1974
STF	Stofbergsfontein, w Cape	Robertshaw 1978a
T	Tshangula, Zimbabwe	Cooke, C.K. 1963, 1979:146
TAB	Tsisab Open T30, Namibia	Vogel & Visser 1981
TAF	Twyfelfontein Affenfelsen, Namibia	Jacobson 1978; Freundlich *et al.* 1980
THA	Twyfelfontein Hasenbild, Namibia	Wendt 1972; Jaconbson 1978; Freundlich *et al.* 1980; Vogel & Visser 1981
TIA	Tiara Shelter, Namibia	Jacobson 1978; Vogel & Visser 1981
TIE	Tienfontein 2, 4A & 7, Orange Free State	Brooker 1980
TIR	Tiras 5, Namibia	Freundlich *et al.* 1980; Vogel & Vissar 1981
TZS	Twyfelfontein Zwei Schneider Shelter, Namibia	Vogel & Visser 1981
U	Uitkomst, Transvaal	Mason 1962b; Vogel 1970
UH	Uri Hauchab 4, Namibia	Freundlich *et al.* 1980
UM	Umhlanga Rocks, Natal	Vogel & Marais 1971
UNI	Uniondale, e Cape	M. Leslie Brooker, pers. comm.
UNJ	Unjab midden, Namibia	Vogel & Visser 1981
URU	Ururu, Namibia	Vogel & Visser 1981
VEN	Ventershoek, n Cape	Klein 1979b
VP	Voigtspost 1, Orange Free State	Horowitz *et al.* 1978:154
VPO	Voigtspost O, Orange Free State	Horowitz *et al.* 1978:154
WAR	Warmquelle, Namibia	Vogel & Visser 1981
WOR	Wortel, Namibia	Vogel & Visser 1981
WT	Wilton name site, e Cape	Deacon, J. 1969, 1972
WK	Wonderwerk, n-w Cape	Butzer *et al.* 1979; Thackeray, A.I. *et al.* 1981; Beaumont & Thackeray, pers. comm.
Z	Zoo Park, Namibia	MacCalman 1965
ZEB	Zebrarivier, Namibia	Vogel & Visser 1981
ZM	Zerrissene Mountains, Namibia	Carr *et al.* 1978; Vogel & Visser 1981
ZOM	Zombepata, Zimbabwe	Cooke, C.K. 1971
ZY	Zaayfontein, n Cape	Sampson 1972

TIM MAGGS
Natal Museum, Pietermaritzburg, South Africa

8

The Iron Age south of the Zambezi

INTRODUCTION

This review is limited to Africa south of the Zambezi and Kunene rivers and will focus on those areas in South Africa which have been most actively investigated in recent years. The geographical coverage of research is uneven although some previously neglected areas, such as Botswana and Mozambique, are now being investigated. The increased tempo of Iron Age research is a recent phenomenon and therefore much of the new information from current and recently completed work is not yet available.

As a background to the Iron Age settlement of the sub-continent two factors, which are apparent from the preceding chapter, pertaining to Late Stone Age distribution need discussion. The first of these is the evidence of relatively sparse hunter-gatherer occupation of some of the eastern areas, notably the Transvaal, in the millennia preceding the arrival of the Iron Age. The second is the evidence of Stone Age pastoralists with pottery from about the beginning of the first millennium AD in the western parts of the sub-continent–indeed in the very areas which down to colonial times were never occupied by Iron Age peoples (Figure 1). As yet there is no good evidence that Stone Age pastoralists spread further east although this lack of evidence is by no means conclusive. Indeed some authors have already asked why the first herders should have chosen the more arid west rather than the grasslands to the east with their richer potential as pasturage. The apparent absence from the eastern areas may merely reflect inadequate archaeological fieldwork, for it is only in the past 8 years that the early arrival of sheep in the west has been recognized (Schweitzer & Scott 1973). What is more, no trace of Stone Age herders has been recovered from excavations in the relatively well researched areas of Zambia or Zimbabwe through which the domestic sheep probably came southwards. On the other hand, Khoisan hunter-gatherers survived in large areas of what are today the central and eastern Cape Province, Orange Free State, Lesotho, upland Natal and even parts of the Transvaal down to the nineteenth century. This pattern might suggest that there was a relatively stable hunter-gatherer population which, unlike its western neighbours, remained uninfluenced by pastoralism. However, the evidence of an apparently sparse pre-Iron Age population in some of the eastern areas may count against this interpretation. Furthermore Late Stone Age materials have been found on several Early Iron Age (EIA) sites (Mason 1974, Maggs 1980b) suggesting both a

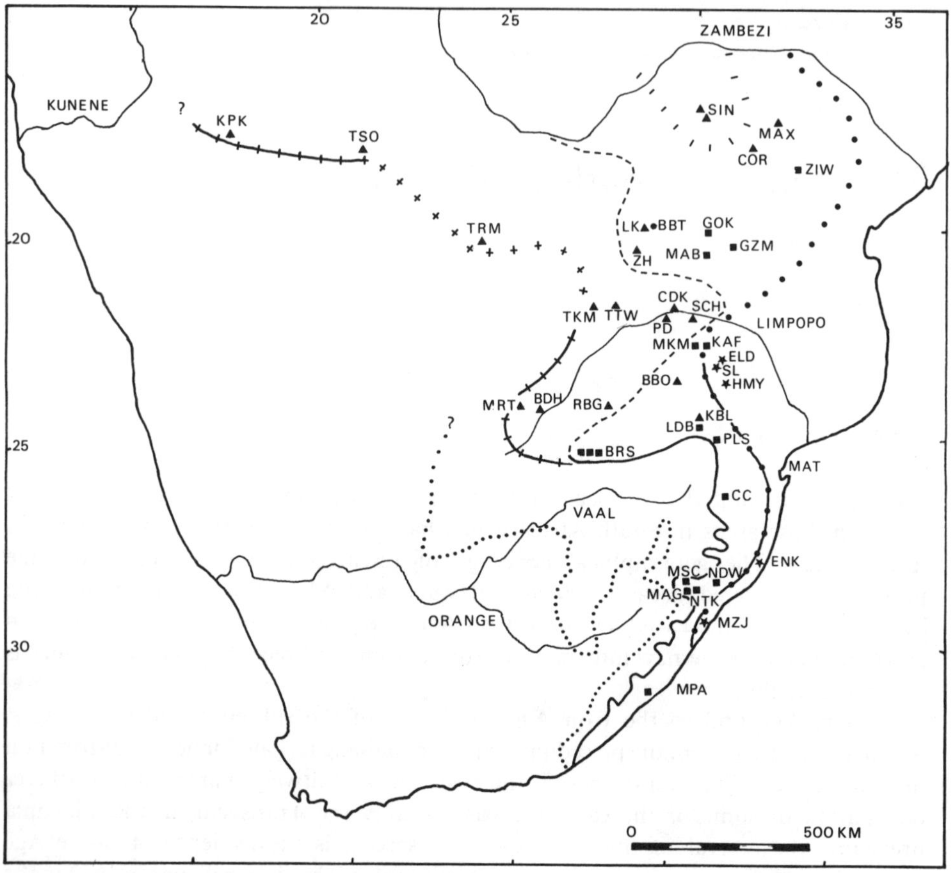

Figure 1.
*Matola Tradition sites around 300 AD with possible limits •—•—•—•—•
■Sites around AD 450-600 of the Lydenburg Cluster, Gokomere-Ziwa etc. with possible limits
△Toutswe, Zhizo, Eiland Coronation, Maxton, Sinoia etc.
Late Iron Age limits . . . .

Key to abbreviations:

BBO Bambo
BBT Bambata
BDH Broadhurst
BRS Broederstroom
CC Castle Cavern
CDK Commando Kop
COR Coronation
ELD Eiland
ENK Enkwazini
GOK Gokomere
GZM Great Zimbabwe
HMY Harmony
KAF Klein Afrika
KBL Klingbeil
KPK Kapako
LDB Lydenburg
LK Leopards Kopje
MAB Mabveni
MAG Magogo
MAT Matola
MAX Maxton
MKM Matakoma
MPA Mpame
MRT Moritsane
MSC Msuluzi Confluence
MZJ Mzonjani
NDW Ndondondwana
NTK Ntshekane
PD Pont Drift
PLS Plaston
RBG Rooiberg
SCH Schroda
SIN Sinoia
SL Silver Leaves
TKM Taukome
TRM Toromoja
TSO Tsodilo
TTW Toutswe
ZH Zhizo
ZIW ZIWA

continuation of the former way of life and perhaps even an increase in the number of its practitioners. But such views are little more than speculations, for the present evidence tells us little about the Stone Age occupation of the eastern areas at the advent of the EIA.

The first major challenge to the Late Stone Age technological system, whether hunter-gatherer or pastoralist, occurred in or before the third century AD with the arrival of EIA communities which had an entirely different way of life and system of beliefs. The main archaeological markers of this are semi-permanent villages which produced ceramics, metals, crops and domestic animals. This cultural and economic 'package' was certainly complete by AD 500, but the evidence from the earlier sites on agriculture, livestock and indeed economy in general is as yet inadequate to show whether the various components came into the region in 'pre-assembled' form or whether some components arrived later in some areas. The inadequacy of the present evidence is partly due to poor preservation of organic remains in tropical and sub-tropical soils and partly due to the relatively few excavations, most of which have been small in scale and have not included techniques, such as flotation, specifically aimed at recovering organic remains.

Research has emphasised ceramic typology in the expectation that it will provide a means of tracing relationships and movements between different parts of sub-equatorial Africa. The relatively distinctive and usually well decorated pottery of the EIA would seem to offer particularly good material for such reconstructions. The widely accepted assumption that the spread of the EIA correlates to the spread of peoples speaking Bantu languages has further encouraged this rather diffusionist preoccupation of EIA studies. However, at the present time, both ceramically and lingustically derived models for this diffusion are highly controversial (e.g. Phillipson 1977, Huffman 1979a, Flight 1980 and Vansina 1980) suggesting that much more research is needed before a generally acceptable solution can be reached.

THE EARLY IRON AGE SEQUENCE

The sequence is best known from Zimbabwe and from recent work in the Transvaal, eastern Botswana and Natal. Other areas have at best a scatter of dated sites (Figure 1). Table 1 shows sites and a tentative scheme of relationships. In the case of Zimbabwe the industries themselves are shown, for these names are well established in the literature. Radiocarbon dates from the sites are give in Table 2.

The clearest combination of early sites includes Silver Leaves (Klapwijk 1974) and Eiland (Evers 1975) from the eastern Transvaal lowveld, Matola (Da Cruz e Silva 1976) in southern Mozambique and Enkwazini (Hall 1980a) and Mzonjani (Maggs 1980a) in Natal. The ceramic link is based on the characteristics of both pots and bowls. The pots have straight necks everted at a distinct angle from the body sometimes with multiple bevelled or fluted lip. The most common decoration takes the form of a groove just below the lip or a band of oblique hatching or oblique rows of impressions on the neck, although a considerable amount of variation occurs especially in the large assemblages. A variety of discontinuous motifs are found on the bodies. Bowls are typically subcarinated in profile, and on the northern sites, particularly Silver Leaves, have horizontal fluting above the carination. Impressed

Table 1. Grouping of Early Iron Age sites.

Date AD	Botswana and Namibia	Zimbabwe	Northern Transvaal	Southern Transvaal and Swaziland	Natal and Transkei	Mozambique
	T } Toutswe					
	T } Toutswe					
	T } Toutswe	MAXTON				
1000	T } Toutswe			?Klingbeil		
	Tsodilo 2		K2 (LIA)			
	T Maiphetwane	Z	Z Commando Kop			
	T Thatswane	Z				
	Tsodilo 1	Z	Z Pont Drift			
	Kapako	Z				?G Massingir
800		Z	Z Schroda	Klingbeil	Ntshekane	?G Hola Hola
		Z CORONATION				
		Z				
	T Taukome	Z			L Ndondondwana	?G Chibuene
600		G SINOIA			L Mpame	
		G		L Plaston	L Ndumu	
		G			L Msuluzi Confl.	
		G		L Lydenburg Heads	L Mhlopeni	
		G ZIWA	Klein Afrika	Broederstroom	L Magogo	
		G		?L Castle Cavern		
400		G				
		?G				
		?BAMBATA?		M Eiland	M Enkwazini	
			?Klein Afrika			
200		?G		M Silver Leaves	M Mzonjani	M Matola IV ?

G: Gokomere
L: Lydenburg cluster
M: Matola Tradition
Z: Zhizo
T: Toutswe

decoration, bevelling and fluting are rare south of Matola but other resemblances are sufficient to justify grouping the sites. Evers (1979) has suggested the name Silver Leaves Tradition to group the Transvaal sites and Matola. However, elsewhere I have argued (Maggs 1980a) that Matola would be more suitable to represent the whole group as it shares characteristics with both the Transvaal and Natal sites. Indeed Silver Leaves, with its predominance of fluting, is rather atypical.

Dated sites of the Matola Tradition all fall within the third and fourth centuries AD with the exception of Matola itself which has produced anomalous readings in the first and ninth centuries (Hall & Vogel 1980). A possible explanation may be contamination from mixing due to biotic activity, a problem that has apparently affected the dating of a number of EIA sites on sandy soils in tropical Africa (Clark & Fagan 1965, Cahen & Moeyersons 1977). It is therefore probable that Matola too belongs to about the third century and, along with its related sites, represents the earliest settled village communities on the eastern seaboard of southern Africa (Figure 1).

Succeeding the Matola tradition south of the Limpopo are a number of sites dated from the fifth to about the eighth centuries. Ceramic resemblance between these sites has been recognized by several workers although no formal nomenclature has yet been agreed upon. Hall and Vogel (1980) have used the informal name Lydenburg cluster which we will adopt here since aspects of the chronology, origins and typological interrelationships are still unclear or controversial.

Typological links derive mainly from the pots which have concave, everted necks typically with several horizontal bands of decoration. Bands or discontinuous motifs may occur on the body as well. A wider range of motifs than in Matola assemblages is common, bands of oblique hatching, opposed hatching (herringbone) and crosshatching being typical. Carinated bowls may still occur but the majority of bowls are of a simpler, hemispherical shape and lack decoration. Regional variation is noticeable between Mpame, of the seventh/eighth century on the Transkei coast (Cronin no date), Msuluzi Confluence of the seventh century (Maggs 1980b) together with similar Natal sites, Lydenburg itself (Inskeep & Maggs 1975, Evers in press) and Plaston (Evers 1977) of the fifth/sixth and seventh century respectively from the eastern Transvaal. Broederstroom in the central Transvaal (Mason 1974) and Castle Cavern in Swaziland (Dart & Beaumont 1969) may also belong within this cluster (the ceramics are unpublished) and because of their fifth century dates may be of value in throwing light on the relationships between the cluster and the Matola Tradition, which they just postdate (Table 1). This point will be discussed below.

At the upper end of the time scale the Lydenburg cluster develops increasingly along regional lines as represented by Ntshekane in Natal (Maggs & Michael 1976) and Klingbeil (Evers 1980) in the eastern Transvaal, dated around the ninth century.

In the northern Transvaal the earliest phase is represented by the sites of Matakoma (= Happy Rest) and Klein Afrika (Prinsloo 1974, 1981). Previously these were regarded as an extension of Gokomere southwards from Zimbabwe (e.g. Fagan 1965, Phillipson 1977). Recently however, it has been suggested that they are part of the 'western stream' of the EIA (Evers 1981, Hanisch 1981a) related on the one hand to Zambian material and on the other to the Lydenburg cluster. While most of the pottery from Klein Afrika (Prinsloo 1981) has the multiple bands of decoration on curved everted necks so typical of the Lydenburg cluster, some vessels are typical of

Table 2. Early Iron Age dates from southern Africa.

South Africa, Swaziland, Mozambique and Namibia	Uncalibrated dates a.d.		Tradition
Matola	R-1327	70 ± 50	(Matola)
(Mozambique)	R-1328	830 ± 50	(Matola)
Enkwazini	Pta-1847	300 ± 50	(Matola)
	Pta-1977	410 ± 50	(Matola)
Mzonjani	Pta-1989	280 ± 40	(Matola)
Silver Leaves	Pta- 901	270 ± 55	(Matola)
	Pta- 914	330 ± 50	(Matola)
	Pta-2360	250 ± 50	(Matola)
	Pta-2459	250 ± 40	(Matola)
Eiland	Pta-1524	270 ± 40	(Matola)
	Pta-1607	390 ± 40	(Matola)
	Pta-1608	320 ± 30	(Matola)
Lydenburg Heads	Pta- 328	490 ± 50	(Lydenburg)
	Pta-1634	540 ± 50	(Lydenburg)
Mpame	Pta-2019	640 ± 60	(Lydenburg)
	Pta-2045	720 ± 40	(Lydenburg)
Msuluzi Confluence	Pta-2195	640 ± 40	(Lydenburg)
	Pta-2197	580 ± 30	(Lydenburg)
Mhlopeni	Pta-2878	550 ± 50	(Lydenburg)
Magogo	Pta-2874	590 ± 50	(Lydenburg)
	Pta-2875	760 ± 50	(Lydenburg)
Ndondondwana	Pta-2388	730 ± 50	(Lydenburg)
	Pta-2389	760 ± 50	(Lydenburg)
Ntshekane	Pta-1057	850 ± 50	(Lydenburg)
	Pta-1058	800 ± 50	(Lydenburg)
Ndumu	Pta- 002	630 ± 40	(Lydenburg)
Plaston	Pta-1635	635 ± 50	(Lydenburg)
Klingbeil	Pta-1633	790 ± 50	(Lydenburg)
	Pta-1747	980 ± 40	(Lydenburg)
	Pta-2160	830 ± 50	(Lydenburg)
Klein Afrika	Pta-1168	330 ± 45	(Matakoma)
	Pta-1320	520 ± 40	(Matakoma)
	Pta-1321	500 ± 40	(Matakoma)
	Pta-2415	415 ± 50	(Matakoma)
	Pta-2420	540 ± 50	(Matakoma)
Schroda	Pta-1819	840 ± 50	(Zhizo)
	Pta-1967	790 ± 50	(Zhizo)
Commando Kop	Pta-1817	835 ± 55	(Zhizo & L.K.)
	Pta-1957	970 ± 40	(Zhizo & L.K.)
Pont Drift 1/1	Pta-1961	835 ± 50	(Zhizo & L.K.)
Pont Drift 1/2	Pta-1818	1110 ± 50	(Zhizo & L.K.)
	Pta-1959	810 ± 50	(Zhizo & L.K.)
Broederstroom	UCLA-1791 A	490 ± 50	
	UCLA-1791 B	460 ± 50	
	RL- 351	430 ± 110	
Castle Cavern	Y-1712	400 ± 60	
(Swaziland)	Gr. N-5022	415 ± 30	
	Gr. N-5315	400 ± 30	
	Y-1995	520 ± 100	
Kapako (Namibia)	Pta- 234	840 ± 50	
Massingir (Mozambique)	Pta-1640	920 ± 40	(? Lydenburg)

Table 2 *(Contd.)*

South Africa, Swaziland, Mozambique and Namibia	Uncalibrated dates a.d.		Tradition
Hola Hola (Mozambique)	R-1326	890 ± 50	(? Gokomere)
Chibuene (Mozambique)	R-1325	770 ± 50	(? Gokomere)
ZIMBABWE			
Mabveni	SR- 43	180 ± 120	(? Bambata)
	SR- 79	585 ± 110	(Gokomere)
	Pta-2105	585 ± 30	(Gokomere)
Great Zimbabwe	M- 913	320 ± 150	(? Bambata)
Gokomere	SR- 26	530 ± 120	(Gokomere)
Kinsale	SR- 117	540 ± 95	(Gokomere)
Sighwa	SR- 119	410 ± 95	(Gokomere)
Malapati	SR- 33	840 ± 100	(Gokomere)
Nyangoni	SR- 32	850 ± 100	(Ziwa)
Leopard's Kopje	SR- 55	700 ± 110	(Zhizo)
	SR- 225	670 ± 90	(Zhizo)
	I-4862A	820 ± 95	(Zhizo)
Taba Zika Mambo	SR- 68	870 ± 100	(Zhizo)
	SR- 512	740 ± 70	(Zhizo)
Makuru	N-1275	690 ± 65	(Zhizo)
Coronation Park	N- 978	710 ± 100	(Coronation)
Coronation Park	N- 979	980 ± 100	(Maxton)
Surtic Farm	Pta-1842	1005 ± 35	(Maxton)
Sinoia Caves	SR- 118	650 ± 95	(Sinoia)
Lanlory	Pta- 606	820 ± 50	(Sinoia)
	SR- 245	1030 ± 50	(Sinoia)
	SR- 248	1180 ± 55	(Sinoia)
Chitope	SR- 163	1110 ± 95	(Sinoia)
Na Ba	Pta-1193	690 ± 50	(Sinoia)
BOTSWANA			
Toutswe	GX3772	1305 ± 95	(Toutswe)
	GX3773	1200 ± 95	(Toutswe)
	GX3774	1090 ± 105	(Toutswe)
	GX3775	1500 ± 95	(Toutswe)
	I-11412	1190 ± 75	(Toutswe)
	I-11413	960 ± 75	(Toutswe)
Thatswane	I-11414	925 ± 80	(Toutswe)
	I-11415	1110 ± 75	(Toutswe)
Taukome	I-11409	955 ± 75	(Toutswe)
	I-11410	710 ± 80	(Toutswe)
	I-11407	685 ± 80	(Toutswe)
Maiphetwane	Pta- 2526	990 ± 50	(Toutswe)
Tsodilo Locality 1	I-11411	850 ± 80	(Other EIA)
Tsodilo Locality 2	Wits- 836	980 ± 50	(Other EIA)
Broadhurst	Wits- 837	1360 ± 50	(Other EIA)

the Matola Tradition. The two elements are evidently contemporary (Prinsloo pers. comm.) but the range of dates, from the fourth to the sixth century (Table 2), make the chronology uncertain although a date of about AD 450 or 500 is most likely.

The distribution of Matakoma-related ceramics is at present limited to the Soutpansberg range and the Kruger National Park in the northern and eastern Transvaal (Hanisch 1981a, Evers 1981). The Limpopo valley, just to the north seems to lack settlements of this early phase although a few sherds attributed to Gokomere have been recovered from the lowest levels of the South Terrace at Mapungubwe (Meyer 1979).

Recent work in Botswana has for the first time proved the existence of EIA settlement in several areas, although at a rather later date than further east. Best known is the Toutswe Tradition which was previously considered Late Iron Age in its affinities and chronology (Lepionka 1979, Huffman 1978a). Denbow (1979, in press) has now shown that it extends back from about AD 1300 to AD 700 and that its occurrence in eastern Botswana is broadly equivalent and closely related to Zhizo of western Zimbabwe and the Limpopo valley.

In south-eastern Botswana a different expression of the EIA has been found at several sites near Gaborone. Broadhurst (Denbow 1981) has pottery that may be broadly related to the Lydenburg cluster and more particularly with Hall's (1981) undated Rooiberg Unit 1 from the west-central Transvaal. The single fourteenth century radiocarbon date from Broadhurst seems very late by comparison with the sequence further east but, if correct, may be explained by different socio-economic conditions in Botswana relating to a greater emphasis on livestock (Denbow *op. cit.*). Moritsane, dated late in the eleventh century, (*op. cit.*) differes somewhat from Broadhurst and seems similar to Hall's (1981) Rooiberg Unit 2 and the Eiland phase in general, which is discussed separately below.

In the Toromoja area on the Botletli River three surface sites have produced sherds, similar to the Bambata ware of Zimbabwe, along with Late Stone Age material which may be associated (Denbow & Campbell 1980). Two EIA sites in the Tsodilo Hills of north-western Botswana have produced dates in the ninth and tenth centuries (Denbow 1980), the earlier being associated with a burial of negro physical type.

From Namibia the only dated site is Kapako in the extreme north on the Okavango River, where the EIA component is dated to the ninth century (Sandelowsky 1979). Overlying this and at a neighbouring site is later material related to the ethnological pottery of the Kavango and Mbukushu peoples. It is of interest that this pottery also has something of an EIA appearance and that the potters among these people are male whereas virtually all other traditional potters in southern Africa are female. A parallel situation occurs in Phillipson's (1974) Lungwebungu ceramic tradition, among the present peoples of western Zambia, where the potters are male and there is an apparent stylistic continuity from the EIA. Though Phillipson acknowledges that the pottery from the two areas is not very similar, he suggests from this evidence that EIA potters in general may have been male, and that it became a predominantly female craft only in the Late Iron Age.

In Zimbabwe the framework of the EIA has been established for a number of years. The basic tradition and, until recently, that considered to be the earliest is the Gokomere Tradition, dated at several central and western sites from the fourth to about the sixth centuries AD with one date in the second century from Mabveni (Robinson, K.R. 1961a). The Ziwa industry, recognised from the eastern highlands (Summers 1958), was intially regarded as distinct, but has more recently been subsumed within the Gokomere Tradition as a barely distinguishable early eastern facies (Huffman 1974a) which is in effect undated and inadequately described. Within the Gokomere Tradition a second phase has been recognised by Huffman (1974a) consisting of two facies, Coronation (Huffman 1971) in the east and Zhizo (Garlake 1967, Huffman 1971, 1974a, Robinson, K.R. 1966a) in the west, both of which date from about the eighth to the tenth centuries. In the Limpopo valley, Hanisch's work has identified Zhizo material at Schroda, Pont Drift and Commando Kop dating essentially to the ninth and tenth centuries (Hall & Vogel 1980, Hanisch 1981 a,b). The two latter sites include an apparent admixture of Late Iron Age pottery of Leopards Kopje affinity which is also the pottery at K2 (Bambandyanalo) dated to the period AD 970-1070 (Meyer 1979, Hall & Vogel 1980). These Zhizo sites are therefore clearly at the interface of the Early and Late Iron Age.

In north-eastern Zimbabwe a further ceramic group, Maxton, dated to the eleventh century, is regarded as a third phase within the Gokomere Tradition (Garlake 1967, Huffman 1974b, Phillipson 1977) despite its relatively late date and the presence of some Late Iron Age features including some vessel shapes and the association with hilltop settlements some even with stone walling.

In the Urungwe district of northern Zimbabwe the Sinoia Tradition assemblages, dated from the seventh century to perhaps as late as the eleventh, are quite distinct from Gokomere and can be linked with assemblages from north of the Zambezi particularly Kapwirimbwe (Garlake 1970, Huffman 1972a, 1974b, Phillipson 1977).

Since Neville Jones (Arnold & Jones 1919) excavations at Bambata Cave near Bulawayo, the Bambata pottery has been something of an enigma in Zimbabwean archaeology. Its context in the upper levels of rock shelters was apparently associated with a Stone Age hunter-gatherer way of life, yet the pottery itself has the appearance of an Iron Age ware (Robinson, K.R. 1966b). The enigma remained until a radically new interpretation by Huffman (1978a) put Bambata forward as a potential ancestor of much of the EIA of South Africa. On typological grounds Huffman argues that Bambata closely resembles both Zambian EIA ceramics from Kalundu and Kapwirimbwe on the one hand and the Lydenburg cluster from the Transvaal and Natal on the other. He therefore proposes a route of diffusion of the EIA into South Africa via Phillipson's (1977) western stream in Zambia to Bambata and finally to the Lydenburg cluster (Huffman 1979a). This diffusion would have taken place during the first half of the first millenium AD before the arrival of Gokomere/Ziwa in Zimbabwe.

Huffman's model opens up interesting new possiblities and represents a complete contrast from Phillipson's eastern and western stream model which places virtually the whole of our region in his eastern stream. However, there are a number of problems to its acceptance. Firstly, the Zambian material that has been compared with Bambata is relatively late in date – not apparently occurring earlier than about the fifth century (Phillipson 1977) and therefore contemporary with the Lydenburg cluster. Material from southern Zambia that is thought to be earlier, namely Situmpa, is still poorly known and its chronology is imprecise because of some anomalously early dates (e.g. Clark & Fagan 1965, Katanekwa 1978). Secondly the typological connection seems questionable since some characteristic features of the Zambian material are rare or absent from Bambata. Thus the Zambian origin of Bambata is uncertain.

Bambata itself, prior to Huffman's (1978a) remodelling, was a poor candidate for an ancestral EIA tradition. Ceramic samples are small and too fragmentary to allow reconstruction of vessel shapes and motif combinations (Robinson, K.R. 1966b, Schofield 1940). There are no radiometric dates and no clear associations with Iron Age economy. Huffman's new concept incorporates some elements previously assigned to Gokomere including part of the Mabveni site (Robinson, K.R. 1961a) together with the earliest of its three radiocarbon dates, a perhaps unreliable date in the second century AD (SR-43) (Hall & Vogel 1980). Therefore, acceptance of this new concept of the Bambata must await a full description of its redefinition, together with new fieldwork to provide adequate ceramic and chronometric samples.

An alternative explanation for the origin of the sixth-seventh century Lydenburg cluster material is that it developed from the third/fourth century Matola tradition. Elsewhere (Maggs 1980a) I have shown that within the Natal EIA sequence most of the attributes characteristic of the seventh century pottery are already present, albeit in low frequencies, by the third century. This explanation avoids the need to propose further extensive EIA migrations into South Africa after the initial settlement by Matola tradition communities.

Present evidence seems inadequate to establish which of these two explanations, or perhaps some combination of them, accounts for the Lydenburg cluster. However, this should become clear once we have material available from the fifth and early sixth centuries. For if the origin is diffusion from Bambata, or some other external source, there should be a sharp break in the sequence, whereas more gradual transition during this period would indicate local development from the Matola tradition. At present we have nothing from this period in Natal, while the pottery from Castle Cavern and Broederstroom (Table 1) is unpublished. However, Lydenburg itself, dated to about AD 500, is relevant to this question. While much of its pottery is quite distinct from the earlier material, an appreciable portion of the pots have straight everted necks with the types and positions of decorative motifs corresponding closely to Matola tradition material, particularly the eastern Transvaal sites of Eiland and Harmony (Evers 1975, 1979), the former dated to about AD 300. Like the Natal evidence, this suggests local evolution of style rather than a break in the sequence as a result of migration.

To explain the other EIA elements in our area Huffman (1979a) proposes two other routes of diffusion. One, moving southwards down the eastern coastal lowlands, deriving our Matola tradition from something like Kwale of East Africa, seems to be generally accepted (Da Cruz e Silva 1976, De Senna Martinez 1976, Evers 1977, Inskeep 1978a, Klapwijk 1974, Maggs 1977 and Phillipson 1977). The main problem against its complete acceptance is the lack of data on the EIA in southern Tanzania and Mozambique north of the Zambezi. The other proposed movement is from Nkope of southern Malawi giving rise to Gokomere/Ziwa of Zimbabwe around 500 AD. Some ceramic elements of Nkope, notably the fluted bowls and the pots with straight, everted necks with a band of oblique hatching or comb-stamping, are more at home with Matola. However, it is probable that Nkope or related developments within the eastern coastal region around the latitude of the

Zambezi gave rise to Gokomere/Ziwa. If these latter ceramics first appeared as late as 500 AD, a period of about two centuries would have elapsed since the beginnings of Nkope, giving plenty of time for both change and movement. Again the problem here is the scarcity of precise information on the first half of the first millennium AD in Zimbabwe.

ECONOMY AND ECOLOGY

Ceramic analysis may eventually give us a clear picture of the geographic expansion of the EIA. But to understand more about these communities and why they came to occupy so much of central and southern Africa we need to look at other aspects, particularly the elements of their economy and how these interacted with the varied environments of the region.

Settlement around AD 300 – the Matola tradition

As yet there is relatively little detailed or direct evidence on economic elements of this period. No adequate faunal samples are available and therefore there is no definite evidence of domestic stock, which could perhaps be absent. Of considerable importance is the identification of cultivated bullrush millet *Pennisetum americanum* at Silver Leaves (Klapwijk 1974). This is the earliest evidence of cultivation in the region and it implies the presence of crops on other Matola sites as well. Iron working is evident on several sites while at Harmony and Eiland (Evers 1975,1979) the manufacture of salt was already developed locally.

The Natal evidence shows that sites are typically 2 hectares or more in extent (Maggs 1980a). Although not as large nor as rich in cultural remains as the 6th/7th century sites, they clearly represent some sort of village rather than a transitory camp. The preference for inland site locations to be in relatively dry savanna country beside rivers seems to be already established, judging by the eastern Transvaal evidence. Another EIA trait, the deliberate burial of pots in pits, is also present.

In Natal 22 sites grouped around Durban and Lake St Lucia (Hall 1980b, Maggs 1980a), and identified from the distinctive pottery, show a very clear pattern of site location in relation to the environment. All but three of the sites are within 3 km of the coastline, the exceptions being within 6 km. Furthermore they are all within the relatively narrow cordon of Pleistocene dunes which were covered by Coastal Forest before clearance by farming communities. This pattern of location applies also to Matola itself.

This environment would have provided very rich agricultural land on initial clearance, because of the high humic content of the forest floor. However, in the hot and humid climate rapid leaching of plant nutrients takes place on the sandy soils, leading to a sharp drop in productivity and the need to clear new plots after a few years.

In addition to this selection for the Coastal Dune Forest environment, the pattern almost certainly reflects the importance of marine protein resources, particularly the highly productive brown mussel *Perna perna* (Hall 1980b). Sites closest to the sea are simply shell middens with no signs of prolonged habitation. They seem to be little more than collecting points for the larger, village sites further inland, which may also include small shell middens.

Detailed, systematic work on site location in the St Lucia area (Hall 1980b) shows even more selectively in site location. Here, settlement is concentrated along the ecotone between the Coastal Dune Forest and the seasonally flooded hygrophilous grasslands on the eastern shores of the lake. This ecotone represents a narrow belt parallel with the coast and has resulted in an almost linear pattern of EIA settlement. Evidently the first villages were established on the forest margin and it seems probable that agricultural plots would have been cut into the forest which would have been replaced by secondary grasslands after abandonment of the plots. Wild edible fruits would have been available and protein from both marine molluscs and wild terrestrial fauna, but domestic stock were probably of little significance since grazing would have been very limited prior to forest clearance.

The ecological pattern of Matola sites in Natal is significant in terms of the present evidence for very rapid initial expansion of the EIA down the east African coastline. Settlements scattered over some 3200 km from Kenya to southern Natal may be within 150 years of each other (Maggs 1980a). Movement at anything approaching this speed would seem to require special economic circumstances, a condition supplied by the Natal ecological model and perhaps applicable to the coastline further north as well. Fairly rapid expansion along inland waterbodies - river and lakes - could also be postulated in view of the preference for waterside site locations. However, it seems very unlikely that rapid linear expansion, of the type proposed on linguistic and ceramic typological evidence, could have taken place away from or cross-cutting linear environmental features such as rivers, shorelines and ecotones.

Settlement from about AD 500 onwards

Taking Namibia, Botswana and South Africa together it seems that the EIA reached western areas considerably later than the eastern coast and lowlands (Figure 1, Table 1). All sites securely dated around AD 300 are east of the Great Escarpment. The Transvaal plateau seems to have been settled only around AD 400-500 with eastern Botswana perhaps as late as the eighth century and northern Namibia the ninth. Little can be said about Zimbabwe before about AD 500 and it is possible that there was little EIA settlement prior to the Gokomere/Ziwa sites of about this time.

Also around AD 500 sites of the Lydenburg cluster extended into higher altitude areas of the Transvaal and Natal than were previously occupied, including much of Natal up to the 1000m contour (Maggs 1980a). Coastal shell middens continued to accumulate and now occurred much further south, as far as the Chalumna River between East London and Port Alfred (Derricourt 1977). The inland sites tend to be larger, richer and more numerous than before.

Much more information on economic activities is available from this period though there are still large gaps. Stock including cattle, small stock and domestic dogs are known from sites where bone preservation is adequate. Judging by the identification work from sites north of the Zambezi, we can expect considerable variation in the ratio between domestic and wild animals at different sites (Plug 1979b, Voigt 1973c). But available evidence from further south indicates that there was already a predominance of domestic stock over wild animals in terms of dietary input (e.g. Maggs & Michael 1976, Maggs 1980b, Voigt 1980, Welbourne 1973).

Futher identified crop plants are relatively few and mainly dated from late in the EIA. Cow peas, *Vigna unguiculata,* ground beans, *Voandzeia subterranea* and sorghum are dated to about the ninth century from Lanlory and Leopards Kopje (Huffman 1972a, 1974a) while sorghum and 'beans' have been identified from a twelfth century Toutswe context at Thatswane in Botswana (Denbow in press) and sorghum also at Schroda (Hanisch 1981b). Earlier material has been identified from flotation samples at two Natal sites. At Magogo, dated to the sixth or seventh century, the millets *Eleusine corocana* and cf. *Pennisetum typhoides* are cultigens and *Citrullus* sp. and cf. *Sorghum* sp. may be likewise. The eighth century site of Ndondondwana has also produced *Pennisetum typhoides,* bullrush millet.

Indirect evidence on the size and location of settlements and the numbers of grindstones support the view that cultigens were staple items of the diet. Grinding technology on many sites differs from the Late Iron Age in that broad oval grooves averaging 7 cm wide, 20 cm long and 2 cm deep were produced, using a relatively small upper stone and often with several rounded facets. No satisfactory explanation for this pattern has been offered, indeed such grooves are often but unjustifiably referred to as spear-sharpening grooves. A possible explanation is that if the small-grained millets were the predominant grain crops the typical EIA grooves were more suitable for grinding than the broader, flatter surfaces typical of the Late Iron Age.

Iron smelting is evident on very many sites, indicating that most villages could produce their own requirements. Often the scale of production, judging by the amount of waste materials, was relatively small but in some cases there is sufficient to suggest a surplus that could have been traded. A surprising fact is that no EIA furnaces, sufficiently complete to indicate shape and size, have yet been reported from south of the Great Lakes region of central Africa. Here the superstructures were frequently built of bricks which would have enabled the upper parts of furnaces to be dismantled and rebuilt (Schmidt 1978, Van Noten 1979). The lack of furnaces from further south, despite the presence of loose chunks broken from their walls at several sites, suggests that they may have been dismantled after each firing.

Copper was also being produced in some areas and there seems to have been some trade in it, but the scale of production and the degree of specialization did not approach that found among some localized Late Iron Age communities. Items are essentially decorative and include techniques as complicated as the making of chains (Mason 1979b, Denbow & Campbell 1980). Other specializations like soapstone working and salt-making occur locally while ivory production, particularly of bangles, seems to have been focused at some sites like Schroda on the Limpopo (Voigt 1981) and Ndondondwana on the Tugela. The clearest evidence of trade over considerable distances is the presence of marine shells even on far interior sites. Clearly trade networks were already in existence linking one settlement with another, but the evidence suggests a very low volume of trade, at least up to the ninth century. In fact there seems to have been a large degree of economic and perhaps also political self-sufficiency at a village level.

Villages are typically around 8 hectares in size, some being perhaps as large as 20 hectares or more. Pits from about 1 m to 2,5 m deep, containing domestic debris and often a carefully buried but bottomless pot, are typical and may be the only 'structures' remaining on eroded sites. Some pits may have been dug for grain storage before being used for refuse. Pieces of burnt daga (adobe) with stick and thatch impressions from many sites suggest a widespread, uniform type of hut construction. At Mabveni (Robinson, K.R. 1961a), two structures were tentatively identified as a hut floor and a granary built of daga and poles raised above ground level by a stone at each corner. More detailed evidence is emerging from several Transvaal sites. Of the more than 40 huts at Broederstroom about half had floors raised on parallel rows of stone slabs or boulders, the superstructures being again of poles and daga (Mason 1979b). Similar huts are reported from Klein Afrika with the caution that they could be misinterpreted as shallow pits (Prinsloo 1981). During occupation hollows form under the hut, perhaps the work of domestic animals. Cultural debris accumulates here to be buried beneath a mass of daga fragments after the hut's destruction. At Schroda (Hanisch 1981b) two huts had carefully prepared sand floors at ground level. The best preserved was oval in shape with a maximum diameter of 2,5 m surrounded by charred posts, some up to 12 cm thick. Beside the entrance was a semi-circular platform supporting a hearth. In front of each hut was an area with prepared earth floor.

Little is yet known about the arrangement of structures within individual sites. However, cultural material tends to occur in concentrations about 5 to 20 m in diameter. Such concentrations are often associated with one or more pits, sometimes with smelting debris and hut daga and may represent individual homesteads. They are scattered about the settlement but are usually not far apart. The impression they give is of a population of a few hundred people per village.

Some pattern of inter-site relationship seems to be appearing in the few areas where fairly concentrated survey work has been undertaken. In the Lydenburg valley (Marker & Evers 1976) five sites, each of about 6 to 8 hectares, are situated about 3 km from each other, although not all are contemporary. Each is located on the colluvial lower slopes beside a tributary stream. A similar pattern occurs in parts of the Tugela Basin where sites are immediately beside rivers or permanent streams, and are built on patches of deep colluvial alkaline soil – the best available land for cultivation. In Natal settlement is apparently limited to the coastal lowlands and the deeply incised valleys of the larger rivers and streams (Maggs 1980c). The latter are relatively hot and dry with a mixed bushveld or savanna vegetation with an understory of 'sweet' grass, which provides relatively sparse but nutritious year-round grazing. By contrast, neighbouring areas of higher altitude and rainfall have a grassland vegetation which is only nutritious in spring and summer.

The high altitude areas of South Africa, notably the Highveld grasslands, were apparently not settled by EIA communities. The Tugela Basin pattern suggests that a mix of fertile soil, waterside location, sweet grazing, adequate rain for crops and good supply of wood were prerequisites for EIA settlement. In South Africa, the zone within which such a mix can be found becomes progressively narrower, its inland margin progressively closer to the east coast, as one moves southwards. The limits are finally reached around latitude S.33°, near modern East London, where the summer rainfall regime gives way to an all season pattern, and further west to winter rainfall, where there is inadequate summer rain to grow tropical African crops such as sorghum and the millets (Maggs 1980c). The distribution of EIA pottery along the coast corresponds closely with the limits of the summer rainfall regime, the southernmost site being close to latitude S.33°.

The ecology of this middle part of the EIA therefore indicates a strong preference for a savanna evironment with settlement concentrated along major water bodies. Intervening watershed areas, especially, those of sufficient altitude to support open grassland, were not settled. They may, however, have been used in spring and summer for transhumant grazing where they were within reach of the permanent settlements.

Like the earlier Matola pattern, here again a selective and discontinuous pattern of site location is evident and relevant to any discussion on population movement. South of about latitude 26° where the savanna starts to be pinched between coastal forest and interior grassland, the relatively narrow corridor suitable for EIA settlement could conceivably have led to rapid population movement in a single direction. However, within the broader savanna regions further north, EIA expansion could have taken place in many directions over a much broader front. Rapid unidirectional movements are unlikely to have taken place in such regions except, as suggested above, along rivers, or along the ecotones between savanna and other ecotypes.

Developments around AD 800-1000

In parts of the Limpopo Basin in particular, economic developments from the basic EIA pattern seem to have been taking place from about the eighth century. Prior to this there is little evidence of EIA settlement on the Limpopo in the northern Transvaal and in eastern Botswana, as we have seen. Now a fairly dense settlement of Zhizo and Toutswe communities takes place.

Many of the Toutswe sites have been found from air photographs, for their characteristic vitrified dung deposits support stands of a particular grass *Cenchrus ciliaris* (Denbow 1979). Over 200 sites have been identified within a relatively small area and some have vitrified dung deposits up to 1,5 m thick (Denbow in press). Taukome, the earliest of the dated sites has preserved the outline of a livestock pen approximately 20 m in diameter, delimited by post holes, beneath dung deposits dated around AD 700.

Denbow (in press) interprets the evidence as indicating a strong pastoral emphasis in this relatively arid area, by contrast with the closely related and contemporary Zhizo sites of south-western Zimbabwe. Evidence of a hierarchical settlement pattern may reflect an increasing inequality of wealth in domestic stock leading to the establishment of larger and more complex socio-economic units.

Both in the Toutswe area and further south in Botswana EIA ceramic traditions continued after the tenth century despite significant changes further east. The relative conservatism, Denbow (1981) suggests, may also have been in part the result of the relative wealth in livestock.

As at Toutswe, the northern Transvaal Zhizo sites demonstrate that livestock, particularly cattle, provided a large proportion of the protein (Welbourne 1975, Voigt 1980). Perhaps similar political and economic developments were taking place here too. Schroda has similar dung accumulations and it is a large settlement (Hanisch 1981b). The several dozen ceramic figurines of stylized human and animal forms also suggest that it was an important centre.

Of even more importance at Schroda are the indications of economic developments presaging those at the neighbouring K2 site 150 years later. It is the earliest site to show the import of glass beads on a considerable scale as well as a volume of ivory working apparently well above the level of local use (Hanisch 1981b, Voigt 1981). The number of carnivore bones suggests hunting for skins, which may also have been exported. Clearly Schroda was in contact with the east coast and it seems to represent 'a large and culturally important Zhizo community which was a forerunner to the Leopards Kopje A community on K2' (Voigt 1981).

The dichotomous system of dividing the Iron Age into Early Iron Age (EIA) and Late Iron Age (LIA) is not entirely successful as neither ceramic typology nor chronology provides a clear cut transition in all regions. Controversy is focused about the dividing line and there are some ceramic units whose position is disputed. To add to the confusion some workers favour the addition of a Middle Iron Age category although this has never been defined.

In regions where there is a distinct break in the sequence around AD 900-1000 the two-fold subdivision seems to work well, although there is no agreement as to the causes of this break. In areas such as eastern Botswana and north-eastern Zimbabwe a continuity in ceramic tradition from the EIA is documented up to the twelfth century or even later, the relevant ceramic units being Toutswe, Maxton and Sinoia. However, sites of the two former units differ in some respects from typical EIA sites. They are commonly on top of relatively steep hills, some have stone walling, while the clear pattern of cattle in the centre surrounded by a ring or arc of huts and supports for grain stores has yet to be demonstrated from early settlements. Glass beads at some sites though few in numbers indicate long-distance trading contacts (Garlake 1967, 1969, Huffman 1971, Denbow in press). Both Zhizo sites on the Limpopo and Toutswe sites may include LIA Leopards Kopje A (or K2) pottery among their assemblages (Hall & Vogel 1980, Denbow in press). This may indicate trade or intermarriage with contemporary LIA communities, but another possiblity that requires examination is that the LIA ceramic elements were developing within settlements which were in transition away from the EIA pattern but which still retained a predominance of terminal EIA ceramics.

Another still unresolved problem which could be examined here or later is that concerning the middle assemblage at the eastern Transvaal site of Eiland. Although a full description has yet to be published, the site, with a predominance of herringbone and crosshatched motifs in its pottery, has given its name to a ceramic unit that is now being found widely distributed in the Transvaal. It is securely dated from around AD 1000 to 1200 at Silver Leaves (Klapwijk 1973) as well as Eiland (Evers 1975, 1981) while contemporary material from the Phalaborwa sites of Loolekop and Kgopolwe may also belong in this ceramic unit rather than the succeeding Phalaborwa or Letaba unit where they were formerly placed (Stuiver & Van der Merwe 1968, Van der Merwe & Scully 1971, Evers 1981). At Mapungubwe and at Toutswe sites it is found as odd imported sherds within the main assemblages (Fouche. 1937, Meyer 1979, Denbow in press). Further afield Eiland pottery has also been found in the Lydenburg area (Collett 1979, Evers 1981) and it can be closely compared if not equated with Rooiberg Unit 2 in the western Transvaal and southern Botswana herringbone pottery such as that from Moritsane (Hall, S. 1981, Denbow 1981). At Bambo, near Pietersburg, the Eiland occupation is on a steep hilltop and associated with a complex of stone-walled enclosures and terraces (Loubser in prep) where cattle were penned in the central area surrounded by the dwellings. This is the first site to provide evidence of settlement pattern and architecture which are essentially LIA in character.

Eiland is thus unusual for the several reasons of its widespread distribution, its dating around the beginning of the LIA and its intrusive occurrence at some sites. Its classification like that of Toutswe is currently in a state of flux. Huffman (1978a) places it along with several early LIA units in his Kutama tradition which is thought to have developed from Klingbeil. Evers, initially unsure (1975) now sees it as EIA (1981) as does S. Hall (1981). It seems likely that with further research it will assume a more important position in the transition from EIA to LIA than it currently enjoys.

THE LATE IRON AGE: ORIGINS

As we have seen there is within the EIA a tendency towards greater regional diversity with time. From about AD 800 elements appear in some areas which seem to herald developments which are more typical of the LIA:-hilltop settlements, stone walling, long distance trading links, the association of political power and wealth in cattle.

It is generally agreed that the LIA has less unity as a concept than the EIA yet the distinct change in many parts of sub-equatorial Africa around the tenth century seems to justify the retention of the two terms even if, as we have just seen, they do not fit all the evidence.

Two main theses have been put forward in recent years to explain the abrupt change: a northern origin by Phillipson (1977) and a southern origin by Huffman (1978a, 1979a). Phillipson attempting to correlate linguistic and ceramic prehistory concludes that at about AD 1000, after population pressure had built up in the Shaba area of southern Zaire, an expansion of Bantu-speaking peoples of the Eastern Highland group introduced the LIA to the eastern half of the subcontinent. This explanation has been criticized by several authors (Garlake 1979, Huffman 1979a, De Maret 1977) on the basis of both the ceramic and other evidence and indeed a wave of migration on the scale envisaged seems unlikely. Using ceramic typology Huffman (1978a), has put forward a scheme in which Klingbeil plays a pivotal role between the EIA Lydenburg cluster and a group of early LIA ceramic units in the Transvaal, Botswana and Zimbabwe namely Gumanye, Harare, Musengezi, Eiland, Toutswe, and the two facies of Leopards Kopje A–K2 and Mambo. These he groups under the term Kutama Tradition and interprets the distribution pattern as a build up of population pressure in the south-eastern part of our region, leading to a movement northwards during the eleventh century, introducing the LIA and the Shona peoples to Zimbabwe and the Limpopo valley (Huffman 1978a, Beach 1980). Some of the evidence on which this is based is rather slender, we have already noted a revision in the status of Toutswe and to some extent Eiland, while the key site of Klingbeil has a relatively small sample and there is a measure of imprecision about its dating (Table 2). But perhaps the greatest obstacle to its acceptance is the evidence of socio-economic developments in the Limpopo valley itself. Huffman (1974a, 1978a) is convinced that the Leopards Kopje ceramic tradition did not develop from its predecessor in the region, Zhizo, yet the new features that characterize the crucial K2 site seem to originate from local Toutswe and Zhizo settlements rather than anything further south, as we have seen. Present evidence is inadequate to resolve this problem and indeed we still know relatively little of the nature of the changes that took place around the eleventh century, let alone the reasons for them.

LATE IRON AGE: THE EARLIER CENTURIES

A convenient point to start this discussion is on the Limpopo at the K2 site, for here we have the most detailed picture of a settlement at the beginning of this period (Eloff & Meyer 1981, Fagan 1964, Fouche 1937, Gardner 1963, Hall & Vogel 1980, Meyer 1979, Voigt 1978, 1980). Furthermore we are able to see some of the ninth century developments reach a new scale here. Ceramically K2 is the type site of the southern facies of Leopards Kopje A and in many ways is characteristic of this unit.

K2 is at least 8 hectares in extent and contains the greatest accumulation of material on any known Iron Age site up to its time. Thanks to the recent excavations by the University of Pretoria the main occupation (Phase 2 of the Greefswald sequence as a whole) is firmly dated to around AD 970-1100. Occupation took place around a large central cattle pen which later was overwhelmed by the accumulation of an enormous midden some 6 metres deep. The living areas show a succession of prepared daga floors and debris from burnt huts. The latter were relatively light structures of thin poles with a sandy plaster. Occupation was intensive and there must have been a large population. Even more than with the Zhizo sites, cattle were the main source of meat, contributing 88% with small stock another 10% and wild animals a negligible amount. Some cattle were humped and while most were of Sanga type some may have been Afrikander. Domestic dogs were also present. Despite the aridity, crops including *Pennisetum typhoides, Sorghum bicolor* and *Vigna unguiculata* (cow peas) were grown and stored in small pole and daga huts. Initial attribution of the large number of human burials to a Khoisan physical type (Galloway 1959) is now discounted and it has been shown that all examples fall within the range of the modern South African Negro (Rightmire 1970). There were also a number of so-called 'beast burials' where a few bovid bones, usually of cattle, had been buried in association with broken pottery and sometimes beads or metal items.

But of even more interest is the evidence for long distance trade on a scale greater than at the earlier sites. Large quantities of glass beads presumably of Indian Ocean origin were imported and some were remelted on site and cast in ceramic moulds to produce the large 'garden roller' beads (Davison 1973). As yet there is little evidence of glassworking from any other southern African site. The main export was evidently ivory for quantities of fragments discarded from working were found but there were only a few finished arm bands. The scale of settlement and of trade indicates that K2 was an important regional centre economically and presumably politically as well.

Contemporary with K2 is the northern or Mambo facies of Leopards Kopje A, best known from Leopards Kopje or Nthabazingwe itself (Robinson, K.R. 1965, 1966a, Huffman 1974a). Huts here are rather more substantial, with thick daga floors, pole and daga walls, while some have exterior poles indicating a surrounding but rather narrow verandah. The interiors are about 3 m in diameter and here again the overall form must have been cone-on-cylinder not hemispherical. Huffman (*op cit.*) stresses both the economic importance of cattle based on faunal remains and dung deposits, as well as their ritual importance based on ceramic cattle figurines and 'beast burials' as at K2. However, in the higher rainfall area of the Matabeleland, cultivation may have been relatively more important and there is ample evidence of it in the form of carbonized seeds of *Eleusine corocana, Sorghum bicolor, Vigna sinensis* (cow peas) and *Voandzeia subterranea* (ground beans), as well as numerous grindstones, pits perhaps for storage and small stone circles that may have supported grain bins. Most Leopards Kopje sites, unlike K2, indicate relatively small communities based on mixed subsistence farming with some metal working and only a small amount of long distance trade as attested by a few glass beads.

The north-south facies subdivision within the Leopards Kopje tradition continues into its second or B phase with Woodlandale in the north and Mapungubwe in the south. While there is evidence of continuity in some spheres there are also some important changes. Ceramically the north and south facies are similar and both have also been compared quite closely with Great Zimbabwe (Robinson, K.R. 1966a, Huffman 1974a). Because the stratigraphic evidence at Great Zimbabwe suggests a gradual change in ceramics and architecture leading up to the Zimbabwe Phase occupation it has been accepted that this phase developed out of the preceding Gumanye occupation (previously called Zimbabwe II or lower Zimbabwe, Robinson 1961b, Garlake 1973, Huffman 1972b, Beach 1980). However the new dating evidence for Mapungubwe necessitates a complete revision of the relationship between Great Zimbabwe and Mapungubwe. This we will discuss in the next section under the Great Zimbabwe Tradition.

Roughly contemporary ceramic units further east include Gumanye, Musengezi and Harare (Figure 2) all of which like Leopards Kopje are part of Huffman's (1978a) Kutama Tradition. Rather less is known about these sites but in many ways they seem to resemble Leopards Kopje if on a smaller and perhaps poorer scale. Gumanye sites are mainly small villages on defendable hilltops (Huffman 1978b) while Harare is known essentially from cemeteries, an unusual occurrence in the Iron Age (Goodall 1962).

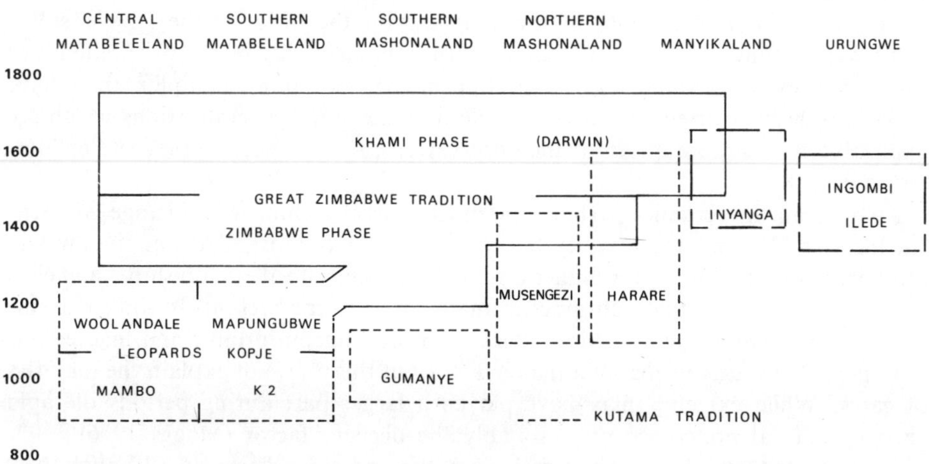

Figure 2. Late Iron Age in Zimbabwe (partly from Huffman 1974).

South of the Limpopo the early centuries of the LIA are known from only a few rather isolated occurrences. In the Olifantspoort area of the central Transvaal Mason's (1971) 'Middle Iron Age' is dated between AD 1000-1500. Sites consist of ten to twenty huts spaced around a central open area which perhaps contained stock pens, the whole covering less than a hectare. There is no stone walling but the daga hut floors are well preserved and of about 2 m diameter, usually with a raised platform to one side. Next to the platform is a hollow for a hearth, while smaller dimples in the floor held pots and wooden mortars. Bones of cattle and small stock, and seeds of millet indicate a typical Iron Age mixed farming economy.

In Natal two small villages on steep hilltops close to the coast have been dated to the eleventh century (Davies 1971, Robey 1980). Huts made of organic materials and of about 5 m diameter are indicated. Pottery is quite distinct from EIA material and shows a break in the sequence, as in so many other areas. The economy is of interest in showing considerable emphasis on shell collecting, hunting and fishing for both marine and fresh-water species. Game consisted chiefly of bushbuck, blue duiker, bushpig and cane rat indicative of coastal forest vegetation. The environmental setting probably explains the absence or paucity of domestic stock and the economy is probably not typical of more open areas further inland. For example Moor Park (Davies 1974) shows considerable emphasis on stock with its stone enclosures, and both bones and figurines of cattle. Dated to the thirteenth and fourteenth centuries Moor Park and several related sites are of interest particularly because of their locations and distribution (Maggs 1980c). They consist of extensive though rough stone walling which augments the natural defences of hill- or spur-top locations and encloses settlements of several hectares. Their distribution extends from the upper fringe of savanna vegetation well into the grasslands of the Natal interior and they are the earliest stone walling and the earliest known Iron Age settlements to occupy the grassland region of South Africa. Like the Olifantspoort sites on the northern fringe of the grasslands they seem to represent early adaptations on the part of some communities to the challenge of the grasslands. Once this challenge had been overcome, the region was opened to Iron Age settlement on a large scale, as we will seen in the final section.

The nature and magnitude of the change with the onset of the LIA is still a controversial topic and the new evidence rather than simplifying the picture tends to show that events around the eleventh century were more complex than have previously been realized. But it is possible to make some generalizations which are applicable to a large number of the observations, though not to every site or local sequence.

Apart from the ceramics perhaps the most general comment on change concerns site location. The waterside, valley-bottom preference of most EIA sites gives way in most areas to a preference for higher ground at the sites of valleys, on slopes and even on hilltops, a pattern that continues in many areas to the present. In some extreme cases, such as Gumanye and Moor Park with their uncomfortable and inaccessible hilltops, defence was probably a major factor but this does not explain the majority of cases. While exigency may have played a larger part during periods of rapid change, cultural preference was probably the decisive factor (Maggs 1976).

Settlements themselves may be rather smaller and more dispersed, although there are some notable exceptions. Structures such as huts show more regional variety, stone walling appears in some areas while pits are much less common. Changes take place in items such as smelting furnaces, grindstones, ceramic sculpture and ornaments, though they are by no means uniform throughout the subcontinent, and from this time on there is increasing geographical variation in most aspects of material culture. Economic specialization appears or increases in many areas although the basic economic elements – the livestock, cultigens and metallurgy – remain. Trade networks seem to become more important though long distance trade probably had a significant affect on only a few communities in the Limpopo valley and further north. The Toutswe and Zhizo evidence shows that economic emphasis on cattle was not unique to the LIA though their socio-religious significance may have increased during this period.

The rate of change was clearly variable, being rapid in Natal and much of Zimbabwe, apparently evolutionary in parts of the northern Transvaal while Botswana remained conservative. A single monolithic explanation, such as a wave of migration, is implausible under the circumstances. What seems more likely is a variety of regional responses to new factors which may include new social-political formations resulting from inequalities in cattle wealth, long distance trade and specialized local production.

THE GREAT ZIMBABWE TRADITION

Following Garlake (1973) I will refer to the archaeological site as Great Zimbabwe to distinguish it from the country. The tradition has also been called the Ruins Tradition (Huffman 1971) but most authors now seem to prefer 'Great Zimbabwe Tradition' as being more appropriate and specific (Garlake 1973, Huffman 1978b, Phillipson 1977). It is normally subdivided into two, a Zimbabwe Phase and a Khami Phase, but because of the new evidence mentioned above I am here adding an initial Mapungubwe Phase as well.

For some years a fourteenth to fifteenth century age was accepted for Mapungubwe which meant that it was regarded as no more than an offshoot, and a rather provincial one at that, from Great Zimbabwe (e.g. Beach 1980, Garlake 1973, Phillipson 1977). This evidence was based on two radiocarbon dates run some 20 years ago on carbonized sorghum seeds (Gardner 1963). More recently it has been learnt that sorghum, along with many other tropical grasses, uses the Hatch-Slack photosynthetic pathway (Vogel *et al.* 1978 and references) which necessitates corrections to ^{14}C readings in order to obtain correct dates. The new evidence based on a long series of dates in careful stratigraphic order fix the floruit of Mapungubwe in the period AD 1050-1150, with a continuation under relatively impoverished circumstances until AD 1200 (Eloff & Meyer 1981, Hall & Vogel 1980).

The subsistence elements of the economy remained the same as at K2 and again there is a very high incidence of cattle (Voigt 1980). The pottery has been seen as a development within the Leopards Kopje Tradition, whether locally or from elsewhere is uncertain although the most recent excavators prefer the latter interpretation (Eloff & Meyer 1981). But in several other important aspects Mapungubwe can be linked with and indeed seen as the origin of the Great Zimbabwe Tradition.

The dual nature of the settlement pattern with an elite hilltop area and a less rich but nevertheless important lower area – the southern terrace – is repeated at Great Zimbabwe and some of the other large sites. Access to the upper area is limited to a few possible routes of which two are the most important. Stone walling appears here and at several other Leopards Kopje B sites, though it is not definitely associated with any Leopards Kopje A sites. The use of walling is rather different from that found in other LIA traditions where it chiefly serves the practical purposes of controlling the movement of livestock. Here, on the other hand, both free-standing and terrace retaining walls serve to divide up and control pedestrian movement within the living area. For example, walls and openings within them serve to define what appears to have been the main route to the hilltop across the southern terrace and up the western ascent, a narrow cleft with rows of foot-holds hollowed out of the rock. At the top is a small wall that could have had no more than a symbolic function. Indeed the purpose of the walling in general seems to have been to define status within the site. Huts are still built of poles and daga but floors are more solid and daga curbs occur on some verandahs.

In view of the vast amount of building material used and the difficulty of carrying it up to the hilltop the community must have had control of a considerable surplus of labour beyond the needs of subsistence activities.

Glass beads in large numbers and sherds of Chinese celadon attest to a flourishing trade, and the key to this was gold. The site is the earliest to have produced evidence of gold, which is also known from some other Leopards Kopje B deposits including Thaba Zikamambo (Robinson, K.R. 1966a) and the infills of several gold mines which yielded sherds of this phase (Summers 1969).

The impressive gold items from graves on the hilltop have not been equalled at other sites, but the techniques of goldworking continue throughout the tradition. Beadmaking and wound wire merely translate into gold, techniques seen earlier in iron and copper. But the larger items – rhinoceros, bowl, 'sceptre', etc. – introduce sheet goldwork which is attached to a carved wooden core with gold tacks.

Spindle whorls appear at Mapungubwe in some numbers indicating the local manufacture of cloth. Finer quality cloth was proabably imported. There seems to have been no ivory working, by contrast with K2, but the number of carnivore bones has led to the suggestion that pelts may have been exported (Voigt 1978).

At this point we need to consider historic documents describing trade on the east African coast. The earlier writers before AD 1000 stress ivory as the major export; thus a ninth century Chinese source, Tuan Ch'eng-Shih (Freeman-Grenville 1962) has ivory along with ambergris and slaves, while al-Mas'ūdi in the earlier tenth century describes the ivory trade in detail and also mentions leopard skins and tortoise-shell. He does mention gold from the land of Sofala but it seems to have been of relatively little importance (Chittick 1977). By the time of the next references in the thirteenth century, however, gold has become a major item and Abu'l-Fidā has gold and iron mining as the chief means of subsistence (Freeman-Grenville 1962). As early as about AD 1120 Kilwa obtained control of the gold trade from Sofala (Freeman-Grenville 1973) and at Ibn Battūta's visit in 1331 he mentions gold dust exported from the interior via Sofala.

The replacement of ivory by gold as the major export and the corresponding growth of the more southerly Islamic coastal settlements can be placed in the twelfth century, contemporary with the floruit of Mapungubwe. The Greefswald sequence

(K2 plus Mapungubwe) therefore clearly documents, from the supply end, the shift in trade emphasis from ivory to gold, as seen in the written sources.

Being dependent on elephant hunting, the ivory trade would clearly have been more widely dispersed than gold mining. Yet it seems that some communities, as at Schroda and especially K2, were able to control sufficient of it to become centres of considerable importance locally. With the beginnings of gold export, Mapungubwe was able to take this process much further and develop new cultural, social and no doubt political innovations which were to be followed by successive phases of the Tradition. Indirect evidence that Mapungubwe was part of a fair-sized political unit is inherent in the fact that the nearest known areas of Iron Age gold mining are at least 120 km distance to the north and west (Fouche 1937, Summers 1969) although there are possible sources closer to the south.

Support for the view put forward above, that Mapungubwe represents the first phase of the Great Zimbabwe Tradition comes from the related site of Mapela, 75 km to the northwest (Figure 3) which is dated to about the end of the twelfth century (Garlake 1968, 1973). It too is a hilltop site with limited access but it has far more extensive stone terracing to provide both defence and level platforms for huts. Again the scale of construction implies considerable labour resources. A hierarchical society is evident in the contrast between the core area, where huts are built with thick floors, walls and mouldings of solid daga, presaging the elite dwellings of later phases, and the rest of the settlement with its pole and daga huts.

While we have concentrated on the large hilltop sites with their connotations of extensive economic and political control, many Leopards Kopje B settlements are not of this kind. Most are much smaller, with little or no stonework and probably of short duration. They were presumably typical LIA villages of subsistence farmers, though no doubt many were under some sort of political authority based on the larger sites.

Many characteristics of the Mapungubwe Phase continue and are further developed in the subsequent phase named after Great Zimbabwe itself. Great Zimbabwe is an urban area far larger than the extent of the impressive stone walling suggests (Caton-Thompson 1931, Garlake 1973, Huffman 1972a, 1974c, Summers 1963, 1971, Summers *et al.* 1961). The main stone structures mark the elite area and can be divided into two main groups, the hilltop (Acropolis) and the valley. The former has two main access routes each approached through a stone wall and one of which goes under a large boulder. There are also short sections of covered passage between enclosures. This was evidently the most important religious area, with finds like the famous stone birds, and it was probably also the seat of political and economic power. Most impressive and probably the longest occupied of the valley ruins is the elliptical building with its conical tower. This is the largest Iron Age structure in sub-equatorial Africa. The scale of walling tends to increase with time and the later, Q walling is well coursed and sometimes carries decoration built into the dry stonework in the form of chevrons, dentelle and dark bands. Other architectural features include stepped platforms, rounded entrances, non-functional rounded buttresses some with a vertical slot and monoliths set upright on top of walls. Walling is largely freestanding but also occurs as terracing to provide hut platforms, especially on the south side of the hill facing the valley ruins. A long, low wall separates the ruins from the western part of the settlement.

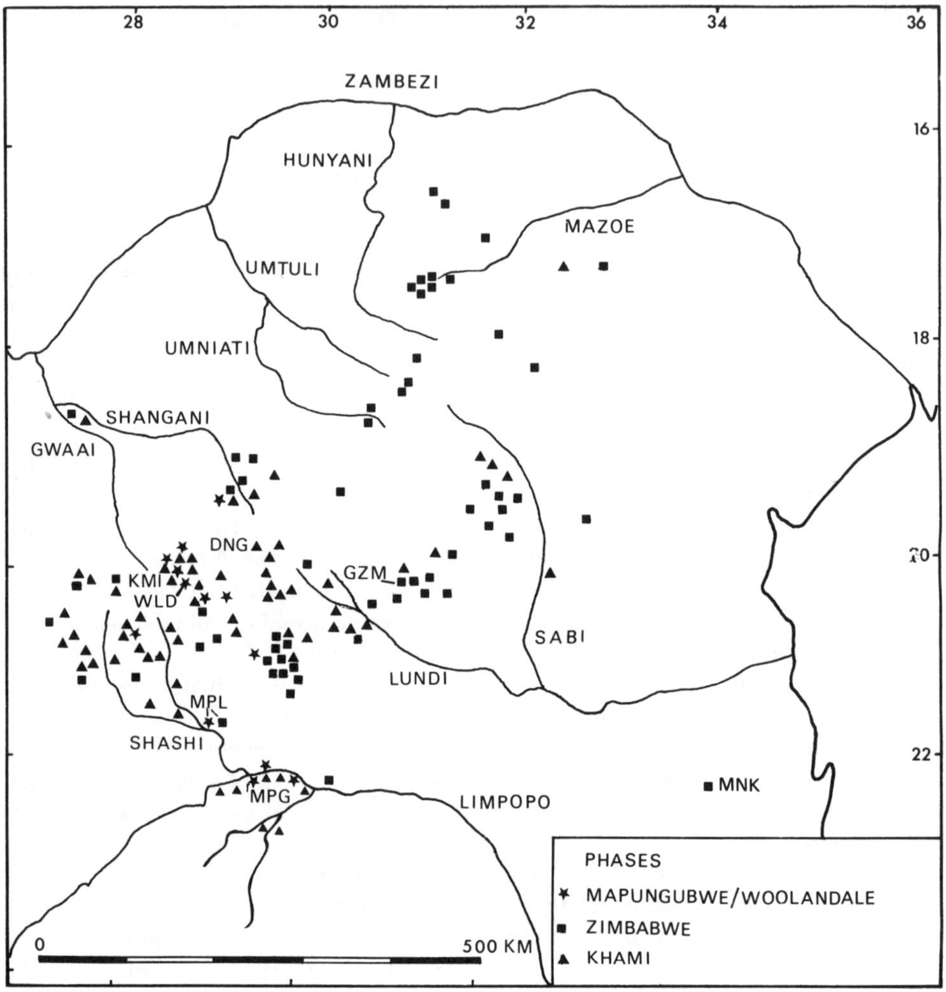

Figure 3. Tentative arrangement of Great Zimbabwe Tradition sites into phases based on Fouche (1937), Garlake (1970b) and Summers (1971).

Abbreviations:

DNG Danangombe
GZM Great Zimbabwe
KMI Khami
MNK Manekweni
MPG Mapungubwe
MPL Mapela
WLD Woolandale

Areas of commoner occupation occur both within and outside the long wall and are characterized by high density, cramped conditions often with a pair of sleeping huts and a kitchen hut grouped around a courtyard – presumably a family unit (Huffman 1977). Commoner areas are concentrated within a radius of 1 km from the hill, while beyond these are a series of smaller occupied areas, each consisting of huts around a small stone enclosure on a hill, which form a ring of outposts around the periphery of the settlement. While the elite areas could have contained a few hundred people at the most, the whole settlement is estimated to have contained some 11,000 people – an urban centre of some magnitude.

The floruit of Great Zimbabwe (Periods III and IV of Summers *et al.* 1961) is firmly dated by radiocarbon and imports to the period AD 1250-1450, but it may also extend back earlier and perhaps a little later. A single radiocarbon date of AD 1130 ±40 (Hall & Vogel 1980) is appreciably earlier than the other recent dates and, if accepted it would support the view that the development of the Great Zimbabwe state caused the eclipse of Mapungubwe in the second half of the twelfth century (Eloff & Meyer 1981).

Numerous other sites can be attributed to the Zimbabwe Phase on the basis of stylistic similarities, though few have been dated and many of the attributions are tentative (Figure 3). They are much smaller and most do not seem to have been occupied for very long. They are probably provincial centres and they often seem to be located in proximity to gold-mining areas or trade routes (Summers 1969) though Manekweni, near the coast in southern Mozambique (Garlake 1976, 1978) is clearly an exception. With fluctuations in the strength of the central authority, outlying centres may well have achieved autonomy at times.

Several reasons have been put forward to explain the concentration of power at Great Zimbabwe, including religion, trade and cattle. Present evidence suggests that while religion was almost certainly associated with political power on the hilltop, there was probably no separate centre of religious power (Huffman 1977). Garlake (1978) has questioned the close association between sites and gold producing areas and also doubts the existence of a unitary state. Instead he identifies ten centres which may each have controlled a distinct territory. Each centre is well situated to take advantage of the differing seasonal grazing potential of different ecotypes and he therefore argues that transhumant beef production was a major concern in the location of the centres. Certainly prime beef formed a large part of the elite diet (Brain 1974, Barker 1978). However, it is the export of gold and the resulting scale of imports that differentiate the Great Zimbabwe Tradition economically from all other Iron Age phenomena in southern Africa. It therefore seems that trade should be regarded as the largest single factor, though by no means the only one, in the rise of Great Zimbabwe.

Historical sources tell us little about Great Zimbabwe, indeed we do not know precisely who lived there, other than that they were clearly a Shona group. Reasons for its decline in the fifteenth century are also somewhat uncertain. Exhaustion of the local environment was probably one factor (Huffman 1972b) but political developments may have played a larger part, for, shortly thereafter, the region seems to have been dominated by two dynasties, the Mutapa in the north-east and the Torwa in the west (Beach 1980). Despite the numerous references to Mutapa in the Portuguese documents little is known of it archaeologically. The reverse is true of Torwa which, in its capital at Khami, may be seen as the successor to Great Zimbabwe.

Khami Phase structures are widely distributed (Figure 3) but the largest and most complex are in Matebeleland. While there is a great deal of cultural continuity, sites are distinguished from the preceding phase on the basis of architectural details (elite huts on terraces, although sometimes still in enclosures, squared entrances, more wall decoration especially check, herringbone, cord and two coloured patterns) and the richly decorated band-and-panel pottery (Caton-Thompson 1931, MacIver 1906, Robinson, K.R. 1959). The emphasis on terracing may indeed reflect continuity in the west from sites like Mapungubwe and Mapela. Although elevated site locations are still chosen there seems to be less emphasis on the need to find naturally inaccessible hilltops. Instead, at the large sites of Khami and Danangombe (Dhlo Dhlo), a series of terraces provide a stepped profile to the relatively low hills which form the focus of the sites. The terracing artificially limits access to these areas, and in each case one entrance is via a covered passage.

The Torwa are thought to have ruled from Khami during approximately the period 1450 to 1640 when the dynasty was weakened by a succession dispute in which the two factions obtained support from Islamic and Portuguese interests (Beach 1980). The Portuguese-supported faction won, but the dynasty never fully recovered and was replaced by the Changamire (Roswi) from further east by 1696. Despite the change in dynasty and the destruction of Khami by fire in about 1640, characteristics of the Khami Phase continued. Danangombe may have succeeded Khami as the capital before the Changamire took over, and it seems to have continued to serve this function for some time (Beach 1980). The Changamire-Roswi remained in power, though much weaker in later years, down to the early nineteenth century when events of the Difaqane overtook them.

Small Khami Phase sites, still undated, occur in the northern Transvaal between the Limpopo and the Soutpansberg range (Fouche 1937, Hanisch 1981a). As the tradition of building continued down to historic times among the Venda and as the history of this people records them moving southwards to this area in about the seventeenth century, there is good reason to associate them with the local expression of the Khami Phase.

THE LATER CENTURIES

From about the sixteenth century archaeological evidence of Iron Age occupation is available from virtually all the areas historically occupied by Bantu-speaking peoples in southern Africa. Oral and documentary historical evidence, social anthropology and ethnology provide rich sources of information to supplement the bare bones of archaeology. From this time it is usually possible to correlate archaeological sites with one of the major historical groupings and in some cases with particular communities and even individual rulers. Although the archaeological coverage is still uneven there is too much material for a detailed review here. Instead we will look at some of the main themes that have emerged from research; the basic chronological and typological framework for areas south of the Limpopo being summarized in Table 3.

One recurring theme is the development of specialized economic production at a local level. The EIA pattern of virtual self-sufficiency at the village level gives way to greater regional interdependence and more emphasis on trade, although this must be seen against the continuing background of a basic Iron Age economy of subsistence farming, augmented by hunting and collecting. This growth of specialization can largely be attributed to the uneven distribution of resources over the landscape, but cultural factors such as the skills developed by particular groups also played a part.

Probably the most important specialization was in metallurgy, particularly iron-working. No longer is it usual to find slag on every site, and indeed some extensive areas must have been dependent on outside sources. The southernmost Bantu-speaking peoples, the southern Nguni, evidently did little or no smelting. At least from the sixteenth century, when the first ship-wrecked Portuguese met them, they were very short of both iron and copper, and exchanged livestock for small pieces (Wilson 1959, Derricourt 1977). The double-ended 'Xhosa spade' made of wood (Shaw & Van Warmelo 1981) appears to have been a substitute for the iron hoe, while spears were often just poles with the end sharpened and fire-hardened as were the 'spades'. At some Highveld sites the proliferation of bone tools indicates that bone was substituted for iron in such tasks as scraping and sewing skins (Maggs 1976, White *et al.* 1979).

From both archaeological and ethnological evidence it is clear that in many areas smelting took place away from centres of population, sometimes associated with a single homestead and sometimes well away from all habitation. This again shows a contrast with the typical EIA smelting which took place within the village.

Areas of concentrated iron and steel production occur in parts of the well wooded lowlands of Natal, Zululand and the eastern Transvaal, though they are probably more widely dispersed in Zimbabwe. The obvious requirements of proximity to iron ore and timber for charcoal are not always the decisive factor in site location, for at Square (Van der Merwe & Killick 1979) some 200 tons of ore were carried a distance of 20 km from Lolwe. In some cases local specialization seems to have developed because the area was relatively unsuitable for the normal Iron Age subsistence activities. Examples are the aridity of the area around Phalaborwa (Van der Merwe & Scully 1971), the broken terrain of Mabhija on the Tugela (Maggs, in prep.) and perhaps tsetse at Hluhluwe (Hall 1980c), all of which were centres of metal production.

Furnaces are mainly of the low shaft kind, less than a metre in internal height, about 50-80 cm in diameter and usually partly dug into the ground (Van der Merwe 1980). From the Transvaal northward there is often a small non-functional hole in the base where on ethnographic evidence, 'medicine' was placed to ensure a successful smelt. In Zimbabwe they often have a broad rake hole, which allows for the removal of thc bloom and even the tapping of slag, in addition to two or more smaller holes or slits for tuyeres (Prendergast 1979). Ethnographic evidence shows that smelting was linked symbolically with human reproduction and furnaces sometimes have moulded breasts and other decoration.

In the eastern Transvaal furnaces are sub-triangular to circular in plan with three slits for the tuyeres (Evers 1981, Van der Merwe & Killick 1979). In Zululand and Natal furnaces are elliptical in plan and have an opening at either end for tuyeres (Hall 1980c, Maggs in press). They were usually built in pairs and each pair may have been fired simultaneously. Judging by the quantity of waste material, the eastern Transvaal furnaces were refired many more times than those in Natal.

Iron and steel were mainly used for implements especially hoes, spearheads, leaf-shaped knife blades, axes, adzes, mining gads, razors, awls for sewing and a variety of more specialized items. Iron was also made into ornaments, particularly beads and bangles.

Copper production was more localized than iron and south of the Vaal and

Table 3. The Late Iron Age in South Africa.

	Eastern Cape	Natal	Orange Free State	Southern Transvaal (Taylor)	Southern Transvaal (Mason)	Eastern Transvaal Highveld	Eastern Transvaal Lowveld	Central Transvaal	Northern Transvaal
	Historic	Historic	Historic	Historic	Historic	Historic	Historic	Historic	Historic
		Zulu	Sotho/	Sotho/	Sotho/	Sotho (Pedi)	Sotho/	Ndebele	Venda
1800			Tswana	Tswana	Tswana		Tswan etc.	Sotho etc.	
	Southern								
		Nqabeni	Types V	Group III	Olifants-	Marateng	Letaba		
		Many others			poort	Phase	Phase	Bambo 11	Khami
			and Z —	Group II	Klipriviers-			(? Letaba)	Phase
	Nguni	(2, 3)			berg	(7, 8)	(8, 9)	Rooiberg	
1600	(1)				Many others			U.4	
				Group I				Rooiberg	
			Type N	(5)	Olifants-			U.3	
			(4)		poort			(10, 11)	
1400	Mpame				'M.I.A.'				
					(6)				
									Icon
		Moor Park							
1200							Eiland		
									Mapungubwe
								Rooiberg	
1000		Blackburn			?Melville			U.2	
		Mpambanyoni							K2

Legend:
(1) Derricourt 1977
(2) Hall & Maggs 1979
(3) Hall 1980b
(4) Maggs 1976
(5) Taylor 1979
(6) Mason 1974
(7) Collett 1979
(8) Evers 1981
(9) Van der Merwe & Scully 1971
(10) Loubser in prep.
(11) Hall 1981

Nkomati rivers there was little or no mining. Furnaces were small and of several different shapes (Friede & Steel 1975, Van der Merwe 1980). Tin mining is known only from Rooiberg in the central Transvaal (Friede & Steel 1976). Some communities, like those at Phalaborwa (Van der Merwe & Scully 1971) and Messina (Van Warmelo 1940) may have become heavily dependent on specialized production and trade. Copper and bronze were used essentially for ornamental objects including beads, bangles, earrings and arm and neck bands, some of which had connotations of prestige. Techniques included casting, wire-pulling and sheet-working. Copper was sometimes reworked into different forms far from the source of the ore, for example the wire earrings at 001 in the Orange Free State (Maggs 1976), and the cast beads and bangles at Mgungundlovu, Dingane's capital (Parkington & Cronin 1979).

Metal items were one of the main categories of trade goods. Other items of geological origin include soft stone for carved pipes and ornaments as well as a variety of colouring matter – graphite, ochres, specularite – used on artefacts and as cosmetics. Other ornamental items which often carried great prestige included a variety of feathers, hides, pelts and shells. Livestock, both large and small, would normally have been acceptable in exchange; grain and salt were also traded. Imported objects especially glass beads spread throughout the regions of Iron Age occupation. While a considerable trade network existed, little in the way of detailed archaeological information is yet available from most areas.

Another aspect that is receiving more attention from recent research is the interaction between Iron Age man and the natural environment, particularly vegetation. Farming communities must have had considerable effect on the vegetation through forest clearance, burning, grazing, cultivation and the use of timber for fuel and building. Feely (1980) and Hall (1980b) have shown that in game reserve areas of lowland Zululand, which have been protected from recent agricultural exploitation, the composition of some grassland and woodland plant communities still reflects the impact of Iron Age occupation. Indeed the maintenance of an open environment in some ecotypes may be an artefact of Iron Age settlement rather than the result of natural vegetation succession.

As mentioned above, it is during this period that a large part of the interior grasslands of South Africa, including the Highveld of the southern Transvaal and Orange Free State as well as inland Natal and Transkei, becomes occupied by Iron Age communities. One very common adaptation that these communities made in the essentially tree-less environment was to build in stone. Sites are therefore very easily found today, especially through the use of aerial photographs, which has led to several studies on distribution and typology in grassland areas of the Transvaal (Figure 4, Mason 1968, Taylor 1979, Collett 1979) Orange Free State (Maggs 1976, Jones 1978) and Natal (Hall & Maggs 1979, Hall 1980b). From this work it is clear that in some grassland areas there was a relatively high density of population in the centuries immediately preceding the colonial era. It has also been possible, using this information and historical sources, to plot the limits of Iron Age settlement in South Africa (Figure 1), though this is not yet the case in Botswana and Namibia.

Settlement of the grasslands is by no means uniform over the landscape, but seems rather to reflect a number of ecological factors. Along the cool, humid Drakensberg escarpment and its footslopes acid soils and sour grazing were probably the major constraints (Maggs 1976, 1980c). On the other hand the absolute southern and western limits to the Iron Age in the eastern and northern Cape, Orange Free State, Botswana and Namibia were set by areas too arid or at least with insufficient summer rain to allow for cultivation. Although historic herder groups beyond this margin obtained iron by trade and some, such as the Nama, apparently employed smiths to smelt copper for them, these groups fall outside the conventional Iron Age bracket and they are known mainly from historical rather than archaeological records.

There is considerable evidence that major ecotones within and on the margins of the grasslands, in particular where they interfaced with savanna or at least thornveld, were favoured areas of settlement, and seem to have supported higher population densities than pure grassland areas. West of Pretoria the Magaliesberg range steps down from the Lichtenburg grassland plains in the south to the bushveld of the central Transvaal. The partly wooded steps of the range were densely settled by stone-building communities but the grassland shows few sites (Mason 1972). In a heavily settled area of the Vredefort Dome (Figure 4) the distribution of three distinct types of settlement was influenced by significant changes in vegetation, geology and topography over a distance of only a few kilometres (Taylor 1979). A little further south a low escarpment runs down the centre of the Orange Free State, supporting some woody vegetation, while open grassland stretches to the east and west. This narrow belt of broken country contains the majority of a particular type of settlement (Maggs 1976). Hall (1980b) has shown that in Zululand a more humid grassland area may have some agricultural advantages over a neighbouring dry valley, especially for maize cultivation, but that livestock numbers would have been

Figure 4. Air photograph of Group II structures in the Vredefort Dome. Dated to the eighteenth century they are associated on typological grounds with a Tswana community. Courtesy of M.O.V. Taylor.

severly restricted unless the upland settlements had access to winter grazing in the sweetveld valley.

A clear advantage of settlements on or near such ecotones would be the potential for livestock transhumance, though in such cases most of the people would remain at home. Access to wood for building and fuel is another factor, though the common grassland adapatations of stone for building and cattle dung for fuel provided substitutes. Lowland soils tend to be better but are in more drought-prone areas. Uplands would have been more suited to the moisture requirements of maize which was widespread in Natal-Zululand by the eighteenth century (Hall & Maggs 1979, Hall & Vogel 1980), but somewhat drier areas would favour sorghum as it is more drought resistant and it is susceptible to fungal diseases in moist conditions.

Within the grasslands, away from their ecotonal margins, opportunities for ecological hedging of bets in pastoralism and cultivation would not have offered themselves to anything like the same extent. Yet even here there was intensive settlement in some areas. The landscape tends to be flat or gently rolling with settlement concentrated on hills or at the sides of the broad river valleys where topography is more broken (Mason 1972, Maggs 1976). Mason has argued that, because such topography supports a great number of floral and faunal species, hillslopes have a greater relative productivity which was a factor in attracting Iron Age settlement. However, such areas tend to have shallow, poorly developed soils and therefore offer little potential for producing the agricultural staples of the Iron Age diet. Other factors which need consideration in understanding this pattern of site location include proximity to both good agricultural and grazing lands, stone for building, microclimatic factors and also socio-cultural preference (Maggs 1976).

High population densities would have led to environmental degradation in both grassland and wooded environments. Areas of relatively steep slopes and high rainfall would be among the most susceptible. It is therefore of interest that two such areas, on the eastern Transvaal escarpment and Inyanga in eastern Zimbabwe, with no apparent connections, both contain large areas of stone terracing associated with dense settlement patterns (Summers 1958, Collett 1979). Both areas show evidence of erosion and redeposition of hillwash on a considerable scale during the Iron Age (Summers 1958, Marker & Evers 1976). The large scale construction of terracing and walled tracks for livestock was evidently an independent but parallel response to similar environmental conditions in these two areas.

A problem to some extent relevant to all archaeology but particularly applicable to the LIA is that of the archaeological visibility of sites. This review has been concerned, to a large extent, with stone-built sites both of the Great Zimbabwe Tradition and of subsistence farming communities. Much less is known about other LIA settlements. For example in most of the region occupied historically by the Nguni peoples (from Swaziland to the eastern Cape) their dispersed settlement pattern and preference for organic building materials has meant that archaeological traces are ephemeral. Sites may be no more than a handful of sherds and a grindstone. The potential of such evidence can easily be dismissed, yet with suitable research strategies it can yield valuable information (Hall 1980b).

A superficial view of the current state of LIA research in southern Africa would emphasise grassland and savanna ecotone communities. Yet the end of the Iron Age was marked by a series of upheavals, the Difaqane, which had massive repercussions not only in our region but in areas north of the Zambezi as well. What is most significant in the present context is that these upheavals, which culminated in the establishment of the Zulu kingdom under Shaka, developed out of a struggle for control of *lowland* areas in Zululand. The disorder spread to other areas as groups such as the Hlubi, Ngwane, Shangane, Ngoni and Mzilikazi's Ndebele broke away or were driven away from the epicentre. As we have seen, EIA distributions show a marked correlation with savanna and even forested environments. This preference was probably present throughout the LIA and was certainly reflected in the early nineteenth century struggle for power in Zululand (Daniel 1973). Population densities were almost certainly higher in savanna than grassland areas despite the archaeological indications. The bias introduced by archaeological visibility will therefore have to be taken into account when converting Iron Age data into history.

Bibliography

ALDWORTH, E. 1979. An interpretation of the Upper Pleistocene sequence in the Transvaal and Lesotho. Unpublished paper presented at a workshop on 'Towards a better understanding of the Upper Pleistocene in Sub-Saharan Africa' organized by the Southern African Association of Archaeologists, Stellenbosch, June 1979.

ALESSIO, M., L. ALLEGRI, F. BELLA, S. IMPROTA, G. BELLUOMINI, C. CALDERONI, C. CORTESI, L. MANFRA, V. PETRONE & A. FRUSCALZO 1978. University of Rome Carbon-14 dates XV. *Radiocarbon* 20: 68-78.

ALEXANDER, J.E. 1838. *An expedition of discovery into the interior of Africa*. London: Henry Colburn.

ANDREWS, P., G.E. MEYER, D.R. PILBEAM, J.A. VAN COUVERING & J.A.H. VAN COUVERING 1981. The Miocene fossil beds of Maboko Island, Kenya: geology, age, taphonomy, and palaeontology. *J. Human Evol.* 10: 35-48.

ANSELL, W.F.H. 1971a. Order Artiodactyla. In: J. Meester & H. W. Setzer, (eds.), *The mammals of Africa: an identification manual*: Part 15. Washington, D.C.: Smithsonian Institution Press.

ANSELL, W.F.H. 1971b. Order Perissodactyla. In: J. Meester & H.W. Setzer, (eds.), *The mammals of Africa: an identification manual:* Part 14. Washington, D.C.: Smithsonian Institution Press.

ANSELL, W.F.H. 1971c. Order Proboscidea. In: J. Meester & H.W. Setzer, (eds.), *The mammals of Africa: an identification manual*: Part 11. Washington, D.C.: Smithsonian Institution Press.

ANTHONY, B.W. 1963. Report on excavations at Peers' Cave, Fish Hoek, C.P., March-June 1963. Unpublished manuscript, South African Museum, Cape Town.

ANTHONY, B.W. 1967. Excavations at Peers Cave, Fish Hoek South Africa. *Palaeoecol. Afr.* 2: 58-59.

ANTHONY, B.W. 1972. The Stillbay Question. In: H.J. Hugot (ed.), *Actes du VIe Congrès panafricain de préhistoire* (Dakar, 1967): 80-82, Chambéry: Les Imprimeries Réunies de Chambéry.

ARMSTRONG, A.L. 1931. Rhodesian archaeological expedition (1929): excavations in Bambata Cave and researches on prehistoric sites in Southern Rhodesia. *J.R. Anthrop. Inst.* 61: 239-276.

ARNOLD, G. & N. JONES 1919. Notes on the Bushman Cave at Bambata, Matopos. *Proc. Rhod. Scient. Assoc.* 17: 7-21.

AVERY, D.M. 1981. Holocene micromammalian faunas from the northern Cape Province, South Africa. *S. Afr. J. Sci.* 77: 265-273.

AVERY, D.M. 1982a. Micromammals as palaeoenvironmental indicators and an interpretation of the late Quaternary in the southern Cape Province, South Africa. *Ann. S. Afr. Mus.* 85: 183-374.

AVERY, D.M. 1982b. The micromammalian fauna from Border Cave, Kwazulu, South Africa. *J. archaeol. Sci.* 9:187-204.

AVERY, G. 1974. Open station shell midden sites and associated features from the Pearly Beach area, south-western Cape. *S. Afr. archaeol. Bull.* 29: 104-144.

AVERY, G. 1975. Discussion on the age and use of tidal fish-traps (visvywers). *S. Afr. archaeol. Bull.* 30: 105-113.

AVERY, G. 1976. A systematic investigation of open station shell midden sites along the south-western Cape coast. Unpublished M.A. thesis: University of Cape Town.

AVERY, G. 1977. Report on the marine bird remains from the Paternoster midden. *S. Afr. archaeol. Bull.* 32: 74-76.

AVERY, G. 1981. Late Holocene avian remains from Rooiels Cave, south-western Cape Province, South Africa. *S. Afr. archaeol. Bull.* 36: 84-87.

AVERY, G. & W.R. SIEGFRIED 1980. Food gatherers along South Africa's seashore. *Oceans* 4: 32-37.

AZZAROLI, A. 1975. Late Miocene interchange of terrestrial faunas across the Mediterranean. *Act. I Col. Intern. sobre Biostrat. Cont. Neogeno Sup. Cuat. Inf.* 4: 67-71.

BADA, J.L. & L. DEEMS 1975. Accuracy of dates beyond the 14C dating limit using the aspartic acid racemisation reaction. *Nature (Lond.)* 255: 218-219.

BADA, J.L., R.A. SCHROEDER, R. PROTSCH & R BERGER 1974. Concordance of collagen-based radiocarbon and aspartic-acid racemization ages. *Proc. Nat. Acad. Sci.* 71: 914-917.

BALFOUR, H. 1909. In: H. Tongue, *Bushman Paintings*. London: Clarendon Press.

BALINSKY, B.I. 1962. Patterns of animal distribution on the African continent. *Ann. Cape Prov. Mus.* 2: 199-310.

BARKER, G. 1978. Economic models for the Manekweni *Zimbabwe*, Mozambique. *Azania* 13: 71-100.

BARNARD, A. 1979. Kalahari Bushman settlement patterns. In: P.C. Burnham & R.F. Ellen (eds.), *Social and ecological systems*: 131-144. London: Academic Press.

BARRADAS, L. 1965 Cronologia da Beira-Mar do sul de Mocambique. *Mem. Inst. Invest. Cient. Mocambique. Ser. B.* 7: 23-35.

BARRADAS, L. 1967. Concheiros da antigua baia de Lourenco Marques. *Trab. Inst. Invest. Cient. Mocambique*. 17: 1-75.

BARROW, J. 1801. *An account of travels into the interior of southern Africa in the years 1797 and 1798.* London: Cadell & Davies.

BATTISS, W.W. 1948. *The artists on the rocks*. Pretoria: Red Fawn Press.

BATTISS, W.W. 1971. The aesthetic value of our rock art. *S. Afr. J. Sci. Spec. Publ.* 2.

BEACH, D. 1980. *The Shona and Zimbabwe 900–1850.* Gwelo: Mambo.

BEATER, B.E. 1976. An engraved stone implement from the Upper Mkomanzi River, Natal. *S. Afr. archaeol. Bull*. 22: 157.

BEAUMONT, P.B. 1973a. Border Cave–a progress report. *S. Afr. J. Sci.* 69: 41-46.

BEAUMONT, P.B. 1973b. The ancient pigment mines of southern Africa. *S., Afr. J. Sci.* 69: 140-146.

BEAUMONT, P.B. 1978. Border Cave. Unpublished M.A. thesis: University of Cape Town.

BEAUMONT, P.B. 1979a. The Stone Age cultural stratigraphy of Border Cave. Unpulished paper presented at a workshop on 'Towards a better understanding of the Upper Pleistocene in sub-Saharan Africa' organized by the Southern African Association of Archaeologists, Stellenbosch, June 1979.

BEAUMONT, P.B. 1979b. A first account of recent excavations at Wonderwerk Cave. Unpublished paper presented at the biennial meeting of the Southern African Association of Archaeologists, University of Cape Town, June 1979.

BEAUMONT, P.B. 1980. On the age of Border Cave hominids 1-5. *Palaeont. Afr*. 23: 21-33.

BEAUMONT, P.B. 1981. The Heuningsneskrans shelter. In: E.A. Voigt (ed.), *Guide to archaeological sites in the northern and eastern Transvaal*: 133-145. Pretoria: S.A. Association of Archaeologists.

BEAUMONT, P.B. & A.K. BOSHIER 1974. Report on test excavations in a prehistoric pigment mine near Postmasburg, northern Cape. *S. Afr. archaeol. Bull.* 29: 41-59.

BEAUMONT, P.B., H. DE VILLIERS & J.C. VOGEL 1978. Modern man in sub-Saharan Africa prior to 49,000 years B.P.: a review and evaluation with particular reference to Border Cave. *S. Afr. J. Sci.* 74: 409-419.

BEAUMONT, P.B. & J.C. VOGEL 1972a. On a new radiocarbon chronology for Africa south of the Equator. Part 1. *Afr. Stud.* 31: 66-89.

BEAUMONT, P.B. & J.C. VOGEL 1972b. On a new radiocarbon chronology for Africa south of the Equator. Part 2. *Afr. Stud.* 31: 155-182.

BECK, B.B. 1980. *Animal tool behavior*. New York: Garland STPM Press.

BENNETT, C.L. 1979. Radiocarbon dating with accelerators. *Am. Scient.* 67: 450-457.

BERGGREN, W.A. & J.A. VAN COUVERING 1974. *The Late Neogene.* Amsterdam: Elsevier.

BIESELE, M. 1978. Sapience and scarce resources: communication systems of the !Kung and other foragers. *Soc. Sci. Info*. 17: 921-947.

BIGALKE, R.C. 1978. Mammals. In: M.J.A. Werger (ed.), *Biogeography and ecology of Southern Africa*: 981-1048. The Hague: W. Junk.

BINFORD, L.R. 1981. *Bones: ancient men and modern myths*. New York: Academic Press.

BINNEMAN, J., J. BRINK & A. SCHOLTZ 1979. A Last Interglacial midden at Herolds Bay Cave. Unpublished paper presented at a workshop on 'Towards a better understanding of the Upper

Pleistocene in sub-Saharan Africa' organized by the Southern African Association of Archaeologists, Stellenbosch, June 1979.

BISHOP, W.W. & J.D. CLARK (eds.) 1967. *Background to evolution in Africa.* Chicago: University of Chicago Press.

BISHOP, W.W. & M. PICKFORD 1975. Geology, fauna and palaeoenvironments of the Ngorora Formation, Kenya Rift Valley. *Nature (Lond.)* 254: 185-192.

BLEEK, D. 1928. *The Naron.* Cambridge: Cambridge University Press.

BOAZ, N.T. & F.C HOWELL 1977. A gracile hominid cranium from Upper Member G of the Shungura Formation, Ethiopia. *Am. J. Phys. Anthrop.* 46: 93-108.

BOND, G. & R. SUMMERS 1951. The Quaternary sucession and archaeology of Chelmer near Bulawayo, Southern Rhodesia. *S. Afr. J. Sci.* 47: 200-204.

BONIFAY, M.—F. 1980. Relations entre les données isotopiques océaniques et l'histoire des grandes faunes européenes plio-pléistocénes. *Quatern. Res.* (N. Y.) 14: 251-162.

BORDES, F. 1961. *Typologie du Paléolithique ancien et moyen.* Bordeaux: Delmas.

BORDES, F. 1977. Time and space limits of the Mousterian. In: R.V.S. Wright (ed.), *Stone tools as cultural markers: change, evolution and complexity*: 37-39. Canberra: Autstralian Institute of Aboriginal Studies.

BORDES, F. 1980. Le débitage Levallois et ses variantes. *Bull. Soc. Prèhist. Fr.* 77: 45-49.

BOSHIER, A. & P.B. BEAUMONT 1972. Mining in southern Africa and the emergence of modern man. *Optima* 22: 2-12.

BOTHMA, J. DU P. 1971. Order Hyracoidea. In: J. Meester & H.W. Setzer (eds.), *The mammals of Africa: an identification manual*: Part 12. Washington, D.C.: Smithsonian Institution Press.

BOWKER, J.H. 1884. Other days in South Africa. *S. Afr. J. Phil.* 3: 68-73.

BRAIN, C.K. 1958. The Transvaal ape-man-bearing cave deposits. *Transv. Mus. Mem.* 11: 1-125.

BRAIN, C.K. 1967. The Transvaal Museum's fossil project at Swartkrans. *S. Afr. J. Sci.* 63: 378-384.

BRAIN, C.K. 1969a. Faunal remains from the Bushman Rock Shelter, eastern Transvaal. *S. Afr. archaeol. Bull.* 24: 52-55.

BRAIN, C.K. 1969b. New evidence for climatic change during Middle and Late Stone Age times in Rhodesia. *S. Afr. archaeol. Bull.* 24: 127-143.

BRAIN, C.K. 1970. New finds at the Swartkrans australopithecine site. *Nature (Lond.)* 225: 1112-1119.

BRAIN, C.K. 1974. Human food remains from the Iron Age at Zimbabwe. *S. Afr. J. Sci.* 70: 303-309.

BRAIN, C.K. 1975. An interpretation of the bone assemblage from the Kromdraai australopithecine site, South Africa. In: R.H. Tuttle (ed.), *Palaeoanthropology, morphology and palaeoecology*: 225-243. The Hague: Mouton.

BRAIN, C.K. 1976. A re-interpretation of the Swartkrans site and its remains. *S. Afr. J. Sci.* 72: 141-146.

BRAIN, C.K. 1981. *The hunters or the hunted?* Chicago: University of Chicago Press.

BRAIN, C.K. 1982. The Swartkrans site: stratigraphy of the fossil hominids and a reconstruction of the environment of early Homo. In: *L'Homo erectus et la place de 1'homme de Tautavel parmi les hominidés fossiles*, Colloque International du C.N.R.S., Nice 1982, pp. 617-675.

BRAIN, C.K., C. VAN RIET LOWE & R. DART 1955. Kafuan stone artifacts in the post-australopithecine breccia at Makapansgat. *Nature (Lond.)* 175: 16-18.

BRAIN, C.K. & C.K. COOKE 1967. A preliminary account of the Redcliff Stone Age cave site in Rhodesia. *S. Afr. archaeol. Bull.* 21: 171-182.

BRENTJES, B. 1969. *African rock art.* London: Dent.

BREUIL, H. 1931. *L'Afrique préhistorique: paléolothique ancien, paléolithique moyen, paléolithique supérieur et microlithique, néolithique. L'art rupestre en Afrique.* Paris: Cahiers d'art.

BRICKHILL, J. 1978. Small fossil carnivores from Makapansgat. *S. Afr. J. Sci.* 74: 145.

BRINK, J.S. & H.J. DEACON 1981. A study of a Last Interglacial midden and carnivore lair at Herolds Bay, Cape Province, South Africa. Unpublished paper presented at biennial meeting of the South African Society for Quaternary Research, Pretoria, May 1981.

BROCK, A., P.L. MCFADDEN & T.C. PARTRIDGE 1977. Preliminary palaeomagnetic results from Makapansgat and Swartkrans. *Nature (Lond.)* 266: 249-250.

BROOK, G.A. 1982. Stratigraphic evidence for Quaternary climatic change at Echo Cave, Transvaal. *Catena* (Braunschweig) 9: 343-351.

BROOKER, M. 1980. Rescue excavations at the Welbedacht Dam site. *Humanitas* 6: 35-42.

BROOKS, A.S. & J.E. YELLEN 1977. Archaeological excavations at ≠ Gi: preliminary report on the first two field seasons. *Botswana Notes Rec.* 9: 21-30.

BROOKS, A.S., A.L. CROWELL & J.E. YELLEN 1980. ≠Gi: a Stone Age archaeological site in the northern Kalahari Desert, Botswana. In: R.E.F. Leakey & B.A. Ogot (eds.), *Proceedings of the 8th Panafrican congress of prehistory and Quaternary studies* (Nairobi, 1977): 304-309. Nairobi: The International Louis Leakey Memorial Institute for African Prehistory.

BROWN, F.H., F.C. HOWELL & G.G. ECK 1978. Observations on problems of correlation of late Cenozoic hominid-bearing formations in the North Lake Turkana basin. In: W.W. Bishop (ed.), *Geological background to Fossil Man*: 473-498. Edinburgh: Scottish Academic Press.

BROWN, F. H. & W.P. NASH 1976. Radiometric dating and tuff mineralogy of Omo group deposits. In: Y. Coppens, F.C. Howell, G.Ll. Isaac & R.E.F. Leakey (eds.), *Earliest man and environments in the Lake Rudolf Basin*: 50-63. Chicago: University of Chicago Press.

BROWN, L. 1965. *Africa: a natural history.* New York: Random House.

BUCHANAN, W.F. 1977. Rescue dig at a Late Stone Age cave, Hout Bay, Cape Province. Unpublished Archaeology Additional Project: Department of Archaeology, University of Cape Town.

BUCHANAN, W.F., S.L. HALL, J. HENDERSON, A. OLIVIER, J.M. PETTIGREW, J.E. PARKINGTON & P.T. ROBERTSHAW 1978. Coastal shell middens in the Paternoster area, south-western Cape. *S. Afr. archaeol. Bull.* 33: 89-93.

BUNN, H.T. 1981. Archaeological evidence for meat-eating by Plio-Pleistocene hominids from Koobi Fora and Olduvai Gorge. *Nature (Lond.)* 291: 574-577.

BUNN, H., J.W.K. HARRIS, G. ISAAC, Z. KAUFULU, E. KROLL, K. SCHICK, N. TOTH & A.K. BEHRENSMEYER 1980. FxJi50: an early Pleistocene site in northern Keyna. *Wld. Archaeol.* 12: 109-136.

BURCHELL, W.J. 1822. *Travels in the interior of southern Africa.* London: Batchworth.

BURKITT, M.C. 1928. *South Africa's past in stone and paint.* Cambridge: Cambridge University Press.

BUTLER, P.M. 1978. Insectivora and Chiroptera. In: V.J. Maglio & H.B.S. Cooke (eds.), *Evolution of African mammals:* 56-68. Cambridge (Mass.): Harvard University Press.

BUTZER, K.W. 1971. Fine alluvial fills of the Orange and Vaal River Basins of South Africa. *Proc. Assoc. Amer. Geog.* 3: 41-48.

BUTZER, K.W. 1973a. Geology of Nelson Bay Cave, Robberg, South Africa. *S. Afr. Archaeol. Bull.* 28: 97-110.

BUTZER, K.W. 1973b. Re-evaluation of the geology of the Elandsfontein (Hopefield) site, South-western Cape, South Africa. *S. Afr. J. Sci.* 69: 234-238.

BUTZER, K.W. 1973c. On the geology of a late Pliocene *Mammuthus* site, Virginia, Orange Free State. *Res. Nat. Mus.* (Bloemfontein) 2(II): 386-393.

BUTZER, K.W. 1973d. The question of Pleistocene 'periglacial' phenomena in South Africa. *Boreas* (Uppsala) 2: 1-11.

BUTZER, K.W. 1973e. A provisional interpretation of the sedimentary sequence from Montagu Cave (Cape Province), South Africa. *Univ. Calif. Anthrop. Rec.* 28: 89-92.

BUTZER, K.W. 1974a. Paleoecology of South African australopithecines: Taung revisited. *Curr. Anthrop.* 15: 367-382, 413-416.

BUTZER, K.W. 1974b. Geo-archaeological interpretation of Acheulean calc-pan sites at Doornlaagte and Rooidam (Kimberley, South Africa). *J. archaeol. Sci.* 1: 1-25.

BUTZER, K.W. 1976. Lithostratigraphy of the Swartkrans Formation. *S. Afr. J. Sci.* 72: 136-141.

BUTZER, K.W. 1978a. Climate patterns in an unglaciated continent. *Geog. Mag.* 51: 201-208.

BUTZER, K.W. 1978b. Sediment stratigraphy of Middle Stone Age sequence at Klasies River Mouth. *S. Afr. archaeol. Bull.* 33: 141-151.

BUTZER, K.W. 1979a. Geomorphology and geo-archaeology at Elandsbaai, western Cape, South Africa. *Catena* (Braunschweig) 6: 157-166.

BUTZER, K.W. 1979b. Preliminary notes on the geology of Die Kelders. *Ann. S. Afr. Mus.* 78: 225-228.

BUTZER, K.W. 1980a. The Taung australopithecine: contextual evidence. *Palaeont. afr.* 23: 59-60.

BUTZER, K.W. 1980b. The Holocene lake plain of North Rudolf, East Africa. *Physical Geography* (Washington) 1: 44-58.

BUTZER, K.W. 1982. *Archaeology as human ecology: method and theory for a contextual approach.* New York: Cambridge University Press.

BUTZER, K.W. (n.d.). Comments on the sedimentary profile of Kangkara and Paardeberg Caves, Cape Folded Ranges. Unpublished manuscript.

BUTZER, K.W., P.B. BEAUMONT & J.C. VOGEL 1978. Lithostratigraphy of Border Cave, Kwazulu, South Africa: a Middle Stone Age sequence beginning c. 195,000 B.P. *J. archaeol. Sci.* 5: 317-341.

BUTZER, K.W., J.D. CLARK & H.B.S. COOKE 1974. The geology, archaeology, and fossil mammals of the Cornelia Beds, O.F.S. *Mem. Nas. Mus.* (Bloemfontein) 9: 1-84.

BUTZER, K.W., G.J. FOCK, R. STUCKENRATH & A. ZILCH 1973. Palaeohydrology of late Pleistocene Lake Alexandersfontein, Kimberley, South Africa. *Nature (Lond.)* 243: 328-330.

BUTZER, K.W., G.J. FOCK L. SCOTT & R. STUCKENRATH 1979. Dating and context of rock engravings in southern Africa. *Science* 203: 1201-1214.

BUTZER, K.W. & D. HELGREN, 1972. Late Cenozoic evolution of the Cape coast between Knysna and Cape St. Francis, South Africa. *Quatern. Res.* (N.Y.) 2: 143-169.

BUTZER, K.W., D.M., HELGREN, G.I. FOCK & R. STUCKENRATH 1973. Alluvial terraces of the Lower Vaal River, South Africa: a reappraisal and reinvestigation. *Jour. Geol.* (Chicago) 81: 341-362.

BUTZER, K.W., R. STUCKENRATH A.J. BRUZEWICZ & D.M. HELGREN 1978. Late Cenozoic paleoclimates of the Gaap Escarpment, Kalahari margin, South Africa. *Quatern. Res.* (N.Y.) 10: 310-339.

BUTZER, K.W., R. STUCKENRATH & J.C. VOGEL 1979. The geo-archeological sequence of Wonderwerk Cave, South Africa. Abstract, Society of Africanist Archaeologists in America Meeting, Calgary.

BUTZER, K.W. & J.C. VOGEL 1979. Archeo-sedimentological sequences from the sub-montane interior of South Africa: Rose Cottage Cave, Heuningsneskrans, and Bushman Rock Shelter. Unpublished paper presented at a workshop on 'Towards a better understanding of the Upper Pleistocene in sub-Saharan Africa' organized by the Southern African Association of Archaeologists, Stellenbosch, June 1979.

CABLE, C. 1980. Comment on 'Time and place' by J.E. Parkington. *S. Afr. archaeol. Bull.* 35: 84.

CABLE, J.H.C., K. SCOTT & P.L. CARTER 1980. Excavations at Good Hope Shelter, Underberg District, Natal. *Ann. Natal Mus.* 24: 1-34.

CAHEN, D. 1978a. New excavations at Gombe (ex-Kalina) Point, Kinshasa, Zaire. *Antiquity* 52: 51-56.

CAHEN, D 1978b. Vers une révision de la nomenclature des industries préhistoriques de l'Afrique centrale. *L'Anthropologie* 82: 5-36.

CAHEN, D. & J. MOEYERSONS 1977. Subsurface movements of stone artifacts and their implications for the prehistory of Central Africa. *Nature (Lond.)* 266: 812-815.

CAHEN, D. & J. MOEYERSONS 1979. Contributions to the Upper Pleistocene archaeology and palaeo-environment of southwestern Central Africa. Unpublished paper presented at a workshop on 'Towards a better understanding of the Upper Pleistocene in sub-Saharan Africa' organized by the Southern African Association of Archaeologists, Stellenbosch, June 1979.

CAIRNS, P.A. 1975. A report on circular stone features associated with coastal shell middens at Cape St. Francis. *S. Afr. archaeol. Bull.* 30: 36-39.

CAMPBELL, C.V. 1967. Lamina, laminaset, bed and bedset. *Sedimentology* 8: 7-26.

CARR, M.J., A.C. CARR. & L. JACOBSON 1978. Hut remains and related features from the Zerrissene Mountain area: their typology, distribution and ecology. *Cimbebasia* B2: 235-258.

CARTER, P.L. 1969. Moshebi's Shelter: excavation and exploitation in eastern Lesotho. *Lesotho Notes Rec.* 8: 1-11.

CARTER, P.L. 1970. Late Stone Age exploitation patterns in southern Natal. *S. Afr. archaeol. Bull.* 25: 55-58.

CARTER, P.L. 1976. The effects of climatic change on settlement in eastern Lesotho during the Middle and Later Stone Age. *Wld. Archaeol.* 8: 197-206.

CARTER, P.L. 1978. The prehistory of eastern Lesotho. Unpublished Ph.D. Dissertation: University of Cambridge.

CARTER, P.L. 1980. Comment on 'Time and place' by J.E. Parkington. *S. Afr. archaeol. Bull.* 35: 85.

CARTER, P.L. & J.C. VOGEL 1974. The dating of industrial assemblages from stratified sites in eastern Lesotho. *Man* (N.S.) 9: 557-570.

CATON-THOMPSON, G. 1931. *The Zimbabwe Culture.* Oxford: University Press.

CHAPPEL, C.A. 1968. A Strandloper skeleton found at Cape St. Francis. *Diastema*: 37-39.

CHAVAILLON, J.N. CHAVAILLON F. HOURS & M. PIPERNO 1978 Le dèbut et la fin de l'acheulèen a Melka-Konturé: méthodologie pour l'étude des changements de civilisation. *Bull. Soc. Préhist. Fr.* 75: 105-115.

CHITTICK, H.N. 1977. The east coast, Madagascar and the Indian Ocean. In: R. Oliver (ed.), *The Cambridge history of Africa 3.* Cambridge: Cambridge University Press.

CHURCHER, C.S. 1978. Giraffidae. In: V.J. Maglio & H.B.S. Cooke (eds.), *Evolution of African mammals*: 509-535. Cambridge (Mass.): Harvard University Press.

CHURCHER, C.S. & M.L. RICHARDSON 1978. Equidae. In: V.J. Maglio & H.B.S. Cooke (eds.), *Evolution of African mammals*: 379-422. Cambridge (Mass.): Harvard University Press.

CLARK, J.D. 1942. Further excavations (1939) at the Mumbwa Caves, Northern Rhodesia. *Trans. R. Soc. S. Afr.* 29: 133-201.

CLARK, J.D. 1950. *The Stone Age Cultures of Northern Rhodesia.* Claremont: South African Archaeological Society.

CLARK, J.D. 1955. A note on a wooden implement from the level of Peat I at Florisbad, Orange Free State. *Navors. Nas. Mus.* (Bloemfontein) 1: 135-140.

CLARK, J.D. 1958a. Certain industries of notched and strangulated scrapers in Rhodesia, their time range and possible use. *S. Afr. archaeol. Bull.* 13: 56-66.

CLARK, J.D. 1958b. Some Stone Age woodworking tools in southern Africa. *S. Afr. archaeol. Bull.* 13: 144-152.

CLARK, J.D. 1959. *The prehistory of southern Africa.* Harmondsworth: Penguin.

CLARK, J.D. 1964. The Sangoan culture of Equatoria: the implications of its stone equipment. In: E. Ripoll Perelló (ed.), *Miscelanea en homenaje al Abate Henri Breuil*: 309-323. Barcelona: Instituto de prehistoria y arqueologia.

CLARK, J.D. 1969. *Kalambo Falls prehistoric site, Vol. 1.* Cambridge: Cambridge University Press.

CLARK, J.D. 1970. *The prehistory of Africa.* London: Thames & Hudson.

CLARK, J.D. 1971a. Human behavioral differences in southern Africa during the Later Pleistocene. *Amer. Anthrop.* 73: 1211-1236.

CLARK, J.D. 1971b. Problems of archaeological nomenclature and definition in the Congo Basin. *S. Afr. archaeol. Bull.* 26: 67-78.

CLARK, J.D. 1974. *Kalambo Falls prehistoric site, Vol. 2.* Combridge: University of Cambridge Press.

CLARK, J.D. 1975. Africa in prehistory: peripheral or paramount? *Man* (N.S.) 10: 175-198.

CLARK, J.D. 1977a. Interpretations of prehistoric technology from ancient Egyptian and other sources. Part II: prehistoric arrow forms in Africa as shown by surviving examples of the traditional arrows of San Bushmen. *Paléorient* 3: 127-150.

CLARK, J.D. 1977b. Bone tools of the earlier Pleistocene. In: B. Arensburg & O. Bar-Yosef (eds.), *Moshe Stekelis Memorial Volume*: 23-37. Jerusalem: Israel Exploration Society.

CLARK, J.D. 1981. Prehistory in southern Africa. In: J. Ki-Zerbo (ed.), *General history of Africa I: methodology and African prehistory*: 487-529. Berkeley: University of California Press.

CLARK, J.D., G.H., COLE, G. Ll. ISAAC & M.R. KLEINDIENST 1966. Precision and definition in African archaeology. *S. Afr. archaeol. Bull.* 21: 114-121.

CLARK, J.D. & B.M. FAGAN 1965. Charcoals, sands and channel-decorated pottery from Northern Rhodesia. *Amer. Anthrop.* 67: 354-371.

CLARK, J.D. & C.V. HAYNES 1970. An elephant butchery site at Mwanganda's Village, Karonga, Malawi, and its relevance for Paleolithic archaeology. *Wld. Archaeol.* 1: 390-411.

CLARK, J.D., C.V., HAYNES, J.E. MAWBY & A. GAUTIER 1970. Interim report on palaeo-anthropological investigations in the Lake Malawi Rift. *Quaternaria* 13: 305-354.

CLARK, J.D. & H. KURASHINA 1979. Hominid occupation of the east-central highlands of Ethiopia in the Plio-Pleistocene. *Nature (Lond.)* 282: 33-39.

CLARK, J.D., K.P. OAKLEY, L.H. WELLS & J.A.C. MCCLELLAND , 1950. New studies on Rhodesian Man. *J.R. Anthrop. Inst.* 77: 7-32.

CLARK, J.D. & E.M. VAN ZINDEREN BAKKER, 1964. Prehistoric culture and Pleistocene vegetation at Kalambo Falls, Northern Rhodesia. *Nature (Lond.)* 201: 971-975.

CLARK, J.D. & J. WALTON 1962. A Late-Stone Age site in the Erongo Mountains, South West Africa. *Proc. Prehist. Soc.* 28: 1-16.

CLARKE, D.L. 1968. *Analytical archaeology.* London: Methuen.

CLARKE, R.J. & F.C. HOWELL 1972. Affinities of the Swartkrans 847 hominid cranium. *Am. J. Phys. Anthrop.* 37: 319-336.

CLARKE, R.J., F.C. HOWELL & C.K. BRAIN 1970. More evidence of an advanced hominid at Swartkrans. *Nature (Lond.)* 225: 1219-1222.

CLOSE, A.E. 1980. Current research and recent radiocarbon dates from northern Africa. *J. Afr. Hist.* 21: 145-167.

COETZEE, C.G. 1971. Order Carnivora, main text. In: J. Meester & H.W. Setzer (eds.), *The mammals of Africa: an identification manual*: Part 8. Washington, D.C. Smithsonian Institution Press.

COETZEE, J.A. 1967. Pollen analytical studies in East and Southern Africa. *Palaeoecol. Afr.* 3: 1-146.

COETZEE, J.A. 1976. A report on an pollen analytical investigation of recent river mouth sediments on the south-west African coast. *Palaeoecol. Afr.* 9: 131-135.

COETZEE, J.A. 1978a. Climatic and biological changes in south-western Africa during the late Cainozoic. *Palaeoecol. Afr.* 10: 13-29.

COETZEE, J.A. 1978b. Late Cainozoic palaeoenvironments of southern Africa. In: E.M. Van Zinderen Bakker Sr. (ed.), *Antarctic glacial history and world palaeoenvironments*: 115-127. Rotterdam: Balkema.

COETZEE, J.A. & E.M. VAN ZINDEREN BAKKER 1952. Pollen spectrum of the southern middleveld of the Orange Free State. *S. Afr. J. Sci.* 48: 275-281.

COLLETT, D.P. 1979. The archaeology of the stone-walled settlements in the eastern Transvaal, South Africa. Unpublished M.Sc. thesis: Department of Archaeology, University of the Witwatersrand.

COLLINGS, G.E., A.R.I. CRUIKSHANK, J.M. MAGUIRE & R.M. RANDALL 1976. Recent faunal studies at Makapansgat Limeworks, Transvaal, South Africa. *Ann. S. Afr. Mus.* 71: 153-165.

COOKE, C.K. 1957. The Waterworks site at Khami, Southern Rhodesia: Stone Age and Proto-historic. *Occ. Pap. natn. Mus. S. Rhod. A.* 3(21): 1-43.

COOKE, C.K. 1963. Report on excavations at Pomongwe and Tshangula Caves, Matopos Hills, Southern Rhodesia. *S. Afr. archaeol. Bull.* 18: 73-151.

COOKE, C.K. 1966a. Re-appraisal of the industry hitherto named the Proto-Stillbay. *Arnoldia (Rhod.)* 2(22): 1-14.

COOKE, C.K. 1966b. The archaeology of the Mafungabusi area, Gokwe, Rhodesia. *Proc. Trans. Rhod. Scient. Ass.* 51: 51-78.

COOKE, C.K. 1969. *Rock art of southern Africa.* Cape Town: Books of Africa.

COOKE, C.K. 1971. Excavations in Zombepata Cave, Sipolilo District, Mashonaland, Rhodesia. *S. Afr. archaeol. Bull.* 26: 104-126.

COOKE, C.K. 1975. Two parts of arrowshafts: Pomongwe Cave excavations 1960-61. *S. Afr. archaeol. Bull.*30: 41.

COOKE, C.K. 1978. The Redcliff Stone Age site, Rhodesia. *Occ. Pap. natn. Mus. Rhod. A* 1: 45-73.

COOKE, C.K. 1979a. Excavations at Diana's Vow Rock Shelter, Makoni District, Zimbabwe Rhodesia. *Occ. Pap. natn. Mus. Rhod. A* 4: 115-148.

COOKE, C.K. 1979b. The industries of the Upper Pleistocene in Rhodesia. Unpublished paper presented at a workshop on 'Towards a better understanding of the Upper Pleistocene in sub-Saharan Africa' organized by the Southern African Association of Archaeologists, Stellenbosch, June 1979.

COOKE, C.K. 1980. Wooden and bone artefacts: Pomongwe Cave, Matobo District, Zimbabwe. *S. Afr. archaeol. Bull.* 35: 25-29.

COOKE, C.K. & K.R. ROBINSON 1954. Excavations at Amadzimba Cave, located in the Matopo Hills, Southern Rhodesia. *Occ. Pap. natn. Mus. S. Rhod.* 2(19): 699-728.

COOKE, C.K. & H.A.B. SIMONS 1969. Mpato Shelter: Sentinel Ranch, Limpopo River, Beitbridge, Rhodesia: excavation results. *Arnoldia (Rhod.)* 4(18): 1-9.

COOKE, C.K., R. SUMMERS & K.R. ROBINSON 1966. Rhodesian prehistory reexamined. *Arnoldia (Rhod.)* 2(12) 1-8.

COOKE, H.B.S. 1949. Fossil mammals of the Vaal River deposits. *Mem. Geol. Surv. S. Afr.* 35: 1.117.

COOKE, H.B.S. 1950. Quaternary fossils from Northern Rhodesia. In: J.D. Clark, *The Stone Age cultures of Northern Rhodesia*: 137-142. Claremont: South African Archaeological Society.

COOKE, H.B.S. 1962. Notes on the faunal material from the Cave of Hearths and Kalkbank. In: R.J. Mason, *Prehistory of the Transvaal*: 447-453. Johannesburg: University of the Witwatersrand Press.

COOKE, H.B.S. 1963. Pleistocene mammal faunas of Africa, with particular reference to southern Africa. In: F.C. Howell, & F. Bourliere (eds.), *African ecology and human evolution*: 65-116. Aldine: Chicago.

COOKE, H.B.S. 1967. The Pleistocene sequence in South Africa and problems of correlation. In: J.D. Clark & W.W. Bishop (eds.), *Background to evolution in Africa*: 175-184. Chicago: University of Chicago Press.

COOKE, H.B.S. 1970. Untitled communication. *News Bull. Soc. Vert. Paleontol.* 90: 2

COOKE, H.B.S. 1978a. Faunal evidence for the biotic setting of early African hominids. In: C. Jolly (ed.), *Early hominids of Africa*: 267-281. New York: St. Martin's Press.

COOKE, H.B.S. 1978b. Africa: the physical setting. In: V.J. Maglio & H.B.S. Cooke (eds.), *Evolution of African mammals*: 17-45. Cambridge (Mass.): Harvard University Press.

COOKE, H.B.S. (n.d.). The faunal remains from Witkrans 2. Unpublished manuscript.

COOKE, H.B.S., B.D. MALAN & L.H. WELLS 1945. Fossil man in the Lebombo Mountains, South Africa: the 'Border Cave', Ingwavuma District, Zululand. *Man* 45: 6-13.

COOKE, H.B.S. & L.H. WELLS 1951. Fossil remains from Chelmer, near Bulawayo, Southern Rhodesia. *S. Afr. J. Sci.* 47: 205-209.

COOKE, H.B.S. & A.F. WILKINSON 1978. Suidae and Tayassuidae. In: V.J. Maglio & H.B.S. Cooke (eds.), *Evolution of African mammals*: 435-482. Cambridge (Mass.): Harvard University Press.

COOKE, H.J. 1975. The palaeoclimatic significance of caves and adjacent landforms in western Ngamiland, Botswana, *Geog. Jour.* 141: 430-444.

COOKE, H.J. & B.T. VERHAGEN 1977. The dating of cave development – an example from Botswana. *Proc. 7th Intern. Speleolog. Cong.*, Sheffield, U.K.: 122-124.

COPPENS, Y. 1980. The differences between *Australopithecus* and *Homo*: prelliminary conclusions from the Omo research expedition's studies. In: L.-K. Konigsson (ed.), *Current argument on Early Man*: 207-225. Oxford: Pergamon Press.

COPPENS, Y., V.J. MAGLIO, C.T. MADDEN & M. BEDEN 1978. Proboscidea. In: V.J. Maglio & H.B.S. Cooke (eds.), *Evolution of African mammals*: 336-367. Cambridge (Mass.): Harvard University Press.

CORRUCINI, R.S. 1975. Morphometric assessment of australopithecine postcranial affinities. *Syst. Zool.* 24: 226-233.

CORRUCINI, R.S. & R.L. CIOCHON 1979. Primate facial allometry and interpretations of australopithecine variation. *Nature (Lond.)* 281: 62-64.

CORVINUS, G. & Q.B. HENDEY 1978. A new Miocene vertebrate locality at Arrisdrift in South West Africa. *N. Jb. Geol. Palaeont. Mb.* (Stuttgart) 1978(4): 193-205.

CORYNDON, S.C. 1978. Hippopotamidae. In: V.J. Maglio & H.B.S. Cooke (eds.), *Evolution of African mammals*: 483-495. Cambridge (Mass.): Harvard University Press.

CRAMB, G. 1961. A second report on work at the Holley Shelter. *S. Afr. J. Sci.* 57: 45-48.

CROSBY, E. 1977. An archaeologically oriented classification of ethnographic material culture. In: R.V.S Wright (ed.), *Stone tools as cultural markers*: 83-96. Canberra: Australian Institute of Aboriginal Studies.

DA CRUZ E SILVA, T.M. 1976. A preliminary report on an Early Iron Age site: Matola IV 1/68. In: T.M. Da Cruz e Silva *et al.*, *Iron Age Research in Mozambique. Collected Preliminary Reports.* Maputo: Universidade Eduardo Mondlane.

DANIEL, J.B. Mc. I. 1973. A geographical study of pre-Shakan Zululand. *S. Afr. geog. J.* 55: 23-31.

DART, R.A. 1925. *Australopithecus africanus*: the man-ape of South Africa. *Nature (Lond.)* 115: 195-199.

DART, R.A. 1957. The osteodontokeratic culture of *Australopithecus prometheus. Transv. Mus. Mem.* 10: 1-105.

DART, R.A. 1967. The antiquity of miniing in southern Africa. *S. Afr. J. Sci.* 63: 264-267.

DART, R.A. & P.B. BEAUMONT 1968. Ratification and retrocession of earlier Swaziland iron ore mining radiocarbon dates. *S. Afr. J. Sci.* 64: 241-246.

DART, R.A. & P.B. BEAUMONT 1969. Iron Age radiocarbon dates from Western Swaziland. *S. Afr. archaeol. Bull.* 24: 71.

DART, R.A. & P.B. BEAUMONT 1971. On a further radiocarbon date for ancient mining in southern Africa. *S. Afr. J. Sci.* 67: 10-11.

DART, R.A. & N. DEL GRANDE 1931. The ancient iron-smelting cavern at Mumbwa. *Trans R. Soc. S. Afr.* 19: 379-427.

DAVIES, O. 1971. Excavations at Blackburn. *S. Afr. archaeol. Bull.* 26: 165-178.

DAVIES, O. 1974. Excavations at the walled early Iron Age site in Moor Park near Estcourt, Natal. *Ann. Natal Mus.* 22: 289-323.

DAVIES, O. 1975. Excavations at Shongweni south cave. *Ann. Natal Mus.* 22: 627-662.

DAVISON, C.C. 1973. Chemical resemblance of garden roller and Ml glass beads. *Afr. Stud.* 32: 247-257.

DAY, M.H., M.D. LEAKEY & C. MAGORI 1980. A new hominid fossil skull (L.H. 18) from the Ngaloba Beds, Laetoli, northern Tanzania. *Nature (Lond.)* 284: 55-56.

DAY, M.H., M.D. LEAKEY & T.R. OLSON 1980. Comment on the status of *Australopithecus afarensis.* Science 207: 1102-1103.

DAY, M.H. & E.H. WICKENS 1980. Laetoli Pliocene hominid footprints and bipedalism. *Nature (Lond.)* 286: 385-387.

DEACON, H.J. 1966. Note on the X-ray of two mounted implements from South Africa. *Man* (N.S.) 1: 87-90.

Deacon, H.J. 1967. Two radiocarbon dates from Scott's Cave, Gamtoos Valley. *S. Afr. archaeol. Bull.* 22: 51-52.

Deacon, H.J. 1969. Melkhoutboom Cave, Alexandria District, Cape Province: a report on the 1967 occurrences investigation. *Ann. Cape Prov. Mus.* 6: 141-169.

Deacon, H.J. 1970a. Two shell midden occurrences in the Tsitsikama National Park, Cape Province: a contribution to the study of the ecology of the Strandlopers. *Koedoe* 13: 37-49.

Deacon, H.J. 1970b. The Acheulean occupation at Amanzi Springs, Uitenhage District, Cape Province. *Ann. Cape Prov. Mus. (Nat. Hist.)* 8: 89-189.

Deacon, H.J. 1972. A review of the post-Pleistocene in South Africa. *S. Afr. archaeol. Soc. Goodwin Ser.* 1: 26-45.

Deacon, H.J. 1975. Demography, subsistence and culture during the Acheulean in southern Africa. In: K.W. Butzer & G.Ll. Isaac (eds.), *After the australopithecines*: 543-569. The Hague: Mouton.

Deacon, H.J. 1976. *Where hunters gathered: a study of Holocene stone age people in the eastern Cape.* S. Afr. archaeol. Soc. monogr. Ser. 1: 1-231.

Deacon, H.J. 1978. The Langkloof archaeological project (Cango Valley). *Palaeoecol. Afr.* 10: 135-137.

Deacon, H.J. 1979. Excavations at Boomplaas Cave – a sequence through the Upper Pleistocene and Holocene in South Africa. *Wld. Archaeol.* 10: 241-257.

Deacon, H.J. (in press). The ecology of hunter-gatherers in South Africa. *Proc. IXe Congres de l'Union Internationale des Sciences Prè-Et Proto-historiques, Nice 1976.*

Deacon, H.J. & M. Brooker 1976. The Holocene and Upper Pleistocene sequence in the southern Cape. *Ann. S. Afr. Mus.* 71: 203-214.

Deacon, H.J. & J. Deacon 1963. Scott's Cave: a late Stone Age site in the Gamtoos Valley. *Ann. Cape Prov. Mus.* 3: 96-121.

Deacon, H.J. & J. Deacon 1980. The hafting, function and distribution of small convex scrapers with an example from Boomplaas Cave. *S. Afr. archaeol. Bull.* 35: 31-37.

Deacon, H.J., J. Deacon & M. Brooker 1976. Four painted stones from Boomplaas Cave, Oudtshoorn District. *S. Afr. archaeol. Bull.* 31: 141-145.

Deacon, H.J., J. Deacon, M. Brooker & M.L. Wilson 1978. The evidence for herding at Boomplaas Cave in the southern Cape, South Africa. *S. Afr. archaeol. Bull.* 33: 39-65.

Deacon, H.J. & R.G. Klein (n.d.). Report on the excavation of the Kangkara Cave and the Paardeberg Cave. Unpublished manuscript.

Deacon, J. 1966. An annotated list of radiocarbon dates for sub-Saharan Africa. *Ann. Cape Prov. Mus.* 5: 5-84.

Deacon, J. 1969. Re-excavation and description of the Wilton type-site, Albany District, eastern Cape. Unpublished M.A. thesis: University of Cape Town.

Deacon, J. 1972. Wilton: an assessement after 50 years. *S. Afr. archaeol. Bull.* 27: 10-45.

Deacon, J. 1974. Patterning in the radiocarbon dates for the Wilton/Smithfield complex in southern Africa. *S. Afr. archaeol. Bull.* 29: 3-18.

Deacon, J. 1976a. Report on stone artefacts from Duinefontein 2, Melkbosstrand, *S. Afr. archaeol. Bull.* 31: 21-25.

Deacon, J. 1976b. Review article: The Stone Age Archaeology of Southern Africa by C. G. Sampson. *S. Afr. Archaeol. Bull.* 31: 58-63.

Deacon, J. 1978. Changing patterns in the late Pleistocene/early Holocene prehistory of southern Africa, as seen from the Nelson Bay Cave stone artifact sequence. *Quatern. Res.* (N. Y.) 10: 84-111.

Deacon, J. 1979a. Guide to archaeological sites in the southern Cape. *Occ. Publ. Dept. Archaeol. Univ. Stellenbosch* 1: 1-149.

Deacon, J. 1979b The Howieson's Poort and related industries in southern Africa with special reference to the name site collection. Unpublished paper presented at a workshop on 'Towards a better understanding of the Upper Pleistocene in sub-Saharan Africa' organized by the Southern African Association of Archaeologists, Stellenbosch, June 1979.

De Bruiyn, H. 1972. Pans in the western Orange Free State. *Geol. Surv. S. Afr. Ann.* 9: 121-124.

De Graaff, G. 1960. A preliminary investigation of the mammalian microfauna in Pleistocene deposits of caves in the Transvaal System. *Palaeont. afr.* 7: 59-118.

De Graaff, G. 1971. Family Bathyergidae. In: J. Meester & H.W. Setzer (eds.), *The mammals of Africa: an identification manual*: Part 6.9. Washington, D.C.: Smithsonian Institution Press.

DELSON, E., N. ELDREDGE & I. TATTERSALL 1977. Reconstruction of hominid phylogeny: a testable framework based on cladistic analysis. *J. Hum. Evol.* 6: 263-278.

DE MARET, P. 1977. Sanga: new excavations, more data and some related problems. *J. Afr. Hist.* 18: 321-337.

DE MUIZON, C. & Q.B. HENDEY 1980. Late Tertiary seals of the South Atlantic Ocean. *Ann. S. Afr. Mus.* 82: 91-128.

DENBOW, J.R. 1973. Malowa Rock Shelter: archaeological report. *Occ. Pap. Malawi Dept. Antiq.* 14: 5-59.

DENBOW, J.R. 1979. *Cenchrus ciliarus*: an ecological indicator of Iron Age middens using aerial photography in eastern Botswana. *S. Afr. J. Sci.* 75: 405-408.

DENBOW, J.R. 1980. Early Iron Age remains from the Tsodilo Hills, northwestern Botswana. *S. Afr. J. Sci.* 76: 474-475.

DENBOW, J.R. 1981. Broadhurst: a 14th century A.D. expression of the Early Iron Age in south-eastern Botswana. *S. Afr. archaeol. Bull.* 36: 66-74.

DENBOW, J.R. (in press). The Toutswe Tradition: a study in socio-economic change. In: Botswana Society (ed.), *Settlement in Botswana.* London: Heinemann.

DENBOW, J.R. & A.C. CAMPBELL 1980. National Museum of Botwana–archaeological research program. *Nyame Akuma* 17.

DENNINGER, E. 1971. The use of paper chromatography to determine the age of albuminous binders and its application to rock paintings. *S. Afr. J. Sci. spec. Publ.* 2: 80-84.

DERRICOURT, R.M. 1973. Radiocarbon chronology of the Late Stone Age and Iron Age in South Africa. *S. Afr. J. Sci.* 69: 280-284.

DERRICOURT, R.M. 1977. *Prehistoric man in the Ciskei and Transkei.* Cape Town: Struik.

DE SENNA MARTINEZ, J.C. 1976. A preliminary report on two Early Iron Age pottery traditions from southern Mozambique coastal plain. In: Da Cruz e Silva *et al., Iron Age research in Mozambique. Collected Preliminary Reports.* Maputo: Universidade Eduardo Mondlane.

DE VILLIERS, H. 1973. Human skeletal remains from Border Cave, Ingwavuma District, KwaZulu, South Africa. *Ann. Transv. Mus.* 28: 229-256.

DE VILLIERS, H. 1974. Human skeletal remains from Cape St. Francis, Cape Province. *S. Afr. archaeol. Bull.* 29: 89-91.

DE VILLIERS, H. 1976a. A second adult human mandible from Border Cave, Ingwavuma District, KwaZulu, South Africa. *S. Afr. J. Sci.* 72: 212-215.

DE VILLIERS, H. 1976b. Reports on skeletal remains. In: D.W. Phillipson, *The prehistory of eastern Zambia.* Nairobi: British Institute in Eastern Africa.

DE VOS, A. 1975. Africa, the devastated continent? The Hague: W. Junk.

DINGLE, R.V. & J. ROGERS 1972. Effects of sea-level changes on the Pleistocene palaeoecology of the Agulhas Bank. *Palaeoecol. Afr.* 6: 55-58.

DORNAN, S. 1925. *Pygmies and Bushmen of the Kalahari.* London: Seeley, Service.

DORST, J. & P. DANDELOT 1970. *A field guide to the larger mammals of Africa.* London: Collins.

DOS SANTOS, J.R. & C.M.N. ERVEDOSA 1970. A estacao arqueologica de Benfica. *Ciencias Biol. Univ. Luanda* 1: 31-51.

DREYER, T.F. 1933. The archaeology of the Matjes River rock shelter. *Trans. R. Soc. S. Afr.* 21: 187-209.

DREYER, T.F. 1938. The archaeology of the Florisbad deposits. *Argeol. Navors. Nas. Mus.* (Bloemfontein) 1: 65:77.

DUNN, E.J. 1931. *The Bushman.* London: Griffin.

EBERT, J.I. 1978. Comparability between hunter-gatherer groups in the past and present: modernization versus explanation. *Botswana Notes Rec.* 10: 19-36.

ELDREDGE, N. & I. TATTERSALL 1975. Evolutionary models, phylogenetic reconstruction, and another look at hominid phylogeny. In: F.S. Szalay (ed.), *Approaches to Primate paleobiology. Contributions to Primatology* 5: 218-242. Basel: Karger.

ELLENBERGER, D.F. & S.C. MACGREGOR 1912. *The history of the Basuto: ancient and modern.* London: Caxton.

ELLERMAN, J.R., T.C. MORRISON-SCOTT & R.W. HAYMAN 1953. *Southern African mammals 1758 to 1951: a reclassification.* London: British Museum (Natural History).

ELOFF, J.F. 1969. Bushman Rock Shelter, eastern Transvaal: excavations 1967-8. *S. Afr. archaeol. Bull.* 24: 60.

ELOFF, J.F. & A. MEYER 1981. The Greefswald sites In: E.A. Voigt (ed.), *Guide to archaeological sites in the northern Transvaal.* Pretoria: Transvaal Museum.

ELPHICK, R. 1977. *Kraal and castle: Khoikhoi and the founding of white South Africa.* New Haven: Yale University Press.

ENGELBRECHT, J.A. 1936. *The Korana: an account of their customs and their history.* Cape Town: Maskew Miller.

EPSTEIN, H. 1937. Animal husbandry of the Hottentots. *Onderstepoort J. vet. Sci.* 9: 631-666.

EPSTEIN, H. 1971. *The origin of the domestic animals of Africa.* Vols. 1 & 2 New York: Africana Publishing Corp.

ERIKSSON, P.B. 1978. An investigation of Quaternary aeolian-lacustrine sediments in Namaqualand. *Palaeoecol. Afr.* 10-11: 41-46.

ESTES, R. 1977. Relationships of the South African fossil frog *Eoxenopides reuningi* (Anura, Pipidae). *Ann. S. Afr. Mus.* 73: 49-80.

EVERS, T.M. 1975. Recent Iron Age research in the eastern Transvaal, South Africa. *S. Afr. archaeol. Bull.* 30: 71-83.

EVERS, T.M. 1977. Plaston Early Iron Age site, White River District, eastern Transvaal, South Africa. *S. Afr. archaeol. Bull.* 32: 170-178.

EVERS, T.M. 1979. Salt and soapstone factories at Harmony, Letaba District, northeast Transvaal. *S. Afr. archaeol. Soc. Goodwin Ser.* 3: 94-107.

EVERS, T.M. 1980. Klingbeil Early Iron Age sites, Lydenburg, eastern Transvaal, South Africa. *S. Afr. archaeol. Bull.* 35: 46-57.

EVERS, T.M. 1981. The Iron Age in the eastern Transvaal, South Africa. In: E.A. Voigt, (ed.), *Guide to archaeological sites in the northern and eastern Transvaal.* Pretoria: Transvaal Museum.

EWER, R.G. 1954. The fossil carnivores of the Transvaal Caves. The Hyaenidae of Kromdraai. *Proc. Zool. Soc. Lond.* 124: 565-585.

EWER, R.F. 1955a. The fossil carnivores of the Transvaal Caves. The Hyaenidae, other than Lycyaena, of Swartkrans and Sterkfontein. *Proc. Zool. Soc. Lond.* 124: 815-837.

EWER, R.F. 1955b The fossil carnivores of the Transvaal Caves. The Lycaenas of Sterkfontein and Swartkrans, together with some general considerations of the Transvaal fossil hyaenids. *Proc. Zool. Soc. Lond.* 124: 839-857.

EWER, R.F. 1955c. The fossil carnivores of the Transvaal Caves: Machairodontinae. *Proc. Zool. Soc. Lond.* 125: 587-615.

EWER, R.F. 1956a. The fossil carnivores of the Transvaal Caves: Felinae. *Proc. Zool. Soc. Lond.* 126: 83-95.

EWER, R.F. 1956b. The fossil carnivores of the Transvaal Caves: Canidae. *Proc. Zool. Soc. Lond.* 126: 97-119.

EWER, R.F. 1956c. The fossil carnivores of the Transvaal Caves: two new viverrids, together with some general considerations. *Proc. Zool. Soc. Lond.* 126: 259-274.

EWER, R.F. 1956d. Some fossil carnivores from the Makapansgat Valley. *Palaeont. afr.* 4: 57-67.

EWER, R.F. 1973. *The Carnivores.* Ithaca (N.Y.): Cornell University Press.

EWER, R.F. & H.B.S. COOKE 1964. The Pleistocene mammals of South Africa. In: D.H.S. Davis (ed.), *Ecological studies in Southern Africa*: 35-48. The Hague: W. Junk.

FAGAN, B.M. 1960. The Glentyre Shelter and Oakhurst re-examined. *S. Afr. archaeol. Bull.* 15: 80-94.

FAGAN, B.M. 1964. The Greefswald sequence: Mapungubwe and Bambandyanalo. *J. Afr. Hist.* 5: 337-361.

FAGAN, B.M. 1965. *Southern Africa during the Iron Age.* London: Thames & Hudson.

FAGAN, B.M. & F. VAN NOTEN 1971. *The hunter-gatherers of Gwisho.* Tervuren: Musèe Royal de l'Afrique centrale.

FAIRHALL, A.W., A.W. YOUNG & J.L. ERICKSON 1976. University of Washington dates IV. *Radiocarbon* 18: 221-239.

FEELY, J.M. 1980. Did Iron Age man have a role in the history of Zululand's wilderness landscape? *S. Afr. J. Sci.* 76: 150-152.

FLETEMEYER, J.R. 1977. Age determination in the teeth of the Cape fur seal and its bearing on the seasonal mobility hypothesis proposed for the western Cape, South Africa. *S. Afr. archaeol. Bull.* 32: 146-149.

FLIGHT, C. 1980. Malcolm Guthrie and the reconstruction of Bantu Prehistory. *History in Africa* 7: 81-116.

FOCK, G.J. 1968. Rooidam, a sealed site of the First Intermediate. *S. Afr. J. Sci.* 64: 153-159.

FOCK, G.J. 1979. *Felsbilder in Südafrika Teil 1: Die gravierungen auf Klipfontein, Kaapprovinz*. Köln: Böhlau Verlag.

FOCK, G.J., K.W. BUTZER, L. SCOTT & R. STUCKENRATH 1980. Rock engravings and the Later Stone Age, northern Cape Province, South Africa: a multidisciplinary study. In: R.E.F. Leakey & B.A. Ogot (eds.), *Proceedings of the 8th Panafrican Congress of prehistory and Quaternary studies (Nairobi, 1977)*: 311-313. Nairobi: The International Louis Leakey Memorial Institute for African Prehistory.

FOLK, R.L. 1966. A review of grain-size parameters. *Sedimentology* (Amsterdam) 6: 73-93.

FOUCHE, L. 1937. *Mapungubwe: ancient Bantu civilization on the Limpopo*. Cambridge: University Press.

FOURIE, G.P. 1970. Die geologie van Florisbad. *Geol. Surv. S. Afr. Rep. D.* 144: 1-55.

FRANCA, J.C. 1964. Nota preliminar sobre uma gruta prè-historica do planalto da Humpata (Angola). *Mem. Junta Invest. Ultram. 2 Ser.* 50: 59-67.

FREEMAN, L.G. 1975. Acheulean sites and stratigraphy in Iberia and the Maghreb. In: K.W. Butzer & G. Ll. Issac (eds.), *After the australopithecines*. 661-744. The Hague: Mouton.

FREEMAN-GRENVILLE, G.S.P. 1962. *The East African coast: select documents*. Oxford: Oxford University Press.

FREEMAN-GRENVILLE, G.S.P. 1973. *Chronology of African history*. Oxford: Oxford University Press.

FREUNDLICH, J.C., H. SCHWABEDISSEN & W.E. WENDT 1980. Köln radiocarbon measurements II. *Radiocarbon* 22: 68-81.

FRIEDE, H. & J.N. PIENAAR 1974. Application of the froth flotation process to Kruger Cave deposit. *S. Afr. J. Sci.* 70: 375-376.

FRIEDE, H.M. & R.H. STEEL 1975. Notes on Iron Age copper-smelting technology in the Transvaal. *Jour. S. Afr. Inst. Min. Metall.* 75: 221-231.

FRIEDE, H.M. & R.H. STEEL 1976. The mining and smelting in the Transvaal during the Iron Age. *Jour. S. Afr. Inst. Min. Metall.* 76: 461-470.

FROBENIUS, L. 1931. *Madsimu Dsangara*. Berlin: Atlantis-Verlag.

FROBENIUS, L. & L.R. VON WILM (eds.) 1921. *Atlas Africanus* I. Munchen: Beck.

GABEL, C. 1963. Lochinvar mound: a Later Stone Age camp-site in the Kafue Basin. *S. Afr. archaeol. Bull.* 18: 40-48.

GALLAGHER, J.P. 1977. Contemporary stone tools in Ethiopia: implications for archaeology. *J. field Archaeol.* 4: 407-414.

GALLOWAY, A. 1959. The skeletal remains from Bambandyanalo. Johannesburg: Witwatersrand University.

GARDNER, G.A. 1963. *Mapungubwe Vol. 2*. Pretoria: Van Schaik.

GARLAKE, P.S. 1967. Excavations at Maxton Farm, near Shamva Hill, Rhodesia. *Arnoldia (Rhod.)* 3(9): 1-15.

GARLAKE, P.S. 1968. Test excavations at Mapela Hill, near the Shashi River, Rhodesia. *Arnoldia (Rhod.)* 3(34): 1-29.

GARLAKE, P.S. 1969. Chitope: an Early Iron Age village in northern Mashonaland. *Arnoldia (Rhod.)* 4(19): 1-14.

GARLAKE, P.S. 1970a. Iron Age sites in the Urungwe District of Rhodesia. *S. Afr. archaeol. Bull.* 25: 25-44.

GARLAKE, P.S. 1970b. Rhodesian ruins – a preliminary assessment of their styles and chronology. *J. Afr. Hist.* 11: 495-513.

GARLAKE, P.S. 1973. *Great Zimbabwe*. London: Chatto & Windus.

GARLAKE, P.S. 1979. Review of D.W. Phillipson, The Earlier prehistory of eastern and southern africa. *J. Afr. Hist.* 20: 457-459.

GATES, W.L. 1976. The numerical simulation of ice-age climate with a global general circulation model. *J. atmospher. Sci.* 33: 1844-1873.

GENTRY, A.W. 1971. Genus *Gazella*. In: J. Meester & H.W. Setzer (eds.), *The mammals of Africa: an identification manual*: Part 15.1 Washington, D.C.: Smithsonian Institution Press.

GENTRY, A.W. 1974. A new genus and species of Pliocene boselaphine (Bovidae, Mammalia) from South Africa. *Ann. S. Afr. Mus.* 65; 145-188.

GENTRY, A.W. 1978. Bovidae. In: V.J. Maglio & H.B.S. Cooke (eds.), Evolution of African mammals: 540-572. Cambridge (Mass.): Harvard University Press.

GENTRY, A.W. 1980. Fossil Bovidae (Mammalia) from Langebaanweg, South Africa. *Ann. S. Afr. Mus.* 79: 213-337.

GENTRY, A.W. & A. GENTRY 1978. Fossil Bovidae (Mammalia) of Olduvai Gorge, Tanzania. *Bull. Brit. Mus. (Nat. Hist.) (Geol. Ser.)* 29: 289-446 and 30: 1-83.

GOOCH, W.D. 1881. The Stone Age of South Africa. *J.R. Anthrop. Inst.* 11: 124-182.

GOODALL, E. 1962. Report on an ancient burial ground, Salisbury. In: G. Mortelmans & J. Nenquin (eds.), *Actes du IVe Congrès panafrican de préhistoire de l'ètude du Quaternaire* (Leopoldville, 1959). Part IV: 315-322. Tervuren: Musée Royale de l'Afrique Centrale.

GOODWIN, A.J.H. 1926. South African stone implement industries. *S. Afr. J. Sci.* 23: 784-788.

GOODWIN, A.J.H. 1928. An introduction to the Middle Stone Age in South Africa. *S. Afr. J. Sci.* 25: 410-418.

GOODWIN, A.J.H. 1929a. The Middle Stone Age. *Ann. S. Afr. Mus.* 29: 95-145.

GOODWIN, A.J.H. 1929b. The Montagu Cave: a full report of the investigation of the Montagu Rock-shelter. *Ann. S. Afr. Mus.* 24: 1-16.

GOODWIN, A.J.H. 1930. A new variation of the Smithfield Culture from Natal. *Trans. R. Soc. S. Afr.* 19: 7-14.

GOODWIN, A.J.H. 1934. Some developments in technique during the Earlier Stone Age. *Trans. R. Soc. S. Afr.* 21: 109-123.

GOODWIN, A.J.H. 1938. Archaeology of the Oakhurst Shelter, George. *Trans. R. Soc. S. Afr.* 25: 229-324.

GOODWIN, A.J.H. 1946a. *The loom of prehistory*. Claremont: South African Archaeological Society Handbook 2.

GOODWIN, A.J.H. 1946b. Prehistoric fishing methods in South africa. *Antiquity* 20: 134-141.

GOODWIN, A.J.H. 1947. The bored stones of South Africa. *Ann. S. Afr. Mus.* 37: 1-210.

GOODWIN, A.J.H. 1953a. *Method in prehistory*. Claremont: South African Archaeological Society Handbook 1.

GOODWIN, A.J.H. 1953b. Two caves at Kalk Bay, Cape Peninsula. *S. Afr. archaeol. Bull.* 8: 59-77.

GOODWIN, A.J.H. & B.D. MALAN 1935. Archaeology of the Cape St. Blaize Cave and raised beach, Mossel Bay. *Ann. S. Afr. Mus.* 24: 111-140.

GOODWIN, A.J.H. & C. VAN RIET LOWE 1929. The Stone Age cultures of South Africa. *Ann. S. Afr. Mus.* 27: 1-289.

GOWLETT, J.A.J. 1978. Kilombe - an Acheulean site complex in Kenya. In: W.W. Bishop (ed.), *Geological background to fossil man*: 337-360. Edinburgh: Scottish Academic Press.

GREENMAN, L. 1966. The geology of Area 2615C, Luderitz, South West Africa. Unpublished thesis: University of Cape Town.

GREY, D.R. & H.J. COOKE 1977. Some problems in the Quaternary evolution of the landforms of northern Botswana. *Catena* (Braunschweig) 4: 123-135.

GRINDLEY, J.R., E. SPEED & T. MAGGS 1970. The age of the Bonteberg Shelter deposits, Cape Peninsula. *S. Afr. archaeol. Bull.* 25: 24.

GRINE, F.E. 1981. Trophic differences between 'gracile' and 'robust' australopithecines: a scanning electron microscope analysis of occlusal events. *S. Afr. J. Sci.* 77: 203-230.

GROBELAAR, C.S. & A.J.H, GOODWIN 1952. Report on the skeletons and implements in association with them from a cave near Bredasdorp, Cape Province. *S. Afr. archaeol. Bull.* 7: 95-107.

HAGERSTRAND, T. 1967. *Innovation diffusion as a spatial process.* Chicago: University of Chicago Press.

HALL, M.J. 1980a. Enkwazini, an Iron Age site on the Zululand coast. *Ann. Natal Mus.* 24: 97-109.

HALL, M.J. 1980b The ecology of the Iron Age in Zululand. Unpublished Ph.D. thesis: Department of Archaeology, University of Cambridge.

HALL, M.J. 1980c. An iron-smelting site in the Hluhluwe Game Reserve, Zululand. *Ann. Natal Mus.* 24: 165-175.

HALL, M.J. & T. MAGGS 1979. Ngabeni, a Late Iron Age site in Zululand. *S. Afr. archaeol. Soc. Goodwin Ser.* 3: 159-175.

HALL, M.J. & J.C. VOGEL 1980. Some recent radiocarbon dates from Southern Africa. *J. Afr. Hist.* 21: 431-455.

HALL, S.L. 1981. Iron Age sequence and settlement in the Rooiberg, Thabazimbi area. Unpublished M.A. thesis: Department of Archaeology, University of the Witwatersrand.

HAMILTON, W.R. 1978. Cervidae and Palaeomerycidae. In: V.J. Maglio & H.B.S. Cooke (eds.), *Evolution of African mammals:* 496-508. Cambridge (Mass.): Harvard University Press.

HAMILTON, W.R. & J.A. VAN COUVERING 1977. Lower Miocene mammals of South West Africa. *Namib. Bull (Suppl. 2, Transv. Mus. Bull.)*: 9-11.

HANISCH, E.O.M. 1981a. The northern Transvaal: environment and archaeology. In: E.A. Voigt (ed.), *Guide to archaeological sites in the northern and eastern Transvaal.* Pretoria: Transvaal Museum.

HANISCH, E.O.M. 1981b. Schroda: a Zhizo site in the northern Transvaal. In: E.A. Voigt (ed.), *Guide to archaeological sites in the northern and eastern Transvaal.* Pretoria: Transvaal Museum.

HARE, E.P. 1980. Organic geochemistry of bone and its relation to the survival of bone in the natural environment. In: A.K. Behrensmeyer & A.P. Hill (eds.), *Fossils in the making*: 208-219. Chicago: University of Chicago Press.

HARINCK, G. 1969. Interaction between Xhosa and Khoi: emphasis on the period 1620-1750. In: L. Thompson (ed.), *African societies in southern Africa*: 145-170 London: Heinemann.

HARRIS, J.M. 1976. Pliocene Giraffoidea (Mammalia, Artiodactyla) from the Cape Province. *Ann. S. Afr. Mus.* 69: 325-353.

HARRIS, J.M. 1977. Deinotheres from southern Africa. *S. Afr. J. Sci.* 73: 281-282.

HARRIS, J.M. & T.D. WHITE 1979. Evolution of the Plio-Pleistocene African Suidae. *Trans. Am. Phil. Soc.* 69: 1-128.

HAUGHTON, S.H. 1932. The fossil Equidae of South Africa. *Ann. S Afr. Mus.* 28: 407-427.

HAUSMAN, A.J. 1980. Holocene human evolution in southern Africa: the bio-cultural development of the Khoisan. Unpublished Ph.D. thesis: State University of New York At Binghamton.

HAUSMAN, A.J. 1983.. Holocene human evolution in the Cape Province, South Africa. In: J.D. Clark & S. Brandt (eds.), *From hunters to farmers: considerations of the causes and consequences of food production in Africa* (in press). Berkeley: University of California Press.

HAY, R.L. 1976. *Geology of Olduvai Gorge.* Berkeley: University of California Press.

HAYDEN, B. (ed.) 1979. *Lithic use-wear analysis.* New York: Academic Press.

HEINE, K. 1978a. Jungquartäre pluviale und interpluviale in der Kalahari (Südliches Afrika). *Palaeoecol. Afr.* 10: 31-39.

HEINE, K. 1978b. Radiocarbon chronology of late Quaternary lakes in the Kalahari, southern Africa. *Catena* (Braunschweig) 5: 145-149.

HEINE, K. 1979. Reply to Cooke's discussion of K. Heine. *Catena (Braunschweig)* 6: 259-266.

HEINE, K. 1981. Aride und pluviale Bedingungen während der letzen Kaltzeit in der Südwest-Kalahari (südliches Afrika). Zeit. *Geomorph. (Berlin) Suppl.* Vol. 38: 1-37.

HEINZ, H.J. & B. MAGUIRE 1974. *The ethno-biology of the !ko Bushmen.* Gaberone: The Botswana Society Occ. Pap. 1.

HESSIG, K. 1971. *Brachypotherium* aus dem Miozän von Südwestafrikas. *Mitt. bayer. Staatssamml. Paläont. hist. Geol.* 11: 125-128.

HELGREN, D.M. 1977. Geological context of the Vaal River faunas. *S. Afr. J. Sci.* 73: 303-307.

HELGREN, D.M. 1978. Acheulean settlement along the lower Vaal River, South Africa. *J. archaeol. Sci.* 5: 39-60.

HELGREN, D.M. 1979a. Rivers of diamonds: an alluvial history of the Lower Vaal Basin, South Africa. *Univ. Chicago Dept. Geog. Res. Pap.* 185: 1-389.

HELGREN, D.M. 1979b. River channelways of the Middle Orange River. *S. Afr. J. Sci.* 75: 462-463.

HELGREN, D.M. & A.S. BROOKS 1983. Geoarchaeology at Gi, a Middle Stone Age and Later Stone Age site in the northwest Kalahari. *J. archaeol. Sci.* 10: 181-197.

HELGREN, D.M. & K.W. BUTZER 1977. Paleosols of the southern Cape coast, South Africa. *Geog. Rev.* 67: 430-445.

HENDEY, Q.B. 1973. Carnivore remains from the Kromdraai australopithecine site (Mammalia: Carnivora). *Ann. Transv. Mus.* 28: 99-112.

HENDEY, Q.B. 1974a. The Late Cenozoic Carnivora of the South-Western Cape Province. *Ann. S. Afr. Mus.* 63: 1-369.

HENDEY, Q.B. 1974b. Faunal dating of the Late Cenozoic of southern Africa, with special reference to the Carnivora. *Quatern. Res.* (N.Y.) 4: 149-161.

HENDEY, Q.B. 1974c. New fossil carnivores from the Swartkrans australopithecine site (Mammalia: Carnivora). *Ann. Transv. Mus.* 29: 27-49.

HENDEY, Q.B. 1976a. The Pliocene fossil occurrences in 'E' Quarry, Langebaanweg, South Africa. *Ann. S. Afr. Mus.* 69: 215-247.

HENDEY, Q.B. 1976b. Fossil peccary from the Pliocene of South Africa. *Science* 192: 787-789.

HENDEY, Q.B. 1978a. Late Tertiary Hyaenidae from Langebaanweg, South Africa, and their relevance to the phylogeny of the family. *Ann. S. Afr. Mus.* 76: 265-297.

HENDEY, Q.B. 1978b. The age of the fossils from Baard's Quarry, Langebaanweg, South Africa. *Ann. S. Afr. Mus.* 75: 1-24.

HENDEY, Q.B. 1978c. Preliminary report on the Miocene vertebrates from Arrisdrift, South West Africa. *Ann. S. Afr. Mus.* 76: 1-41.

HENDEY, Q.B. 1978d. Late Tertiary Mustelidae (Mammalia, Carnivora) from Langebaanweg, South Africa. *Ann. S. Afr. Mus.* 76: 329-357.

HENDEY, Q.B. 1980a. *Agriotherium* (Mammalia, Ursidae) from Langebaanweg, South Africa, and relationships of the genus. *Ann. S. Afr. Mus.* 81: 1-109.

HENDEY, Q.B. 1980b. Origin of the giant panda. *S. Afr. J. Sci.* 76: 179-180.

HENDEY, Q.B. 1981a. Palaeoecology of the late Tertiary fossil occurrences in 'E' Quarry, Langebaanweg, South Africa, and a reinterpretation of their geological context. *Ann. S. Afr. Mus.* 84: 1-104.

HENDEY, Q.B. 1981b. Geological succession at Langebaanweg, Cape Province, and global envents of the late Tertiary. *S. Afr. J. Sci.* 77: 33-38.

HENDEY, Q.B. & H. HENDEY 1968. New Quaternary fossil sites near Swartklip, Cape Province. *Ann. S. Afr. Mus.* 52: 43-73.

HENDEY, Q.B. & R. SINGER 1965. The faunal assemblages from the Gamtoos Valley Shelters. *S. Afr. archaeol. Bull.* 20: 206-213.

HEWITT, J. 1921. On several implements and ornaments from Strandloper sites in the eastern Province. *S. Afr. J. Sci.* 18: 454-467.

HEWITT, J. 1931a. Artefacts from Melkhoutboom *S. Afr. J. Sci.* 28: 540-548.

HEWITT, J. 1931b. Discoveries in a Bushman Cave at Tafelberg Hall *Trans. R. Soc. S. Afr.* 19: 185-196.

HILL, W.C. OSMAN & J. MEESTER 1971. Infraorder Lemuriformes. In: J. Messter & H.W. Setzer (eds.), *The mammals of Africa: an identification manual:* Part 3.2. Washington, D.C.: Smithsonian Institution Press.

HITZEROTH, L. 1973. Results of a Stone Age excavation at Duncombe Farm, Concession. *Rhod. Prehist.* 10: 14-18.

HODDER, I. 1977. The distribution of material culture items in the Baringo district, western Kenya. *Man* (N.S.) 12: 239-269.

HODDER, I. 1980. Interpretation of spatial variation in material culture: some suggestions and hypotheses. In: R.E.F. Leakey & B.A. Ogot (eds.), *Proceedings of the 8th Panafrican congress of prehistory and Quaternary studies* (Nairobi, 1977): 29-30 Nairobi: The International Louis Leakey Memorial Institute for African Prehistory.

HOERNLE, W. 1925. The social organization of the Namaqua Hottentots of Southwest Africa. *Amer. Anthrop.* 27: 1-24.

HOFFMAN, A.C. 1955. Some important contributions of the Orange Free State to our knowledge of Primitive Man. *S. Afr. J. Sci.* 51: 163-168.

HOLM, E. 1961. The rock art of South Africa. In: H.G. Bandi (ed.), *The art of the Stone Age.* London: Methuen.

HOOIJER, D.A. 1972. A late Pliocene rhinoceros from Langebaanweg, Cape Province. *Ann. S. Afr. Mus.* 59 : 151-191.

HOOIJER, D.A. 1976. The late Pliocene Equidae of Langebaanweg, Cape Province, South Africa. *Zool. Verh.* (Leiden) 148: 1-39.

HOOIJER, D.A. 1978. Rhinocerotidae. In: V.J. Maglio & H.B.S. Cooke (eds.), *Evolution of African mammals*: 371-378. Cambridge (Mass.): Harvard University Press.

HOPWOOD, A.T. 1928. New and little known mammals from the Miocene of Africa. *Am. Mus. Novit.* 344: 1-9.

HOROWITZ, A. 1975. Preliminary palaeoenvironmental implications of pollen analysis of Middle Breccia from Sterkfontein. *Nature (Lond.)* 258: 417-418.

HOROWITZ, A., C.G. SAMPSON, L. SCOTT & J.C. VOGEL. 1978. Analysis of the Voigtspost site, O.F.S., South Africa. *S. Afr. archaeol. Bull.* 33: 152-159.

HORWITZ, L. 1979. From materialism to middens: a case study at Elandsbay, western Cape, South Africa. Unpublished B.A. (Hons.) thesis: Department of Archaeology, University of Cape Town.

HOWELL, F.C. 1978a. Hominidae. In: V.J. Maglio & H.B.S. Cooke (eds.), *Evolution of African mammals*: 154-248. Cambridge (Mass.): Harvard University Press.

HOWELL, F.C. 1978b. Overview of the Pliocene and earlier Pleistocene of the lower Omo Basin, southern Ethiopia. In: C. Jolly (ed.), *Early Hominids of Africa*: 85-130. New York: St. Martin's Press.

HOWELL, F.C., S.L. WASHBURN & R.L. CIOCHON 1978. Relationships of *Australopithecus* and *Homo*. *J. Hum. Evol.* 7: 127-131.

HOWELLS, W.W. 1973. Cranial Variation in Man. *Pap. Peabody Mus.* 67. Cambridge (Mass.): Harvard University.

HOWELLS, W.W. 1976. Explaining modern man: evolutionists versus migrationists. *J. Hum. Evol.* 5: 477-495.

HOWELLS, W.W. 1980. *Homo erectus*-who, when and where: a survey. *Yearbook Phys. Anthrop.*: 1-23.

HSU, K.J. *et al.* 1977. History of the Mediterranean salinity crisis. *Nature (Lond.)* 267: 399-403.

HUBLIN, J.J. 1978. Quelques caractères apomorphes due crâne néandertalien et leur interpretation phylogènique. *C.R. Acad. Sci. Ser. D* 287: 923-926.

HUFFMAN, T.N. 1971. A guide to the Iron Age of Mashonaland. *Occ. Pap. natn. Mus. S. Rhod.* 4: 20-44.

HUFFMAN, T.N. 1972a. The Early Iron Age in the Urungwe District, Rhodesia. *Rhod. Prehist.* 8: 12-16.

HUFFMAN, T.N. 1972b. The rise and fall of Great Zimbabwe. *J. Afr. Hist.* 13: 353-366.

HUFFMAN,T.N. 1973. Test excavation at Makuru, Rhodesia. *Arnoldia (Rhod.)* 5(39): 1-21.

HUFFMAN, T.N. 1974a. The Leopard's Kopje tradition. *Mem. natl. Mus. Rhod.* 6.

HUFFMAN, T.N. 1974b. The linguistic affinities of the Iron Age in Rhodesia. *Arnoldia (Rhod.)* 7(7): 1-12.

HUFFMAN, T.N. 1974c. Ancient mining and Zimbabwe. *Jour. S. Afr. Inst. Min. Metall.* 74: 238-242.

HUFFMAN, T.N. 1977. Zimbabwe: southern Africa's first town *Rhod. Prehist.* 7(15): 9-14.

HUFFMAN, T.N. 1978a. The origins of Leopard's Kopje: an 11th century Difaquane. *Arnoldia (Rhod.)* 8(23): 1-23.

HUFFMAN, T.N. 1978b. The Iron Age of the Buhwa District, Rhodesia. *Occ. Pap. natn. Mus. Rhod.* 4(3): 81-100.

HUFFMAN, T.N. 1979a. African origins. *S. Afr. J. Sci.* 75: 233-237.

HUFFMAN, T.N. 1979b. Test excavations at NaBa and Lanlory, northern Mashonaland. *S. Afr. archaeol. Soc. Goodwin Ser.* 3: 14-46.

HUGHES, A.G. & P.V. TOBIAS 1977. A fossil skull probably of the genus *Homo* from Sterkfontein, Transvaal. *Nature (Land.)* 265: 310-312.

HUMPHREYS, A.J.B. 1970. The remains from Koffiefontein burials excavated by W. Fowler and preserved in the McGregor Museum, Kimberley. *S. Afr. archaeol. Bull.* 25: 104-115.

HUMPHREYS, A.J.B. 1972a. The Type R settlements in the context of the later prehistory and early history of the Riet River valley. Unpublished M.A. thesis: University of Cape Town.

HUMPHREYS, A.J.B. 1972b. Comments on aspects of raw material usage in the Later Stone Age of the Middle Orange River area. *S. Afr. archaeol. Soc. Goodwin Ser.* 1: 46-53.

HUMPHREYS, A.J.B. 1973. Report on some collections of Middle Stone Age artefacts from Riverton, Kimberley District, South Africa. *Ann. Cape Prov. Mus.* 9: 177-185.

HUMPHREYS, A.J.B. 1974a. A preliminary report on test excavations at Dikbosch Shelter I, Herbert District, northern Cape. *S. Afr. archaeol. Bull.* 29: 115-119.

HUMPHREYS, A.J.B. 1974b. Note on a date for a burial from the Riet River. *S. Afr. J. Sci.* 70: 271.

HUMPHREYS, A.J.B. 1979. The Holocene sequence in the northern Cape and its position in the prehistory of South Africa. Unpublished Ph.D. thesis: University of Cape Town.

HUMPHREYS, A.J.B. 1980. Comment on 'Time and place' by J.E. Parkington. *S. Afr. archaeol. Bull.* 35: 93-94.

HUMPHREYS, A.J.B. & T.M. O'C MAGGS 1970. Further graves and cultural material from the banks of the Riet River. *S. Afr. archaeol. Bull.* 25: 116-126.

INSKEEP, R.R. 1962. The age of the Kondoa rock paintings in the light of recent excavations at Kisese II rock shelter. In: G. Mortelmans & J. Nenquin (eds.), *Actes du IVe Congrès panafricain de préhistoire et de l'ètude du Quaternaire* (Leopoldville, 1959). Part III: 249-256. Tervuren: Musée Royal de l'Afrique Centrale.

INSKEEP, R.R. 1965. University of Cape Town excavations at Plettenberg Bay. *Scient. S. Afr.* 2: 575-577.

INSKEEP, R.R. 1967. The Late Stone Age. In: W.W. Bishop & J.D. Clark (eds.), *Background to evolution in Africa:* 557-582. Chicago: University of Chicago Press.

INSKEEP, R.R. 1969. The archaeological background. In: M. Wilson & L. Thompson (eds.), *The Oxford history of South Africa I: South Africa to 1870*: 1-39. Oxford: Oxford University Press.

INSKEEP, R.R. 1972. Nelson's Bay Cave, Robberg Peninsula Plettenberg Bay. *Palaeoecol. Afr.* 6: 247-249.

INSKEEP, R.R. 1978a. *The peopling of southern Africa.* Cape Town: David Philip.

INSKEEP, R.R. 1978b. The Bushmen in prehistory. In: P.V. Tobias (ed.), *The Bushmen: San hunters and herders of southern Africa*: 33-56. Cape Town: Human & Rousseau.

INSKEEP, R.R. & T.M. O'C MAGGS 1975. Unique art objects in the Iron Age of the Transvaal. *S. Afr. archaeol. Bull.* 30: 114-138.

ISAAC, G. Ll. 1967. The stratigraphy of the Peninj Group – early Middle Pleistocene formations west of Lake Natron, Tanzania. In: W.W. Bishop & J.D. Clark (eds.), *Background to evolution in Africa:* 229-257. Chicago: University of Chicago Press.

ISAAC, G. Ll. 1969. Studies of early culture in East Africa. *Wld. Archaeol.* 1: 1-28.

ISAAC, G. Ll. 1972. Early phases of human behavior: models in Lower Palaeolithic archaeology. In: D.L. Clarke (ed.), *Models in archaeology*: 167-199. London: Methuen.

ISAAC, G. Ll. 1975. Stratigraphy and cultural patterns in East Africa during the middle ranges of Pleistocene time. In: K.W. Butzer & G. Ll. Isaac (eds.), *After the australopithecines*: 495-542. The Hague: Mouton.

ISAAC, G. Ll. 1977. *Olorgesailie: archaeological studies of a Middle Pleistocene lake basin in Kenya.* Chicago: University of Chicago Press.

ISAAC, G. Ll. 1978. The food-sharing behavior of protohuman hominids. *Sci. Am.* 238: 90-108.

ISAAC, G. Ll. & G.H. CURTIS 1974. Age of early Acheulean industries from the Peninj Group, Tanzania. *Nature (Lond.)* 249: 624-627.

JACOBSON, L. 1975. The gemsbok creation myth and Brandberg rock art. *S. Afr. J. Sci.* 71: 314.

JACOBSON, L. 1976. A critical reivew of the Damaraland Cutlure. *Cimbebasia* B2: 203-208.

JACOBSON, L. 1978. A study of functional variability in the Later Stone Age of western Damaraland, Namibia. Unpublished B.A. (Hons.) thesis: University of Cape Town.

JACOBSON, L. 1980. The final stone-tool-using peoples of western Damaraland. In: R.E.F. Leakey & B.A. Ogot (eds.), *Proceedings of the 8th Panafrican Congress of prehistory and Quaternary studies* (Nairobi, 1977): 310. Nairobi: The International Louis Leakey Memorial Institute for African Prehistory.

JACOBSON, L. & J.C. VOGEL 1975. Recent radiocarbon dates from the Brandberg. *S. Afr. J. Sci.* 71: 349.

JAEGER, J.-J. & H.B. WESSELMAN 1976. Fossil remains of micromammals from the Omo Group deposits. In: Y. Coppens, F.C. Howell, G. Ll. Isaac & R.E.F. Leakey (eds.), *Earliest man and environments in the Lake Rudolf Basin*: 351-360. Chicago: University of Chicago Press.

JANOSSY, D. 1975. Mid-Pleistocene microfaunas of continental Europe and adjoining areas. In: K.W. Butzer & G. Ll. Isaac (eds.), *After the australopithecines*: 375-398. The Hague: Mouton.

JARVIS, J.V.M, 1969. The breeding season and litter size of African mole rats. *J. Reproduct. Fertil. Suppl.* 6: 237-248.

JOHANSON, D.C., M. TAIEB, Y. COPPENS & H. ROCHE 1978. Expèdition international de l'Afar, Ethiopie (4e et 5e campagnes 1975-1977): nouvelles dècouvertes d'hominidés et découvertes d'industries lithiques pliocénes á Hadar. *C.R. Acad. Sci. Paris Ser. D:* 287: 237-240.

JOHANSON, D.C. & T.D. WHITE 1979. A systematic assessment of early African hominids. *Science* 203: 321-330.

JOHANSON, D.C., T.D. WHITE & Y. COPPENS 1978. A new species of the genus *Australopithecus* (Primates: Hominidae) from the Pliocene of eastern Africa. *Kirtlandia* 28: 1-14.

JOHNSON, T., H. RABINOWITZ & P. SIEFF 1959. *Rock paintings of the south-west Cape.* London: Longmans & Green.

JOLLY, K. 1947. Preliminary note on new excavations at Skildergat, Fish Hoek. S. Afr. archaeol. Bull. 2: 11-12.

JOLLY, K. 1948. The development of the Cape Middle Stone Age in the Skildegat Cave, Fish Hoek. *S. Afr. archaeol. Bull.* 3: 106-107.

JONES, N. 1926. *The Stone Age in Rhodesia.* Oxford: Oxford University Press.

JONES, N. 1933. Excavations at Nswatugi and Madiliyangwa, and notes on new sites located and examined in the Matopo Hills, Southern Rhodesia, 1932. *Occ. Pap. Rhod. Mus.* 1(2): 1-44.

JONES, 1940. Bambata Cave: a reorientation *Occ.Pap. natn. Mus. S. Rhod.* 1(9): 11-28.

JONES, N. 1949. *The prehistory of Southern Rhodesia.* Cambridge: Cambridge University Press.

JONES, P. 1978. An approach to stone settlement typology of the Late Iron Age: stone walling on the Klip River. *Afr. Stud.* 37: 83-97.

JONES, P.R. 1979. Effects of raw materials on biface manufacture. *Science* 204: 835-836.

JONES, P.R. 1980. Experimental butchery with modern stone tools and its relevance for Palaeolithic archaeology. *Wld. Archaeol.* 12: 153-175.

JOUBERT, E. & P.K.N. MOSTERT 1975. Distribution patterns and status of some mammals in South West Africa. *Madoqua* 9: 5-44.

KANNEMEYER, D.R. 1890. Stone implements: with a description of Bushman stone implements and relics: their names, uses, mode of manufacture and occurrence. *Cape Illus. Mag.* 1: 120-130.

KATANEKWA, N.M. 1978. Some Early Iron Age sites from the Machili Valley of south-western Zambia. *Azania* 13: 135-166.

KAUFULU, Z. E.S. VRBA & T.D. WHITE 1981. Age of the Chiwondo Beds, northern Malawi. *Ann. Transv. Mus.* 33: 1-7.

KEELEY, L.H. & N. TOTH 1981. Microwear polishes on early stone tools from Koobi Fora, Kenya. *Nature (Lond.)* 293: 464-465.

KEITH, A. 1931. *New discoveries relating to the antiquity of man.* New York: W.W Norton.

KELLER, C.M. 1969. Mossel Bay: a redescription. *S. Afr. archaeol. Bull.* 23: 131-140.

KELLER, C.M. 1970a. Montagu Cave: a preliminary report. *Quaternaria* 13: 187-203.

KELLER, C.M. 1970b. ^{14}C dates – Montagu Cave. *S. Afr. archaeol. Bull.* 25: 47.

KELLER, C.M. 1973. Montagu Cave in prehistory. *Univ. Calif. Anthrop. Rec.* 28: 1-150.

KENDALL, R.L. 1966. An ecological history of the Lake Victoria basin. *Ecol. Monogr.* 39: 121-176.

KENNETT, J.P. 1980. Paleoceanographic and biogeographic evolution of the Southern Ocean during the Cenozoic, and Cenozoic microfossil datums. *Palaeogeogr., Palaeoclimatol., Palaeoecol.* 31: 123-152.

KERR, R.A. 1980. Changing global sea levels as a geologic index. *Science* 209: 483-486.

KLAPWIJK, M. 1973. An Early Iron Age site near Tzaneen, N.E. Transvaal. *S. Afr. J. Sci.* 69: 324.

KLAPWIJK, M. 1974. A preliminary report on pottery from the north-eastern Transvaal, South Africa. *S. Afr. archaeol. Bull.* 29: 19-23.

KLEIN, R.G. 1972a. Preliminary report on the July through September 1970 excavations at Nelson Bay Cave, Plettenberg Bay (Cape Province, South Africa). *Palaeoecol. Afr.* 6: 177-208.

KLEIN, R.G. 1972b. The late Quaternary mammalian fauna of Nelson Bay Cave (Cape Province, South Africa): its implications for megafaunal extinctions and environmental and cultural change. *Quatern. Res.* (N.Y.) 2: 135-142.

KLEIN, R.G. 1973. Geological antiquity of Rhodesian Man. *Nature (Lond.)* 244: 311-312.

KLEIN, R.G. 1974. Environment and subsistence of prehistoric man in the southern Cape Province, South Africa. *Wld. Archaeol.* 5: 249-284.

KLEIN, R.G. 1975a. Middle Stone Age man-animal relationships in southern Africa: evidence from Die Kelders and Klasies River Mouth. *Science* 190: 265-267.

KLEIN, R.G. 1975b. Paleoanthropological implications of the non-archaeological bone assemblage from Swartklip I, South-Western Cape Province, South Africa. *Quatern. Res.* (N.Y.) 5: 275-288.

KLEIN, R.G. 1975c. Ecology of Stone Age man at the southern tip of Africa. *Archaeology* 28: 238-247.

KLEIN, R.G. 1976a. The mammalian fauna of the Klasies River Mouth sites, South Africa. *S. Afr. archaeol. Bull.* 31: 75-98.

KLEIN, R.G. 1976b. A preliminary report on the 'Middle Stone age' open-air site of Duinefontein 2 (Melkbosstrand, south-western Cape Province, South Africa). *S. Afr. archaeol. Bull.* 31: 12-20.

KLEIN, R.G. 1977a. The ecology of early man in southern Africa. *Science* 197: 115-126.

KLEIN, R.G. 1977b. The mammalian fauna from the Midde and Later Stone Age (later Pleistocene) levels of Border Cave, Natal Province, South Africa. *S. Afr. archaeol. Bull.* 32: 14-27.

KLEIN, R.G. 1978a. Preliminary analysis of the mammalian fauna from the Redcliff Stone Age cave site, Rhodesia. *Occ. Pap. natn. Mus. Rhod. A* 4: 74-80.

KLEIN, R.G. 1978b. A preliminary report on the larger mammals from the Boomplaas Stone Age cave site, Cango Valley, Oudtshoorn District, South Africa. *S. Afr. archaeol. Bull.* 33: 66-75.

KLEIN, R.G. 1978c. The fauna and overall interpretation of the 'Cutting 10' Acheulean site at Elandsfontein (Hopefield), southwestern Cape Province, South Africa. *Quatern. Res.* (N.Y.) 10: 69-83.

KLEIN, R.G. 1978d. Stone Age predation on large African bovids. *J. archaeol. Sci.* 5: 195-217.

KLEIN, R.G. 1979a. Stone age exploitation of animals in southern Africa. *Am. Scient.* 67: 151-160.

KLEIN, R.G. 1979b. Palaeoenvironmental and cultural implications of late Holocene archaeological faunas from the Orange Free State and north-central Cape Province, South Africa. *S. Afr. archaeol. Bull.* 34: 34-49.

KLEIN, R.G. 1980. Environmental and ecological implications of large mammals from Upper Pleistocene and Holocene sites in southern Africa. *Ann. S. Afr. Mus.* 81: 223-283.

KLEIN, R.G. 1981a. Later Stone Age subsistence at Byneskranskop Cave, South Africa. In: R.S.O.

Harding & G. Teleki (eds.), *Omnivorous Primates: gathering and hunting in human evolution*: 166-190. New York: Columbia University Press.

KLEIN, R.G. 1981b. Ungulate mortality and sedimentary facies in the late Tertiary Varswater Formation, Langebaanweg, south-western Cape Province, South Africa. *Ann. S. Afr. Mus.* 84: 233-254.

KLEIN, R.G. 1981c. Stone Age predation on small African bovids. *S. Afr. archaeol. Bull.* 36: 55-65.

KLEIN, R.G. 1982a. Age (mortality) profiles as a means of distinguishing hunted species from scavenged ones in Stone Age archaeological sites. *Paleobiology* 8: 151-158.

KLEIN, R.G. 1982b. Patterns of ungulate mortality and ungulate mortality profiles from Langebaanweg (early Pliocene) and Elandsfontein (Middle Pleistocene), south-western Cape Province, South Africa. *Ann. S. Afr. Mus.* 90: 49-94.

KLEIN, R.G. (in press a). The Stone Age prehistory of Southern Africa. *Ann. Rev. Anthrop.* 12.

KLEIN, R.G. (in press b). Mammalian extinctions and stone age people in Africa. In: P.S. Martin & R.G. Klein (eds.), *Quaternary extinctions: a prehistoric revolution.* Tucson: University of Arizona Press.

KLEIN, R.G. (n.d.). The mammalian fauna from Equus Cave: carnivore and ungulate ecology in the southern Kalahari during the Last Interglacial. Unpublished manuscript.

KLEIN, R.G. & K. SCOTT 1974. The fauna of Scotts Cave, Gamtoos Valley, south-eastern Cape Province. *S. Afr. J. Sci.* 70: 186-187.

KOLBEN, P. 1731. *The present state of the Cape of Good Hope.* Transl. Guido Medley. London: W. Innys.

KRONFELD, J. & J.C. VOGEL 1980. U fingerprints in groundwater. *Nuclear Active* 22: 30-32.

KRUGER, F.J. 1977. Ecology of Cape fynbos in relation to fire. *USDA Forest Service General Technical Report* WO-3: 230-244.

KRUGER, F.J. 1979. Fire. In: J. Day, W.R. Siegfried, G.N. Louw & M.L. Jarman (eds.), Fynbos ecology: a preliminary synthesis. *S. Afr. natl. Scient. Programmes Rep.* 40: 43-57.

KURTEN, B. 1959. Rates of evolution in fossil mammals. *Cold Spring Harbor Symposia in Quantitative Biology* 24: 205-215.

KURTEN, B. 1965. The carnivores of the Palestine caves. *Acta zool. Fenn.* 107: 1-74.

KURTEN, B. 1968. *Pleistocene mammals of Europe.* London: Weidenfeld & Nicolson.

KURTEN, B. & E. ANDERSON 1980. *Pleistocene mammals of North America.* New York: Columbia University Press.

LAIDLER, P.W. 1933. Dating evidence concerning the Middle Stone Age, and a Capsio-Wilton site in the south-east Cape. *S. Afr. J. Sci.* 30: 530-542.

LAITMAN, J.T., R.C. HEIMBUCH & E.S. CRELIN 1979. The basicranium of fossil hominids as an indicator of their upper respiratory systems. *Am. J. Phys. Anthrop.* 51: 15-34.

LANCASTER, I.N. 1979a. Quaternary environments in the arid zone of southern Africa. *Dept. Geogr. Environ. Stud. Univ. Witwatersrand Occ. Pap.* 22: 1-77.

LANCASTER, I.N. 1979b. Evidence for a widespread late Pleistocene humid period in the Kalahari. *Nature (Lond.)* 279: 145-146.

LANCASTER, I.N. 1981a. Formation of the Holocene Lake Chilwa sand bar, southern Malawi. *Catena* (Braunschweig) 8: 369-382.

LANCASTER, I.N. 1981b. Paleoenvironmental implications of fixed dune systems in southern Africa. *Palaeogeog, Palaeoclimat., Palaeoecol.* 33: 327-346.

LAVOCAT, R. 1961. Le gisement des vertébrés Micocène des Beni Mellal (Maroc). Etude systématique de la fauna des mammifères et conclusions generales. *Notes Mem. Serv. Carte geol. Maroc* 155: 29-54.

LAVOCAT, R. 1973. Les rongeurs du Miocène d'Afrique orientale. 1. Miocène inferieur *Mem. Trav. Inst. Montpellier* 1: 1-284.

LAVOCAT, R. 1978. Rodentia and Lagomorpha. In: V.J. Maglio & H.B.S. Cooke (eds.), *Evolution of African mammals*: 69-89. Cambridge (Mass.): Harvard University Press.

LAWTON, A.C. 1962. Bantu pottery of southern Africa. *Ann. S. Afr. Mus.* 49(1).

LEAKEY, M.D. 1970. Stone artifacts from Swartkrans. *Nature (Lond.)* 225: 1222-1225.

LEAKEY, M.D. 1971. *Olduvai Gorge Vol. 3: Excavations in Beds I and II, 1960-1963.* Cambridge Cambridge University Press.

LEAKEY, M.D. 1975. Culture patterns in the Olduvai sequence. In: K.W. Butzer & G. Ll. Isaac (eds.), *After the australopithecines:* 477-493. The Hague: Mouton.

LEAKEY, M.D. 1979. *Olduvai Gorge: my search for Early Man.* London: Collins.

LEAKEY, M.D. 1981. Tracks and tools. *Phil. Trans. R. Soc. Lond. B* 292: 95-102.

LEAKEY, M.D. & R.L. HAY 1979. Pliocene footprints in the Laetolil Beds at Laetoli, northern Tanzania. *Nature (Lond.)* 278: 317-323.

LEAKEY, M.G. & R.E.F. LEAKEY 1978. *Koobi Fora Research Project. Volume 1: The fossil hominids and an introduction to their context, 1968-1974.* Oxford: Oxford University Press.

LEAKEY, R.E.F. 1980. How many species of hominids at Lake Turkana? In: L.K. Königsson (ed.), *Current argument on early man*: 29-30. Oxford Pergamon Press.

LEAKEY, R.E.F. & A.C. WALKER 1980. Comment on the status of *Australopithecus afarensis. Science* 207: 1103.

LEE, D.N. & H.C. WOODHOUSE 1970. *Art on the rocks of southern Africa.* Cape Town: Purnell.

LEE, R.B. 1965. Subsistence ecology of !Kung Bushmen. Unpublished Ph.D. thesis: University of California, Berkeley.

LEE, R.B. 1968. What hunters do for a living, or, how to make out on scarce resources. In: R.G. Lee & I. DeVore (eds.), *Man the hunter*: 30-48. Chicago: Aldine.

LEE, R.B. 1969. !Kung Bushman subsistence: an input-output analysis. In: A.P.Vayda (ed.), *Environment and cultural behavior*: 47-69. New York: Natural History Press.

LEE, R.B. 1972. The !Kung Bushmen of Botswana. In: M. Bicchieri (ed.), *Hunters and gatherers today*: 327-368. New York: Holt, Rinehart & Winston.

LEE, R.B. 1979. *The !Kung San: men, women and work in a foraging society.* Cambridge: Cambridge University Press.

LEITH, G. 1898. On the caves, shell-mounds and stone implements of South Africa. *J.R. Anthrop. Inst.* 28: 258-274.

LEPIONKA, L. 1979. Ceramics at Toutswemogala, Botswana. *S. Afr. archaeol. Soc. Goodwin Ser.* 3: 62-71.

LE ROUX, H.J. 1945. Die toestand, verspreiding en verbrokkeling van die Hottentotstamme in Suid-Afrika 1672-1713. Unpublished M.A. thesis: University of Stellenbosch.

LESER, H. 1971. Landschaftsökogische Studien in Kalahari-Sandgebiet um Auob und Nossob (östliches Südestafrica). *Erdwiss. Forsch.* (F. Steiner, Wiesbaden) 3: 1-243.

LEWIN, R. 1981a. Ethiopian stone tools are world's oldest. *Science* 211: 806-807.

LEWIN, R. 1981b. Protohuman activity etched in fossil bones. *Science* 213: 123-124.

LEWIS-WILLIAMS, J.D. 1972. The syntax and function of the Giant's Castle rock paintings. *S. Afr. archaeol. Bull.* 27: 49-65.

LEWIS-WILLIAMS, J.D. 1974a. Superpositioning in a sample of rock paintings from the Barkly East District. *S. Afr. archaeol. Bull.* 29: 93-103.

LEWIS-WILLIAMS, J.D. 1974b. Re-thinking the southern African rock paintings. *Origini* (Roma) 8: 229-257.

LEWIS-WILLIAMS, J.D. 1977. Believing and seeing San rock interpretation of symbolic meanings in southern San rock paintings. Unpublished Ph.D. thesis University of Natal, Durban.

LEWIS-WILLIAMS, J.D. 1980. Ethnography and iconography: aspects of southern San thought and art. *Man* (N.S.) 15: 467-482.

LEWIS-WILLIAMS, J.D. 1981. *Believing and seeing.* New York: Academic Press.

LINICK, T.W. 1977. La Jolla natural radiocarbon measurements VII. *Radiocarbon* 19: 19-48.

LIVINGSTONE, D.A. 1967. Postglacial vegetation of the Ruwenzori Mountains in Equatorial Africa. *Ecol Monogr.* 37: 25-52.

LIVINGSTONE, D.A. 1971. A 22,000 year pollen record from the plateau of Zambia. *Limnol Oceanogr.* 16: 349-356.

LOUBSER, J. (in prep.). Ndebele archaeology of the Pietersburg area. M.A. thesis: Department of Archaeology, University the Witwatersrand.

LOUW, A.W. 1969. Bushman Rock Shelter, Ohrigstad, eastern Transvaal: a preliminary investigation, 1965. *S. Afr. archaeol. Bull.* 24: 39-51.

LOUW, J.T. 1960. Prehistory of the Matjes River shelter. *Mem. natn. Mus. (Bloemfontein)* 1: 1-143.

LOVEJOY, C.O. 1975. Biomechanical perspectives on the lower limb of early hominids. In: R.H. Tuttle (ed.), *Primate functional morphology and evolution*: 291-326. The Hague: Mouton.

LOVEJOY, C.O. 1978. A biomechanical review of the locomotor diversity of early hominids. In: C.J. Jolly (ed.), *Early hominids of Africa*: 403-429. London: Duckworth.

MABERLEY, C.T. ASTLEY 1967. *The game animals of Southern Africa.* Cape Town: Nelson.

MACCALMAN, R. 1965. Radiocarbon dates from South West Africa. *S. Afr. archaeol. Bull.* 20: 215.

MACFARLANE, D.R. 1945. The Orange River high level gravels at Aliwal North in relation to crustal movement and the Stone Age. *S. Afr. J. Sci.* 41: 415-428.

MACIVER, D.R. 1906. *Mediaeval Rhodesia*. London: Macmillan.

MAGGS, T.M. O'C 1967. A quantitative analysis of the rock art from a sample area in the Western Cape. *S. Afr. J. Sci.* 63: 100-104.

MAGGS, T.M. O'C. *Iron age communities of the southern Highveld.* Pietermaritzburg: Natal Museum.

MAGGS, T.M. O'C. 1977. Some recent radiocarbon dates from eastern and southern Africa. *J. Afr. Hist.* 18: 161-191.

MAGGS, T.M. O'C. 1980a. Mzonjani and the beginning of the Iron Age in Natal. *Ann. Natal Mus.* 24: 71-96.

MAGGS, T.M. O'C. 1980b. Msuluzi Confluence: a seventh century Early Iron Age site on the Tugela River. *Ann. Natal Mus.* 24: 111-146.

MAGGS, T.M. O'C. 1980c. The Iron Age sequence south of the Vaal and Pongola rivers: some historical implications. *J. Afr. Hist.* 21: 1-15.

MAGGS, T.M.O'C. (in press). Mabhija: pre-colonial south industrial development in the Tugela Basin. *Ann. Natal Mus.* 25:

MAGGS, T.M. O'C & M.A. MICHAEL 1976. Ntshekane: an Early Iron Age site in the Tugela Basin, Natal. *Ann. Natal Mus.* 22: 705-740.

MAGGS, T.M. O'C & E. SPEED 1967. Bonteberg Shelter. *S. Afr. archaeol. Bull.* 22: 80-93.

MAGGS, T. & V. WARD 1980. Driel Shelter: rescue at a Late Stone Age site on the Tugela River. *Ann. Natal Mus.* 24: 35-70.

MAGLIO, V.J. 1975. Pleistocene faunal evolution in Africa and Eurasia. In: K.W. Butzer & G. Ll. Isaac (eds.), *After the australopithecines:* 419-476. The Hague: Mouton.

MAGLIO, V.J. 1978. Patterns of faunal evolution. In: V.J. Maglio & H.B.S. Cooke (eds.), *Evolution of African mammals*: 603-619. Cambridge (Mass.): Harvard University Press.

MAGURIE, B. 1980. Further observations on the nature and provenance of the lithic artifacts from the Makapansgat Limeworks. *Palaeont. afr.* 23: 127-151.

MAGUIRE, J.M. 1978. Southern African fossil porcupines. *S. Afr. J. Sci.* 74: 144

MAGUIRE, J.M., D. PEMBERTON & M.H. COLLETT 1980. The Makapansgat Limeworks grey breccia: hominids, hyaenas, hystricids or hillwash? *Palaeont. afr.* 23: 75-98.

MAINGARD, L.F. 1931. The lost tribes of the Cape. *S. Afr. J. Sci.* 28: 487-504.

MALAN, B.D. 1946. The distribution and chronology of the Modderpoort Cutlure. *S. Afr. J. Sci.* 42: 254-260.

MALAN, B.D. 1952. The final phase of the Middle Stone Age in South Africa. *Proceedings of the Pan-African Congress on prehistory, 1947*: 188-194. New York: Philosophical Library.

MALAN, B.D. 1955. The archaeology of the Tunnel Cave and Skildergat Kop, Fish Hoek, Cape of Good Hope. *S. Afr. archaeol. Bull.* 10: 3-9.

MALAN, B.D. 1956. An engraved flake from Muden, Natal. *S. Afr. archaeol. Bull.* 11: 63-65.

MALAN, B.D. & H.B.S. COOKE 1941. A preliminary account of the Wonderwerk Cave, Kuruman District. *S. Afr. J.Sci.* 37: 300-312.

MALAN, B.D. & L.H. WELLS 1943. A further report on the Wonderwerk Cave, Kuruman. *S. Afr. J. Sci.* 40: 258-270.

MANABE, S. & D.G. HAHN 1977. Simulation of the tropical climate of an ice age. *J. geophys. Res.* 82: 3889-3911.

MANKINEN, E.A. & G.B. DALRYMPLE 1979. Revised geomagnetic polarity time scale for the interval 0-5m.y. B.P. *J. geophys. Res.* 84: 615-626.

MARKER, M.E. 1972. Karst landform analysis as evidence for climatic change in the Transvaal. *S. Afr. geog. Jour.* 54: 152-162.

MARKER, M.E. & T.M. EVERS 1976. Iron Age settlement and soil erosion in the eastern Transvaal, South Africa. *S. Afr. archaeol. Bull.* 31: 153-165.

MARKS, S. 1972. Khoisan resistance to the Dutch in the seventeenth and eighteenth centuries. *J.Afr. Hist.* 13: 55-80.

MARSHALL, L. 1976. *The !Kung of Nyae Nyae.* Cambridge (Mass.): Harvard University Press.

MARTIN, A.R.H. 1959. Stratigraphy and history of Groenvlei, a South African coastal fen. *Austr. J. Bot.* 7: 142-167.

MARTIN, A.R.H. 1960. The ecology of Groenvlei, a South African coastal fen. *J. Ecol.* 48: 307-329.

MARTIN, A.R.H. 1962. Evidence relating to the Quaternary history of the Wilderness Lakes. *Trans. Geol. Soc. S. Afr.* 64: 19-42.

MARTIN, A.R.H. 1968. Pollen analysis of Groenvlei lake sediments, Knysna (South Africa). *Rev. Palaeobotan. Palynol.* 7: 107-144.

MARTIN, C. 1938. A rock shelter on Nyazongo Mountain, Penhalonga District, Southern Rhodesia. *Occ. Pap. Queen Vic. Mem. Lib. Salisbury* 1: 1-18.

MARTIN, H. & R.J. MASON 1954. The test trench in the Phillips Cave, Ameib, Erongo Mountains, South West Africa. *S. Afr. archaeol. Bull.* 9: 148-151.

MASAO, F.T. 1979. *The Later Stone Age and the rock paintings of central Tanzania.* Wiesbaden: Franz Steiner Verlag.

MASON, I.C. & J.P. MAULE 1960. *The indigenous livestock of eastern and southern Africa.* Farnham Royal, England: Commonwealth Agricultural Bureaux.

MASON, R.J. 1951. The excavation of four caves near Johannesburg. *S. Afr. archaeol. Bull.* 6: 71-79.

MASON, R.J. 1957. The Transvaal Middle Stone Age and statistical analysis. *S. Afr. archaeol. Bull.* 12: 119-137.

MASON, R.J. 1962a. The Sterkfontein stone artefacts and their maker. *S. Afr. archaeol. Bull.* 17: 109-125.

MASON, R.J 1962b. *Prehistory of the Transvaal.* Johannesburg: University of the Witwatersrand Press.

MASON, R.J. 1965. Makapansgat Limeworks fractured stone objects and natural fracture in Africa. *S. Afr. archaeol. Bull.* 20: 3-16.

MASON, R.J. 1966. The excavation of Doornlagte Earlier Stone Age camp, Kimberley District. In: L.D. Cuscoy (ed.), *Actas del V Congress Panafricano de prehistoria y de estudio de Cuaternario* (Tenerife, 1963): 187-188. Santa Cruz de Tenerife: Museo Arqueologico de Tenerife.

MASON, R.J. 1967a. Analytical procedures in the Earlier and Middle Stone Age cultures in southern Africa. In: W.W. Bishop & J.D. Clark (eds.), *Background to evolution in Africa*: 737-764. Chicago: University of Chicago Press.

MASON, R.J. 1967b. Prehistory as a science of change – new research in the South Arican interior. *Occ. Pap. archaeol. Res. Unit Univ. Witwatersrand* 1: 1-19.

MASON, R.J. 1968. Transvaal and Natal Iron Age settlement revealed by aerial photography and excavation. *Afr. Stud.* 27: 167-180.

MASON, R.J. 1969a. Tentative interpretations of new radiocarbon dates for stone artefact assemblages from Rose Cottage Cave, O.F.S. and Bushman Rock Shelter, Transvaal. *S. Afr. archaeol. Bull.* 24: 57-59.

MASON, R.J. 1969b. The Oppermansdrif Dam archaeological projects – Vaal Basin. *S. Afr. archaeol. Bull.* 24: 182-192.

MASON, R.J. 1971. Multivariate analysis of Cave of Hearths Middle Stone Age artefact assemblages. *Occ. Pap. Dept. Archaeol. Univ. Witwatersrand* 7: 1-34.

MASON, R.J. 1972. Locational models of Transvaal Iron Age settlements. In: D.L. Clarke (ed.), *Models in archaeology.* London: Methuen.

MASON, R.J. 1974a. The last Stone Age San (Bushmen) of the Vaal-Limpopo Basin. *S. Afr. J. Sci.* 70: 375.

MASON, R.J. 1974b. Background to the Transvaal Iron Age – new discoveries at Olifantspoort and Broederstroom. *Jour S. Afr. Inst. Min. Metall.* 74: 211-216.

MASON, R.J. 1976. The earliest artefact assemblages in South Africa. *IXe Congr. Union Int. Sci. prehist. protohist. Nice, 1976. Colloq. V* (preprint): 140-156.

MASON, R.J. 1979a. The Kalkbank Middle Stone Age site. Unpublished paper presented at a workshop on 'Towards a better understanding of the Upper Pleistocene in sub-Saharan Africa' organized by the Southern African Association of Archaeologists, Stellenbosch, June 1979.

MASON, R.J. 1979b. Archaeology and early farmers on the south bank of Hartebeespoort Dam, Transvaal. *Education and Culture* 2(8): 28-29.

MASON, R.J., R.A. DART & J.W. KITCHING 1958. Bone tools at the Kalkbank Middle Stone Age site and the Makapansgat australopithecine locality, central Transvaal. *S. Afr. archaeol. Bull.* 13: 85-116.

MASON, R.J. & J.C. VOGEL 1974. Later Stone Age plant remains from Kruger Cave, Olifantsnek, Rustenburg District. *S. Afr. J.Sci.* 70: 376.

MAZEL, A. 1980. Comment on 'Time and place' by J.E. Parkington. *S. Afr. archaeol. Bull.* 35: 100-102.

MAZEL, A. & J. PARKINGTON 1978. Sandy Bay revisited: variability among Late Stone Age tools. *S. Afr. J. Sci.* 74: 381-382.

MAZEL, A. & J. PARKINGTON 1981. Stone tools and resources: a case study from southern Africa. *Wld. Archaeol.* 13: 16-30.

MEESTER, J. 1965. The origins of the southern African mammal fauna. *Zool. Afr.* 1: 87-93.

MEIRING, A.J.D. 1956. The macrolithic culture of Florisbad. *Navors. Nas. Mus.* (Bloemfontein) 1: 205-237.

MERRICK, H.V. 1975. Change in later Pleistocene lithic industries in eastern African. Unpublished Ph.D. thesis: University of California, Berkeley.

MEYER, A. 1979. Verslag oor die keramiek van Greefswald. Unpublished M.A. thesis: University of Pretoria.

MEYER, G.E. 1978. Hyracoidea. In: V.J. Maglio & H.B.S. Cooke (eds.), *Evolution of African mammals*: 284-314. Cambridge (Mass.): Harvard University Press.

MILLER, S.F. 1969. The Nachikufan industries of the Later Stone Age in Zambia. Unpublished Ph.D. thesis: University of California, Berkeley.

MILLER, S.F. 1971. The age of the Nachikufan industries in Zambia. *S. Afr. archaeol. Bull.* 26: 143-146.

MOEYERSONS, J. 1978. The behavior of stones and stone implements, buried in consolidating and creeping Kalahari Sands. *Earth Surface Processes* 3: 115-128.

MOFFETT, R.O. & H.J. DEACON 1977. The flora and vegetation in the surrounds of Boomplaas Cave, Cango Valley. *S. Afr. archaeol. Bull.* 32: 127-145.

MOODIE, D. 1841. *Specimens of the authentic records of the colony of the Cape of Good Hope.* Cape Town: A.S. Robertson.

MOORIS, A.G. 1981. Copper discolouration of bone and the incidence of copper artefacts with human burials in South Africa. *S. Afr. archaeol. Bull.* 36: 36-42.

MORRISON, M.E.S. 1968. Vegetation and climate in the uplands of south-western Uganda during the later Pleistocene period. I Muchoya Swamp, Kigezi District. *J. Ecol.* 56: 363-384.

MCFADDEN, P.L., A. BROCK & T.C. PARTRIDGE 1979. Palaeomagnetism and the age of the Makapansgat hominid site. *Earth Plan. Sci. Lett.* 44: 373-382.

MCHENRY, H.M. & R.S. CORRUCINI 1975. Multivariate analysis of early hominid pelvic bones. *Am. J.Phys. Anthrop.* 43: 263-270.

MCHENRY H.M. & L.A. TEMERIN 1979. The evolution of hominid bipedalism: evidence from the fossil record. *Yearbook Phys. Anthrop.* 22: 105-131.

MCKEE, E.D. & G.W. WEIR 1953. Terminology for stratification and cross-stratification in sedimentary rocks. *Geol. Soc. Amer. Bull.* 64: 381-390.

NELSON, C.M. 1976. Flaked stone tool variations in the Later Stone Age of eastern and southern Africa. In: B. Abebe, J. Chavaillon & J.E.G. Sutton (eds.), *Proceedings of the VIIth Panafrican congress of prehistory and Quaternary Studies* (Addis Ababa, 1971): 131-152.

NETTERBERG, F. 1974. Observations on calcretes at some archaeological and palaeontological sites in South Africa. *S. Afr. archaeol. Soc. Goodwin Ser.* 2: 20-24.

NORTHEY, L.D. 1979. An analysis of non-archaeological faunal assemblages from southern Africa, with particular reference to the discrimination of hominid and non-hominid agents of bone accumulation. Unpublished M.A. thesis: University of Chicago.

OAKLEY, K.P. 1954. Study tour of early hominid sites in southern Africa, 1953. *S. Afr. archaeol. Bull.* 9: 75-87.

OBERMAIER, H. & H. KUHN 1930. *Bushmen art: rock paintings of South-West Africa*. Oxford: Oxford University Press.

OLSON, T.R. 1978. Hominid: phylogenetics and the existence of *Homo* in Member I of the Swartkrans Formation, South Africa. *J. Hum. Evol.* 7: 159-178.

OPPEERMAN, H. 1978. Excavations in the Buffelskloof Rock Shelter near Calitzdorp, southern Cape. *S. Afr. archaeol. Bull.* 33: 18-28.

ORPEN, J.M. 1908. *Reminiscences of life in South Africa from 1846 to the present day, with historical researches. Vol. 1.* Durban: P. Davis & Sons.

OXNARD, C.E. 1975a. The place of the australopithecines in human evolution: grounds for doubt? *Nature (Lond.)* 258: 389-395.

OXNARD, C.E. 1975b. *Uniqueness and diversity in human evolution.* Chicago: University of Chicago Press.

PAGER, H. 1971. *Ndedema*. Graz: Akademische Druck.

PAGER, H. 1975. *Stone Age myth and magic.* Graz: Akademische Druck.

PARKINGTON, J.E. 1976. Coastal settlement between the mouths of the Berg and Olifants Rivers, Cape Province. *S. Afr. archaeol. Bull.* 31: 127-140.

PARKINGTON, J.E. 1977a. Follow the San. Unpublished Ph.D. thesis: University of Cambridge.

PARKINGTON, J.E. 1977b. Soaqua: hunter-fisher-gatherers of the Olifants Rivers, western Cape. *S. Afr. archaeol. Bull.* 32: 150-157.

PARKINGTON, J.E. 1979. Reports on reasearch into the Late Stone Age of the Western Cape. Cape Town: Department of Archaeology, University of Cape Town.

PARKINGTON, J.E. 1980. Time and place: some observations on spatial and temporal patterning in the Later Stone Age sequence in southern Africa. *S. Afr. archaeol. Bull.* 35: 73-83.

PARKINGTON, J.E. 1981. The effects of environmental change on the scheduling of visits to the Elands Bay Cave, Cape Province, S.A. In: I. Hodder, G. Ll. Isaac & N. Hammond (eds.), *Patterns in the past: studies in honour of David Clarke:* 341-359. Cambridge: Cambridge University Press.

PARKINGTON, J. & M. CRONIN 1979. The size and layout of Mgungundlovu. *S. Afr. archaeol. Soc Goodwin Ser.* 3: 133-148.

PARKINGTON, J. & C. POGGENPOEL 1971a. Excavations at De Hangen 1968. *S. Afr. archaeol. Bull.* 26: 3-36.

PARKINGTON, J. & C. POGGENPOEL 1971b. A Late Stone Age burial from Clanwilliam. *S. Afr. archaeol. Bull.* 26: 82-84.

PARTRIDGE, T.C. 1978. Re-appraisal of the lithostratigraphy of the Sterkfontein hominid site. *Nature (Lond.)* 275: 282-287.

PARTRIDGE, T.C. 1979. Re-appraisal of the lithostratigraphy of the Makapansgat Limeworks hominid site. *Nature (Lond.)* 279: 484-488.

PARTRIDGE, T.C. 1982. The chronological position of the fossil hominids of southern Africa. In: *L'Homo erectus et la place de l'homme de Tautavel parmi les hominidés fossiles,* Colloque International du C.N.R.S., Nice 1982, pp. 617-675.

PARTRIDGE, T.C. 1983. Some preliminary observations on the stratigraphy and sedimentology of the Kromdraai B hominid site. *Palaeoecol. Afr.* 15: 3-12.

PATERSON, W. 1790. *A narrative of four journeys into the country of the Hottentots, and Caffraria, in the Years 1777, 1778, 1779.* Second Edition. London: J. Johnson.

PATTERSON, B. 1965. The fossil elephant shrews (Family Macroscelididae). *Bull. Mus. comp. Zool. Harv.* 133: 295-335.

PATTERSON, B. 1978. Pholidota and Tubulidentata. In: V.J. Maglio & H.B.S. Cooke (eds.), *Evolution of African mammals*: 268-278. Cambridge (Mass.): Harvard University Press.

PEABODY, F.E. 1954. Travertines and cave deposits of the Kaap Escarpment of South Africa and the type locality of *Australopithecus africanus. Bull. Geol. Soc. Am.* 65: 671-706.

PEERS, B. 1927. Excavation report of 1927 on Peers Cave. Unpublished manuscript, South African Museum, Cape Town.

PEERS, B. 1928. Ancient Fish Hoek and man his home. Unpublished manuscript, University of Cape Town Libraries.

PEERS, B. 1929. Preliminary report on the archaeology: the Fish Hoek – Noord Hoek Valley. Unpublished manuscript, University of Cape Town Libraries.

PERINGUEY, L. 1911. The Stone Ages of South Africa. *Ann. S. Afr, Mus.* 8: 1-218.

PETTER, F. 1971. Order Lagomorpha. In: Meester & H.W. Setzer (eds.), *The mammals of Africa: an identification manual:* Part 5. Washington, D.C.: Smithsonian Institution Press.

PHILLIPSON, D.W. 1974. Iron Age history and archaeology in Zambia. *J. Afr. Hist.* 15: 1-25.

PHILLIPSON, D.W. 1976. The prehistory of eastern Zambia. *Mem. Br. Inst. East Afr.* 6: 1-229.

PHILLIPSON, D.W. 1977. *The later prehistory of eastern and southern Africa.* London: Heinemann.

PICKFORD, M. 1975. New fossil Orycteropodidae (Mammalia: Tubulidentata) from East Africa. *Orycteropus minutus* sp. nov. and *Orycteropus chemeldoi* sp. nov. *Neth. J. Zool.* 25: 57-88.

PICKFORD, M. 1981. Preliminary Miocene mammalian biostratigraphy for western Kenya. J. Hum. Evol. 10: 73-97.

PILBEAM, D. & S.J. GOULD 1974. Size and scaling in human evolution. *Science* 186: 892-901.

PLUG, I. 1978. Die Latere Steentydperk van die Boesmanrotsskuiling in Oos Transvaal. Unpublished M.A. thesis: University of Pretoria.

PLUG, I. 1979a. Striped Giraffe Shelter faunal report. *Cimbebasia* B 3(1): 71-72.

PLUG, I. 1979b. Namakala and Nanga: faunal report on two Early Iron Age sites, Zambia. *S. Afr. archaeol. Bull.* 34: 123-126.

PLUG, I. 1981. Some research results on the late Pleistocene and early Holocene deposits of Bushman Rock Shelter, eastern Transvaal. *S. Afr. archaeol. Bull.* 36: 14-21.

PLUG, I. 1982. Bone tools and shell, bone and ostrich egg shell beads from Bushman Rock Shelter (BRS), eastern Transvaal. *S. Afr. archaeol. Bull.* 37: 57-62.

PÓCOCK, T.N. 1976. Pliocene mammalian microfauna from Langebaanweg: a new fossil genus linking the Otomyinae with the Murinae. *S. Afr. J. Sci.* 72: 58-60.

POGGENPEOL, C.A. & P.T. ROBERTSHAW 1981. The excavation of Smitswinkelbaai Cave, Cape Peninsula. *S. Afr. archaeol. Bull.* 36: 29-35.

POTTS, R. & P. SHIPMAN 1981. Cutmarks made by stone tools on bones from Olduvai Gorge, Tanzania. *Nature (Lond.)* 291: 577-580.

POWER, J.H. 1955. Power's site, Vaal River. *S. Afr. archaeol. Bull.* 10: 96-101.

PRENDERGAST, M.D. 1979. Chisvingo Hill furnace site, northern Mashonaland. *S. Afr. archaeol. Soc. Goodwin Ser.* 3: 47-51.

PRICE WILLIAMS, D. 1981. A preliminary report on recent excavations of Middle and Late Stone Age levels at Sibebe Shelter, north-west Swaziland. *S. Afr. archaeol. Bull.* 36: 22-28.

PRICE WILLIAMS, D. & A. WATSON 1982. New observations on the prehistory and palaeoclimate of the Late Pleistocene in southern Africa. *Wld. Archaeol.* 13: 372-381.

PRICE WILLIAMS, D., A. WATSON & A.S. GOUIDE 1982. Quaternary colluvial stratigraphy, archaeological sequences and palaeoenvironment in Swaziland, southern Africa. *Geog. Jour.* 148: 50-67.

PRINSLOO, H.P. 1974. Early Iron Age site at Klein Afrika near Wylliespoort, Soutpansberg Mountains, South Africa. *S. Afr. J. Sci.* 70: 271-273.

PRINSLOO, H.P. 1981. The Early Iron Age at Klein Afrika. Unpublished conference paper. Southern African Association of Archaeologists, July 1981.

PROTSCH, R. 1974. The Fish Hoek hominid: another member basic *Homo sapiens afer. Anthropolog. Anz.* 34: 241-249.

PROTSCH, R. 1977. Mumbwa: its absolute chronology and archaeology. *Z. Morph. Anhrop.* 68: 1-7.

PROTSCH, R. & H. DE VILLIERS 1974. Bushman Rock Shelter, Ohrigstad, eastern Transvaal, South Africa. *J. Hum. Evol.* 3: 387-396.

QUIN, P.J. 1959. *Foods and feeding habits of the Pedi.* Johannesburg: Witwatersrand University Press.

RAVEN-HART, R. 1967. *Before Van Riebeeck.* Cape Town: Struik.

RAVEN-HART, R. 1971. *The Cape of Good Hope 1652-1702. The first fifty years of Dutch colonization as seen by callers.* Vols. I & II. Cape Town. Balkema.

REPENNING, C.A. 1967. Palearctic-Nearctic mammalian dispersal in the Late Cenozoic. In: D.M. Hopkins (ed.), *The Bering Land Bridge*: 288-311. Stanford: Stanford University Press.

REPENNING, C.A. 1980. Faunal exchanges between Siberia and North America. *Canad. J. Anthrop.* 1: 37-44.

RICH, P.V. 1980. Preliminary report on the fossil avian remains from late Tertiary sediments at Langebaanweg (Cape Province), South Africa. *S. Afr. J. Sci.* 76: 166-170.

RIGHTMIRE, G.P. 1970. Iron age skulls from southern Africa re-assessed by multiple discriminant analysis. *Am. J. Phys. Anthrop.* 33: 147-167.

RIGHTMIRE, G.P. 1975. Problems in the study of later Pleistocene man in Africa. *Amer. Anthrop.* 77: 28-52.

RIGHTMIRE, G.P. 1976. Relationships of Middle and Upper Pleistocene hominids from sub-SaharanAfrica. *Nature (Lond.)* 260: 238-240.

RIGHTMIRE, G.P. 1978a. Florisbad and human population succession in southern Africa. *Am J. Phys. Anthrop.* 48: 475-486.

RIGHTMIRE, G.P. 1978b. Human skeletal remains from the southern Cape Province and their bearing on the Stone Age prehistory of South Africa. *Quatern. Res.* (N.Y.) 9: 219-230.

RIGHTMIRE, G.P. 1979. Implications of Border Cave skeletal remains for later Pleistocene human evolution. *Curr. Anthrop.* 20: 23-35.

RIGHTMIRE, G.P. 1981. Later Pleistocene hominids of eastern and southern Africa. *Anthropologie* (Brno) 19: 15-26.

ROBERTS, A. 1951. *The mammals of South Africa.* Johannesburg: Central News Agency.

ROBERTSHAW, P.T. 1977. Excavations at Paternoster, south-western Cape. *S. Afr. archaeol. Bull.* 32: 63-73.

ROBERTSHAW, P.T. 1978a. Archaeological investigations at Langebaan Lagoon, Cape Province. *Palaeoecol. Afr.* 10: 139-148.

ROBERTSHAW, P.T. 1978b. The origin of pastoralism in the Cape. *S. Afr. hist. J.* 10: 117-133.

ROBERTSHAW, P.T. 1979. Excavations at Duiker Eiland, Vredenburg district, Cape Province. *Ann. Cape Prov. Mus. (Hum. Sci.)* 1: 1-26.

ROBEY, T. 1980. Mpambanyoni: a Late Iron Age site on the Natal south Coast. *Ann. Natal Mus.* 24: 147-164.

ROBINSON, J.T. 1954a. The genera and species of the Australopithecinae. *Am. J. Phys. Anthrop.* 12: 181-200.

ROBINSON, J.T. 1954b. Prehominid dentition and hominid evolution. *Evolution* 8: 324-334.

ROBINSON, J.T. 1956. *The dentition of the Australopithecinae.* Pretoria: Transvaal Museum.

ROBINSON, J.T. 1959. A bone implement from Sterkfontein. *Nature (Lond.)* 184: 583-585.

ROBINSON, J.T. 1962. Sterkfontein stratigraphy and the significance of the Extension site. *S. Afr. archaeol. Bull.* 17: 87-107.

ROBINSON, J.T. 1967. Variation and the taxonomy of the early hominids. In: T. Dobzhansky, M.K. Hecht & W.C. Steere (eds.), *Evolutionary biology* 1: 69-100. New York: Appleton, Century, Crofts.

ROBINSON, J.T. 1972. *Early hominid posture and locomotion.* Chicago: University of Chicago Press.

ROBINSON, J.T., L. FRELDMAN & B.A. SIGMON 1972. Some aspects of pongid and hominid bipedality. *J. Hum. Evol.* 1: 361-369.

ROBINSON, J.T. & R.J. MASON 1957. Occurrence of stone artefacts with *Australopithecus* at Sterkfontein. *Nature (Lond.)* 180: 521-524.

ROBINSON, K.R. 1952. Excavations in two rock shelters near the Rusawi River, central Mashonaland. *S. Afr. archaeol. Bull.* 7: 108-129.

ROBINSON, K.R. 1958. Some Stone Age sites in Inyanga District. In: R. Summers, *Inyanga: prehistoric settlements in Southern Rhodesia*: 270-309. Cambridge: Cambridge University Press.

ROBINSON, K.R. 1959. *Khami Ruins.* Cambridge: Cambridge University Press.

ROBINSON, K.R. 1961a. An Early Iron Age site from the Chibi District, Southern Rhodesia. *S. Afr. archaeol. Bull.* 16: 75-102.

ROBINSON, K.R. 1961b. Excavations on the Acropolis Hill. *Occ. Pap. natn. Mus. S. Rhod.* 3(23A): 159-192.

ROBINSON, K.R. 1964. Dombozanga Rock Shelter, Mtetengwe River, Beit Bridge, Southern Rhodesia; excavation results. *Arnoldia (Rhod.)* 1(17): 1-14.

ROBINSON, K.R. 1965. The Leopard's Kopje Culture: a preliminary report on recent work. *Arnoldia (Rhod.)* 1(25): 1-6.

ROBINSON, K.R. 1966a. The Leopard's Kopje Culture: its position in the Iron Age of Southern Rhodesia. *S. Afr. archaeol. Bull.* 21: 5-51.

ROBINSON, K.R. 1966b. Bambata ware: its position in the Rhodesian Iron Age in the light of recent evidence. *S. Afr. archaeol. Bull.* 21: 81-85.

ROBINSON, K.R. 1975. Note on an engraved stone from the Marandellas District, Rhodesia. *S. Afr. archaeol. Bull.* 30: 42-43.

ROBINSON, K.R. & C.K. COOKE 1950. Some unusual elements in the Wilton Industry in the Matopo area of Southern Rhodesia. *S. Afr. archaeol. Bull.* 5: 108-114.

ROCHE, H. & J.J. TIERCELIN 1977. Découverte d'une industrie lithique ancienne *in situ* dans la formation d'Hadar. *C.R. Acad. Sci. Paris Ser. D.*, 284: 1871-1874.

RUDNER, I. & J. RUDNER 1954. A local Later Stone Age development. *S. Afr. archaeol. Bull.* 9: 103-107.

RUDNER, J. 1957. The Brandberg and its archaeological remains. *Jour. S.W.A. scient. Soc.* 12: 7-44.

RUDNER, J. 1968. Strandloper pottery from South and South West africa. *Ann. S. Afr. Mus.* 49: 441-663.

RUDNER, J. 1971a. Ostrich eggshell flasks and soapstone objects from the Gordonia District, north-western Cape. *S. Afr. archaeol. Bull.* 26: 139-142.

RUDNER, J. 1971b. Painted burial stones from the Cape. *S. Afr. J. Sci. spec. Publ.* 2: 54-61.

RUDNER, J. 1972. Radiocarbon dates from the Brandberg in South West Africa. *S. Afr. archaeol. Bull.* 27: 164-165.

RUDNER, J. 1979. The use of stone artefacts and pottery among the Khoisan peoples in historic and protohistoric times. *S. Afr. archaeol. Bull.* 34: 3-17.

RUDNER, J. & I. RUDNER 1970. *The hunter and his art: a survey of rock art in southern Africa.* Cape Town: Struik.

RUDNER, J. & I. RUDNER 1973. A note on early excavations at Robberg. *S. Afr. archaeol. Bull.* 28: 94-96.

SACKETT, J.R. 1977. The meaning of style in archaeology: a general model. *Am. Antiq.* 42: 369-380.

SAMPSON, C.G. 1976a. Excavations at Glen Elliot Shelter, Colesberg District, northern Cape. *Res. natn. Mus.* (Bloemfontein) 2: 125-210.

SAMPSON, C.G. 1976b. Zeekoegat 13: a Later Stone Age open-site near Venterstad, Cape. *Res. natn. Mus.* (Bloemfontein) 2: 211-237.

SAMPSON, C.G. 1967c. Excavations at Zaayfontein Shelter, Norvalspont, northern Cape. *Res. natn. Mus.* (Bloemfontein) 2: 41-124.

SAMPSON, C.G. 1968a. The Middle Stone Age industries of the Orange River Scheme area. *Mem. natn. Mus.* (Bloemfontein) 4: 1-111.

SAMPSON, C.G. 1968b. Two analyses of the Middle Stone Age industries from the Cave of Hearths and Olieboompoort Shelter, northern Transvaal-method B. *Occ. Pap. Archaeol. Res. Unit Univ. Witwatersrand* 2: 1-11.

SAMPSON, C.G. 1970. The Smithfield Industrial Complex: further field results. *Mem. natn. Mus.* (Bloemfontein) 5: 1-172.

SAMPSON, C.G. 1972. The Stone Age industries of the Orange River scheme and South Africa. *Mem. natn. Mus.* (Bloemfontein) 6: 1-288.

SAMPSON, C.G. 1974. *The Stone Age archaeology of Southern Africa.* New York: Academic Press.

SAMPSON, C.G. 1980. Archaeological distribution of the Zeekoe Valley Bushmen. Unpublished manuscript.

SAMPSON, C.G. & M. SAMPSON 1967 Riversmead Shelter: excavations and analysis. *Mem. natn. Mus.* (Bloemfontein) 3: 1-111.

SAMPSON, C.G. & M.D. SOUTHARD 1973. Variability and change in the Nachikufan Industry of Zambia. *S. Afr. archaeol. Bull.* 28: 78-89.

SANDELOWSKY, B.H. 1971. Ostrich eggshell caches from South West Africa. *S. Afr. archaeol. Bull.* 26: 153.

SANDELOWSKY, B.H. 1977. Mirabib-an archaeological study in the Namib. *Madoqua* 10: 221-283.

SANDELOWSKY, B.H. 1979. Kapako and Vungu Vungu: Iron Age sites on the Kavango River. *S. Afr. archaeol. Soc. Goodwin Ser.* 3: 52-61.

SANDELOWSKY, B.H., J.H. VAN ROOYEN & J.C. VOGEL, 1979. Early evidence for herders in the Namib. *S. Afr. archaeol. Bull.* 34: 50-51.

SANDELOWSKY, B.H. & A. VIERECK 1969. Supplementary report on the archaeological expedition of 1962 to the Erongo mountains of South West Africa. *Cimbebasia* Bl (1): 4-43.

SANTA LUCA, A.P. 1978. A re-examination of presumed Neandertal-like fossils. *J. Hum. Evol.* 7: 619-636.

SAVAGE, R.J.G. & W.R. MAMILTON 1973. Introduction to the Miocene mammal faunas of Gebel Zelten, Libya. *Bull. Br. Mus. nat. Hist., Geol.* 22: 514-527.

SCHALKE, H.J.W.G. 1973. The upper Quaternary of the Cape Flats area (Cape Province, South Africa). *Scripta Geol.* 15: 1-57.

SCHAPERA, I. 1926. A preliminary consideration of the relationship between Hottentots and Bushmen. *S. Afr. J. Sci.* 23: 833-866.

SCHAPERA, I. 1930. *The Khoisan peoples of South Africa.* London: Routledge & Kegan Paul.

SCHERZ, E. 1970. *Felsbilder in Südwest-Afrika.* Köln: Böhlau-Verlag.

SCHMIDT, P.R. 1978. Historical archaeology-a structural approach in an African culture. *Contributions in intercultural and comparative studies* 3. Westport: Greenwood.

SCHOFIELD, J.F. 1940. A report on the pottery from Bambata Cave. *S. Afr. J. Sci.* 37: 61-372.

SCHOONRAAD, M. 1980. Sommige petrogliewe van die Magaliesmoot. *Humanitas* 6(3).

SCHRIRE, C. 1962. Oakhurst: a re-examination and vindication. *S. Afr. archaeol. Bull.* 17: 181-195.

SCHRIRE, C. 1980. An inquiry into the evolutionary status and apparent identity of San hunter-gatherers. *Hum. Ecol.* 8: 9-32.

SCHULTZE, L. 1907. *Aus Namaland und Kalahari.* Jena: Gustav Fischer.

SCHWEITZER, F.R. 1970. A preliminary report of excavations of a cave at Die Kelders. *S. Afr. archaeol. Bull.* 25: 136-138.

SCHWEITZER, F.R. 1974. Archaeological evidence for sheep at the Cape. *S. Afr. archaeol. Bull.* 29: 75-82.

SCHWEITZER, F.R. 1975. The ecology of post-Pleistocene peoples on the Gansbaai coast, south-western Cape. Unpublished M.A. thesis: University of Stellenbosch.

SCHWEITZER, F.R. 1979. Excavations at Die Kelders, Cape Province, South Africa. *Ann. S. Afr. Mus.* 78: 101-233.

SCHWEITZER, F.R. & K.J. SCOTT 1973. Early occurrence of domestic sheep in sub-Saharan Africa. *Nature (Lond.)* 241: 547.

SCHWEITZER, F.R. & M.L. WILSON 1978. A preliminary report on excavations at Byneskranskop, Bredasdorp District, Cape. *S. Afr. archaeol. Bull.* 33: 134-140.

SCOTT, L. 1976. Preliminary palynological results from the Alexandersfontein basin near Kimberley. *Ann. S. Afr. Mus.* 71: 193-199.

SCOTT, L. 1982a. A late Quaternary pollen record from the Transvaal bushveld, South Africa. *Quatern. Res.* (N.Y.) 17: 339-370.

SCOTT, L. 1982b. A 5,000-year old pollen sequence from spring deposits in the bushveld at the north of the Soutpansberg, South Africa. *Palaeoecol. Afr.* 14: 45-56.

SCOTT, L. 1982c. Late Quaternary fossil pollen grains from the Transvaal, South Africa. *Rev. Palaeobotan. Palynol.* 36: 241-278.

SCOTT, L. 1982d. Pollen analysis of Late Cainozoic deposits in the Transvaal, South Africa, and their bearing on palaeoclimates. *Palaeoecol. Afr.* 15: 101-107.

SCOTT, L. 1982e. Palynological evidence for vegetation patterns in the Transvaal (South Africa) during the Late Pleistocene and Holocene. *Bothalia* (Proceedings of the AETFAT congress, Pretoria, 1982) (in press).

SCOTT, L. & R.G. KLEIN 1981. A hyena-accumulated bone assemblage from Late Holocene deposits at Deelpan, Orange Free State. *Ann. S. Afr. Mus.* 86: 217-227.

SCOTT, L. & J.C. VOGEL (n.d.) Late Quaternary pollen profile from the Transvaal Highveld, South Africa. Unpublished manuscript.

SEELY, M.K. & B.H. SANDELOWSKY 1974. Dating the regression of a river's end point. *S. Afr. archaeol. Soc. Goodwin Ser.* 2: 61-64.

SELBY, M.J., C.H. HENDY & M.K. SEELY 1979. A late Quaternary lake in the central Namib Desert, southern Africa, and some implications. *Palaeogeog., Palaeoclimat., Palaeoecol.* 26: 37-41.

SHACKLETON, N.J. 1973. Oxygen-isotope analysis as a means of determining season of occupation of prehistoric midden sites. *Archaeometry* 15: 133-141.

SHACKLETON, N.J. 1982. Stratigraphy and chronology of the Klasies River Mouth deposits: oxygen isotope evidence. In: R. Singer, & J.J., Wymer *The Middle Stone Age at Klasies River Mouth in South Africa*: 194-199. Chicago: University of Chicago Press.

SHACKLETON, N.J. & N.D. OPDYKE 1973. Oxygen isotope and palaeomagnetic stratigraphy of Equatorial Pacific Core V28-238: oxygen isotope temperatures and ice volumes on a 10^5 and 10^6 year scale. *Quatern. Res.* (N.Y.) 3: 39-55.

SHACKLETON, N.J. & N.D. OPDYKE, 1976. Oxygen-isotope and paleomagnetic stratigraphy of Pacific Core V28-239. Late Pliocene to latest Pleistocene. *Geol. Soc. Amer. Mem.* 145: 449-464.

SHACKLETON, N.J. & N.D. OPDYKE 1977. Oxygen isotope and palaeomagnetic evidence for early Northern Hemisphere glaciation. *Nature (Lond.)* 270: 216-219.

SHACKLEY, M. 1980. An Acheulean industry with *Elephas recki* fauna from Namib IV, South West Africa (Namibia). *Nature (Lond.)* 284: 340-341.

SHAW, E.M. & N.J. VAN WARMELO 1981. The material culture of the Cape Nguni 3. *Ann. S. Afr. Mus.* 58(3).

SHAW, T. 1979. Holocene adaptations in West Africa: The Late Stone Age. *Early Man News* 3/4: 51-82.

SHAW, T. 1981. Letter to the Editor. *S. Afr. archaeol. Bull.* 36: 54.

SHEPPARD, J.G. & E.R. SWART 1973. Rhodesian radiocarbon measurements V. *Radiocarbon* 15: 545-553.

SHIPMAN, P., A. WALKER, J.A. VAN COUVERING, P.J. HOOKER & J.A. MILLER, 1981. The Fort Ternan hominoid site, Kenya: Geology, age, taphonomy and palaeoecology. *J. Hum. Evol.* 10: 49-72.

SHORTRIDGE, G.C. 1934. *The mammals of South-West Africa.* London: Heinemann.

SIDNEY 1965. The past and present distribution of some African ungulates. *Trans. Zool. Soc. Lond.* 30: 1-396.

SIESSER, W.G. 1980. Late Miocene origin of the Benguela upwelling system off northern Namibia. *Science* 208: 283-285.

SIESSER, W.G. & R.V. DINGLE 1981. Tertiary sea-level movements around southern Africa. *J. Geol.* 89: 83-96.

SILBERBAUER, F.B. 1979. Stable carbon isotopes and prehistoric diets in the eastern Cape Province, South Africa. Unpublished M.A. thesis: University of Cape Town.

SILBERBAUER, G. 1965. *Bushman survey report.* Gaberone: Bechuanaland Government.

SILBERBAUER, G. 1972. The G/wi Bushmen. In: M. Bicchieri (ed.), *Hunters and gatherers today*: 271-325. New York: Holt, Rinehart & Winston.

SIMONS, E.L., P. ANDREWS & D.R. PILBEAM 1978. Cenozoic apes. In: V.J. Maglio & H.B.S. Cooke, (eds.), *Evolution of African mammals*: 120-146. Cambridge (Mass.): Harvard University Press.

SIMONS, E.L.& E. DELSON 1978. Cercopithecidae and Parapithecidae. In: V.J. Maglio & H.B.S. Cooke, (eds.), *Evolution of African mammals*: 100-119. Cambridge (Mass.): Harvard University Press.

SIMPSON, C.G. 1973. Tertiary penguins (Sphenisciformes, Spheniscidae) from Ysterplaats, Cape Town, South Africa. *S. Afr. J. Sci.* 69: 342-344.

SIMPSON, C.G. 1979. Tertiary penguins from the Duinefontein site, Cape Province, South Africa. *Ann. S. Afr. Mus.* 79: 1-7.

SINGER, R. & J.R. CRAWFORD 1958. The significance of the archaeological discoveries at Hopefield, South Africa. *J.R. Anthrop. Inst.* 88: 11-19.

SINGER, R. & H. LEHMAN 1963. The hemoglobins of Africander cattle. *Occ. pap. R. anthrop. Inst.* 18: 119-125.

SINGER, R. & J. WYMER 1968. Archaeological investigations at the Saldanha Skull site in South Africa. *S. Afr. archaeol. Bull.* 23: 63-74.

SINGER, R. & J. WYMER 1969. Radiocarbon date for two painted stones from a coastal cave in South Africa. *Nature (Lond.)* 244: 508-510.

SINGER, R. & J. WYMER 1982. *The Middle Stone Age at Klasies River Mouth in South Africa*. Chicago: University of Chicago Press.

SMITH, A.B. 1981. An archaeological investigation of Holocene deposits at Rooiels Cave, south-western Cape. *S. Afr. archaeol. Bull.* 36: 75-83.

SMITH, A.B. & M.R. RIPP 1978. An archaeological reconnaissance of the Doorn/Tanqua Karoo. *S. Afr. archaeol. Bull.* 33: 118-133.

SMITH, C.A. 1966. Common names of South African plants. *Mem.bot. Surv. S. Afr.* 35: 1-642.

SMITHERS, R.H.N. 1966. *The mammals of Rhodesia, Zambia, and Malawi.* London: Collins.

SMITHERS, R.H.N. 1971a. The mammals of Botswana. *Nat. Mus. Rhod. Mem.* 4: 1-340.

SMITHERS, R.H.N. 1971b. Family Felidae. In: J. Meester & H.W. Setzer (eds.), *The mammals of Africa: an identification manual:* Part 8.1. Washington. D.C.: Smithsonian Institution Press.

SMITHERS, R.H.N. & J.L.P. LOBAO TELLO 1976. Check list and atlas of the mammals of Mozambique. *Nat. Mus. Rhod. Mem.* 8: 1-184.

SOUTH AFRICAN WEATHER BUREAU 1965. *Climate of South Africa, Parts 8-9.* Pretoria.

SPARRMAN, A. 1785. *A voyage to the Cape of Good Hope towards the Antarctic polar circle and round the world, but chiefly into the country of the Hottentots and Caffres, from the year 1772 to 1776.* Vol. 1. London: Robinson.

STAMELMAN, A. 1948. A short archaeological excursion down the valley of the Lower Vaal. *S. Afr. archaeol. Bull.* 3: 71-73.

STAPLETON, P. J. HEWITT 1927. Stone implements from a rock-shelter at Howieson's Poort near Grahamstown. *S. Afr. J. Sci.*24: 574-587.

STAPLETON, P. & J. HEWITT 1928. Stone implements from Howieson's Poort, near Grahamstown. *S. Afr. J. Sci.* 25: 399-409.

STEUDEL, K. 1978. A multivariate analysis of the pelvis of early hominids. *J. Hum. Evol.* 7: 583-595.

STEUDEL, K. 1980. New estimates of early hominid body size. *Am. J. Phys. Anthrop.* 52: 63-70.

STEUDEL, K. & J.T. ROBINSON (n.d.). Allometry, diet and the early hominids. In: P.V. Tobias & Y. Coppens (eds.), *Les Plus Anciens Hominidés.* Paris: CNRS (in press).

STEYN, H.P. 1971. Aspects of the economic life of some nomadic Nharo Bushman groups. *Ann. S. Afr. Mus.* 56: 275-322.

STEYN, H.P. 1981. Nharo plant utilization: an overview. *Khoisis, Occ. pap. Dept. Anthrop. & Archaeol. Univ. Stellenbosch* 1: 1-31.

STILES, D.H. & T.C. PARTRIGE 1979. Results of recent archaeological and palaeoenvironmental studies at the Sterkfontein Extension Site. *S. Afr. J. Sci.* 75: 346-352.

STORRAR, P. 1978. *Portrait of Plettenberg Bay.* Cape Town: Purnell.

STORY, R. 1958. Some plants used by the Bushmen in obtaining food and water. *Mem. bot. Surv. S. Afr.* 30. Pretoria: Government Printer.

STORY, R. 1964. Plant lore of the Bushmen. In: D.H.S. Davis (ed.), *Ecological studies in southern Africa.* The Hague: W. Junk.

STOW, G.W. 1905. *The native races of South Africa.* London: Swan Sonneschein.

STRINGER, C.B. 1974. Population relationships of later Pleistocene hominids: a multivariate study of available crania. *J. archaeol. Sci.* 1: 317-342.

STRINGER, C.B. 1978. Some problems in Middle and Upper Pleistocene hominid relationships. In: D.J. Chivers & K.A. Joysey (eds.), *Recent advances in primatology, 3, Evolution*: 395-418. London: Academic Press.

STRINGER, C.B., F.C. HOWELL & J.K. MELENTIS 1979. The significance of the fossil hominid skull from Petralona, Greece. *J. archaeol. Sci.* 6: 235-253.

STROMER, E. 1926. Reste land- und süsswasser-bewohender Wirbeltiere aus den Diamantenfeldern Deutsch-Südwestafrikas. In: E. Kaiser (ed.), *Die Diamantenwüste Südwestafrikas* 2: 107-153. Berlin: D. Reimer.

STROMER, E. 1931a. Reste süsswasser- und land-bewohnender Wirbeltiere aus den Diamantenfeldern Klein-Namaqualandes (Südwes-Afrika). *Sber. bayer. akad. Wiss.* 1931. 17-47.

STROMER, E. 1931b. *Palaeothentoides africanus,* nov. gen., nov. spec., ein erstes Beuteltier aus Afrika. *Sber. bayer. Akad. Wiss.* 1931: 177-190.

STUIVER, M. 1969. Yale natural radiocarbon measurements IX. *Radiocarbon* 11: 545-658.

STUIVER, M. & N.J. VAN DER MERWE 1968. Radiocarbon chronology of the Iron Age in sub-Saharan Africa. *Curr. Anthrop.* 9: 54-58.

SUMMERS, R. 1958. *Inyanga: prehistoric settlements in Southern Rhodesia.* Cambridge University Press.

SUMMERS, R. 1963. *Zimbabwe: a Rhodesian mystery.* Johannesburg; Nelson.

SUMMERS, R. 1969. Ancient mining in Rhodesia and adjacent areas. *Nat. Mus. Rhod. Mem.* 3.

SUMMERS, R. 1971. *Ancient ruins and vanished civilisations of southern Africa.* Cape Town: Bulpin.

SUMMERS, R., K.R. ROBINSON & A. WHITTY 1961. Zimbabwe excavations. *Occ. pap. natn. Mus. S. Rhod.* 3(23A).

SWITSUR, V.R. & R.G. WEST 1973. University of Cambridge natural radiocarbon measurements XII. *Radiocarbon* 15: 534-544.

SYDOW. W. 1967. The pre-European pottery of South West Africa. *Cimbebasia Mem.* 1: 7-74.

SZABO, B.J. 1979. Dating fossil bone from Cornelia, Orange Free State, South Africa. *J. archaeol. Sci.* 6: 201-203.

SZABO, B.J. & K.W. BUTZER 1979. Uranium-series dating of lacustrine limestones from pan deposits with Final Acheulian assemblage at Rooidam, Kimberley District, South Africa. *Quatern. Res.* (N.Y.) 11: 257-260.

TAIEB, M., D.C. JOHANSON, Y. COPPENS & J.J. TIERCELIN 1978. Expédition internationale de l' Afar, Ethiopie (4e et 5e campagnes 1975-1977): chronostratigraphie des gisements à hominidés pliocènes d' Hadar et corrélations avec les sites préhistoriques du Kada Gona. *C.R. Acad. Sci. Paris Ser. D.* 287: 459-461.

TALMA, A.S., J.C. VOGEL & T.C. PARTRIDGE 1974. Isotopic contents of some Transvaal speleothems and their palaeoclimatic significance. *S. Afr. J. Sci.* 70: 135-140.

TANAKA, J. 1969. The ecology and social structure of Central Kalahari Bushmen: a preliminary report. *Kyoto Univ. Afr. Stud.* 3; 1-26.

TANAKA, J. 1976. Subsistence ecology of Central Kalahari San. In: R.B. Lee & I. DeVore (eds.), *Kalahari hunter-gatherers: studies of the !Kung San and their neighbours:* 98-119. Cambridge (Mass.): Harvard University Press.

TANKARD, A.J. 1975. The marine Neogene Saldanha Formation. *Trans. geol. Soc. S. Afr.* 78: 257-264.

TANKARD, A.J. 1976. The stratigraphy of a coastal cave and its palaeoclimatic significance. *Palaeoecol. Afr.* 1: 151-159.

TANKARD, A.J. & J. ROGERS 1978. Late Cenozoic palaeoenvironments on the west coast of southern Africa. *J. Biogeogr.* 5: 319-337.

TANKARD, A.J. & F.R. SCHWEITZER 1974. The geology of Die Kelders Cave and environs: a palaeoenvironmental study. *S. Afr. J. Sci.* 70: 365-369.

TANKARD, A.J. & F.R. SCHWEITZER 1976. Textural analysis of cave sediments: Die Kelders, Cape Province, South Africa. In: D.A. Davidson & M.L. Shackley (eds.), *Geoarchaeology:* 289-316. London: Duckworth.

TAYLOR, M.O.V. 1979. Late Iron Age settlements on the northern edge of the Vredefort dome. Unpublished M.A. thesis. University of the Witwatersrand.

THACKERAY, A.I. 1977. Stone artefacts from Klipfonteinrand. Unpublished B.A. (Hons.) thesis: Department of Archaeology, University of Cape Town.

THACKERAY, A.I. 1981. The Holocene cultural sequence in the northern Cape Province, South Africa. Unpublished Ph.D. thesis: Yale University.

THACKERAY, A.I., J.F. THACKERAY, P.B. BEAUMONT & J.C. VOGEL 1981. Dated rock engravings from Wonderwerk Cave, South Africa. *Science* 214: 64-67.

THACKERAY, J.F. 1979. An analysis of faunal remains from archaeological sites in southern South West Africa (Namibia). *S. Afr. archaeol. Bull.*34: 18-33.

THACKERAY, J.F. 1980. New approaches in interpreting archaeological faunal assemblages with examples from southern Africa. *S. Afr. J. Sci.* 76: 216-223.

THACKERAY, J.F. & E.C. FEAST 1974. A midden burial from Cape St. Fancis, eastern Cape Province. *S. Afr. archaeol. Bull.* 29: 92.

THEAL, G. McC. 1896. *The Portuguese in South Africa, with a description of the native races between the River Zambesi and the Cape of Good Hope during the sixteenth century.* London: T. Fisher Unwin.

THIERGART, F., U. FRANTZ & K. RAUKOPE 1962. Palynologische Untersuchungen von Tertiarkohlen und einer Oberflächen probe nahe Knysna, Südafrika. *Adv. Front. Pl. Sci.* 4: 151-178.

THOM, H.B. (ed.) 1952. *Journal of Van Riebeeck.* Vol. 1. Cape Town: Balkema.

THOM, H.B. (ed.) 1954. *Journal of Van Riebeeck.* Vol. 2. Cape Town: Balkema.

THOM, H.B. (ed.) 1958. *Journal of Van Riebeeck.* Vol. 3. Cape Town: Balkema.

THOMAS, D.H. 1974. *Predicting the past: and introduction to anthropological archaeology.* New York: Holt, Rinehart & Winston.

TOBIAS, P.V. 1949. The excavation of Mwulu's Cave, Potgietersrust District. *S. Afr. archaeol. Bull.* 4: 2-13.

TOBIAS, P.V. 1967. *Olduvai Gorge Vol. 2: The cranium and maxillary dentition of Australopithecus (Zinjanthropus) boisei.* Cambridge: Cambridge University Press.

TOBIAS, P.V. 1972. Recent human biological studies in southern Africa, with special reference to Negroes and Khoisans. *Trans. R. Soc. Afr.* 40: 109-133.

TOBIAS, P.V. 1978a. The place of *Australopithecus africanus* in hominid evolution. In: D.J. Chivers & K.A. Joysey (eds.), *Recent advances in Primatology, 3, Evolution*: 373-394. London: Academic Press.

TOBIAS, P.V. 1978b. The South African australopithecines in time and hominid phylogeny, with special reference to the dating and affinities of the Taung skull. In: C.J. Jolly (ed.), *Early hominids of Africa*: 45-84. London: Duckworth.

TOBIAS, P.V.1978c. The earliest Transvaal members of the genus *Homo* with another look at some problems of hominid taxonomy and systematics. *Z. Morph. Anthrop.* 69: 225-265.

TOBIAS, P.V. 1978d. The San: an evolutionary perspective. In: P.V. Tobias (ed.), *The Bushmen*: 16-23. Cape Town: Human & Rousseau.

TOBIAS, P.V. 1980a. A survey and synthesis of the African hominids of the late Tertiary and early Quaternary periods. In: L.K. Konigsson (ed.), *Current argument on early man*: 86-113. Oxford: Pergamon Press.

TOBIAS, P.V. 1980b. The natural history of the helicoidal occlusal plane and its evolution in early *Homo*. *Am J. Phys. Anthrop.* 53: 173-187.

TOBIAS, P.V. & A.R. HUGHES 1969. The new Witwatersrand University excavation at Sterkfontein. *S. Afr. archaeol. Bull.* 24: 158-169.

TOMLINSON, R.W. 1974. Preliminary biogeographical studies on the Inyanga mountains, Rhodesia. *S. Afr. Geog. Jour.* 56: 15-26.

TUFFREAU, A. 1979. Les débuts du paléolithique moyen dans la France septentrionale. *Bull. Soc. Préhist. Fr.* 76: 140-142.

TURNER, B.R. 1980. Sedimentological characteristics of the 'Red Muds' at Makapansgat Limeworks. *Palaeont. Afr.* 23: 51-58.

VAIL, P.R., R.M. MITCHUM & S. THOMPSON 1977. Seismic stratigraphy and global changes in sea level. Part 4. Global cycles of relative changes of sea level. *Mem. Am. Ass. petrol. Geol.* 26: 83-97.

VAIL, P.R. & J. HARDENBOL 1979. Sea-level changes during the Tertiary. *Oceanus* 22: 71-79.

VAN COUVERING, J.A.H. & J.A. VAN COUVERING 1976. Early Miocene mammal fossils from East Africa. In: G. Ll. Isaac & E.R. McCown (eds.), *Human origins*: 155-207. Menlo Park (Calif.): W.A. Benjamin.

VAN DER MERWE, N.J. 1980. The advent of iron in Africa. In: T.A. Wertime & J.D. Muhly (eds.), *The coming of the age of iron.* New Haven: Yale University Press.

VAN DER MERWE, N.J. & D.J. KILLICK 1979. Square: an iron smelting site near Phalaborwa. *S. Afr. archaeol. Soc. Goodwin Ser.* 3: 86-93.

VAN DER MERWE, N.J. & R.T.K. SCULLY 1971. The Phalaborwa story: archaeological and ethnographic investigation of a South African Iron Age group. *Wld. Archaeol.* 3: 178-196.

VAN NOTEN, F. 1974. Excavations at the Gordons Bay shell midden, south-western Cape. *S. Afr. archaeol. Bull.* 29: 122-142.

VAN NOTEN, F. 1977. Excavations at Matupi cave. *Antiquity* 51: 34-40.

VAN NOTEN, F. 1979. The Early Iron Age in the Interlacustrine Region: the diffusion of iron technology. *Azania* 14: 61-80.

VAN RIET LOWE, C. 1936. The Smithfield 'N' Culture. *Trans. R. Soc. S. Afr.* 23: 367-372.

VAN RIET LOWE, C. 1945. The evolution of the Levallois technique in South Africa. *Man* 45: 49-59.

VAN RIET LOWE, C. 1952. The Vaal River chronology: an up-to-date summary. *S. Afr. archaeol. Bull.* 7: 135-149.

VAN RIJSSEN, W. 1980. Ways of seeing: some aspects of the interpretation of rock paintings. Unpublished B.A. (Hons.) thesis: Department of Archaeology, University of Cape Town.

VANSINA, J. 1980. Bantu in the crystal ball. *History in Africa* 6: 287-333 and 7: 293-325.

VAN WARMELO, N.J. 1940: The copper miners of Musina. *S. Afr. Dept. Native Affairs Ethnol. Publ.* 8.

VAN ZINDEREN BAKKER, E.M. 1953. *South African pollen grains and spores I.* Cape Town: Balkema.

VAN ZINDEREN BAKKER, E.M. 1955. A preliminary survey of the peat bogs of the Alpine belt of northern Basutoland. *Acta Geographica* 14: 413-422.

VAN ZINDEREN BAKKER, E.M. 1956. *South African pollen grains and spores II.* Cape Town: Balkema.

VAN ZINDEREN BAKKER, E.M. 1957. A pollen analytical investigation of the Florisbad deposits (South Africa). In: J.D. Clark (ed.), *Proceedings of the 3rd Panafrican congress on prehistory* (Livingstone, 1955): 56-67. London: Chatto & Windhus.

VAN ZINDEREN BAKKER, E.M. 1963. Analysis of pollen samples from northeast Angola. In: J.D. Clark *Prehistoric cultures of northeast Angola and their significance in tropical Africa I.* Museo do Dundo Publ. Cult. 62: 213-217.

VAN ZINDEREN BAKKER, E.M. 1964. A pollen diagram from equatorial Africa, Cherangani, Kenya. *Geol. Mijnbouw* 43: 123-128.

VAN ZINDEREN BAKKER, E.M. 1969. The Pleistocene vegetation and climate of the basin. In: J.D. Clark (ed.), *Kalambo Falls prehistoric site Vol. 1*: 57-84. Cambridge University Press.

VAN ZINDEREN BAKER, E.M. 1982. Pollen analytical studies of the Wonderwerk Cave, South Africa. *Pollen et Spores* (Paris) 24: 235-250.

VAN ZINDEREN BAKKER, E.M. & K.W. BUTZER 1973. Quaternary environmental changes in southern Africa. *Soil Sci.* 116: 236-248.

VAN ZINDEREN BAKKER, E.M. & J.D. CLARK 1962. Pleistocene climates and cultures in north-eastern Angola. *Nature (Lond.)* 196: 639-642.

VAN ZINDEREN BAKKER, E.M. & M.J.A. WERGER 1974. Environment, vegetation and phytogeography of the high-altitude bogs of Lesotho. *Vegetatio* 29: 37-49.

VINNICOMBE, P. 1965. Bushman fishing as depicted in rock paintings. *Scient. S. Afr.* 2: 578-581.

VINNICOMBE, P. 1967. Rock painting analysis. *S. Afr. archaeol. Bull.* 22: 129-141.

VINNICOMBE, P. 1971. A Bushman hunting kit from the Natal Drakensberg. *Ann. Natal Mus.* 20: 611-625.

VINNICOMBE, P. 1972a. Motivation in African rock art. *Antiquity* 46: 124-133.

VINNICOMBE, P. 1972b. Myth, motive, and selection in southern African rock art. *Africa* 42: 192-204.

VINNICOMBE, P. 1976. *People of the Eland.* Pietermaritzburg: University of Natal Press.

VISSER, D.J.L. & C. VAN RIET LOWE 1956. Die geologie en argeologie van die Klein-Caledonriviervallei. *Geol. Surv. S. Afr. Mem.* 47: 1-48.

VOGEL, J.C. 1969. Radiocarbon dating of Bushman Rock Shelter, Ohrigstad District. *S. Afr. archaeol. Bull.* 24: 56.

VOGEL, J.C. 1970. Groningen radiocarbon dates IX. *Radiocarbon* 12: 444-471.

VOGEL, J.C. 1979. Problems in dating the Middle Stone Age. Unpublished paper presented at a workshop on 'Towards a better understanding of the Upper Pleistocene in sub-Saharan Africa' organized by the Southern African Association of Archaeologists, Stellenbosch, June 1979.

VOGEL, J.C. 1983. The age of the Kuiseb river silt terrace at Homeb. *Palaeoecol. Afr.* 15: 201-209.

VOGEL, J.C., A. FULS & R.P. ELLIS 1978. The geographical distribution of Kranz grasses in South Africa. *S. Afr. J. Sci.* 74: 209-215.

VOGEL, J.C. & M. MARAIS 1971. Pretoria radiocarbon dates. *Radiocarbon* 13: 378-394.

VOGEL, J.C. & E. VISSER 1981. Pretoria radiocarbon dates II. *Radiocarbon* 23: 43-80.

VOIGT, E.A. 1973a. Stone age molluscan utilization at Klasies River Mouth caves. *S. Afr. J. Sci.* 69: 306-309.

VOIGT, E.A. 1973b. Klasies River Mouth: an exercise in shell analysis. *Transv. Mus. Bull.* 14: 14-15.

VOIGT, E.A. 1973c. Faunal remains from the Iron Age sites of Matope Court, Namichimba and Chikumba, Southern Malawi. *Malawi Dept. Antiq. Publ.* 13: 135-167.

VOIGT, E.A. 1975. Studies of marine mollusca from archaeological sites: dietary preferences, environmental reconstructions and ethnological parallels. In: A. Clason (ed.), *Archaeozoological studies:* 87-98. Amsterdam: North Holland.

VOIGT, E.A. 1978. The faunal remains from Greefswald as a reflection of Iron Age economic and cultural activities. Unpublished M.A. thesis: University of Pretoria.

VOIGT, E.A. 1980. Reconstructing Iron Age economies of the northern Transvaal: preliminary report. *S. Afr. archaeol. Bull.* 35: 39-45.

VOLMAN, T.P. 1978. Early archaeological evidence for shellfish collecting. *Science* 201: 911-913.

VOLMAN, T.P. 1981. The Middle Stone Age in the southern Cape. Unpublished Ph.D. thesis: University of Chicago.

VON BOECKMANN, K. 1921. Stoffe der tracht. In: L. Frobenius & L.R. VON Wilm (eds.), *Atlas Africanus I.* Munich: Beck.

VRBA, E.S. 1974. Chronological and ecological implications of the fossil Bovidae at the Sterkfontein australopithecine site. *Nature (Lond.)* 250: 19-23.

VRBA, E.S. 1975. Some evidence of chronology and palaeoecology of Sterkfontein, Swartkrans, and Kromdraai from the fossil Bovidae. *Nature (Lond.)* 254: 301-304.

VRBA, E.S. 1976. The fossil Bovidae of Sterkfontein, Swartkrans and Kromdraai. *Transv. Mus. Mem.* 21: 1-116.

VRBA, E.S. 1977. New species of *Parmularius* Hopwood and *Damaliscus* Sclater (Alcelaphini, Bovidae, Mammalia) from Makapansgat, and comments on faunal chronological correlation. *Palaeont. Afr.* 20: 137-151.

VRBA, E.S. 1978. Problematical alcelaphine fossils from the Kromdraai Faunal Site (Mammalia: Bovidae). *Ann. Transv. Mus.* 31: 21-28.

VRBA, E.S. 1979. Phylogenetic analysis and classification of fossil and recent Alcelaphini (Mammalia: Bovidae). *Biol. J. Linn. Soc.* 11: 207-208.

VRBA, E.S. 1980. The significance of bovid remains as indicators of environment and predation patterns. In: A.K. Behrensmeyer & A.P. Hill (eds.), *Fossils in the making*: 247-271. Chicago: University of Chicago Press.

VRBA, E.S. 1981. The Kromdraai australopethecine site revisted in 1980: recent investigations and results. *Ann. Transv. Mus. (Pretoria) 33 (3): 17-60.*

VRBA, E.S. 1982. Biostratigraphy and chronology, based on Bovidae, of southern African hominid-associated assemblages. In: *L' Homo erectus et la place de l'homme de Tautavel parmi les hominides* fossiles. Colloque International du C.N.R.S., Nice 1982, pp. 707-752.

WADLEY, L. 1983. Results of a Stone Age excavation from Duncombe Farm shelter, Concession, 1970. *Rhod. Prehist.* 10: 14-18.

WADLEY, L. 1976. Radiocarbon dates from Big Elephant Shelter, Erongo Mountains, South West Africa. *S. Afr. archaeol. Bull.* 31: 146.

WADLEY, L. 1979. Big elephant shelter and its role in the Holocene prehistory of central South West Africa. *Cimbebasia* B 3(1) 1-75.

WALKER, A.C. 1977. Functional anatomy of oral tissues: mastication and deglutition. In: J.H. Shaw, E.A. Sweeney, C.C. Capuccino & S.M. Meller (eds.), Textbook of oral biology. Philadelphia: Saunders.

WALKER, A.C. 1978. Prosimian primates. In: V.J. Maglio & H.B.S. Cooke (eds.), *Evolution of African mammals*: 90-99. Cambridge (Mass.): Harvard University Press.

WALKER, A.C. 1980. Functional anatomy and taphonomy. In: A.K. Behrensmeyer & A.P. Hill (eds.), *Fossils in the making*: 182-196. Chicago: University of Chicago Press.

WALKER, N.J. 1978. The Later Stone Age of the Matopos. *Rhod. Preshist.* 16: 2-7.

WALKER, N.J. 1980. Later Stone Age research in the Matopos. *S. Afr. archaeol. Bull.* 35: 19-24.

WALTER, R.C. & J.L. ARONSON 1982. Revisions of K/Ar ages for the Hadar hominid site, Ethiopia. *Nature (Lond.)* 296: 122-127.

WEBB, S.D. 1969. Extinction-origination equilibrium in Late Cenozoic land mammals of North America. *Evolution* 23: 688-702.

WEBB, S.D. 1977. A history of savanna vertebrates in the New World. Part 1: North America. *Ann. Rev. Ecol. Syst.* 8: 355-380.

WELBOURNE, R.G. 1971. A study of prehistoric relationships between environment, animals and man at one Earlier Stone Age site, one Middle Stone Age site and eight Iron Age sites in the Vaal-Limpopo Basins. Unpublished M.A. thesis: University of the Witwatersrand.

WELBOURNE, R.G. 1973. Identification of animal remains from the Broederstroom 24/73 Early Iron Age site. *S. Afr. J. Sci.* 69: 325.

WELBOURNE, R.G. 1975. Toutswe Iron Age site: its yield of bones. *Botswana Notes Rec.* 7: 1-16.

WELLS, L.H. 1962. Pleistocene faunas and the distribution of mammals in southern Africa. *Ann. Cape Prov. Mus.* 2: 37-40.

WELLS, L.H. 1964. The Vaal River 'Younger Gravels' faunal assemblage: a revised list. *S. Afr. J. Sci.* 60: 92-94.

WELLS, L.H. 1967 Antelopes in the Pleistocene of southern Africa. In: J.D. Clark & W.W. Bishop (eds.), *Background to evolution in Africa*: 99-105. Chicago: University of Chicago Press.

WELLS, L.H. & H.B.S. COOKE 1956. Fossil Bovidae from the Limeworks Quarry, Makapansgat, Potgietersrus. *Palaeont. Afr.* 4: 1-55.

WELLS, L.H., H.B.S. COOKE & B.D. MALAN 1942. The associated fauna and culture of the Vlakkraal thermal springs, O.F.S. *Trans. R. Soc. S. Afr.* 29: 203-233.

WELLS, M.J. 1965. An analysis of plant remains from Scotts Cave in the Gamtoos Valley. *S. Afr. archaeol. Bull.* 20: 79-84.

WELLS, M.J. 1970. Plant remains from Amanzi Springs. *Ann. Cape Prov. Mus. (Nat. Hist.)* 8: 191-194.

WENDORF, F., R.L. LAURY, C.C. ALBRITTON, R. SCHILD, C.V. HAYNES, P.E. DAMON, M. SHAFIQULLAH & R. SCARBOROUGH 1975. Dates for the Middle Stone Age of East Africa. *Science* 187: 740-742.

WENDT, W.E. 1972. Preliminary report on an archaeological research programme in South West Africa. *Cimbebasia* B 2(1): 1-61.

WENDT, W.E. 1974. 'Art mobilier' aus der Apollo 11-Grotte in Südwest-Afrika: die ältesten datierten Kunstwerke Afrikas. *Acta Praehist. Archaeol.* 5: 1-42.

WENDT, W.E. 1976. 'Art mobilier' from the Apollo 11 Cave, South West Africa: Africa's oldest dated works of art. *S. Afr. archaeol. Bull.* 31: 5-11.

WENDT, W.E. 1977/78. Felsgravierungen im Südlichen Südwestafrika. *Jour S.W.A. Scient. Soc.* 32: 7-68.

WERGER, M.J.A. (ed.) 1978. *Biogeography and ecology of Southern Africa.* The Hague: W. Junk.

WESTPHAL, E.O.J. 1963. The linguistic prehistory of southern Africa: Bush, Kwadi, Hottentot and Bantu linguistic relationships. *Africa* 33: 253-256.

WHITE, D.A., M. WHITE & O. BOURQUIN 1979. The restoration of an ancient Bantu settlement at Bekkersberg, Orange Free State. O.F.S. Provincial Administration *Nature Conservation Miscellaneous Publ.* 5.

WHITE, J.P. & D.S. THOMAS 1972. What mean these stones? Ethnotaxonomic models and archeological interpretations in the New Guinea Highlands. In: D.L. Clarke (ed.), *Models in archaeology* : 275-308. London: Methuen.

WHITE, L.A. 1949. *The science of culture.* New York: Farrar, Straus & Giroux.

WHITE, R. 1982. Rethinking the Middle/Upper Paleolithic transition. *Curr. Anthrop.* 23: 169-192.

WHITE, T.D. 1977. The anterior mandibular corpus of early African Hominidae: functional significance of shape and size. Unpublished Ph.D. thesis: University of Michigan.

WHITE, T.D. 1980. Evolutionary implications of Pliocene hominid footprints. *Science* 208: 175-176.

WHITE, T.D. & J.M. HARRIS 1977. Suid evolution and correlation of African hominid localities. *Science* 198: 13-21.

WHITE, T.D., D.C. JOHANSON & W.H. KIMBALL 1981. *Australopithecus africanus.*: its phyletic position reconsidered. *S. Afr. J. Sci.* 77: 445-470.

WILKINSON, M.J. 1977. Evolution of the australopithecine-bearing deposits of Sterkfontein Cave. Unpublished manuscript.

WILLIAMS, J. 1978. A brief comparison of model simulations of glacial period maximum atmospheric circulation. *Palaeogeogr., Palaeoclimatol., Palaeoecol.* 25: 191-198.

WILLCOX, A.R. 1963. *The rock art of South Africa.* London: Nelson.

WILLCOX, A.R. 1965. Archaeological notes from the northern Cape. *S. Afr. archaeol. Bull.* 20: 139-140.

WILLCOX, A.R. 1973. *Rock paintings of the Drakensberg.* Second Edition. Cape Town: Struik.

WILSON, M. 1959. The early history of the Transkei and Ciskei. *Afr. Stud.* 18: 167-180.

WILSON, M. 1969. The hunters and herders. In: M. Wilson & L. Thompson (eds.), *The Oxford history of South Africa. I*: 40-74. Oxford: Oxford University Press.

WILSON, M.C. 1982. Cut marks and early hominids: evidence for skinning. *Nature (Lond.)* 298: 303.

WILSON, M.L. & HALKETT, D. 1981. The use of marine shell for decorating coastal (Khoisan) pottery. *S. Afr. archaeol. Bull.* 36: 43.

WOOD, B.A. 1976. Remains attributable to *Homo* in the East Rudolf succession. In: Y. Coppens, F.C. Howell, G. Ll. Isaac & R.E.F. Leakey (eds.), *Earliest man and environments in the Lake Rudolf Basin*: 490-506. Chicago: University of Chicago Press.

WOOD, B.A. & C.G. STACK 1980. Does allometry explain the differences between 'gracile' and 'robust' australopithecines? *Am. J. Phys. Anthrop.* 52: 55-62.

WRIGHT, J.B. 1971. *Bushman raiders of the Drakensberg 1840-1870.* Pietermaritzburg: University of Natal Press.

WYMER, J. & R. SINGER 1972. Middle Stone Age occupational settlementson the Tzitzikama coast, eastern Cape Province, South Africa. In: P.J. Ucko, R. Tringham & G.W. Dimbleby (eds.), *Man, settlement and urbanism*: 207-210. London: Duckworth.

YELLEN, J.E. 1977. *Archaeological approaches to the present: models for reconstructing the past.* New York: Academic Press.

YELLEN, J.E. & H. HARPENDING 1972. Hunter-gatherer populations and archaeological inference. *Wld. Archaeol.* 4: 244-253.

YELLEN, J.E. & R.B. LEE 1976,. The Dob-/du/da environment: considerations for a hunting and gathering way of life. In: R.B. Lee & I. DeVore (eds.), *Kalahari hunter-gatherers*: 27-46. Combridge (Mass.): Harvard University Press.

Index